VOLUME ONE HUNDRED AND FIFTY SEVEN

PROGRESS IN MOLECULAR BIOLOGY AND TRANSLATIONAL SCIENCE

Epigenetics and Psychiatric Disease

VOLUME ONE HUNDRED AND FIFTY SEVEN

PROGRESS IN MOLECULAR BIOLOGY AND TRANSLATIONAL SCIENCE

Epigenetics and Psychiatric Disease

Edited by

DENNIS R. GRAYSON
Center for Alcohol Research in Epigenetics, The Psychiatric Institute, Department of Psychiatry, University of Illinois, Chicago, IL, United States

ELSEVIER

ACADEMIC PRESS
An imprint of Elsevier

Academic Press is an imprint of Elsevier
125 London Wall, London EC2Y 5AS, United Kingdom
525 B Street, Suite 1650, San Diego, CA 92101-4495, United States
50 Hampshire Street, 5th Floor, Cambridge, MA 02139, United States
The Boulevard, Langford Lane, Kidlington, Oxford OX5 1GB, United Kingdom

First edition 2018

ISBN: 978-0-12-813565-5
ISSN: 1877-1173

For information on all Academic Press publications
visit our website at https://www.elsevier.com/books-and-journals

Publisher: Zoe Kruze
Acquisition Editor: Ashlie M. Jackman
Editorial Project Manager: Joanna Collett
Production Project Manager: James Selvam
Cover Designer: Miles Hitchen

Typeset by Thomson Digital

CONTENTS

CONTRIBUTORS

Schahram Akbarian
Icahn School of Medicine at Mount Sinai, New York, NY, United States

Daniel Almeida
Douglas Mental Health University Institute, McGill University, Montreal, QC, Canada

Reema Abdulrahman S. Alyamani
School of Healthcare Science, Manchester Metropolitan University, Manchester, United Kingdom

Sarah Barnett Burns
Douglas Mental Health University Institute, McGill University, Montreal, QC, Canada

Dae D. Chung
Texas A&M University Health Science Center, Bryan, TX, United States

Rochelle L. Coulson
Genome Center, MIND Institute, UC Davis, CA, United States

Tiffany S. Doherty
University of Delaware, Newark, DE, United States

Yogesh Dwivedi
University of Alabama at Birmingham, Birmingham, AL, United States

Dennis R. Grayson
Center for Alcohol Research in Epigenetics, The Psychiatric Institute, Department of Psychiatry, University of Illinois, Chicago, IL, United States

Alessandro Guidotti
Center for Alcohol Research in Epigenetics, The Psychiatric Institute, Department of Psychiatry, University of Illinois, Chicago, IL, United States

Yan Jiang
Icahn School of Medicine at Mount Sinai, New York, NY, United States

Chuan Jiao
School of Life Science, Central South University, Changsha, China

Bibi S. Kassim
Icahn School of Medicine at Mount Sinai, New York, NY, United States

Aniket Kumar
Christian Medical College, Vellore, India

Janine M. LaSalle
Genome Center, MIND Institute, UC Davis, CA, United States

Elad Lax
Department of Pharmacology and Therapeutics, McGill University Medical School, Montreal, QC, Canada

Chunyu Liu
University of Illinois, Chicago, IL, United States
School of Life Science, Central South University, Changsha, China

Amanda H. Mahnke
Texas A&M University Health Science Center, Bryan, TX, United States
Women's Health in Neuroscience Program, Texas A&M University Health Science Center, Bryan, TX, United States

Rajesh C. Miranda
Texas A&M University Health Science Center, Bryan, TX, United States
Women's Health in Neuroscience Program, Texas A&M University Health Science Center, Bryan, TX, United States

Chris Murgatroyd
School of Healthcare Science, Manchester Metropolitan University, Manchester, United Kingdom

Nagalakshmi B.
Savitribai Phule Pune University, Pune, Maharashtra, India

Eric J. Nestler
Icahn School of Medicine at Mount Sinai, New York, NY, United States

Jacob Peedicayil
Christian Medical College, Vellore, India

Catherine J. Peña
Icahn School of Medicine at Mount Sinai, New York, NY, United States

Prashanth Rajarajan
Icahn School of Medicine at Mount Sinai, New York, NY, United States

Tania L. Roth
University of Delaware, Newark, DE, United States

Bhaskar Roy
University of Alabama at Birmingham, Birmingham, AL, United States

Sneha Sagarkar
Savitribai Phule Pune University, Pune, Maharashtra, India

Amul J. Sakharkar
Savitribai Phule Pune University, Pune, Maharashtra, India

Nihal A. Salem
Texas A&M University Health Science Center, Bryan, TX, United States

Moshe Szyf
Department of Pharmacology and Therapeutics, McGill University Medical School, Montreal, QC, Canada

Alexander M. Tseng
Texas A&M University Health Science Center, Bryan, TX, United States

Gustavo Turecki
Douglas Mental Health University Institute, McGill University, Montreal, QC, Canada
McGill University, Montreal, QC, Canada

Kangli Wang
School of Life Science, Central South University, Changsha, China

Ning Yuan
Hunan Brain Hospital, Changsha, China

PREFACE

This issue of *Progress in Molecular Biology and Translational Science* (PMBTS) focuses on epigenetics in psychiatry. The field is still quite young, and having started some 12–13 years ago when reports in the literature first started to appear on the methylation status of candidate genes thought to be involved in the etiopathogenesis of schizophrenia and other psychiatric disorders. Strictly speaking, epigenetics refers to heritable modifications in gene expression that arise by factors that do not alter the genetic code. However, in a more general sense, epigenetics refer to the mechanisms by which external/environmental factors influence profiles of gene expression in cell populations of an individual. Some common examples include maternal infection or other complications during pregnancy, physical, and/or emotional abuse during early childhood development, as well as periods of drug use in adolescents. These examples that are highly variable in intensity, have been shown to impact neurodevelopment in a large number of both preclinical and clinical psychiatric studies. Epigenetic mechanisms at the molecular level include the regulation of cytosine methylation, modifications of histone proteins, and expression of noncoding RNAs, each of which influence transitions in chromatin architecture and hence alter patterns of gene expression. The regulation of microRNA production and other RNAs are often included in this mechanistic category.

The field of epigenetics has undergone several methodological transitions; the most seminal of which was the switch from single gene candidate studies to genome-wide analyses that examine the distribution of epigenetic marks across the entire genome. This has forever changed the epigenetics research landscape as investigators search for effects that affect the entire genome. This includes not only whole genome bisulfite sequencing but also chromatin immunoprecipitation sequencing with antibodies directed at specific epigenetic marks. A second discovery that changed how many researchers carry out epigenetics research was the discovery of the DNA demethylation pathway, which starts with the oxidation of 5-methylcytosine to form 5-hydroxymethylcytosine (5hmC). The existence of 5hmC was known since the 1970s to be a constituent of DNA particularly in the context of viral and phage DNAs. However, in 2009, important contributions to the field cemented 5hmC as the first intermediate in DNA demethylation initiated by the ten–eleven translocation methylcytosine dioxygenase enzyme family of proteins. 5hmC accumulates in cells indicating that it serves a specific epigenetic function aside from being an intermediate in demethylation. In addition, further oxidation of 5hmC produces 5-formylcytosine and 5-carboxycytosine both of which accumulate in cells and comprise subsequent steps in the cascade.

In the context of this collection of papers that comprise this book, adversity during either embryonic or early development represents a focal point. However, each chapter then diverges and highlights some of the more recent findings in a particular area of interest to researchers in the field of psychiatry. In the Chapter 1, Doherty and Roth describe recent progress in the area of early life adversity and epigenetic regulation particularly as it relates to stress. They describe changes in methylation and hydroxymethylation in response to stress and analyze these changes in the context of gender differences.

Very little is currently known regarding how the 3D genome is structured and how the 3D architecture changes following alterations in histone modifications or DNA methylation. In Chapter 2, Rajarajan *et al.* review early studies on the spatial organization of the genome in the schizophrenia brain and discuss how integrated approaches can be used to examine chromosomal conformations and epigenetic regulation.

In Chapter 3, Peña and Nestler describe recent progress in the epigenetics of depression. They discuss numerous studies reporting on both candidate genes and genome-wide changes in epigenetic marks in both preclinical rodent and postmortem human brain. Moreover, they provide a summary of recent studies of changes occurring in DNA methylation, histone modifications, chromatin remodeling, and noncoding RNAs in depression.

Coulson and LaSalle describe recent findings on how DNA methylation and long noncoding RNAs interact with circadian rhythms at selected imprinted loci in Chapter 4. They also discuss various disorders associated with these imprinted regions. Moreover, they provide insight into mechanisms involved in imprinting and the effects of circadian rhythms during neurodevelopment.

In Chapter 5, Lax and Szyf provide an overview of global and site-specific changes in DNA methylation in drug addiction. They describe recent findings in animal models of drug addiction and alterations in methylation in humans addicted to drugs. Finally, they provide insight into DNA methylation as a potential biomarker in lymphocytes in human studies.

In Chapter 6, Grayson and Guidotti examine epigenetics in the context of schizophrenia and bipolar disorder. They describe two epigenetic mouse models of psychosis. The first is the methionine mouse model in which large doses of methionine are administered to mice facilitating increased DNA methylation. The second is a stress model that involves prenatal restraint of pregnant dams and the consequences of this stress on the adult offspring.

In Chapter 7, Alymani and Murgatroyd discuss early life programming in the neuroendocrine regulation of stress. More specifically, they describe epigenetic mechanisms that translate early-life conditions into gene expression changes underpinning stress-related behaviors. Moreover, they discuss multiple venues of stress during pre- and postnatal development and the consequences in terms of epigenetic changes in the animal and the clinic.

In Chapter 8, Peedicayil and Kumar describe epigenetic changes associated with mood disorders. They also describe epigenetic and nutritional drugs for treating mood disorders, including major depressive disorder and bipolar disorder. The prospect of epigenetic therapies and drugs to treat mood disorders is making considerable progress in recent years.

In Chapter 9, Liu *et al.* provide a detailed synopsis of the current status of DNA methylation in psychiatric diseases. Moreover, they highlight large numbers of gene targets that are methylate in psychiatric diseases, as well as numerous epigenomic-wide sites identified by association studies. Finally, they integrate current advances in the field with a focus on future perspectives and challenges.

In Chapter 10, Roy and Dwivedi describe recent studies on epigenetic perturbations and the etiopathogenesis of major depressive disorder. They describe recent work showing how genetic susceptibility coupled with environmental stimuli can increase risk for suicidal ideation. They also discuss epigenetic marks as potential biomarkers for suicidal behavior.

In Chapter 11, Nagalakshimi B. *et al.* discuss recent progress in preclinical and clinical studies on epigenetic mechanisms associated with traumatic brain injuries. They cover a range of topics from DNA methylation and demethylation to histone modifications and microRNAs in different animal models of traumatic brain injury. Finally, they finish their chapter with a discussion of recent work in mitoepigenetics and traumatic brain injury.

In Chapter 12, Mahnke *et al.* describe recent advances in the epigenetics of fetal alcohol spectrum disorders (FASD). They summarize studies of microRNAs (miRNA) in different FASD models. Moreover, they describe recent advances in long noncoding RNAs (lncRNAs) in fetal alcohol spectrum disorders, as well as alcohol use disorders. Finally, they provide a look at future directions in the field and discuss the utility of lncRNAs as therapeutic targets.

In Chapter 13, Burns *et al.* provide a detailed analysis of the epigenetics of early life adversity. They discuss various animal models and human studies on the impact of early life adversity. Finally, they describe how individual variation impacts the consequences of early life adversity in humans and end with a look at current technical limitations in the field.

Collectively, this volume of *PMBTS* provides a comprehensive look at how a variety of epigenetic mechanisms impact brain function leading to psychiatric disorders. I am pleased to have worked with all of the authors and coworkers in bringing this volume to completion. I would also like to acknowledge the staff members at Elsevier who have worked hard in bringing this project to fruition.

Dennis R. Grayson

CHAPTER ONE

Epigenetic Landscapes of the Adversity-Exposed Brain

Tiffany S. Doherty, Tania L. Roth[1]
University of Delaware, Newark, DE, United States
[1] Corresponding author. E-mail address: troth@psych.udel.edu

Contents

Abstract

It is understood that adversity during development has the power to alter behavioral trajectories, and the role of the epigenome in that relationship is currently under intense investigation. Several studies in both nonhuman animals and humans have established a link between early adversity and epigenetic regulation

Progress in Molecular Biology and Translational Science, volume 157
ISSN 1877-1173
http://dx.doi.org/10.1016/bs.pmbts.2017.11.025

of genes heavily implicated in the stress response, plasticity and cognition, and psychiatric disorders such as depression and anxiety. Thus the relatively recent surge of studies centering on the epigenetic outcomes of stress has great potential to inform treatments and interventions for psychiatric disorder precipitated by early adversity. Here we review what we know and what we do not know, and suggest approaches to help further elucidate the relationship between early adversity, epigenetics, and behavior.

1. INTRODUCTION

In the following sections, we discuss ongoing considerations in the field of neuroepigenetics. Discussion will begin with current knowledge of each topic, focusing on DNA methylation. We will then narrow the focus to where we are at with that knowledge at the intersection of early adversity, the epigenome, and behavior. Lastly, discussion will turn to what we do not know within these areas, the direction the literature suggests we go next, and some specific steps necessary to address outstanding issues.

1.1 Early Adversity, Epigenetics, and Psychiatric Disorder

Adversity in early life, particularly in relation to the caregiver, is a major predisposing factor in the development of psychiatric disorder.[1–3] Investigation into the mechanisms of this relationship reveals epigenetic alterations as promising candidates. Exposure to early adversity elicits changes in the epigenome that can vary by gene, brain region, sex, and across time (see Table 1).[4–11] In animal models, these changes are associated with behavioral alterations reflective of psychological dysfunction.[7,9,12] In humans, they are present in individuals with a history of early adversity and in those with psychiatric disorder.[4,13–17] Taken together, these data present a strong argument for the functional role of epigenetics in adverse psychiatric outcomes induced by disruption of the early environment.

1.2 DNA Methylation

Methylation of DNA is an epigenetic modification whereby methyl groups are added to cytosine bases in the DNA sequence. This is generally associated with repression of gene transcription (for review see Ref. 22) though it is known in some cases to give rise to enhanced transcriptional states.[23] While

Table 1 Models of Early Adversity and Their Associated Epigenetic Outcomes, Including the Brain Region and Time Point at Which They Were Found and Any Sex Differences Reported.

Type of Early Adversity	Epigenetic Outcome	Brain Region/ Peripheral Measure	Age	Sex Differences	Reference
	Increased methylation *GR*	Hippocampus	Adult	Only male brains analyzed	7
Low levels of licking/grooming	Decreased methylation *GR*	Hippocampus	Adult	Only female brains analyzed	18
	Increased methylation *ERα*	Medial preoptic area	Adult	Only female brains analyzed	11
	Decreased methylation *Avp*	Paraventricular nucleus	Adult	Only male brains analyzed	9
Maternal separation	Increased methylation *GR*	Hippocampus	Adolescent	Present only in males	19
	Increased methylation *Ntsr1*	Amygdala	Adult	Only male brains analyzed	12
	Decreased methylation *Pomc*	Pituitary	Adult	Only male brains analyzed	20
	Increased methylation *Bdnf* IX	Whole prefrontal cortex	Infant, adolescent, adult	Males and females	10
	Increased methylation *Bdnf* I	Medial prefrontal cortex	Adolescent	Present only in males	5
	Decreased methylation *Bdnf* IV	Medial prefrontal cortex	Adolescent	Present only in females	5
	Decreased methylation *Bdnf* I	Medial prefrontal cortex	Adult	Males and females	5

(Continued)

Table 1 Models of Early Adversity and Their Associated Epigenetic Outcomes, Including the Brain Region and Time Point at Which They Were Found and Any Sex Differences Reported.—cont'd.

Type of Early Adversity	Epigenetic Outcome	Brain Region/ Peripheral Measure	Age	Sex Differences	Reference
	Increased methylation *Bdnf* IV	Medial prefrontal cortex	Adult	Present only in females	5
Scarcity-adversity model of low nesting resources	Decreased methylation *Bdnf* I	Amygdala	Infant	Present only in females	6
	Increased methylation *Bdnf* IV	Amygdala	Adolescent	Present only in females	6
	Decreased methylation *Bdnf* I	Amygdala	Adult	Present only in males	6
	Decreased methylation *Bdnf* IV	Amygdala	Adult	Males and females	6
	Decreased methylation *Bdnf* I, IV	Dorsal hippocampus	Adult	Present only in males	6
	Increased methylation *Bdnf* I	Ventral hippocampus	Adult	Present only in females	6
	Increased methylation *Bdnf* IV	Ventral hippocampus	Infant, adolescent	Present only in females	6
	Increased methylation *NR3C1*	Hippocampus	Adult	Only male brains analyzed	4
Childhood abuse/ neglect (humans)	Site-specific differential methylation, *BDNF*, *NR3C1* , *FKBP5*	Saliva	Child/adolescent, 5–14 years	Males and females	13
	Increased methylation, *BDNF*, *OXTR*	Blood	Adult	Males and females	21
	Differential methylation Genome-wide	Saliva	Child, average age nine	Gene-dependent sex differences	8

BDNF, Brain-derived neurotrophic factor; *GR* , glucocorticoid receptor; *OXTR* , oxytocin receptor.

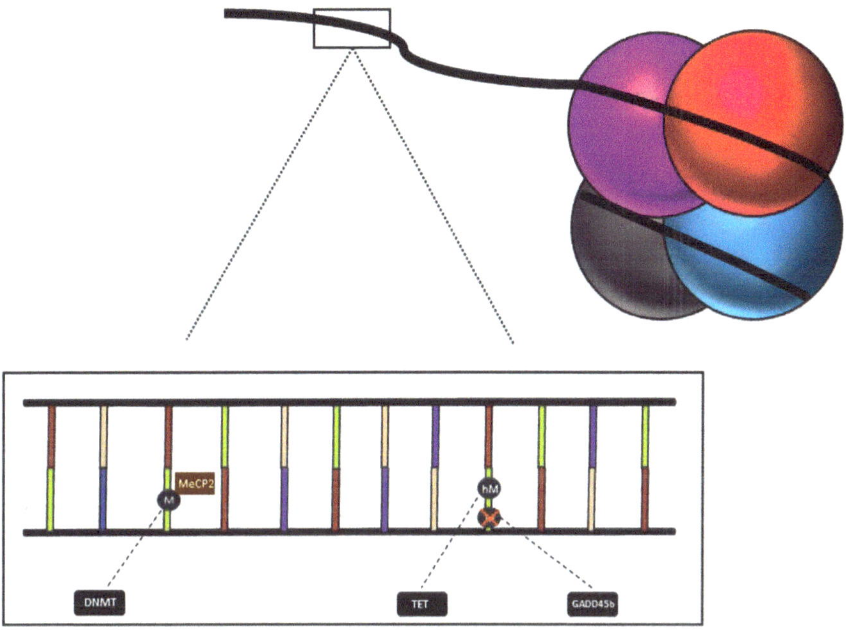

Fig. 1 Schematic of DNA wrapped around histones. Zoom-in of DNA strand demonstrates the addition of methyl groups (M) to cytosines by DNMT enzymes, association of methyl-CpG binding protein 2 with methyl groups, oxidation of methyl groups by ten-eleven translocation enzymes to create hydroxymethylcytosines, and demethylation activity of GADD45b.

traditionally discussed as occurring at cytosine-guanine (CG) dinucleotides, non-CG methylation is also known to occur and have considerable implications for gene expression.[24–26] Methyl group addition is accomplished by DNA methyltransferases DNMT1, DNMT3a, or DNMT3b depending on whether the mark is new or being maintained during replication.[27] DNA demethylation may occur through oxidation of methyl marks[28] or via enzymes such as Gadd45b.[29] The effects of methylation are partly accomplished via methyl-CpG binding protein 2 (MeCP2), which binds to methylated DNA and facilitates gene silencing or expression via recruitment of corepressors[30] or coactivators[23] (see Fig. 1 for schematic of epigenetic enzymes and their roles). DNA methylation is the most heavily studied epigenetic alteration in the realm of early stress and psychiatric disorder and is thus the focus of the ensuing discussion.

2. SPECIFICITY OF EXPERIENCE-INDUCED METHYLATION CHANGES

2.1 Tissue Specificity

Though some postmortem brain studies have been published, human studies of early-stress, psychiatric outcome, and the epigenome generally rely on peripheral measures such as saliva and blood. Use of these peripheral measures has greatly informed the literature on stress-induced methylation statuses of genes critical to mental health, such as brain-derived neurotrophic factor (*BDNF*),[13,31] serotonin transporter (*SLC6A4*),[16] and oxytocin receptor (*OXTR*).[21] However, the question remains of whether the epigenetic state of these peripheral tissues accurately reflects what is going in the brain, the target of interest in the study of psychiatric disorder.

Examination of human samples of various tissue types (i.e. brain, blood, saliva) has revealed baseline levels of epigenetic variation between tissues, but has also revealed some promising correlations that prompt further investigation. A 2012 study examining methylated regions of the human genome in multiple brain regions and in whole blood reported high variation between tissues, especially at genes critical to developmental processes. However, a correlation between brain and blood was reported for some of the differences found between individual subjects.[32] Importantly, this report shows tissue-specific differences in methylation to occur primarily in intragenic locations, with much less between-tissue variation occurring in promoter regions.[32] Given that promoter and intragenic methylation have distinct effects on transcriptional activity (for review see Ref. 33), each of these findings bear important implications. Another study comparing methylation between brain and blood in humans found blood to be a poor predictor of brain and in contrast to the previously mentioned study, found that the majority of the few sites exhibiting correlation with brain tissue were located in gene bodies.[34] A 2014 study revealed that though overall positively correlated, methylation within CpG islands is highly dissimilar between saliva and blood in humans, with saliva being a closer match to methylation patterns found in the brain.[35] Blood is also a poor predictor of human brain *BDNF* exon I methylation, though muscle tissue methylation at this locus exhibits a high correlation to brain.[36] Interestingly, methylation in brain and blood have been found similar in an animal model. A study examining correlations between brain and blood in a mouse model of Cushing's disease found the two to be positively correlated for methylation of *Fkbp5*, though methylation changes occurred at different regions in each tissue type.[37] However, the disease variable must be taken into account when interpreting these results in terms of tissue specificity as disease could have multiple epigenetic effects between tissues that may not be seen in a normal state. The results of cross-tissue comparison studies in humans should also be interpreted with caution given that, due to complications typically associated with postmortem tissue collection, it is not often possible to obtain central and peripheral samples from the same subjects (though some of the aforementioned studies were able to do so). Even so, they provide a valuable foundation for exploring the issue of intra-tissue epigenetic variation in relation to the study of psychiatric disorder.

Despite methodological complications, studies of the epigenome following early stress in humans report findings that are promising in terms of the predictive value of peripheral measures. For example, increased methylation of the glucocorticoid receptor (*GR*) gene following early life stress has been reported in both whole blood[14] and saliva.[13] This is consistent with a 2009 report of increased *GR* methylation in postmortem brain tissue taken from suicide victims with a history of child abuse.[4] In an interesting cross-species comparison, researchers examined white blood cell progenitor cells in human and nonhuman primate cord blood alongside brain tissue (prefrontal cortex) from rats, all following exposure to early adversity. They reported similar early stress-induced epigenetic patterns between these species and their respective tissue for a handful of genes, one of which is associated with major depressive disorder in humans.[38]

2.2 Cell-type Specificity

Specificity of methylation patterns also occurs between cell types. We know that in the central nervous system neurons and glia display distinct patterns of methylation. Compared to glia, neurons exhibit increased levels of both CG and non-CG methylation as well as enriched methylation at CG island shores, enhancers, and intergenic regions.[39–42] Methylation patterns also differ by neuron type[43,44] in what appears to be a functionally significant manner, especially for developmental processes.[24] These cells also exhibit more variability in their epigenetic state than nonneurons,[45] suggesting that the neuronal epigenome responds dynamically to experience. This has proven true in response to early stress exposure: neurons in the rodent medial prefrontal cortex (mPFC) exhibit increased levels of *Bdnf* methylation in response to adverse caregiving when compared to nonneurons.[46]

Beyond the importance of understanding precisely where in the genome epigenetic modifications are taking place in response to adversity, taking into account cell-type specificity in peripheral samples may be critical for drawing parallels to the brain. For example, saliva samples containing a higher proportion of leukocytes are more positively correlated with blood methylation patterns whereas saliva samples containing a higher proportion of epithelial cells are more positively correlated with brain methylation patterns.[35]

3. BEYOND METHYLATION

3.1 DNA Hydroxymethylation

Once established, a methyl mark on a cytosine can be oxidized into a hydroxyl group, creating a hydroxymethylated cytosine. This is accomplished by the family of ten-eleven translocation (TET) proteins.[47] This recently discovered mark may play multiple roles in the epigenome. Evidence supports a transient role critical to the process of active demethylation[28] as well as a stable role in neuronal function[48,49] and behavioral outcomes.[50] The relationship of this mark to gene expression appears to be quite dynamic[51,52] and may interact with methylation, changing gene expression outcomes.[53]

3.2 Early Adversity and the Hydroxymethylome

Significantly different patterns of hydroxymethylation are seen in nonhuman primates exposed to early-life adversity (i.e. peer-reared) when compared to their maternally-reared counterparts.[53] Our lab has also found significant changes in this mark using a rodent model of adverse caregiving. Specifically, male rats exposed to our model of early adversity (i.e. the scarcity-adversity model of low nesting resources) exhibit decreased levels of hydroxymethylation within amygdala tissue in adolescence.[54]

The recent discovery of this mark means there has been much less time to address it in studies of early adversity. In fact, methods most commonly used in the field up to this point are unable to distinguish between methylation and hydroxymethylation, though methods do now exist for this purpose[55–57] and are certain to become more commonly used. For example, oxidative bisulfite sequencing (oxBS-seq) entails the oxidation of hydroxymethylcytosines followed by conversion to uracil in a bisulfite treatment, allowing for sole detection of methylated cytosines.[56] Another option is Tet-assisted bisulfite sequencing (TAB-seq) wherein hydroxymethylcytosines are preserved via glucosylation and methylated cytosines are oxidized to carboxylcytosines via Tet1. The carboxylcytosines are then converted to thymine in a bisulfite treatment, allowing for sole detection of hydroxymethylated cytosines.[57]

To the best of our knowledge, no studies yet exist that examine hydroxymethylation at any time point following developmental stress exposure in humans. However, this modification has been studied in both the developing and adult human brain[58–60] as well as in human peripheral samples (i.e. blood and saliva),[61] and we know it is responsive to environmental conditions.[62] These factors comprise the necessary foundation to move forward with examination of this mark within the realm of early adversity in humans.

4. SEX DIFFERENCES

4.1 Animal Models

The epigenome and its many mechanisms are unarguably sexually dimorphic in development and throughout the lifespan. Sex differences have been reported in MeCP2, Gadd45b, and DNMT3a expression in the developing rat amygdala,[63–65] in MeCP2 expression in the developing rat ventromedial hypothalamus and preoptic area,[64] and in methylation patterns of steroid receptors (e.g. estrogen receptor α) in sexually dimorphic regions of the brain (e.g. mediobasal hypothalamus, preoptic area).[66,67] Postdevelopment, sex-specific histone modifications have been reported in the bed nucleus of the stria terminalis and preoptic area in adult mice[68] and again in methylation patterns of steroid receptors in sexually dimorphic brain regions.[66,67] In humans, baseline (i.e. in nonstress/non-disease samples) sex differences in methylation have been reported.[69,70] Bear in mind that as the axis of this review spins on early adversity, these lists of baseline epigenetic differences are for illustration and thus are not exhaustive.

Though foundational studies of the epigenome in the early stress literature were conducted only in males, a large amount of data have since been produced to support the sex-specific nature of epigenetic modifications following early adversity. One of these foundational studies found that natural variations in maternal care (i.e. high versus low licking and grooming by a dam) resulted in altered methylation patterns and anxiety-like behavior in adult male offspring.[7] A study published approximately a decade later examined outcomes in female offspring and found interesting differences (i.e. different in many respects from the male data) in both epigenetic and behavioral responses.[18] Though methodological differences between these studies restrict a direct comparison, the collective results nevertheless point to the need for sex-specific examination.

Using a version of the scarcity-adversity model of low nesting resources, wherein pups experience daily exposure to adverse conditions outside the home cage during the first week of life, our lab has observed sex-specific alterations in epigenetic regulators in the mPFC as late as postnatal day 90.[71] We have also reported numerous sex differences in adversity-induced *Bdnf* methylation patterns. We find that females exhibit changes in *Bdnf* methylation that very rarely match male patterns.[5,6,54] We also observe significant, brain-region specific differences in genome-wide DNA methylation and hydroxymethylation in adversity-exposed adolescent males, an effect not observed in females.[54] These alterations are accompanied by sex-specific deficits in select behavioral tasks,[72] data which support the hypothesis that altered epigenetic patterns are behaviorally functional. Exposure to maternal separation, another commonly used model of early adversity, results in similar sex-specific outcomes. For example, in a mouse model of maternal separation, altered *Nr3c1* (*GR*) methylation is seen in the adolescent male but not female hippocampus of C57BL/6J mice, a change accompanied by various male-specific behavioral alterations.[19]

4.2 Humans

Interestingly, methylation of *NR3C1* in humans exposed to early adversity is not reported to differ between males and females in adulthood.[15] The same appears to be true for early stress-induced methylation changes in genes whose methylation status significantly predicts depression.[13] However, sex differences in methylation have been reported in children[8] and adolescents[73] exposed to early stress. Many variables could account for these opposing results. Tissue types less representative of the brain (e.g. blood) may not pick up sex differences that would normally be seen in the brain. Differing types, intensities, and timing of adversity may have more or less of a sex effect. Lastly, gene-specific studies not reporting a sex difference may simply be examining a gene that does not exhibit sex differences in methylation in response to adversity.

5. METHYLATION ACROSS THE LIFESPAN

We now know that methylation across the genome is tied to age in both nonhuman animals[9,74] and humans.[70,75,76] In addition, data suggest that the effect of early environmental alterations on health at a later age is of an epigenetic nature. For example, prenatal exposure to famine is related to altered methylation of growth- and metabolic-related genes more than half a century after exposure.[77,78] Adversity in the context of caregiving in humans and nonhumans is known to leave its epigenetic mark early on[14,20,79] and/or in adulthood.[4,9,20] That is, in some cases these marks are seen persistently across development and into adulthood while in other cases they are dynamic across the lifespan, appearing in some life stages and not in others. These stress-induced changes in methylation have even been reported to override changes typically seen in the aging methylome.[9]

In nonhuman primates, serotonin transporter methylation appears to be elevated in both infancy[79] and adulthood[80] following early stress, though it is important to note that different types of early stress were employed in these two studies. In a mouse model of early stress, *Avp* hypomethylation was detected at six weeks of age and persisted up to the last measurement at one year of age.[9] In a similar model, *Pomc* hypomethylation is maintained within the same time frame.[20] In contrast to these sustained marks, our lab has found developmental changes in methylation of the *Bdnf* gene in our adversity-exposed rats. In infancy we find that adversity-exposed females exhibit a brain region-specific decrease in methylation of *Bdnf* exon I alongside increased methylation of exon IV, the reverse of which is then found in adulthood.[6] In males, the only developmental change detected in *Bdnf* methylation is at exon I in adolescence.[5] Not only is that mark found to be reversed (i.e. opposite direction in methylation) in adulthood, six additional alterations in *Bdnf* methylation emerge at this time point.[5,6] We also find several significant, maltreatment-specific alterations in epigenetic regulators in the mPFC of our adult animals that are not detected in infancy.[71] Taken together, these data underline the importance of longitudinal studies of environmental effects upon the epigenome, a type of study that is somewhat lacking in the field at present.

6. WHAT WE DO NOT KNOW AND HOW WE GET THERE

6.1 Tissue- and Cell-type Specificity

Currently, one of the largest issues in studying epigenetic signatures following adversity in humans is the lack of understanding of the relationship between central and peripheral epigenetic marks. Though we may not necessarily expect similar marks between central and peripheral measures, thorough investigation and comparison of the epigenome between tissue types will aid in understanding if there is a relationship between them, and if there is, how we can use that relationship to understand what is happening in the brain when only peripheral measures are available. An example of the utility of this approach can be seen in work by Walton and colleagues.[34] Following initial examination of tissue-specific methylation in samples obtained from epileptic patients during neurosurgery, these researchers isolated a small number of sites in which significant correlations existed between methylation in blood and brain. Using these sites they examined methylation in blood samples of schizophrenia subjects, an approach that led them to discover an association between schizophrenia and methylation of a gene (*AVP*) and its receptor (*AVPR1A*) previously associated with this disorder.[34]

Collection of more biosamples in rodent studies will be incredibly useful in this endeavor. Previous reports of correlation of tissue-specific methylation between-species (i.e. mouse and human)[43,81] support the examination of peripheral tissues in animal models to draw parallels in humans, though more studies are needed to know if correlations also exist for adversity-induced methylation patterns. Peripheral and central tissue collection in rodent studies can be easily accomplished within subjects and will greatly contribute to the quest of understanding the peripheral-central relationship of the psychiatric epigenome. However, there are important caveats to consider in making cross-species comparisons. For example, the previously discussed correlation between methylation patterns in mouse and human brain is more highly conserved in inhibitory than excitatory neurons.[43]

When possible, collecting multiple types of peripheral samples within human subjects will help us better understand intra-tissue epigenetic variation in humans. Given that methylation patterns exhibit brain region-specific changes in response to adversity, it will also be necessary to examine multiple regions individually to understand which of them exhibit a relationship to peripheral measures and what each relationship looks like.

Though access to human brain tissue is difficult to obtain (especially when specific conditions, like adversity, are required), the collection of more brain regions from the same subject is warranted.

Currently, there are few studies on cell-type specific methylation following early stress. However, existing literature clearly demonstrates stress-induced epigenetic responses and the field would greatly benefit from expanding these studies to understand more precisely where these changes are occurring. Given the known mental health outcomes of developmental stress exposure, this understanding is critical to the intersection of epigenetics and psychopathology.

6.2 Hydroxymethylation and Non-CG Methylation

Another necessary step in the field will be increased utilization of the techniques referenced in Section 3.2 in order to distinguish between methylated and hydroxymethylated cytosines. Considering what is known at this point about the distinct effects of these two modifications on gene expression, it will be critical to parse them apart if we wish to understand epigenetic regulation of behavior following early adversity. Additionally, investigation of hydroxymethylation in human early adversity studies will be essential in moving the field forward.

Another component of the epigenome requiring more attention is non-CG methylation. Examining methylation at CG sites is the dominant practice at present. However, non-CG methylation is not only known to occur (as referenced in Section 1.2) but is reportedly the preeminent form of methylation in neurons.[40] Furthermore, when occurring in the gene body, it has been reported to be more predictive of neuronal gene expression than both CG methylation and measures of chromatin accessibility.[24] Given this powerful relationship to functional outcome, examination of non-CG methylation is paramount to understanding how epigenetic alterations subserve the relationship between early adversity and psychiatric dysfunction.

6.3 Sex Differences

In terms of understanding sex differences in the epigenome following early life stress, female offspring need to be included more consistently in animal research, something that is already occurring based on recently introduced NIH policies on balancing sex in cell and animal studies. Something else to consider is that studies in postmortem brains suffer from a lack of female tissue, making it difficult to verify sex differences in human studies using peripheral samples. Sex differences are prevalent in psychiatric disorders,[82,83] particularly in disorders for which stress is a precipitating factor.[84] Given these facts alongside our knowledge of a sexually dimorphic epigenome in individuals exposed to early adversity and in those with psychiatric disorder, it is highly plausible that epigenetic mechanisms are a causal factor in these sex-specific outcomes. Failing to consistently study sex differences will greatly undermine our understanding of and ability to treat psychiatric dysfunction.

6.4 Lifespan Methylation

Concerning aging, we know the human epigenome bears the marks of early stress at individual time points (e.g. adolescence, adulthood) but we have very little information on how methylation changes across the lifespan in these individuals. The literature would greatly benefit from longitudinal studies in early-stress cohorts (in both animal models and humans) that report epigenetic states at various age points following the initial exposure. Understanding the state of the epigenome at a given point in time has great potential to inform the medical community based on the age a given treatment or intervention might be administered.

7. CONCLUDING REMARKS

For a young field, great strides have certainly been made. The foundation that exists in the current literature gives us every opportunity to address these challenges and others that will surely arise. Given the nature of these studies, some obstacles will be more difficult to address than others. For example, obtaining postmortem brain tissue under specific conditions (i.e. from subjects with known histories of early stress) is difficult, as is obtaining peripheral samples from the same subjects due to timing (of death and of the research) as well as the conditions of death. Longitudinal studies also present unique challenges due to issues such as subject attrition. However, the potential of this work to improve the human condition definitively outweighs the cost. The implications are far-reaching, with the potential to shape both policy and medicine in a manner that will meaningfully affect the lives of many in terms of development and/or in the prevention and treatment of disease.

ACKNOWLEDGEMENTS

Work on this article was supported by a grant from The Eunice Kennedy Shriver National Institute of Child Health and Human Development (R01HD087509) to TLR and a University Dissertation Fellowship from the University of Delaware to TSD.

REFERENCES

1. Vachon DD, Krueger RF, Rogosch FA, Cicchetti D. Assessment of the harmful psychiatric and behavioral effects of different forms of child maltreatment. *JAMA Psychiatry.* 2015;72(11):1135–1142.
2. Carr CP, Martins CMS, Stingel AM, Lemgruber VB, Juruena MF. The role of early life stress in adult psychiatric disorders: a systematic review according to childhood trauma subtypes. *J Nerv Mental Dis.* 2013;201(12):1007–1020.
3. Heim C, Nemeroff CB. The role of childhood trauma in the neurobiology of mood and anxiety disorders: preclinical and clinical studies. *Biol Psychiatry.* 2001;49(12):1023–1039.
4. McGowan PO, Sasaki A, D'Alessio AC, et al. Epigenetic regulation of the glucocorticoid receptor in human brain associates with childhood abuse. *Nat Neurosci.* 2009;12(3):342–348.
5. Blaze J, Scheuing L, Roth TL. Differential methylation of genes in the medial prefrontal cortex of developing and adult rats following exposure to maltreatment or nurturing care during infancy. *Dev Neurosci.* 2013;35(4):306–316.
6. Roth TL, Matt S, Chen K, Blaze J. Bdnf DNA methylation modifications in the hippocampus and amygdala of male and female rats exposed to different caregiving environments outside the homecage. *Dev Psychobiol.* 2014;56:1755–1763.
7. Weaver IC, Cervoni N, Champagne FA, et al. Epigenetic programming by maternal behavior. *Nat Neurosci.* 2004;7(8):847–854.
8. Cicchetti D, Hetzel S, Rogosch FA, Handley ED, Toth SL. An investigation of child maltreatment and epigenetic mechanisms of mental and physical health risk. *Dev Psychopathol.* 2016;28(4pt2):1305–1317.
9. Murgatroyd C, Patchev AV, Wu Y, et al. Dynamic DNA methylation programs persistent adverse effects of early-life stress. *Nat Neurosci.* 2009;12(12):1559–1566.
10. Roth TL, Lubin FD, Funk A, Sweatt JD. Lasting epigenetic influence of early-life adversity on the BDNF gene. *Biol Psychiatry.* 2009;65(9):760–769.
11. Champagne FA, Weaver IC, Diorio J, Dymov S, Szyf M, Meaney MJ. Maternal care associated with methylation of the estrogen receptor-α1b promoter and estrogen receptor-α expression in the medial preoptic area of female offspring. *Endocrinology.* 2006;147(6):2909–2915.
12. Toda H, Boku S, Nakagawa S, et al. Maternal separation enhances conditioned fear and decreases the mRNA levels of the neurotensin receptor 1 gene with hypermethylation of this gene in the rat amygdala. *PLOS ONE.* 2014;9(5):e97421.
13. Weder N, Zhang H, Jensen K, et al. Child abuse, depression, and methylation in genes involved with stress, neural plasticity, and brain circuitry. *J Am Acad Child Adolesc Psychiatry.* 2014;53(4). 417–24.e415.
14. Romens SE, McDonald J, Svaren J, Pollak SD. Associations between early life stress and gene methylation in children. *Child Dev.* 2015;86(1):303–309.
15. Perroud N, Paoloni-Giacobino A, Prada P, et al. Increased methylation of glucocorticoid receptor gene (NR3C1) in adults with a history of childhood maltreatment: a link with the severity and type of trauma. *Transl Psychiatry.* 2011;1(12):e59.
16. Vijayendran M, Beach S, Plume JM, Brody G, Philibert R. Effects of genotype and child abuse on DNA methylation and gene expression at the serotonin transporter. *Front Psychiatry.* 2012;3:55.

17. Klengel T, Mehta D, Anacker C, et al. Allele-specific FKBP5 DNA demethylation mediates gene-childhood trauma interactions. *Nat Neurosci.* 2013;16(1):33–41.
18. Pan P, Fleming AS, Lawson D, Jenkins JM, McGowan PO. Within- and between-litter maternal care alter behavior and gene regulation in female offspring. *Behav Neurosci.* 2014;128(6):736.
19. Kundakovic M, Lim S, Gudsnuk K, Champagne FA. Sex-specific and strain-dependent effects of early life adversity on behavioral and epigenetic outcomes. *Front Psychiatry.* 2013;4:78.
20. Wu Y, Patchev AV, Daniel G, Almeida OF, Spengler D. Early-life stress reduces DNA methylation of the pomc gene in male mice. *Endocrinology.* 2014;155(5):1751–1762.
21. Unternaehrer E, Meyer AH, Burkhardt SC, et al. Childhood maternal care is associated with DNA methylation of the genes for brain-derived neurotrophic factor (BDNF) and oxytocin receptor (OXTR) in peripheral blood cells in adult men and women. *Stress.* 2015;18(4):451–461.
22. Bird A. DNA methylation patterns and epigenetic memory. *Genes Dev.* 2002;16(1):6–21.
23. Chahrour M, Jung SY, Shaw C, et al. MeCP2, a key contributor to neurological disease, activates and represses transcription. *Science.* 2008;320(5880):1224–1229.
24. Mo A, Mukamel Eran A, Davis Fred P, et al. Epigenomic signatures of neuronal diversity in the mammalian brain. *Neuron.* 2015;86(6):1369–1384.
25. Lister R, Pelizzola M, Dowen RH, et al. Human DNA methylomes at base resolution show widespread epigenomic differences. *Nature.* 2009;462(7271):315–322.
26. Xie W, Barr CL, Kim A, et al. Base-resolution analyses of sequence and parent-of-origin dependent DNA methylation in the mouse genome. *Cell.* 2012;148(4):816–831.
27. Bestor TH. The DNA methyltransferases of mammals. *Hum Mol Genet.* 2000;9 (16):2395–2402.
28. Ito S, Shen L, Dai Q, et al. Tet proteins can convert 5-methylcytosine to 5-formylcytosine and 5-carboxylcytosine. *Science.* 2011;333(6047):1300–1303.
29. Ma DK, Jang M-H, Guo JU, et al. Neuronal activity-induced Gadd45b promotes epigenetic DNA demethylation and adult neurogenesis. *Science.* 2009;323 (5917):1074–1077.
30. Nan X, Ng H-H, Johnson CA, et al. Transcriptional repression by the methyl-CpG-binding protein MeCP2 involves a histone deacetylase complex. *Nature.* 1998;393 (6683):386–389.
31. Perroud N, Salzmann A, Prada P, et al. Response to psychotherapy in borderline personality disorder and methylation status of the BDNF gene. *Transl Psychiatry.* 2013;3(1): e207.
32. Davies MN, Volta M, Pidsley R, et al. Functional annotation of the human brain methylome identifies tissue-specific epigenetic variation across brain and blood. *Genome Biol.* 2012;13(6):R43.
33. Jones PA. Functions of DNA methylation: islands, start sites, gene bodies and beyond. *Nat Rev Genet.* 2012;13(7):484.
34. Walton E, Hass J, Liu J, et al. Correspondence of DNA methylation between blood and brain tissue and its application to schizophrenia research. *Schizophr Bull.* 2015;42 (2):406–414.
35. Smith AK, Kilaru V, Klengel T, et al. DNA extracted from saliva for methylation studies of psychiatric traits: evidence tissue specificity and relatedness to brain. *Am J Med Genet B Neuropsychiatr Genet.* 2015;168(1):36–44.
36. Stenz L, Zewdie S, Laforge-Escarra T, et al. BDNF promoter I methylation correlates between post-mortem human peripheral and brain tissues. *Neurosci Res.* 2015;91:1–7.
37. Ewald ER, Wand GS, Seifuddin F, et al. Alterations in DNA methylation of Fkbp5 as a determinant of blood–brain correlation of glucocorticoid exposure. *Psychoneuroendocrinology.* 2014;44:112–122.

38. Nieratschker V, Massart R, Gilles M, et al. MORC1 exhibits cross-species differential methylation in association with early life stress as well as genome-wide association with MDD. *Transl Psychiatry.* 2014;4(8):e429.
39. Li X, Baker-Andresen D, Zhao Q, Marshall V, Bredy TW. Methyl CpG Binding Domain Ultra-Sequencing: a novel method for identifying inter-individual and cell-type-specific variation in DNA methylation. *Genes Brain Behav.* 2014;13(7):721–731.
40. Lister R, Mukamel EA, Nery JR, et al. Global epigenomic reconfiguration during mammalian brain development. *Science.* 2013;341(6146):1237905.
41. Guintivano J, Aryee MJ, Kaminsky ZA. A cell epigenotype specific model for the correction of brain cellular heterogeneity bias and its application to age, brain region and major depression. *Epigenetics.* 2013;8(3):290–302.
42. Kozlenkov A, Roussos P, Timashpolsky A, et al. Differences in DNA methylation between human neuronal and glial cells are concentrated in enhancers and non-CpG sites. *Nucleic Acids Res.* 2013;42(1):109–127.
43. Luo C, Keown CL, Kurihara L, et al. Single-cell methylomes identify neuronal subtypes and regulatory elements in mammalian cortex. *Science.* 2017;357(6351):600–604.
44. Kozlenkov A, Wang M, Roussos P, et al. Substantial DNA methylation differences between two major neuronal subtypes in human brain. *Nucleic Acids Res.* 2015;44(6):2593–2612.
45. Iwamoto K, Bundo M, Ueda J, et al. Neurons show distinctive DNA methylation profile and higher interindividual variations compared with non-neurons. *Genome Res.* 2011;21(5):688–696.
46. Blaze J, Roth TL. Caregiver maltreatment causes altered neuronal DNA methylation in female rodents. *Dev Psychopathol.* 2017;29(2):477–489.
47. Pastor WA, Aravind L, Rao A. TETonic shift: biological roles of TET proteins in DNA demethylation and transcription. *Nat Rev Mol Cell Biol.* 2013;14(6):341.
48. Szulwach KE, Li X, Li Y, et al. 5-hmC-mediated epigenetic dynamics during postnatal neurodevelopment and aging. *Nat Neurosci.* 2011;14(12):1607–1616.
49. Kato T, Iwamoto K. Comprehensive DNA methylation and hydroxymethylation analysis in the human brain and its implication in mental disorders. *Neuropharmacology.* 2014;80:133–139.
50. Li X, Wei W, Zhao Q-Y, et al. Neocortical Tet3-mediated accumulation of 5-hydroxymethylcytosine promotes rapid behavioral adaptation. *Proc Natl Acad Sci USA.* 2014;111(19):7120–7125.
51. Robertson J, Robertson AB, Klungland A. The presence of 5-hydroxymethylcytosine at the gene promoter and not in the gene body negatively regulates gene expression. *Biochem Biophys Res Commun.* 2011;411(1):40–43.
52. Valinluck V, Tsai H-H, Rogstad DK, Burdzy A, Bird A, Sowers LC. Oxidative damage to methyl-CpG sequences inhibits the binding of the methyl-CpG binding domain (MBD) of methyl-CpG binding protein 2 (MeCP2). *Nucleic Acids Res.* 2004;32(14):4100–4108.
53. Massart R, Suderman M, Provencal N, et al. Hydroxymethylation and DNA methylation profiles in the prefrontal cortex of the non-human primate rhesus macaque and the impact of maternal deprivation on hydroxymethylation. *Neuroscience.* 2014;268:139–148.
54. Doherty TS, Forster A, Roth TL. Global and gene-specific DNA methylation alterations in the adolescent amygdala and hippocampus in an animal model of caregiver maltreatment. *Behav Brain Res.* 2016;298:55–61.
55. Kawasaki Y, Kuroda Y, Suetake I, Tajima S, Ishino F, Kohda T. A Novel method for the simultaneous identification of methylcytosine and hydroxymethylcytosine at a single base resolution. *Nucleic Acids Res.* 2017;45(4):e24.
56. Booth MJ, Ost TW, Beraldi D, et al. Oxidative bisulfite sequencing of 5-methylcytosine and 5-hydroxymethylcytosine. *Nat Protoc.* 2013;8(10):1841.

57. Yu M, Hon GC, Szulwach KE, et al. Tet-assisted bisulfite sequencing of 5-hydroxymethylcytosine. *Nat Protoc.* 2012;7(12):2159.
58. Wang T, Pan Q, Lin L, et al. Genome-wide DNA hydroxymethylation changes are associated with neurodevelopmental genes in the developing human cerebellum. *Hum Mol Genet.* 2012;21(26):5500–5510.
59. Wen L, Li X, Yan L, et al. Whole-genome analysis of 5-hydroxymethylcytosine and 5-methylcytosine at base resolution in the human brain. *Genome Biol.* 2014;15(3):R49.
60. Gross JA, Pacis A, Chen GG, Barreiro LB, Ernst C, Turecki G. Characterizing 5-hydroxymethylcytosine in human prefrontal cortex at single base resolution. *BMC Genomics.* 2015;16(1):672.
61. Godderis L, Schouteden C, Tabish A, et al. Global methylation and hydroxymethylation in DNA from blood and saliva in healthy volunteers. *BioMed Res Int.* 2015;2015:845041.
62. Tellez-Plaza M, Tang W-Y, Shang Y, et al. Association of global DNA methylation and global DNA hydroxymethylation with metals and other exposures in human blood DNA samples. *Environ Health Perspect.* 2014;122(9):946–954.
63. Kolodkin MH, Auger AP. Sex difference in the expression of DNA methyltransferase 3a in the rat amygdala during development. *J Neuroendocrinol.* 2011;23(7):577–583.
64. Kurian JR, Forbes-Lorman RM, Auger AP. Sex difference in Mecp2 expression during a critical period of rat brain development. *Epigenetics.* 2007;2(3):173–178.
65. Kigar SL, Chang L, Hayne MR, Karls NT, Auger AP. Sex differences in Gadd45b expression and methylation in the developing rodent amygdala. *Brain Res.* 2016;1642:461–466.
66. Nugent BM, Schwarz JM, McCarthy MM. Hormonally mediated epigenetic changes to steroid receptors in the developing brain: implications for sexual differentiation. *Horm Behav.* 2011;59(3):338–344.
67. Schwarz JM, Nugent BM, McCarthy MM. Developmental and hormone-induced epigenetic changes to estrogen and progesterone receptor genes in brain are dynamic across the life span. *Endocrinology.* 2010;151(10):4871–4881.
68. Shen EY, Ahern TH, Cheung I, et al. Epigenetics and sex differences in the brain: a genome-wide comparison of histone-3 lysine-4 trimethylation (H3K4me3) in male and female mice. *Exp Neurol.* 2015;268:21–29.
69. El-Maarri O, Becker T, Junen J, et al. Gender specific differences in levels of DNA methylation at selected loci from human total blood: a tendency toward higher methylation levels in males. *Hum Genet.* 2007;122(5):505–514.
70. Boks MP, Derks EM, Weisenberger DJ, et al. The relationship of DNA methylation with age, gender and genotype in twins and healthy controls. *PLoS ONE.* 2009;4(8):e6767.
71. Blaze J, Roth T. Exposure to caregiver maltreatment alters expression levels of epigenetic regulators in the medial prefrontal cortex. *Int J Dev Neurosci.* 2013;31(8):804–810.
72. Doherty TS, Keller S, Blaze J, Roth TL. Phenotypic outcomes in adolescence and adulthood in the scarcity-adversity model of low nesting resources outside the home cage. *Dev Psychobiol.* 2017;59(6):703–714.
73. Essex MJ, Thomas Boyce W, Hertzman C, et al. Epigenetic vestiges of early developmental adversity: childhood stress exposure and DNA methylation in adolescence. *Child Dev.* 2013;84(1):58–75.
74. Maegawa S, Hinkal G, Kim HS, et al. Widespread and tissue specific age-related DNA methylation changes in mice. *Genome Res.* 2010;20(3):332–340.
75. Horvath S. DNA methylation age of human tissues and cell types. *Genome Biol.* 2013;14(10):3156.
76. Horvath S, Zhang Y, Langfelder P, et al. Aging effects on DNA methylation modules in human brain and blood tissue. *Genome Biol.* 2012;13(10):R97.

77. Heijmans BT, Tobi EW, Stein AD, et al. Persistent epigenetic differences associated with prenatal exposure to famine in humans. *Proc Natl Acad Sci USA.* 2008;105 (44):17046–17049.
78. Tobi EW, Lumey L, Talens RP, et al. DNA methylation differences after exposure to prenatal famine are common and timing- and sex-specific. *Hum Mol Genet.* 2009;18 (21):4046–4053.
79. Kinnally EL, Capitanio JP, Leibel R, et al. Epigenetic regulation of serotonin transporter expression and behavior in infant rhesus macaques. *Genes Brain Behav.* 2010;9 (6):575–582.
80. Kinnally EL, Feinberg C, Kim D, et al. DNA methylation as a risk factor in the effects of early life stress. *Brain Behav Immun.* 2011;25(8):1548–1553.
81. Irizarry RA, Ladd-Acosta C, Wen B, et al. The human colon cancer methylome shows similar hypo- and hypermethylation at conserved tissue-specific CpG island shores. *Nat Genet.* 2009;41(2):178–186.
82. Dohrenwend BP, Dohrenwend BS. Sex differences and psychiatric disorders. *Am J Sociol.* 1976;81(6):1447–1454.
83. Earls F. Sex differences in psychiatric disorders: origins and developmental influences. *Psychiatr Dev.* 1987;5(1):1–23.
84. Bangasser DA, Valentino RJ. Sex differences in stress-related psychiatric disorders: neurobiological perspectives. *Front Neuroendocrinol.* 2014;35(3):303–319.

CHAPTER TWO

Chromosomal Conformations and Epigenomic Regulation in Schizophrenia

Prashanth Rajarajan[1], Yan Jiang, Bibi S. Kassim, Schahram Akbarian
Icahn School of Medicine at Mount Sinai, New York, NY, United States
[1] Corresponding author. E-mail address: prashanth.rajarajan@icahn.mssm.edu

Contents

Abstract

Chromosomal conformations, including promoter-enhancer loops, provide a critical regulatory layer for the transcriptional machinery. Therefore, schizophrenia, a common psychiatric disorder associated with broad changes in neuronal gene expression in prefrontal cortex and other brain regions implicated in psychosis, could be associated with alterations in higher-order chromatin. Here, we review early studies on spatial genome organization in the schizophrenia postmortem brain and discuss how integrative approaches using cell culture and animal model systems could gain deeper insight into the potential roles of higher-order chromatin for the neurobiology of and novel treatment avenues for common psychiatric disease.

Progress in Molecular Biology and Translational Science, volume 157
ISSN 1877-1173
http://dx.doi.org/10.1016/bs.pmbts.2017.11.022

1. INTRODUCTION TO THE EPIGENOMICS OF SCHIZOPHRENIA

Schizophrenia is a major psychiatric disorder often with onset in adolescence or young adulthood, with a wide range of symptoms ranging from delusions and hallucinations to severe social withdrawal and apathy, along with reduced lifespan primarily due to cardiovascular disease and suicide.[1–3] Antipsychotic medications are widely prescribed and mostly target dopaminergic and serotonergic receptor systems,[4,5] but often fail to deliver a satisfactory therapeutic response.[6,7] Cognitive symptoms in particular are severe, chronically disabling, and often persistent during the course of illness.[8] Regrettably, these symptoms are ineffectively treated with antipsychotic medication.[9] The genetic risk architecture of schizophrenia is exceedingly complex. For example, in a study involving 150,000 subjects, the Psychiatric Genomics Consortium identified altogether 108 haplotypes that by individual small effect contribute to the heritable risk for schizophrenia.[10] In addition, there are rare mutations and variants discovered by comprehensive sequencing of all protein coding genes (the "exome," which comprises approximately 1% of total genome sequence),[11] some of which may be causal to the disease etiology in some of the cases with schizophrenia.[12]

Importantly, there can be little doubt that dysregulation of neuronal gene expression in prefrontal cortex (PFC) and various other brain regions implicated in the neural circuitry of psychosis contributes to the pathophysiology of schizophrenia, broadly affecting excitatory and inhibitory neurotransmission, metabolism, myelination, and immune signaling.[13–19] With the transcriptional process intimately connected to chromatin structure and function in human cells and model organisms alike,[20,21] one would therefore expect that epigenomic markers associated with open (active, loose) chromatin permissive for gene expression, versus repressed and silenced chromatin, will show significant alterations in brain tissue from subjects diagnosed with schizophrenia. Indeed, work conducted over the course of the last 15 years has provided first insights into epi-(Greek for *over* and *above*)-genomic aberrations encountered in brain tissue from subjects diagnosed with schizophrenia. Such types of epigenomic explorations in the postmortem brain initially focused on DNA methylation, one of the key epigenetic mechanisms involved in the regulation of gene expression.[22] Methylation at the cytosine C5 position, primarily in the context of cytosine-guanine (CpG) dinucleotides, when located in gene promoters often is implicated in gene repression by directly impeding the binding of transcription factors and by inducing repressive chromatin structure nonpermissive to transcription.[23] Early studies, examining the DNA methylation status of candidate genes affected by dysregulated expression in brains of schizophrenia reported differential DNA methylation profiles in diseased cerebral cortex for key regulators of neuronal connectivity, such as *REELIN* (*RELN*)[24,25] and key transcription factors, such as sex-determining region Y-box containing gene 10 (SOX10),[26] to mention just two examples. Recent genome-wide mapping of the DNA methylome[27,28] reported DNA methylation changes for various genes implicated in excitatory or inhibitory neurotransmission, with one study exploring the DNA methylome in the prefrontal cortex of 191 subjects with schizophrenia in comparison to 335 controls collected across the lifespan identifying >2000 CpG sites with altered methylation levels in diseased tissue.[29] However, the functional implications of the overall extremely subtle methylation differences (on average, 1.3% difference between schizophrenia and control brains for significantly affected CpG sites) in the aforementioned study[29] remain unclear.

Future studies, exploring the DNA methylome of specific brain cell populations in diseased versus control brains[30,31] as opposed to the earlier studies which utilized tissue homogenate or correlational analyses between the brain's DNA methylomes and structural or functional defects in neurons,[32,33] may provide deeper insight into the role of epigenetic dysregulation affecting the brains of subjects with schizophrenia. Of note, some of the epigenomic determinants of chromatin structure and function in fetal and adult human brain, including histone methylation and acetylation markings associated with *cis*-regulatory sequences, such as gene promoters, enhancers, and repressors, show an impressive, up to 26-fold enrichment for single nucleotide polymorphisms associated with heritable risk for schizophrenia.[34] These effects were highly tissue-specific, because brain histone methylation landscapes showed no enrichment for polymorphisms associated with rheumatoid arthritis and other common disorders usually not affecting the central nervous system.[34] Therefore, given that: (1) DNA methylation alterations have been reported in postmortem brain of subjects with schizophrenia, and (2) the significant association of histone modification landscapes from brain cells with the genetic risk architecture of the disease, it is very likely that "epigenomic dysregulation"—including alterations in the expression of specific genes or disruptions in the coordinated regulation of multiple transcriptional units—could be key drivers underlying cortical dysfunction in schizophrenia.

2. REGULATION OF CHROMOSOMAL CONFORMATIONS IN NEURODEVELOPMENT

As discussed in the introduction, evidence arising both from candidate gene studies and genome-scale mappings of DNA methylation and histone modifications provided strong support for the hypothesis that alterations in chromatin structure and function could contribute to transcriptional dysregulation in brains of subject with schizophrenia. However, as discussed in a recent review,[35] comprehensive exploration of the epigenome in schizophrenia and other cognitive disease has to go far beyond DNA methylation, histone modifications, and other regulators of gene expression that are typically "charted" as epigenetic signals directly onto and above the "linear" genome in two dimensions (Fig. 1). Instead, it is increasingly recognized that the spatial configuration and packaging of interphase chromosomes, the equivalent of 250 cm of DNA from a single nucleus when unwound, involves myriads of nonrandomly occurring chromosomal loop structures and DNA–DNA proximity conformations that bypass the linear genome across a wide range of genomic distance, from hundreds of base pairs (such as a short loop formation connecting a transcription start site and proximal gene promoter with a nearby enhancer) to many megabases of interspersed sequence (as in the case of some long-range contacts on the inactive X chromosome in female somatic cells). Proper regulation of these types of

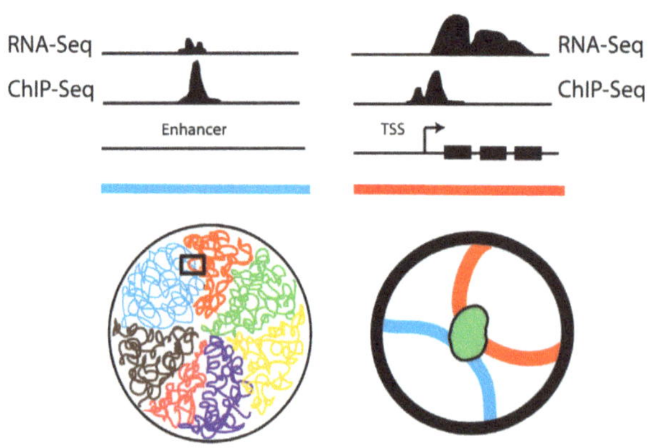

Fig. 1 *(Top)* Conventional ChIP-Seq and RNA-seq tracks on the linear genome demonstrate enhancer and promoter features. *(Bottom)* However, these traditional approaches miss the fact that the two elements, although quite distal in the linear genome, are actually in close physical proximity through looping interactions.

3D genome structures are considered critical for the regulation of gene expression, maintenance of genome integrity and stability, control of growth and differentiation, among many other functions. Surprisingly, however, there is extremely little knowledge about the regulation of the three-dimensional (3D) genome in developing or mature brain, including potential alterations in neuropsychiatric disease, such as schizophrenia.

2.1 Building Blocks of the 3D Genome

We will start our discussion of the 3D genome and its implications for schizophrenia research with a brief overview of its basic principles and organization. Nuclei, separated by a nuclear membrane from the cytoplasm, contain the genome packaged into chromatin fibers as nucleosomal arrays. Nucleosomes are comprised of 146 bp DNA wrapped around a core histone octamer, and interconnected by linker DNA and linker histones. Chromatin can exist in different "states," including "open" (eu-) and condensed (hetero-) chromatin. These are differentially defined by three characteristics: (1) loose or dense nucleosomal packaging, (2) specific types of posttranslational histone modifications, and (3) presence or absence of various chromatin regulatory proteins that either facilitate or repress transcription. For example, actively expressed genes in open chromatin show high levels of histone acetylation, with nucleosome-free intervals occupied by activator proteins (transcription factors) and the RNA polymerase II initiation complex (Fig. 2). Superimposed upon this nucleosomal organization is the 3D conformation of chromatin fibers and entire chromosomes, often described in terms, such as "loops" or "globules" and in toto referred to as the "3D genome." These chromosomal conformations, to a certain extent, reflect the two alternative chromatin states mentioned earlier. For example, euchromatic and heterochromatic sequences tend to assemble into alternating regions of approximately ~5 megabases (Mb). These "compartments," positioned along the same chromosome, could interact with compartments from different chromosomes.[36] Furthermore, A or B compartment designations are correlated with features, such as DNA accessibility and transcriptional activity. Thus, A compartments are enriched for euchromatic regions with much higher overall levels of transcription; whereas B compartments primarily harbor inactive and heterochromatic sequences[37] (Fig. 2). Some of the largest clusters of heterochromatin are microscopically visible and enriched at the nuclear periphery and around nucleolar membranes.[38]

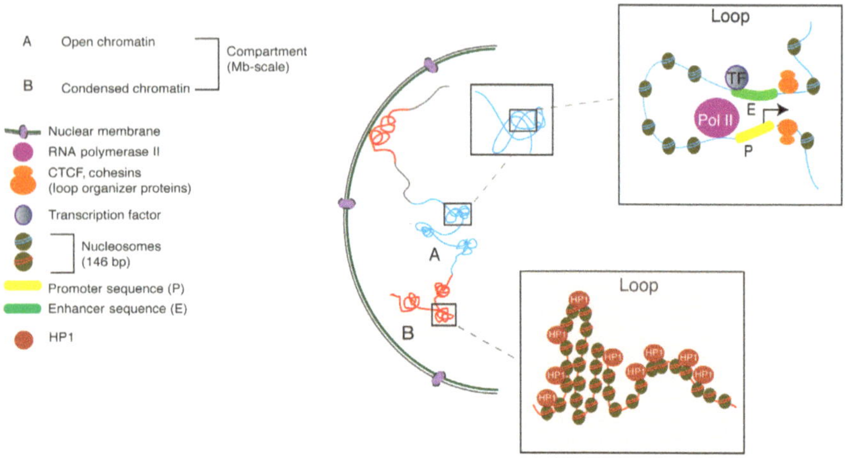

Fig. 2 Within the nucleus, chromatin is organized into megabase-scale active or inactive compartments (*blue* and *red circles*, respectively), which contain many loops that are facilitative and allow promoter-enhancer contacts *(top box)* or are repressive and make heterochromatic the contained regions *(bottom box)*. Loops are often scaffolded by key proteins, CTCF and cohesin (*orange circles*).

Studies exploring scaffolding and regulatory proteins with a critical role for the spatial conformations of the chromosomal materials have primarily focused on the cohesin complex and the CCCTC-binding factor/zinc finger protein CTCF. Cohesins are comprised of five core subunits SMC1, SMC3, RAD21/REC8, and STAG1-3 in humans.[39] In addition, accessory proteins load or release the complex onto chromosomes.[39] Cohesins were initially explored in the context of sister chromatid cohesion and segregation during cell division (mitosis).[39] However, these proteins maintain a heavy presence in the nuclear proteome of postmitotic cells, including neurons.[40] Cohesins form ring-like structures, literally entrapping distant chromatin fibers into chromosomal loops[39] (Fig. 2). Cohesins are highly enriched at actively expressed genes in a tissue- and cell-type specific manner.[39] In contrast, CTCF, while dispensable for cohesin loading onto DNA, orchestrates cohesin enrichment at select binding sites.[40] As a result, chromosomal loops cooccupied by cohesins and CTCF at both ends often associate with broader stretches of regulatory domains, marking the coregulated repression or expression of groups of genes in a cell-type specific manner.[41] The CTCF binding sites are thought to often (but not exclusively) be positioned in inward/convergent and, to a somewhat lesser degree, tandem orientation at the two contact sites of the loop.[37,42] CTCF directionally recognizes binding sites via an 11 zinc finger array, while cohesin undergoes assembly from the CTCF's C-terminal end, often resulting in higher-order chromatin with loop-bound head-to-head CTCF configurations.[43]

2.2 Chromosomal Scaffolding Protein Mutations and Neuropsychiatric Disease

Importantly, a rapidly increasing list of deleterious mutations in genes encoding scaffolding proteins for the 3D genome is linked to neuropsychiatric disease. These include neurodevelopmental disorders, such as Cornelia de Lange syndrome (CdLS)[39,44,45] and adult-onset progressive demyelination syndromes.[46] Neurodevelopmental disease phenotypes in CdLS include intellectual disability, psychosis, and other psychiatric maladies. The underlying genetic defect includes microdeletions and copy number variations affecting core members of the cohesin complex, including *SMC1A* and *SMC3*, and the accessory subunit *NIPBL*.[44] The neurological manifestations could be due to 3D genome disorganization in brain cells and decompaction of chromatin, albeit the precise molecular mechanisms remain to be elucidated.[47] In addition, genetic mutations in *CTCF*, as a key organizer for chromosomal loops, have been linked to monogenic causes of microcephaly and cognitive disorder.[48,49] Consistent with these findings from clinical genetics, selective ablation of *Ctcf* in postnatal mouse brain causes behavioral alterations and dysregulated transcription of hundreds of neuronal transcripts.[50] It remains to be shown, however, whether the neurological manifestations of *Cohesin* gene mutations or *CTCF* are associated with widespread 3D genome alterations in brain. In addition to the aforementioned CTCF and cohesin complex, other examples of neurodevelopmental disease resulting from mutations in genes encoding 3D genome organizer proteins have been identified. These include *Special AT-rich Sequence Binding Proteins 1* and *2* (*SATB1*, *SATB2*) that govern chromosomal territories extending across hundreds of kilobases[51] and anchor chromatin fibers into the nuclear matrix.[52] Of note, *SATB2* is essential for craniofacial development and proper differentiation of transcallosal cortical projection neurons.[53,54] The gene has also been linked to some cases of Glass syndrome (*OMIM 612313*) and mental retardation.[53,55] The related protein, *SATB1*, is essential for connectivity and maturation of the GABAergic interneuron population in the cerebral cortex.[56,57] These findings are of particular interest to the field, given that epigenomic dysregulation of GABAergic gene expression has been reported both for schizophrenia postmortem brain and in the animal and cell culture models.[33,58–61]

3. LOOP DISRUPTIONS POTENTIALLY INVOLVED IN SCHIZOPHRENIA: EARLY FINDINGS

3.1 Promoter-Enhancer Loops and Transcriptional Regulation

Because, as discussed earlier, the molecular pathology of schizophrenia includes transcriptional dysregulation in cerebral cortex and other brain regions, chromosomal conformations associated with the regulation of gene expression are of particular interest to the field. To this end, promoter-enhancer loops represent one type of chromosomal interaction that is becoming increasingly understood. Promoters are often defined as *cis*-regulatory sequences within 1000 bp from the next gene transcription start site. In contrast, enhancers are a type of *cis*-regulatory sequence positioned >1 kb from the nearest transcription start site.[62] Promoters (but not enhancers) typically include a core promoter as docking site for general transcription factors (TFIIA/B/D/E/F/H) and components of RNA polymerase II holoenzyme, as part of the preinitiation complex.[62] These core promoters drive low levels of basal transcription. However, gene expression is heavily stimulated by "activators" or transcription factors that bind, in sequence-specific fashion, at the site of promoters and enhancers.[62] The transcription factors bind to nucleosome-free intervals in open chromatin of promoters and enhancer sequences and recruit coactivator complexes, such as Mediator and CREB-binding protein (CPB)/p300,[62] to mention a few examples. Promoters in contrast to enhancers are often CpG rich,[62] which is of interest, given that alterations in promoter CpG DNA methylation have been reported in the brains of schizophrenia patients.[24–27,33] Importantly, enhancers, as distal regulatory elements, while critically important for transcriptional regulation are separated from their target gene often by many hundreds of base pairs, and in some cases many kilo- or even mega-bases of interspersed linear genome.[62] Various mechanisms have been proposed by which enhancer chromatin could regulate the expression of target genes from distant chromosomal locations. Such mechanisms include sliding along the chromosome to "track" promoters,[63] or alternatively, a physical "bridge" built via protein-protein interactions.[64] Presently, however, most studies implicate the promoter-enhancer (chromosomal) loop model, which involves physical interaction, or at least spatial proximity, between the enhancer chromatin and the promoter target.[65]

3.2 Evidence for Loop Disruption in Brains of Subjects With Schizophrenia

Importantly, chromosome conformation capture (3C), a widely used DNA–DNA proximity assay on chromatin preparations treated by restriction digest followed by religation, is applicable to postmortem tissue, with many chromosomal loop formations showing some level of preservation in brain tissue collected several hours after death.[66] One interesting and early example of 3C applied on human brain involved *GAD1*,[67] encoding the 67 kDa glutamic acid decarboyxlase GABA synthesis enzyme, a gene frequently showing dysregulated expression in multiple areas of the forebrain of subjects with schizophrenia.[18,68–86] One of the loop formations that were altered in a small pilot cohort of diseased prefrontal cortex[67] also existed in neuronal cultures derived from induced pluripotent stem cells. Furthermore, this type of chromatin loop, which bypassed approximately 50 kb of linear genome, was conserved in mouse and human brain,[67] which could indicate an important regulatory function for GAD1 expression across multiple mammalian lineages. In any case, according to the aforementioned study,[67] promoter-enhancer loops are potentially disrupted in diseased brain, thereby contributing to dysregulated gene expression. In the case of this 50 kb GAD1 promoter-enhancer conformation, it has been suggested that the distal sequence (the enhancer) carries, via the loop, a cargo of transcription factors into close spatial proximity with the target gene promoter[67](Fig. 3A).

Another interesting example of higher-order chromatin relevant for schizophrenia was recently described for the *GRIN2B* gene locus, encoding an NMDA glutamate receptor subunit.[87] Importantly, deleterious *GRIN2B* mutations rank prominently in exome sequencing studies capturing rare monogenic forms of neuropsychiatric disease, including intellectual disability, epilepsy, autism, and psychosis spectrum disorders, including schizophrenia.[88–92] Multiple loop-bound intronic and intergenic DNA sequences, up to 450 kb downstream from the *GRIN2B/Grin2b* transcription start site (TSS), compete for access to the TSS.[87] These sequences are loaded with multiple transcription factors and, via chromosomal loop formations, are in physical proximity to the *GRIN2B/Grin2b* transcription start site.[87] However, in addition to such long-range promoter-enhancer loops, the *GRIN2B/Grin2b* promoter also interacts with intragenic repressive chromatin embedded in intron sequences.[87] Therefore, it was proposed that transcriptional regulation at the *GRIN2B/Grin2b* locus involves a dynamic and competitive interplay of multiple loop formations, each of which could engage with the *GRIN2B* promoter[87] (Fig. 3B). However, this process is

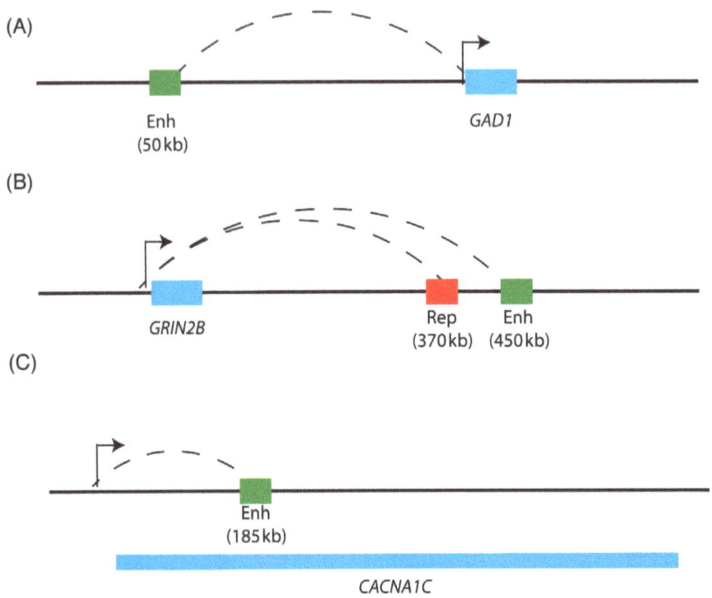

Fig. 3 Depictions of distal promoter-enhancer loop interactions on the linear genome. (A) *GAD1* (*blue box*) and its enhancer (*green box*) that lies 50 kb upstream. (B) *GRIN2B* (*blue box*) and its enhancer (*green box*) that lies 450 kb downstream and its repressor that lies 370 kb downstream. (C) *CACNA1* (introns and exons span the region demarcated by *blue box*) and its enhancer, which lies in an intronic sequence 185 kb downstream from the promoter. HP1, Heterochromatin-associated protein 1.

counterbalanced by promoter-bound higher-order chromatin involving repressive intronic sequences 30 kb downstream from the transcription start site[87,93] (Fig. 3B). Interestingly, multiple loop-bound sequences interacting with the *GRIN2B* promoter harbor single nucleotide polymorphisms (SNP) implicated with liability for working memory,[87] and schizophrenia and personality traits associated with schizophrenia.[87] Notably, one these risk-associated SNP alleles was associated with a promoter-enhancer loop formation and thought to convey decreased nucleoprotein binding and motif loss for the CCAAT/Enhancer Binding Protein CEBPB (C/EBPβ).[87] CEBPB is a transcription factor implicated in consolidation of cortical and hippocampal learning and memory.[94–96] Because many enhancer elements are defined by sequential linear alignment of multiple transcription factors within short distances,[97,98] additional activator proteins may synergistically cooperate with loop-bound CEBPB to regulate *GRIN2B* expression.[87] Taken together, these findings point to a complex and multilayered regulation of chromosomal conformations across at least 1 Mb of sequence surrounding the *GRIN2B* locus. Therefore, loop-bound DNA targeting the *GRIN2B* promoter and carrying disease-associated sequence polymorphisms, could either facilitate or repress expression, depending on the protein "cargo," with multiple distal loop formations competing in a highly dynamic and activity-dependent manner for access to the *GRIN2B* promoter sequences.[87] Importantly, the DNA sequence variants affecting *GRIN2B* enhancer function, and working memory and liability for schizophrenia are located nearly half a megabase downstream from the *GRIN2B* transcription start site.[87] Therefore, these *GRIN2B* higher-order chromatin studies provide the first example how chromosome conformation capture approaches could be harnessed to assign neurological function to risk-associated noncoding sequences that on the linear genome appear to be too far removed from gene promoters to impact expression.

Similarly, risk-associated sequences positioned in enhancer elements within the *CACNA1C* calcium channel gene body (a gene locus, which ranks prominently in the polygenic risk maps of common psychiatric disease, including schizophrenia and depression)[99] were recently identified as potent modulators of reporter gene activity.[34] Remarkably, however, these *CACNA1C* intragenic enhancer sequences were, in human prefrontal cortex, physically bound to the gene transcription start site via a 185 kb spanning chromosomal loop formation,[34] providing yet another example how the study of chromosomal conformations in human brain tissue could help to illuminate the role of noncoding DNA that, on the linear genome, is positioned far away from transcription start sites (Fig. 3C).

However, there is evidence pointing to the importance of chromosomal conformations and the "3D genome" for human cognition and schizophrenia far beyond the aforementioned candidate gene examples. For example, significant overrepresentation of enhancer sequences has been observed within the pool of polymorphisms and haplotypes associated with genetic risk for schizophrenia.[34,100] Furthermore, postmortem studies using more conventional epigenomic assays, including genome-scale DNA methylation survey, have suggested that epigenetic dysregulation of enhancer sequences could contribute to the neurobiology of mood and psychosis spectrum disorders, including astrocyte dysfunction in depression,[101] and broad changes in neuronal gene expression in schizophrenia.[29,102] Finally, recent genome-wide chromosomal conformation mapping in neural progenitor cells and fetal brain tissues reveals many additional examples of promoter-enhancer loops at the sites of schizophrenia relevant genes.[103]

4. (EPI)GENOMIC EDITING OF LOOP-BOUND REGULATORY SEQUENCES IN THE PRECLINICAL MODEL

In the paragraphs aforementioned, we provided specific examples of regulatory noncoding DNA that is associated with the genetic risk architecture of schizophrenia and via chromosomal loops physically associates with promoters and transcription start sites of important neuronal genes, such as *GAD1*, *GRIN2B*, and *CACNA1C*. One could argue that such types of promoter-enhancer loops provide unique opportunities for highly locus- and sequence-selective interventions, targeted against individual risk polymorphisms associated with schizophrenia, and aimed at conveying a therapeutic effect by affecting target gene expression. Thus, the field will soon be challenged with the task to "convert" this molecular information into testable hypotheses aimed at gaining deeper insights into the neurobiology of schizophrenia. Perhaps more importantly, such newly gained information could be harnessed to develop novel therapies aimed at improving cognitive dysfunction. As efforts are currently being made to translate these findings from schizophrenia (epi)genomics into clinical applications and discovery,[104] we predict that intermediate preclinical steps will involve molecular and cellular exploration of cells in vitro[105] and behavioral studies in mice and other small animals.

Importantly, the molecular toolboxes to test this approach, at least for the preclinical model, already exist. Specifically, genome editing strategies via RNA-guided nucleases, including the clustered regularly interspaced short palindromic repeat (CRISPR)-CRISPR-associated protein systems (CRISPR-Cas), introduced to the field only a few years ago,[106] have now been widely adopted in all areas of genomic medicine, including the neurosciences.[107] Mutations in and CRISPR-mediated disruption of enhancer sequences have been observed and accomplished for several neuropsychiatric risk genes, including the NMDA receptor gene *GRIN2B*[87] and the *FOXG1* transcription factor.[103] Strikingly, mutations and manipulations in the enhancers have been linked with changes in the target gene expression even though the enhancer sequences lie distal to the target gene promoters on the linear genome. While the precise molecular mechanisms underlying these phenomena, such as allele-specific binding of specific transcription factors, often remain incompletely understood, it is noteworthy that such experiments have been conducted to date primarily in cell culture systems. Typically, transcriptional changes are explored after targeted mutations in the regulatory element, or by allele-specific comparisons of reporter gene expression activity. Therefore, the next phase of experiments should include in vivo genomic editing of risk-associated promoter and enhancer sequences, and then test for changes in cognition and behavior in the animal. In doing this, we must keep in mind that such genomic editing may carry drawbacks given that mutagenic interventions are likely to be irreversible. However, CRISPR-Cas and other RNA-guided nuclease systems can easily be converted into epigenomic editing tools (by using mutant protein with inactivated nuclease function, fused to a transcriptional activator, such as VP64 or P300,[108] or a repressor, such as KRAB,[109] even with multilocus manipulation.[110] With the underlying DNA sequence left intact, it will be interesting to explore whether simultaneous epigenomic targeting of enhancer and promoter sequences within multiple risk haplotypes could offer a promising approach to effectively alter cognition and behavior. Just as in the aforementioned examples of genomic editing, proof-of-principle studies for epigenomic editing of schizophrenia risk loci by CRISPR-Cas-mediated loading of loop-bound sequences with artificial transcriptional activators and repressors already have been published. These experiments, performed on the clustered *PROTOCADHERIN* locus (regulating neuronal connectivity), have one caveat: they were conducted in cell culture and not in the animal.[111]

Obviously, preclinical assessment of cognitive and behavioral changes after genomic and epigenomic editing of psychiatric risk haplotypes is only feasible for genomic sites that show at least some degree of conservation between human and mouse (or other laboratory animals') genomes. Such an "epigenomic conservation" could include similarities in sequential arrangements of genes and transcriptional units at the locus of interest, conservation of histone modification landscapes, and similarities in chromosomal conformations, including promoter-enhancer loops important for transcriptional regulation. While a more detailed investigation on the epigenomic conservation across species for genomic sites harboring schizophrenia risk haplotypes awaits further investigation, genome-scale[112,113] studies suggest that chromatin structure and function is conserved in human and mouse for a large number of regulatory noncoding sequences, even if these are not necessarily accompanied by DNA sequence conservation. This general observation also holds for brain chromatin, and there are striking similarities in locus-specific higher-order chromatin landscapes in human and mouse for several of the aforementioned neuronal genes and loci, including *GAD1*, *GRIN2B*, and the clustered protocadherins.[67,87,111] With region-specific multiplex gene editing in adult mouse brain in vivo[114] now possible, we predict that the genomic and epigenomic editing of loop-bound regulatory DNA sequences will soon become an important avenue for preclinical schizophrenia research.

5. FUTURE DIRECTIONS

The application of 3D genome mapping technologies in the field of neuropsychiatry is still a relatively new endeavor. A few studies, as enumerated earlier, have delved into locus-specific dynamics in chromatin conformations and their potential impact in schizophrenia etiology. However, using agnostic, genome-wide approaches, we need to map the baseline 3D interaction landscape across various brain-relevant cell types, from neurons to microglia, to truly understand the nuanced ways in which disease risk variants impact loops in one or some but not in all cell types. Such an approach would not only elucidate any fundamental differences in overall genome topologies across cell types but would also allow us as a field to parse out cell-type-specific impacts of noncoding variants on promoter-enhancer loops and, as a result, variation in gene expression programs. Especially considering that various neuropsychiatric disorders are suspected to have neurodevelopmental origins, mapping chromatin states across differentiation from neural progenitor stage to other cell fates could prove fruitful.

With the advent of newer iterations in genome-wide, as well as locus-targeted high throughput chromatin conformation mapping, paired with the trend in decreasing sequencing costs, the field is poised to uncover elements in the relatively underexplored epigenetic layer of 3D regulation. Moreover, innovations in CRISPR-Cas technologies, allowing for fine-tuned gene activation/inactivation, locus deletion, and multiplexing to target various disease-relevant loci at once will further enhance our abilities to tease out the functionality of newly discovered looping interactions, with the potential of uncovering novel therapeutic targets for complex neuropsychiatric illness.

ACKNOWLEDGMENT

This work is supported in part by 1F30MH113330-01 to P.R.

Author contributions: SA wrote the text. PR provided the figures. YJ and BSK contributed intellectual input on the technical aspects of 3D genomics and the epigenomic editing assays.

REFERENCES

1. Hennekens CH, Hennekens AR, Hollar D, Casey DE. Schizophrenia and increased risks of cardiovascular disease. *Am Heart J.* 2005;150:1115–1121.
2. Laursen TM, Munk-Olsen T, Vestergaard M. Life expectancy and cardiovascular mortality in persons with schizophrenia. *Curr Opin Psychiatry.* 2012;25:83–88.
3. Saha S, Chant D, McGrath J. A systematic review of mortality in schizophrenia: is the differential mortality gap worsening over time? *Arch Gen Psychiatry.* 2007;64: 1123–1131.
4. Taly A. Novel approaches to drug design for the treatment of schizophrenia. *Expert Opin Drug Discov.* 2013;8:1285–1296.
5. Kim DH, Stahl SM. Antipsychotic drug development. *Curr Top Behav Neurosci.* 2010;4:123–139.
6. Lieberman JA, Stroup TS, McEvoy JP, et al. Effectiveness of antipsychotic drugs in patients with chronic schizophrenia. *N Engl J Med.* 2005;353:1209–1223.
7. Swartz MS, Perkins DO, Stroup TS, et al. Effects of antipsychotic medications on psychosocial functioning in patients with chronic schizophrenia: findings from the NIMH CATIE study. *Am J Psychiatry.* 2007;164:428–436.
8. Ibrahim HM, Tamminga CA. Schizophrenia: treatment targets beyond monoamine systems. *Annu Rev Pharmacol Toxicol.* 2011;51:189–209.
9. Green MF. Impact of cognitive and social cognitive impairment on functional outcomes in patients with schizophrenia. *J Clin Psychiatry.* 2016;77(Suppl 2):8–11.
10. Schizophrenia Working Group of the Psychiatric Genomics C.Biological insights from 108 schizophrenia-associated genetic loci. *Nature.* 2014;511:421–427.
11. Ruderfer DM, Hamamsy T, Lek M, et al. Patterns of genic intolerance of rare copy number variation in 59,898 human exomes. *Nat Genet.* 2016;48:1107–1111.
12. Genovese G, Fromer M, Stahl EA, et al. Increased burden of ultra-rare protein-altering variants among 4,877 individuals with schizophrenia. *Nat Neurosci.* 2016;19:1433–1441.
13. Arion D, Corradi JP, Tang S, et al. Distinctive transcriptome alterations of prefrontal pyramidal neurons in schizophrenia and schizoaffective disorder. *Mol Psychiatry.* 2015;20:1397–1405.
14. Zhao Z, Xu J, Chen J, et al. Transcriptome sequencing and genome-wide association analyses reveal lysosomal function and actin cytoskeleton remodeling in schizophrenia and bipolar disorder. *Mol Psychiatry.* 2015;20:563–572.
15. Horvath S, Mirnics K. Schizophrenia as a disorder of molecular pathways. *Biol Psychiatry.* 2015;77:22–28.
16. Middleton FA, Mirnics K, Pierri JN, Lewis DA, Levitt P. Gene expression profiling reveals alterations of specific metabolic pathways in schizophrenia. *J Neurosci.* 2002;22:2718–2729.
17. Vawter MP, Shannon Weickert C, Ferran E, et al. Gene expression of metabolic enzymes and a protease inhibitor in the prefrontal cortex are decreased in schizophrenia. *Neurochem Res.* 2004;29:1245–1255.
18. Mirnics K, Middleton FA, Marquez A, Lewis DA, Levitt P. Molecular characterization of schizophrenia viewed by microarray analysis of gene expression in prefrontal cortex. *Neuron.* 2000;28:53–67.
19. Volk DW, Chitrapu A, Edelson JR, Roman KM, Moroco AE, Lewis DA. Molecular mechanisms and timing of cortical immune activation in schizophrenia. *Am J Psychiatry.* 2015;172:1112–1121.
20. Brown JB, Celniker SE. Lessons from modENCODE. *Annu Rev Genomics Hum Genet.* 2015;16:31–53.
21. Lundberg SM, Tu WB, Raught B, Penn LZ, Hoffman MM, Lee SI. ChromNet: learning the human chromatin network from all ENCODE ChIP-seq data. *Genome Biol.* 2016;17:82.

22. Klose RJ, Bird AP. Genomic DNA methylation: the mark and its mediators. *Trends Biochem Sci.* 2006;31:89–97.
23. Bock C, Beerman I, Lien WH, et al. DNA methylation dynamics during in vivo differentiation of blood and skin stem cells. *Mol Cell.* 2012;47:633–647.
24. Abdolmaleky HM, Cheng KH, Russo A, et al. Hypermethylation of the reelin (RELN) promoter in the brain of schizophrenic patients: a preliminary report. *Am J Med Genet B Neuropsychiatr Genet.* 2005;134B:60–66.
25. Grayson DR, Jia X, Chen Y, et al. Reelin promoter hypermethylation in schizophrenia. *Proc Natl Acad Sci USA.* 2005;102:9341–9346.
26. Iwamoto K, Bundo M, Yamada K, et al. DNA methylation status of SOX10 correlates with its downregulation and oligodendrocyte dysfunction in schizophrenia. *J Neurosci.* 2005;25:5376–5381.
27. Mill J, Tang T, Kaminsky Z, et al. Epigenomic profiling reveals DNA-methylation changes associated with major psychosis. *Am J Hum Genet.* 2008;82:696–711.
28. Wockner LF, Noble EP, Lawford BR, et al. Genome-wide DNA methylation analysis of human brain tissue from schizophrenia patients. *Transl Psychiatry.* 2014;4:e339.
29. Jaffe AE, Gao Y, Deep-Soboslay A, et al. Mapping DNA methylation across development, genotype and schizophrenia in the human frontal cortex. *Nat Neurosci.* 2016;19:40–47.
30. Siegmund KD, Connor CM, Campan M, et al. DNA methylation in the human cerebral cortex is dynamically regulated throughout the life span and involves differentiated neurons. *PLoS One.* 2007;2:e895.
31. Jiang Y, Matevossian A, Huang HS, Straubhaar J, Akbarian S. Isolation of neuronal chromatin from brain tissue. *BMC Neurosci.* 2008;9:42.
32. McKinney B, Ding Y, Lewis DA, Sweet RA. DNA methylation as a putative mechanism for reduced dendritic spine density in the superior temporal gyrus of subjects with schizophrenia. *Transl Psychiatry.* 2017;7:e1032.
33. Ruzicka WB, Subburaju S, Benes FM. Circuit- and diagnosis-specific DNA methylation changes at gamma-aminobutyric acid-related genes in postmortem human hippocampus in schizophrenia and bipolar disorder. *JAMA Psychiatry.* 2015;72:541–551.
34. Roussos P, Mitchell AC, Voloudakis G, et al. A role for noncoding variation in schizophrenia. *Cell Rep.* 2014;9:1417–1429.
35. Rajarajan P, Gil SE, Brennand KJ, Akbarian S. Spatial genome organization and cognition. *Nat Rev Neurosci.* 2016;17:681–691.
36. Lieberman-Aiden E, van Berkum NL, Williams L, et al. Comprehensive mapping of long-range interactions reveals folding principles of the human genome. *Science.* 2009;326:289–293.
37. Rao SS, Huntley MH, Durand NC, et al. A 3D map of the human genome at kilobase resolution reveals principles of chromatin looping. *Cell.* 2014;159:1665–1680.
38. Padeken J, Heun P. Nucleolus and nuclear periphery: velcro for heterochromatin. *Curr Opin Cell Biol.* 2014;28:54–60.
39. Nasmyth K, Haering CH. Cohesin: its roles and mechanisms. *Annu Rev Genet.* 2009;43:525–558.
40. Wendt KS, Yoshida K, Itoh T, et al. Cohesin mediates transcriptional insulation by CCCTC-binding factor. *Nature.* 2008;451:796–801.
41. Dowen JM, Fan ZP, Hnisz D, et al. Control of cell identity genes occurs in insulated neighborhoods in mammalian chromosomes. *Cell.* 2014;159:374–387.
42. Vietri Rudan M, Barrington C, Henderson S, et al. Comparative Hi-C reveals that CTCF underlies evolution of chromosomal domain architecture. *Cell Rep.* 2015;10:1297–1309.
43. Guo Y, Xu Q, Canzio D, et al. CRISPR inversion of CTCF sites alters genome topology and enhancer/promoter function. *Cell.* 2015;162:900–910.

44. Boyle MI, Jespersgaard C, Brondum-Nielsen K, Bisgaard AM, Tumer Z. Cornelia de Lange syndrome. *Clin Genet.* 2015;88:1–12.
45. Yan J, Zhang F, Brundage E, et al. Genomic duplication resulting in increased copy number of genes encoding the sister chromatid cohesion complex conveys clinical consequences distinct from Cornelia de Lange. *J Med Genet.* 2009;46:626–634.
46. Finnsson J, Sundblom J, Dahl N, Melberg A, Raininko R. LMNB1-related autosomal-dominant leukodystrophy: clinical and radiological course. *Ann Neurol.* 2015;78: 412–425.
47. Nolen LD, Boyle S, Ansari M, Pritchard E, Bickmore WA. Regional chromatin decompaction in Cornelia de Lange syndrome associated with NIPBL disruption can be uncoupled from cohesin and CTCF. *Hum Mol Genet.* 2013;22:4180–4193.
48. Watson LA, Wang X, Elbert A, Kernohan KD, Galjart N, Berube NG. Dual effect of CTCF loss on neuroprogenitor differentiation and survival. *J Neurosci.* 2014;34:2860–2870.
49. Gregor A, Oti M, Kouwenhoven EN, et al. De novo mutations in the genome organizer CTCF cause intellectual disability. *Am J Hum Genet.* 2013;93:124–131.
50. Hirayama T, Tarusawa E, Yoshimura Y, Galjart N, Yagi T. CTCF is required for neural development and stochastic expression of clustered Pcdh genes in neurons. *Cell Rep.* 2012;2:345–357.
51. Cai S, Lee CC, Kohwi-Shigematsu T. SATB1 packages densely looped, transcriptionally active chromatin for coordinated expression of cytokine genes. *Nat Genet.* 2006;38:1278–1288.
52. Britanova O, Akopov S, Lukyanov S, Gruss P, Tarabykin V. Novel transcription factor Satb2 interacts with matrix attachment region DNA elements in a tissue-specific manner and demonstrates cell-type-dependent expression in the developing mouse CNS. *Eur J Neurosci.* 2005;21:658–668.
53. Docker D, Schubach M, Menzel M, et al. Further delineation of the SATB2 phenotype. *Eur J Hum Genet.* 2014;22:1034–1039.
54. Alcamo EA, Chirivella L, Dautzenberg M, et al. Satb2 regulates callosal projection neuron identity in the developing cerebral cortex. *Neuron.* 2008;57:364–377.
55. Leoyklang P, Suphapeetiporn K, Siriwan P, et al. Heterozygous nonsense mutation SATB2 associated with cleft palate, osteoporosis, and cognitive defects. *Hum Mutat.* 2007;28:732–738.
56. Close J, Xu H, De Marco Garcia N, et al. Satb1 is an activity-modulated transcription factor required for the terminal differentiation and connectivity of medial ganglionic eminence-derived cortical interneurons. *J Neurosci.* 2012;32:17690–17705.
57. Denaxa M, Kalaitzidou M, Garefalaki A, et al. Maturation-promoting activity of SATB1 in MGE-derived cortical interneurons. *Cell Rep.* 2012;2:1351–1362.
58. Labouesse MA, Dong E, Grayson DR, Guidotti A, Meyer U. Maternal immune activation induces GAD1 and GAD2 promoter remodeling in the offspring prefrontal cortex. *Epigenetics.* 2015;10:1143–1155.
59. Guidotti A, Ruzicka W, Grayson DR, et al. S-adenosyl methionine and DNA methyltransferase-1 mRNA overexpression in psychosis. *Neuroreport.* 2007;18:57–60.
60. Kundakovic M, Chen Y, Guidotti A, Grayson DR. The reelin and GAD67 promoters are activated by epigenetic drugs that facilitate the disruption of local repressor complexes. *Mol Pharmacol.* 2009;75:342–354.
61. Costa E, Grayson DR, Mitchell CP, Tremolizzo L, Veldic M, Guidotti A. GABAergic cortical neuron chromatin as a putative target to treat schizophrenia vulnerability. *Crit Rev Neurobiol.* 2003;15:121–142.
62. Vernimmen D, Bickmore WA. The hierarchy of transcriptional activation: from enhancer to promoter. *Trends Genet.* 2015;31:696–708.
63. Hatzis P, Talianidis I. Dynamics of enhancer-promoter communication during differentiation-induced gene activation. *Mol Cell.* 2002;10:1467–1477.

64. Mueller-Storm HP, Sogo JM, Schaffner W. An enhancer stimulates transcription in trans when attached to the promoter via a protein bridge. *Cell.* 1989;58:767–777.
65. Gorkin DU, Leung D, Ren B. The 3D genome in transcriptional regulation and pluripotency. *Cell Stem Cell.* 2014;14:762–775.
66. Mitchell AC, Bharadwaj R, Whittle C, et al. The genome in three dimensions: a new frontier in human brain research. *Biol Psychiatry.* 2014;75:961–969.
67. Bharadwaj R, Jiang Y, Mao W, et al. Conserved chromosome 2q31 conformations are associated with transcriptional regulation of GAD1 GABA synthesis enzyme and altered in prefrontal cortex of subjects with schizophrenia. *J Neurosci.* 2013;33: 11839–11851.
68. Volk DW, Austin MC, Pierri JN, Sampson AR, Lewis DA. Decreased glutamic acid decarboxylase67 messenger RNA expression in a subset of prefrontal cortical gamma-aminobutyric acid neurons in subjects with schizophrenia. *Arch Gen Psychiatry.* 2000;57:237–245.
69. Impagnatiello F, Guidotti AR, Pesold C, et al. A decrease of reelin expression as a putative vulnerability factor in schizophrenia. *Proc Natl Acad Sci USA.* 1998;95: 15718–15723.
70. Akbarian S, Kim JJ, Potkin SG, et al. Gene expression for glutamic acid decarboxylase is reduced without loss of neurons in prefrontal cortex of schizophrenics. *Arch General Psychiatry.* 1995;52:258–266.
71. Guidotti A, Auta J, Davis JM, et al. Decrease in reelin and glutamic acid decarboxylase67 (GAD67) expression in schizophrenia and bipolar disorder: a postmortem brain study. *Arch Gen Psychiatry.* 2000;57:1061–1069.
72. Hashimoto T, Volk DW, Eggan SM, et al. Gene expression deficits in a subclass of GABA neurons in the prefrontal cortex of subjects with schizophrenia. *J Neurosci.* 2003;23:6315–6326.
73. Veldic M, Guidotti A, Maloku E, Davis JM, Costa E. In psychosis, cortical interneurons overexpress DNA-methyltransferase 1. *Proc Natl Acad Sci USA.* 2005;102:2152–2157.
74. Torrey EF, Barci BM, Webster MJ, Bartko JJ, Meador-Woodruff JH, Knable MB. Neurochemical markers for schizophrenia, bipolar disorder, and major depression in postmortem brains. *Biol Psychiatry.* 2005;57:252–260.
75. Veldic M, Kadriu B, Maloku E, et al. Epigenetic mechanisms expressed in basal ganglia GABAergic neurons differentiate schizophrenia from bipolar disorder. *Schizophr Res.* 2007;91:51–61.
76. Benes FM, Lim B, Matzilevich D, Walsh JP, Subburaju S, Minns M. Regulation of the GABA cell phenotype in hippocampus of schizophrenics and bipolars. *Proc Natl Acad Sci USA.* 2007;104:10164–10169.
77. Thompson Ray M, Weickert CS, Wyatt E, Webster MJ. Decreased BDNF, trkB-TK+ and GAD67 mRNA expression in the hippocampus of individuals with schizophrenia and mood disorders. *J Psychiatry Neurosci.* 2011;36:195–203.
78. Curley AA, Arion D, Volk DW, et al. Cortical deficits of glutamic acid decarboxylase 67 expression in schizophrenia: clinical, protein, and cell type-specific features. *Am J Psychiatry.* 2011;168:921–929.
79. Sheng G, Demers M, Subburaju S, Benes FM. Differences in the circuitry-based association of copy numbers and gene expression between the hippocampi of patients with schizophrenia and the hippocampi of patients with bipolar disorder. *Arch Gen Psychiatry.* 2012;69:550–561.
80. Volk DW, Matsubara T, Li S, et al. Deficits in transcriptional regulators of cortical parvalbumin neurons in schizophrenia. *Am J Psychiatry.* 2012;169:1082–1091.
81. Gilabert-Juan J, Varea E, Guirado R, Blasco-Ibanez JM, Crespo C, Nacher J. Alterations in the expression of PSA-NCAM and synaptic proteins in the dorsolateral prefrontal cortex of psychiatric disorder patients. *Neurosci Lett.* 2012;530:97–102.

82. Hashimoto T, Bazmi HH, Mirnics K, Wu Q, Sampson AR, Lewis DA. Conserved regional patterns of GABA-related transcript expression in the neocortex of subjects with schizophrenia. *Am J Psychiatry.* 2008;165:479–489.
83. Hashimoto T, Arion D, Unger T, et al. Alterations in GABA-related transcriptome in the dorsolateral prefrontal cortex of subjects with schizophrenia. *Mol Psychiatry.* 2008;13:147–161.
84. Huang HS, Matevossian A, Whittle C, et al. Prefrontal dysfunction in schizophrenia involves mixed-lineage leukemia 1-regulated histone methylation at GABAergic gene promoters. *J Neurosci.* 2007;27:11254–11262.
85. Bullock WM, Cardon K, Bustillo J, Roberts RC, Perrone-Bizzozero NI. Altered expression of genes involved in GABAergic transmission and neuromodulation of granule cell activity in the cerebellum of schizophrenia patients. *Am J Psychiatry.* 2008;165:1594–1603.
86. Fatemi SH, Stary JM, Earle JA, Araghi-Niknam M, Eagan E. GABAergic dysfunction in schizophrenia and mood disorders as reflected by decreased levels of glutamic acid decarboxylase 65 and 67 kDa and Reelin proteins in cerebellum. *Schizophr Res.* 2005;72:109–122.
87. Bharadwaj R, Peter CJ, Jiang Y, et al. Conserved higher-order chromatin regulates NMDA receptor gene expression and cognition. *Neuron.* 2014;84:997–1008.
88. Talkowski ME, Rosenfeld JA, Blumenthal I, et al. Sequencing chromosomal abnormalities reveals neurodevelopmental loci that confer risk across diagnostic boundaries. *Cell.* 2012;149:525–537.
89. Epi KC, Epilepsy Phenome/Genome P, Allen AS, et al. De novo mutations in epileptic encephalopathies. *Nature.* 2013;501:217–221.
90. Hamdan FF, Srour M, Capo-Chichi JM, et al. De novo mutations in moderate or severe intellectual disability. *PLoS Genet.* 2014;10:e1004772.
91. O'Roak BJ, Deriziotis P, Lee C, et al. Exome sequencing in sporadic autism spectrum disorders identifies severe de novo mutations. *Nat Genet.* 2011;43:585–589.
92. Hu C, Chen W, Myers SJ, Yuan H, Traynelis SF. Human GRIN2B variants in neurodevelopmental disorders. *J Pharmacol Sci.* 2016;132:115–121.
93. Jiang Y, Jakovcevski M, Bharadwaj R, et al. Setdb1 histone methyltransferase regulates mood-related behaviors and expression of the NMDA receptor subunit NR2B. *J Neurosci.* 2010;30:7152–7167.
94. Taubenfeld SM, Milekic MH, Monti B, Alberini CM. The consolidation of new but not reactivated memory requires hippocampal C/EBPbeta. *Nat Neurosci.* 2001;4:813–818.
95. Taubenfeld SM, Wiig KA, Monti B, Dolan B, Pollonini G, Alberini CM. Fornix-dependent induction of hippocampal CCAAT enhancer-binding protein [beta] and [delta] Co-localizes with phosphorylated cAMP response element-binding protein and accompanies long-term memory consolidation. *J Neurosci.* 2001;21:84–91.
96. Merhav M, Kuulmann-Vander S, Elkobi A, Jacobson-Pick S, Karni A, Rosenblum K. Behavioral interference and C/EBPbeta expression in the insular-cortex reveal a prolonged time period for taste memory consolidation. *Learn Mem.* 2006;13:571–574.
97. Dickel DE, Visel A, Pennacchio LA. Functional anatomy of distant-acting mammalian enhancers. *Philos Trans R Soc Lond B Biol Sci.* 2013;368:20120359.
98. Stampfel G, Kazmar T, Frank O, Wienerroither S, Reiter F, Stark A. Transcriptional regulators form diverse groups with context-dependent regulatory functions. *Nature.* 2015;528:147–151.
99. Cross-Disorder Group of the Psychiatric Genomics C.Identification of risk loci with shared effects on five major psychiatric disorders: a genome-wide analysis. *Lancet.* 2013;381:1371–1379.
100. Gusev A, Lee SH, Trynka G, et al. Partitioning heritability of regulatory and cell-type-specific variants across 11 common diseases. *Am J Hum Genet.* 2014;95:535–552.

101. Nagy C, Suderman M, Yang J, et al. Astrocytic abnormalities and global DNA methylation patterns in depression and suicide. *Mol Psychiatry.* 2015;20:320–328.
102. Roussos P, Giakoumaki SG, Zouraraki C, et al. The relationship of common risk variants and polygenic risk for schizophrenia to sensorimotor gating. *Biol Psychiatry.* 2015;79(12):988–996.
103. Won H, de la Torre-Ubieta L, Stein JL, et al. Chromosome conformation elucidates regulatory relationships in developing human brain. *Nature.* 2016;538:523–527.
104. Breen G, Li Q, Roth BL, et al. Translating genome-wide association findings into new therapeutics for psychiatry. *Nat Neurosci.* 2016;19:1392–1396.
105. Brennand KJ, Landek-Salgado MA, Sawa A. Modeling heterogeneous patients with a clinical diagnosis of schizophrenia with induced pluripotent stem cells. *Biol Psychiatry.* 2014;75:936–944.
106. Doudna JA, Charpentier E. Genome editing. The new frontier of genome engineering with CRISPR-Cas9. *Science.* 2014;346:1258096.
107. Heidenreich M, Zhang F. Applications of CRISPR-Cas systems in neuroscience. *Nat Rev Neurosci.* 2016;17:36–44.
108. Hilton IB, D'Ippolito AM, Vockley CM, et al. Epigenome editing by a CRISPR-Cas9-based acetyltransferase activates genes from promoters and enhancers. *Nat Biotechnol.* 2015;33:510–517.
109. Thakore PI, D'Ippolito AM, Song L, et al. Highly specific epigenome editing by CRISPR-Cas9 repressors for silencing of distal regulatory elements. *Nat Methods.* 2015;12:1143–1149.
110. Stricker SH, Koferle A, Beck S. From profiles to function in epigenomics. *Nat Rev Genet.* 2017;18:51–66.
111. Jiang Y, Loh YE, Rajarajan P, et al. The methyltransferase SETDB1 regulates a large neuron-specific topological chromatin domain. *Nat Genet.* 2017;49:1239–1250.
112. Xiao S, Xie D, Cao X, et al. Comparative epigenomic annotation of regulatory DNA. *Cell.* 2012;149:1381–1392.
113. Dincer A, Gavin DP, Xu K, et al. Deciphering H3K4me3 broad domains associated with gene-regulatory networks and conserved epigenomic landscapes in the human brain. *Transl Psychiatry.* 2015;5:e679.
114. Zetsche B, Heidenreich M, Mohanraju P, et al. Multiplex gene editing by CRISPR-Cpf1 using a single crRNA array. *Nat Biotechnol.* 2017;35:31–34.

CHAPTER THREE

Progress in Epigenetics of Depression

Catherine J. Peña, Eric J. Nestler[1]
Icahn School of Medicine at Mount Sinai, New York, NY, United States
[1] Corresponding author.E-mail address: eric.nestler@mssm.edu

Contents

Abstract

Depression is a prevalent and complex psychiatric syndrome. Epigenetic mechanisms bridge the genetic and environmental factors that contribute to the pathophysiology of depression. A surge of research over the last decade has identified changes in DNA methylation, histone modifications, histone organization, and noncoding RNAs associated with depression and stress-induced depression-like behavior in animal models. We focus here on associations of epigenetic factors concurrent with depression and depression-like behavior, although risk for depression and some of the associated epigenetic changes are known to have developmental origins. Finally, emerging technology may enable breakthroughs in the ability to rescue depression-associated

Progress in Molecular Biology and Translational Science, volume 157
ISSN 1877-1173
http://dx.doi.org/10.1016/bs.pmbts.2017.12.011

epigenetic modifications at specific genes, greatly enhancing specificity of future potential therapeutic treatments.

1. INTRODUCTION

Major depressive disorder is characterized by persistence of sadness, loss of interest or pleasure, feelings of guilt or low self-worth, and other symptoms, including sleep and circadian changes, changes in weight or appetite, and diminished concentration. More than 300 million people worldwide are living with depression, making it the leading global cause of disability.[1] Depression can affect people at any age, and can persist or reoccur throughout the lifespan. Previous experience of depression increases the likelihood of future episodes. Women are approximately 50% more likely than men to experience depression, with increased risk associated with puberty, peripartum periods, and menopause.

As with other psychiatric disorders, depression is a complex, heterogeneous syndrome. The causes of depression cannot be pinned down to any one biological or external factor, and instead result from the interplay of genetic, neurobiological, environmental, and cultural factors. The genetic heritability of depression is estimated to be from 35%–40%.[2,3] While heritable genetic factors clearly play a role, a significant proportion of depression risk is attributed to life experience and environment, particularly previous and/or recent stress experience. In the last decade, there has been a considerable focus on the role of epigenetics as a possible bridge between genes and experience in the pathophysiology of depression.

The epigenetic landscape, originally envisioned by Waddington more than 60 years ago, sits beautifully at the intersection of genetics, development, environmental influence, and inheritance.[4] The word "epigenetic" means "above the genome" and epigenetic mechanisms refer to structural and molecular factors that alter expression of genes without altering the sequence of DNA. These include many types of chemical modifications to histone proteins, DNA, and RNA, as well as alterations in other regulatory proteins and noncoding RNAs.[5–7] While epigenetic mechanisms were previously defined as stable through cell division, the dynamic nature of epigenetic modifications is now understood. As primary regulators of gene expression, epigenetic mechanisms are ideal molecular candidates to flexibly encode and translate environmental factors that increase risk for depression and other highly complex psychiatric syndromes.

This chapter discusses major advances in the study of epigenetics and depression, with a focus on recent findings. While early life events are known to increase lifelong risk for depression, the epigenetic consequences of early life stress in mediating depression and other psychiatric disease is discussed in other chapters of this text.[8] We focus here on associations of epigenetic factors concurrent with depression and depression-like behavior. Work to date implicating epigenetic regulation in the context of depression and depression-like behavior has come from studies of both postmortem human brain and animal models.

Research discussed in this chapter focuses on a few key regions of the brain (Fig. 1). Impaired motivation and anhedonia are core symptoms of depression, which implicate the brain's limbic circuitry, including ventral tegmental area (VTA) and nucleus accumbens (NAc).[9] Cognitive dysfunction and impaired attention, also common symptoms of depression, implicate the prefrontal cortex (PFC), hippocampus, and amygdala, among other regions. The hippocampus and amygdala are also known to be highly sensitive to the effects of stress and to control the hypothalamic-pituitary-adrenal axis and the secretion of glucocorticoids. The NAc is of particular interest as it is a site of integration for glutamatergic signals from the PFC, hipocampus, and amygdala, as well as dopaminergic projections from VTA.[9]

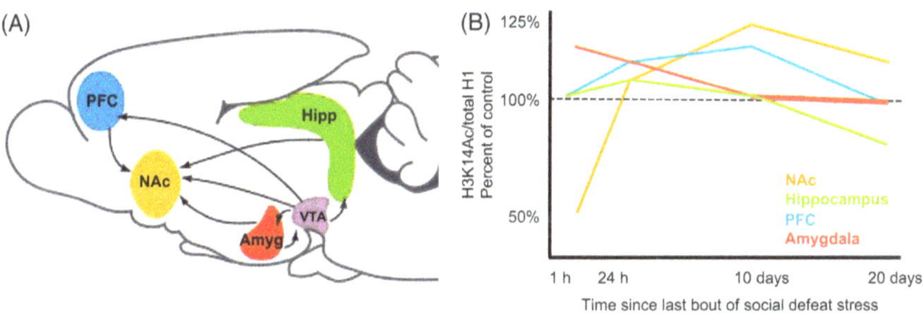

Fig. 1 (A) Brain regions implicated in major depressive disorder and response to stress, shown on sagittal view of mouse brain. (B) H3K14Ac levels in mouse brain after chronic social defeat stress. Chronic fluoxetine treatment normalizes hippocampal H3K14Ac at 20 days (not shown). *Amyg*, Amygdala; *Hipp*, hippocampus; *NAc*, nucleus accumbens; *PFC*, prefrontal cortex; *VTA*, ventral tegmental area. *Adapted from Covington HE, Maze I, LaPlant QC, Vialou VF, Ohnishi YN, Berton O, et al. Antidepressant actions of histone deacetylase inhibitors. J Neurosci. 2009;29(37):11451–60.*[53] *Covington HE, Vialou VF, Laplant Q, Ohnishi YN, Nestler EJ. Hippocampal-dependent antidepressant-like activity of histone deacetylase inhibition. Neurosci Lett. 2011;493(3):122–126.*[54] *Covington HE III, Maze I, Vialou V, Nestler EJ. Antidepressant action of HDAC inhibition in the prefrontal cortex. Neuroscience. 2015;298:329–335.*[56]

2. OVERVIEW OF EPIGENETIC MECHANISMS

The human genome contains approximately 6 billion nucleotides of DNA neatly packaged into 23 pairs of chromosomes.[10] The fundamental unit of chromatin is the nucleosome, which allows the extraordinary organization and compaction of DNA into a microscopic cell nucleus. The nucleosome consists of ~146 bp of DNA wrapped around a core histone octamer (~1.65 turns). Each octamer contains two copies each of the histones H2A, H2B, H3, and H4. Epigenetic mechanisms control the spacing of nucleosomes and the degree to which they are tightly condensed or loosely wound. In simplified terms, chromatin exists across a continuum between an inactivated, condensed state (heterochromatin), which represses gene transcription, and an activated, open state (euchromatin), which is permissive for gene transcription. The three-dimensional (3D) structure and packaging of DNA thus regulates gene expression, and is controlled by complex biochemical processes.

2.1 DNA Methylation

DNA methylation is one way in which the structure of DNA can be altered by covalent attachment of a methyl group to the C5 position of cytosine (5mC) predominantly at cytosine-guanine dyads (CpG sites) to affect transcription without altering the overall sequence.[11] CpG dinucleotides cluster within or around gene promoters, referred to as "CpG islands," and tend to be hypomethylated in normal somatic cells.[11] Gene promoter hypermethylation generally represses gene expression. It can either prevent access of transcriptional activators and RNA polymerase II to the transcription start site, or bind to methyl-CpG-binding proteins to recruit transcriptional corepressors to modify the surrounding chromatin into a silenced state.[11–13] The process of methylation is dependent on the presence of methyl donors (provided by nutrients, such as folic acid, methionine, and choline) and DNA methyltransferases, which catalyze either *de novo* DNA methylation (i.e., DNMT3) or maintenance of methylation across cell division and DNA repair (i.e., DNMT1). More recent findings have indicated that a significant portion of DNA methylation occurs at non-CpG sites and that DNA methylation can either induce or suppress gene expression depending on other factors.[14]

DNA methylation is considered to be a relatively stable epigenetic modification, making it an attractive target in the study of long-lasting epigenetic signatures of psychiatric disease. Until recently, DNA methylation was thought to be a permanent modification that could only be reduced in tissue by passive mechanisms, such as failure to maintain methylation in daughter cells. This view was supported by evidence of an overall accumulation of DNA methylation across the lifespan of an organism. However, DNA methylation is now known to be a reversible process with the discovery of active DNA demethylase enzymes, namely the ten–eleven translocation (TET) family proteins. TET proteins oxidize 5mC into 5-hydroxymethylcytosine (5hmC), and subsequently into 5-formylcytosine and 5-carboxylcytosine.[15] Through deamination, glycosylation, or base excision repair, these newly discovered forms of cytosine modification can then be converted back into an unmethylated state. In contrast to the generally repressive effect of 5mC on gene expression, 5hmC is more correlated with transactivation.[16] Interestingly, 5mC oxidation derivatives are expressed at highest levels in neurons. Most studies of DNA methylation in psychiatric disorders to date have not distinguished between 5mC and 5hmC, which is clearly a major need in the field.

2.2 Chromatin Modifications

Posttranslational modifications to histone proteins, most commonly on amino acid residues within the N-terminal tails that protrude from the nucleosome core, represent a more dynamic process of gene regulation. Such modifications include acetylation, ubiquitination, or SUMOylation at lysine (K) residues, methylation at lysine or arginine residues, and phosphorylation at serine or threonine residues, among many others. Histone acetylation is generally associated with chromatin decondensation and increased transcriptional activity, by negating the positive charge of K residues in histone tails and increases spacing between nucleosomes. The acetylation state of these histones is controlled by histone acetyltransferases (HATs) and histone deacetylases (HDACs). There is a considerable interest in the therapeutic potential of HDAC inhibitors, which effectively increase gene expression by shifting histones to an acetylated state.[17] Histone tails can be methylated by histone methyltransferases (HMTs), and demethylated by histone demethylases (HDMs). The exact location and valence of histone tail methylation (mono-, di- or trimethylation) either promotes euchromatin or heterochromatin, and are thus associated with either activation or repression of gene transcription, respectively.[18] The specificity of numerous HATs and HDACs for specific K residues remains incompletely understood. In contrast, distinct HMTs and HDMs control the methylation of specific K and arginine residues and even the valence of methylation. The complexity of the "histone code" is illustrated by the number of possible posttranslational modifications to H3 lysines alone. Human H3 has 13 lysine residues that can potentially be methylated with one, two, or three methyl groups. This results in huge potential combinatorial power over the level of gene transcription, with more than 67 million (4^13) different H3K-methylation patterns, in one nucleosome.

The functional consequences of histone modifications are mediated in part through "readers"—proteins that bind to specific modified residues and effect transcriptional change.[5,6] For example, different families of *chromatin remodeling proteins*, which use ATP-derived energy to alter nucleosome spacing and condensation, recognize specific forms of modified histones and enhance or repress the activity of nearby genes. The involvement of this diverse family of proteins is just now being studied in the nervous system.[19] Ultimately, hundreds of proteins are thought to be recruited to a gene in concert with its activation or repression, again emphasizing the extraordinary complexity of epigenetic mechanisms. Despite this complexity, strong evidence links chromatin modifications with stress and depression.

2.3 Noncoding RNAs

Only 20% of the human genome is associated with protein-coding genes, with less than 1% transcribed into mRNA, while the rest is thought to contribute to regulation of those genes. Interestingly, higher order species have increasingly greater proportions of noncoding DNA, which includes a range of noncoding RNAs. The diversity of noncoding RNAs and their role in transcriptional regulation is just beginning to be understood. Small noncoding RNAs, including microRNAs (miRNAs), are found in most eukaryotic organisms. They are produced either from their own gene, or commonly from introns of their target or other genes.[7] MicroRNAs are first transcribed from a longer pre-miRNA, including a 5′-cap and poly-A tail, after which a hairpin loop is formed, trimmed, and exported from the cell nucleus to cytoplasm, and processed to a mature single-stranded ~22 nucleotide by a series of enzymes, including DROSHA and DICER. Within an RNA-induced silencing complex (RISC), miRNAs base-pair with complementary mRNA sequences which typically results in silencing by sequestration, degradation, or translational suppression of target mRNAs. Other small noncoding RNAs include short interfering RNAs (siRNAs)

Piwi-interacting RNAs (piRNAs), and small nucleolar RNAs (snoRNAs). Long noncoding RNAs (lncRNAs) are longer than 200 nucleotides and also associated with silencing of target genes. LncRNAs are largely species- and tissue-specific, with increasing proportions of the genome occupied by lncRNAs in higher order species.[20] LncRNAs are also highly expressed in the brain, making them attractive as potential mechanisms of gene regulation in complex neuropsychiatric syndromes, such as depression. Given our relatively recent understanding of noncoding RNAs, a major frontier in the neurobiology of psychiatric disease is to understand the role of specific small and lncRNAs as biomarkers or in the onset and maintenance of depression.[21]

2.4 3D Chromatin Structure

Beyond modifications of DNA, histones, and chromatin remodeling, an important new effort in neuroepigenetics research concerns the 3D organization of the genome in neurons and glia. For example, chromosomal loop formations—which often require CCCTC-binding factor (CTCF), cohesins, and other proteins assembled into scaffolds and anchors—potentially bypass many kilobases, even megabases, of linear genome, thereby repositioning promoter-distal regulatory elements next to their target promoters. Exploration of regulatory DNA elements in the context of chromosomal loopings and higher order chromatin is beginning to assign regulation of genes and complex behavior for some of the noncoding sequences in the human genome.[22]

3. METHODS FOR STUDYING EPIGENETICS AND DEPRESSION

As stated earlier, depression is a heterogeneous disorder, and the difficulty in defining specific subsets of depression has made clear identification of its multiple etiologies very difficult. On top of this, to study the molecular correlates of human depression requires access to relevant biological tissue from human patients. Many studies utilize easily obtainable peripheral tissue, such as blood or saliva. However, the main substrate of depression is the brain; and few studies have provided robust evidence for consistent epigenetic modification of genes between peripheral and central tissues, the main exception being for the gene encoding the glucocorticoid receptor (NR3C1). Studies utilizing human brain tissue samples pose different challenges as tissue is necessarily collected postmortem and studies must control for age, cause of death, variables related to sample handling and quality, and genetic background which must be accounted for to understand the additional contribution of epigenetic modifications. Often, information on life history is absent and at best it is difficult to use such information to infer causality.

Stress paradigms in animals have enabled considerable progress in understanding the neurobiology of depression. The wide use of several standardized rodent chronic stress paradigms, and the objective measurement of depression-like behavior, has also enabled comparison of epigenetic factors in depression across studies. While acute stress paradigms are designed to evaluate an animal's initial coping response, chronic stress paradigms involve prolonged exposure to either physical or psychological stressors, such as social subordination or variable, daily environmental stress. In the chronic social defeat stress paradigm, an experimental mouse is subjected to "bullying" by a new larger more aggressive mouse each day for 10 days.[23] A variation of this paradigm removes the physical component by testing whether depression-like behavior develops in animals that "witness" another mouse being physically aggressed.[24] Such chronic stress paradigms successfully recapitulate certain behavioral features of human depression. For example, chronic social stress in adult male mice results in anhedonia (reduced preference for a sucrose solution), social avoidance, circadian changes, weight changes, and anxiety-like behaviors, such as decreased exploration of a novel environment.[25] Recently, social defeat has been adapted to female mice, which display a similar range of behavioral and neural abnormalities.[26] These depression-like behaviors are effectively reversed by chronic but not acute treatment with existing antidepressant medications, a treatment course comparable to that required in humans.[27] Not all mice that experience chronic social defeat stress succumb to depression-like behavior, as not all humans who experience stress become depressed. Approximately, half to two-thirds of socially stressed mice are "susceptible" and exhibit depression-like behavior, whereas remaining mice are "resilient."[25] This enables researchers to explore molecular and environmental factors that increase or decrease susceptibility and resilience.

4. DNA METHYLATION IN DEPRESSION

Human and animal studies converge to support a role for DNA methylation in mediating part of the impact of stress on the brain. At a broad level, the *de novo* methyltransferase *DNMT3a* is increased in NAc among humans suffering from major depression.[28] These findings are replicated in mouse models of chronic social defeat stress and subchronic variable stress.[28,29] *Dnmt3a* mRNA is higher in female mouse NAc than male, although there are no sex differences in human DNMT3a expression in NAc, and depression—or stress exposure—increases *DNMT3a* in both male and female humans and mice, respectively. Artificially, increasing *Dnmt3a* mRNA expression in mouse NAc increases depression-like behavior in male and female mice.[28,29] Knockout of *Dnmt3a* mRNA in NAc selectively reduces depression-like behavior in female mice after subchronic variable stress,[28] although depression-like behavior was rescued in male mice exposed to chronic defeat stress by infusion of the general DNA methylation inhibitor, RG108, into Nac.[29] Interestingly, NAc *DNMT3a* levels are partially rescued (reduced) with antidepressant treatment in both men and women,[28] although it remains to be seen whether peripheral DNMT3a may be used as a predictive biomarker.

DNMT3b, which exhibits higher expression during embryonic development, and *DNMT1*, which has a higher affinity for hemimethylated DNA and is necessary for maintenance methylation, are not different in depressed NAc. Among male and female suicide completers, however, *DNMT1* mRNA is decreased in frontopolar cortex and amygdala, and *DNMT3b* is increased in frontopolar cortex and decreased in brain stem regions.[30]

Expression of the DNA demethylase *Tet1*, but not *Tet2* or *Tet3*, is downregulated in NAc of mice susceptible to chronic social defeat stress.[31] Paradoxically, knocking out *Tet1* from mouse NAc relieved depression-like behaviors, suggesting a complicated relationship between target gene methylation and behavior. It remains to be seen whether TET enzymes may be involved in broad DNA methylation changes associated with human depression, and whether similar changes are observed in other brain regions.

Increases in DNMTs and decreases in demethylases would suggest an overall hypermethylation state in the brains of depressed individuals and mice. However, genome-wide 5mC and 5hmC analyses instead find regions of both hyper- and hypomethylation in postmortem PFC among depressed compared to control subjects, indicating specificity in DNA methylation changes.[32] 5mC and 5hmC genome-wide profiling indicate enrichment of differentially methylated regions in genes related to nervous system development, mitochondria function, and immune system regulation.[32–34] Genome-wide methylation analyses have yet to be published for other brain regions directly testing association with adult depression, although several important studies have examined the association with history of child abuse, which is positively associated with depression.

DNA methylation at several candidate genes has been examined directly. Among these candidate genes altered in stress and depression are two key neurotrophic factors, brain-derived neurotrophic factor (*BDNF*) and glial cell-derived neurotrophic factor (*GDNF*). At least 20 studies have now examined the association between DNA methylation of BDNF in human blood or saliva samples and depression in humans. A majority of these studies have found increased methylation within *BDNF* promoter I or VI, and three studies found reversal by antidepressant treatment (reviewed by Chen *et al.*[35]). Adult chronic social defeat stress in C57BL/6 mice does not alter *Bdnf* promoter P3 methylation in NAc, although other promoters and brain regions have not been examined.[36] Chronic unpredictable stress also increases *Gdnf* mRNA levels in NAc of C57BL/6 mice, although the effects of *Gdnf* methylation may be mediated by methyl-binding proteins in a complex, strain-specific manner.[37] *BDNF* is also highly regulated in depression by histone modifications (Box 1), and is exemplar of how individual genes are regulated at multiples levels.

Serotonin signaling in the brain is undeniably linked to antidepressant mechanisms and is regulated in part by the serotonin transporter encoded by *SLC6A4*. Methylation of a CpG "island" in the *SLC6A4* promoter in vitro reduces gene expression, after accounting for genetic variation in the serotonin transporter-linked polymorphic region.[44] Nearly 10 studies to date have found a positive association between *SLC6A4* methylation and depression in humans.[35,45] Antidepressant treatment has also been found to reverse *SLC6A4* methylation changes.[35] *SLC6A4* methylation is also positively linked with frontal gray matter brain volume and resting-state connectivity between cortical brain regions.[46] These studies have used peripheral tissue samples, suggesting that *SLC6A4* methylation may be a useful biomarker.[46,47]

Finally, DNA methylation changes are found in several genes critical along the HPA axis response to stress in human depression and animal models. Corticotropin releasing factor (CRF) in the paraventricular nucleus (PVN) of the hypothalamus is increased in mice susceptible to chronic social defeat stress, accompanied by decreased DNA methylation of the *Crf* promoter.[48] Both effects are reversed by chronic imipramine treatment.[48] DNA methylation is also increased at the *Crf* promoter in the PVN of female rats subjected to chronic unpredictable stress, suggesting that DNA methylation may play a role in determining sex-specific regulation of HPA-axis

BOX 1 A closer look at epigenetic regulation of BDNF-TRKB signaling in depression

Many of the histone modifications in stress and depression reviewed in this chapter are at a global rather than gene-specific level. One prominent focus of targeted studies has been BDNF. BDNF is repeatedly found to be decreased in serum of depressed patients and in PFC and hippocampus of depressed subjects and suicide victims examined postmortem.[38] Chronic stress in rodents similarly decreases BDNF in hippocampus.[36] In contrast, chronic social defeat stress in mice increases BDNF in NAc, and removing NAc BDNF prevents development of depression-like behavior in response to this form of stress.[27] Indeed, the effects of manipulating BDNF levels on depression-like behavior are sex-, brain region-, and developmental timing-specific.[39] BDNF is also an attractive target as its levels are responsive to antidepressant treatment by conventional SSRI's and by ketamine.[39] Within the hippocampus, decreased *BDNF* mRNA is accompanied by increased H3K27me2 at two promoters (associated with transcriptional repression).[36] However, H3K27me2 was unaffected by chronic imipramine treatment and therefore unlikely to mediate the antidepressant's effect to normalize BDNF and behavior. Instead, chronic stress and chronic imipramine treatment interact to increase H3 acetylation and H3K4me2 at two *BDNF* promoters (an effect not seen after either stress or imipramine alone).[36]

Consistent with evidence from animal models, postmortem studies in human brains suggest that antidepressants promote open chromatin structures by decreasing H3K27me3 levels at certain *BDNF* promoters in PFC of a depression sample.[40] Follow up studies in peripheral blood reveal higher peripheral *BDNF* expression in treatment responders compared to nonresponders, with H3K27me3 levels being inversely correlated with both *BDNF IV* expression levels and with symptom severity.[41] Decreased expression of BDNF's receptor, tyrosine receptor kinase B (*TRKB*), in the PFC of depressed cases is also associated with an enrichment of H3K27me3 levels in the promoter of both *TRKB* and its astrocytic variant, *TRKB.T1*.[42] The elevated H3K27me3 levels associate with changes in DNA methylation in the promoter suggesting the presence of dual epigenetic control over *TRKB.T1* expression, as reported for many genes in simpler systems. In addition, mice overexpressing *TRKB.T1* are more susceptible to chronic social stress than wild type mice suggesting that epigenetic changes at the *TRKB.T1* promoter could define the vulnerability to chronic social stress and the development of depression.[43]

function.[49] Increased DNA methylation of hippocampal glucocorticoid receptor (*NR3C1*) is also associated with depression and depression-like behaviors in human and animal models, particularly with a history of early life adversity.[50,51]

The examples summarized here are certainly not exhaustive of the work done to uncover how aberrant DNA methylation may either contribute to the pathophysiology of depression, or serve as a biomarker. There is a complementary body of work on epigenetic mechanisms that mediate developmental origins of depression and other psychiatric syndromes (see other Chapters 1, 7, 10 and 13, in this volume). As in genome-wide association studies searching for genetic alterations that increase risk for depression, it is clear that stress effects on DNA methylation are not restricted to a small number of candidate genes. Additional genome-wide DNA methylation studies in humans and animals will help to uncover how changes at dozens or more genes act in concert to promote susceptibility or resilience to environmental stress.

5. HISTONE MODIFICATIONS IN DEPRESSION

5.1 Histone Acetylation

The potential importance of chromatin modifications in depression was initially suggested by observations that HDAC inhibition—alone or in combination with antidepressant treatment—ameliorated depression-like behavior in rodents.[36,52–54] Histone 3 lysine 14 (H3K14) acetylation is decreased in NAc of depressed humans and transiently decreased in NAc of mice that experience chronic social defeat stress.[53] Blocking class I HDACs with the inhibitor MS275, infused directly into mouse NAc, reduces depression-like behavior after chronic stress, suggesting that histone acetylation in NAc mediates the effects of chronic stress on behavior.[53] In support of this, overexpression of a dominant-negative form of the class I *Hdac2* in NAc prevented depression-like behavior in a chronic unpredictable mouse stress paradigm, while expression of a form of *Hdac2* with increased chromatin binding affinity was prodepressant.[37] In contrast, expression of *Hdac5*, a class II HDAC, is decreased in mice susceptible to social defeat stress and increased by chronic imipramine treatment suggesting a proresiliency effect of this HDAC.[55] Furthermore, *Hdac5* knock-out mice display exaggerated depression-like behavior after chronic social defeat stress in adulthood.[55] The opposing effects of HDAC2 and HDAC5 are difficult to reconcile with a simplistic role of histone acetylation in depression and stress adaptation. However, it is conceivable that HDAC2 and HDAC5 may regulate distinct populations of genes. Additionally, it is important to note that HDAC5 could also regulate nonhistone targets due to its cytoplasmic, as well as nuclear localization. Genome-wide studies of NAc gene expression in defeated mice treated systemically with fluoxetine or intra-NAc with MS275 demonstrated that both treatments reverse a large proportion of defeat-induced differential gene expression. Although each treatment also regulate subsets of unique genes, there was also significant overlap, suggesting that antidepressant effects of fluoxetine may in part be mediated by affecting histone acetylation.[53]

Stress does not regulate histone modifications uniformly across brain regions. In contrast to the NAc, chronic social defeat stress transiently increases H3K14Ac in the PFC, amygdala, and hippocampus, 24 h after stress.[54,56] Within the amygdala, H3K14Ac levels normalize within 10 days after stress ceases, while hippocampal levels are persistently decreased thereafter, and PFC levels are persistently increased.[54,57] The long-lasting decrease in hippocampal H3K14Ac likely contributes to depressive-like behaviors, as chronic fluoxetine treatment rescues decreased hippocampal H3K14Ac but has no effect on amygdala levels.[54] Infusion of the HDAC inhibitor MS275 into the hippocampus rescues sucrose preference behavior (anhedonia) but not social avoidance, while infusion into the amygdala rescues social avoidance behavior but not sucrose preference.[54] These results reflect how PFC, amygdala, and hippocampus regulate different aspects of depression-like behavior. Interestingly, intrahippocampal HDAC inhibition by MS275 restores social avoidance if paired with social housing conditions, suggesting that H3K14Ac in the hippocampus facilitates adaptation.[54]

If H3K14Ac levels reflect increased permissiveness for transcriptional changes in response to stimuli, it is possible that the increase in H3K14Ac within NAc, PFC, hippocampus, and amygdala 24 h after stress represents a natural window of behavioral therapeutic intervention. Fluoxetine appears to restore this epigenetic-mediated ability to adapt to environmental experience and update behavioral response, but only within hippocampus. Supplementing traditional serotonin reuptake inhibitor (SSRI)-based antidepressant treatment with HDAC inhibitors increases efficacy in mice.[58] Indeed, the possibility of using HDAC inhibitors in patients with treatment-resistant depression has recently been suggested.[59]

The regional specificity in histone acetylation response to stress may be explained in context of work elucidating the circuit-level relationships of these regions in mice susceptible or resilient to depression-like behavior after chronic social defeat stress. Chronic stress increases measures of regional activity in NAc and ventral hippocampus, but decreases activity measures in mPFC among susceptible mice.[60–62] In support of a causal role for pathway-specific activity in depression-like behavior, optogenetic activation of ventral hippocampus-NAc projections is prodepressive, while activation of mPFC-NAc or basolateral amygdala-NAc projections are antidepressant.[61] Together with the regional acetylation data, a pattern emerges in which the basolateral amygdala works in opposition to NAc during and immediately following stress or artificial activation, and at long-term time points engagement of the hippocampus positively regulates NAc and can either produce prodepressive or proresilient effects if paired with stress or enrichment, respectively.[54,61]

One illustration of how genetics and epigenetic processes work in concert to influence response to stress and depression-like behavior comes from rats that have been selectively bred to be high responders (HR) or low responders (LR) based on their locomotor response in a novel environment. Cross-fostering of pups from HR to LR litters demonstrates that stress-response is genetic. In unstressed conditions, HR rats have high preference for a sucrose solution (low anhedonia-like behavior) compared to LR rats, while after repeated social defeat stress anhedonia increases among HR rats and decreases among LR rats.[63] These behavioral changes are mirrored in global H3 and H2B acetylation levels in hippocampus, which in turn may be mediated by similar changes in the histone acetyl transferase, CREB-binding protein.[63] Thus, epigenetic processes are at the intersection of genetic factors and environmental stress experiences.

5.2 Histone Methylation

Histone methylation is implicated in genome-wide association studies of human depression.[64] Chronic social defeat stress downregulates the histone methyltransferases G9a and G9a-like protein, which catalyze H3K9me2, a major repressive mark, in NAc.[65] Overexpression of *G9a* in NAc is antidepressant-like, and increased H3K9me2 at specific gene promoters is implicated in the antidepressant effect of fluoxetine.[65,66] Indeed, chronic exposure to fluoxetine reduces *Camkii* expression in NAc in association with reduced H3Ac and increased H3K9me2 levels at the *Camkiia* promoter in NAc. Interestingly, these effects are found in the NAc of depressed humans exposed to antidepressants suggesting that the stress-induced loss of repressive methylation is maladaptive and that the reinstatement of these marks at specific gene loci might contribute to the therapeutic effects of antidepressant drugs. Another gene, which illustrates this mode of regulation, is *Ras*. Reduced H3K9me2 at this gene in the NAc of susceptible mice results in increased *Ras* expression, induction of ERK signaling, and ultimately CREB activation which in this brain region acts to induce depression-like behavior.[65]

The observed decreases in H3K9me2 in NAc would be expected to mediate a more permissive transcriptional state similar to the effect of the global increases in H3 acetylation described earlier. Curiously, however, manipulations that decrease repressive methylation induce susceptibility, whereas manipulations that increase acetylation induce resilience. To reconcile and integrate these findings, genome-wide approaches are required to examine regulation of histone modifications at specific gene loci to understand the precise coordinated regulation of depression-related target genes.

Additional methylation at the H3K9 locus, catalyzed by SETDB1 to H3K9me3, is also associated with transcriptional repression. Within the hippocampus, genome-wide profiling of H3K9me3-assocaited genes by ChIP-seq found that restraint stress dramatically enriched this mark at repetitive elements, nontranscriped regions of the genome.[67,68] Such an effect may influence genomic instability. Finally, whole forebrain overexpression of *Setdb1* reduced depression-like behavior, suggesting that the increase in H3K9me3 after acute stress may represent an adaptive response.[69] These modifications are associated with changes in the 3D structure of chromatin as described in a later section.

H3K27 also can be mono-, di-, or trimethylated and these states are associated with activation, bivalency, or repression. ChIP-chip analysis (chromatin immunoprecipitation followed by genome-wide promoter microarrays) examined stress-induced redistribution of H3K27me2 in NAc of mice subjected to chronic social defeat or protracted social isolation. Significant and dynamic changes in repressive histone methylation were observed in upstream regulatory regions in both stress models with more genes implicating increased histone methylation. Interestingly, approximately 20% of genes are similarly regulated in both models H3K27me3 alterations have not been described in rodent stress models at a genome-wide level to date, although this mark is associated with repression of specific genes, including *Rac1*.[70] *Rac1* influences characteristic dendritic spine morphology changes in depressed human brain and socially defeated mice.[71]

Aside from the few examples cited earlier, human postmortem studies examining histone modifications in depression are sparse. Elevated levels of H3K4me3—a mark of gene activation—were reported at the synapsin gene family in PFC of depressed humans.[72] There are also reports of altered H3K4me3 or H3K27me3 in promoter regions of several candidate genes, including *BDNF* (Box 1) and others (e.g., *OAZ*, *SYN2*, *TRKB*) in postmortem PFC.[40] In general, however, there is a major need for genome-wide profiles for prominent histone modifications in depression and in response to antidepressant treatment, across brain regions, in order to define the genes and molecular pathways that mediate these responses. Thus far, there are no genome-wide analyses of histone modifications in depressed human brain. This is a high priority for future research.

6. CHROMATIN REMODELING IN DEPRESSION

Chromatin remodeling, not to be confused with chromatin modifications, refers to the actual movement of nucleosomes along DNA.[73] Chromatin remodeling complexes use the energy of ATP hydrolysis to alter the packing state of chromatin and work in concert with chromatin modifying enzymes to direct nucleosomal dynamics. Very little is known concerning the role of chromatin remodeling complexes in depression or any other psychiatric syndrome. However, our group demonstrated that chronic social defeat stress or other forms of chronic stress induces a repressive chromatin remodeling complex in NAc, which by ChIP-seq was shown to mediate stress-repression of a set of genes important for mediating stress susceptibility.[74] Induction of the same complex was found in NAc of depressed humans, providing translational validation. Induction of this repressive complex at suppressed genes correlates with lower levels of activating histone marks (e.g., H3K4me3 and H4K16ac) and increased levels of certain repressive histone marks (e.g., H3K9me2), thus emphasizing the coordinated nature of epigenetic regulation.[74] These findings underscore the importance of mapping numerous epigenetic mechanisms at a genome-wide scale in order to define the combinatorial code of epigenetic changes associated with depression or antidepressant responses.

Histone variant dynamics may also reflect stress experience across the lifespan and play a role in susceptibility or resilience to depression. The four canonical histone subunits within the nucleosome (H2A, H2B, H3, and H4) also come in several variants. Histone variants, such as H2A.Z and H3.1–3 are incorporated into chromatin independently of DNA replication; this process therefore requires removal of existing nucleosomes. In particular, H3.3-containing nucleosomes at genomic enhancer and promoter regions can undergo rapid turnover, and such turnover is further associated with histone modifications marking active gene expression.[75] It was recently found that H3.3 dynamics are increased in depressed human NAc and in response to chronic social defeat stress in mice, and reversed by antidepressant treatment.[76] Moreover, H3.3 was found to accumulate across development to near-saturation by midadolescence, and chronic stress in a postnatal sensitive period of mouse development accelerates H3.3 accumulation in NAc and increases susceptibility to later stress.[76–78] Preventing H3.3 dynamics promotes resilience to chronic social defeat stress in mice.[76] These findings represent a novel frontier in understanding epigenetic mechanisms of depression.

7. REGULATION OF THE 3D GENOME IN DEPRESSION

There is now an increasing appreciation of the degree to which the 3D structure of chromatin is under dynamic regulation in the developing and adult brain, and this mode of regulation is beginning to be considered in neuropsychiatric disorders.[22] One recent study found that knockout of *Setdb1* (an H3K9me3 HMT) in forebrain neurons induces a dramatic reorganization of ~1.2 Mb of DNA around the protocadherin gene cluster.[79] This reorganization is associated with altered levels of CTCF binding, DNA methylation, and several histone modifications (e.g., H3K27ac) and with altered expression levels of specific protocadherin isoforms. As noted earlier, *Setdb1* knockout mice exhibit increased depression-like behavioral abnormalities, raising the important goal for future research to directly link changes in neuronal function and behavior to the disordered 3D genome at the protocadherin locus. Of note, the 3D structure of this locus as well as patterns of CTCF binding and H3K9me3 enrichment are highly homologous between mouse forebrain neurons and human glutamate-like neurons induced from pluripotent stem cells, supporting the validity of using the mouse for epigenetic explorations of depression mechanisms.[79]

8. NONCODING RNAS IN DEPRESSION

Increasing evidence implicates a role for microRNAs (miRNAs) in depression or as potential biomarkers for diagnosing and targeting treatment for this complex mood disorder.[80] microRNAs are both spatially and temporally regulated, similar to other transcripts. miRNAs are relatively conserved evolutionarily, suggesting that they play an essential role in transcriptional regulation.[7] miRNAs have been estimated to target up to hundreds of genes. As depression and other complex psychiatric syndromes are now accepted to involve many subtle transcriptional changes, miRNAs represent an attractive potential therapeutic tool to achieve broad regulatory ability over entire gene networks.[81]

Although many miRNAs are conserved across species, miR-1202 appears to be primate-specific and enriched within the brain. Depression decreases miR-1202 in postmortem human PFC, and history of antidepressant treatment partially restores its expression.[39] miR-1202 can be measured in human blood and similarly reflects depression status and response to chronic but not acute antidepressant treatment.[79,80] Moreover, peripheral baseline miR-1202 levels predict resting state connectivity between frontal regions, and correlate with activity changes during a response inhibition task in other relevant brain regions.[82] One identified target of miR-1202 is metabotropic glutamate receptor-4 (*GRM4*), which is inversely correlated with miR-1202, increased in depressed postmortem PFC, and rescued with a history of antidepressant treatment.[39] GRM4 pre- and postsynaptically modulates glutamatergic, dopaminergic, GABAergic, and serotonergic signaling throughout the brain. miR-1201 may be a useful biomarker to detect depression and treatment response and altered neurotransmission. Other possible biomarkers of major depression and antidepressant response that were recently identified include miR-146b-5p, miR-425-3p, miR-24-3p, miR-185, and miR-491-3p.[83] Another study found 21 miRNAs downregulated in PFC of depressed suicide cases.[84] These miRNAs are predicted to bind several targets, including *DNMT3B* and *TRKB.T1*. Additionally, miR-218 is reduced in PFC of depressed humans and mice susceptible to chronic social defeat stress.[85] One of its targets is the developmental netrin-1 guidance cue receptor DCC, and both are expressed in pyramidal neurons in humans and mice. *Dcc* overexpression increases vulnerability to chronic stress, thus demonstrating functional relevance for miR-218.[85] Human studies of miRNAs in depression have predominately focused on changes within PFC, and more research is needed to understand how miRNAs in other brain regions contribute to depression.

Let-7a is one of the first two identified miRNAs and is found in a wide range of species. It is thought to regulate genes involved in developmental patterning and cell-cycle regulation. Mice exposed to acute restraint stress have increased expression of Let-7a in PFC and amygdala, but not hippocampus.[86,87] Let-7a increases were maintained in amygdala after chronic stress, but not in PFC. Additional miRNAs, including miR-9 and miR

26-a/b, were increased acutely in PFC, and miR-376b and miR-208 were increased in hippocampus following acute or chronic stress, while miR-9-1 was decreased in hippocampus.[87]

Other reports implicate miRNAs in the posttranscriptional regulation of *Nr3c1* (glucocorticoid receptor). Compared to Sprague-Dawley rats, F344 stress-sensitive rats have lower *Nr3c1* protein but not mRNA levels in the hypothalamic PVN and corresponding increases in miR-18a levels. Overexpression of miR-18a downregulates NR3C1 protein levels in cultured neurons suggesting that the exaggerated HPA stress response in F344 rats may be mediated by miR-18a regulation of *Nr3c1* translation.[88] Interestingly, in amygdala, acute stress induced expression of several miRNAs is implicated in regulation of anxiety through effects on the mineralocorticoid receptor and *Crfr1* expression.[89,90]

A significant downregulation of several miRNAs has been reported in the hippocampus of rats resistant to "learned helplessness"—a subchronic stress model—after inescapable foot shock compared to those rats that develop learned helplessness.[91] These miRNA-mediated effects coincide with robust changes in hippocampus gene expression and are believed to represent a potential coping mechanism that allows the development of physiological responses to overcome the effects of stress.[92]

miRNAs may also be essential components of antidepressant efficacy and of resilience. Chronic fluoxetine treatment in mice increases miR-16 levels in serotonergic raphe nuclei, which downregulates serotonin transporter (*Slc6a4*) levels, thus altering serotonergic activity in the raphe.[93] Intraraphe administration of fluoxetine induces the release of S100β, a known inhibitor of miR-16. Via the reciprocal connections between dorsal raphe and locus coeruleus (LC), S100β is believed to migrate to noradrenergic neurons in the LC and decrease expression of miR-16. By decreasing miR-16 in the LC, S100β turned on the expression of serotonergic functions in noradrenergic neurons. Moreover, intraraphe and -intra-LC miR-16 administration alleviated the behavioral deficits induced by 6 weeks of chronic unpredictable stress suggesting that the therapeutic effects of fluoxetine may be mediated via its actions on miRNA expression in the brain.[93] A different study identified dicer, an miRNA-processing protein, as a major target of the transcription factor, β-catenin, in the NAc, where both proteins contribute to a state of resilience.[94] Small RNA sequencing identified numerous miRNAs whose induction in this region associates with resilience to chronic social defeat stress.

Finally, miRNAs may help explain sex differences in the pathophysiology of depression. Male and female mice exposed to subchronic variable stress each had broad changes in miRNA expression in NAc.[95] In this stress paradigm, only female mice developed depression-like behavior after subchronic variable stress, while male mice developed depression-like behavior after chronic stress. miRNA alterations were also nearly completely nonoverlapping between the sexes.[95] Predicted functions of stress-altered female miRNA-mRNA networks include neurotransmitter receptor activity and binding, gated channel activity, and cytokine activity.[95] More research is needed to understand sex differences in miRNAs in depression in human brain and peripheral tissue.

The earlier findings suggest the involvement of miRNAs in the pathophysiology of depressive disorders. A key challenge moving forward is to directly test the functional relevance of miRNAs in depression by altering specific miRNAs in animal brain and testing effects on susceptibility to stress. In addition, while in silico tools to predict miRNA regulation of mRNA targets have improved in recent years, more work is needed to empirically test predictions in vitro and in vivo.

9. CONCLUSIONS

Looking toward the future, it will be important for studies to integrate peripheral epigenetic modifications—DNA methylation and noncoding RNAs in particular—with changes within the brain, across relevant brain regions. An essential challenge in neuropsychiatry moving forward is to increase the cell-type specificity of studies of epigenetic modifications within the brain. Epigenetic changes are known to be specific to developmental stage, tissue or region, and cell type, and studies are just now beginning to measure and manipulate epigenetic mechanisms with this resolution. Furthermore, in order to push the field forward constant dialogue is needed between human and animal researchers. Model systems enable researchers to test the functional relevance of findings from human studies to determine whether identified epigenetic alterations are correlative or causative. Finally, there is an urgent demand for increased specificity in animal studies testing the causative nature of epigenetic changes. Manipulation of DNMTs or HATs, for example, will have genome-wide impact, yet sequencing studies are increasingly revealing specificity in epigenetic regulation. Recent developments in molecular tools, including zinc finger proteins, transcription activator-like effectors (TALES), and CRISPR now enable researchers to target *individualgenes* for a single type of epigenetic modification.[96–98] These technologies enable brain-region-specific, gene-specific, and potentially, temporal targeting of epigenetic modifiers to determine whether epigenetic regulation at a specific locus is responsible for life-long perturbations in psychiatric disorders. This technology also offers therapeutic promise for the future.

ACKNOWLEDGMENTS

This work is supported in part by K99 MH115096 to C.J.P and P50 MH096890 and Hope for Depression Research Foundation funding to E.J.N.

REFERENCES

1. World Heath Organization (WHO). Depression and Other Common Mental Disorders. Available from: http://apps.who.int/iris/bitstream/10665/254610/1/WHO-MSD-MER-2017.2-eng.pdf
2. Kendler KS, Prescott CA. A population-based twin study of lifetime major depression in men and women. *Arch Gen Psychiatry.* 1999;56(1):39–44.
3. Geschwind DH, Flint J. Genetics and genomics of psychiatric disease. *Science.* 2015;349 (6255):1489–1494.
4. Waddington CH. Canalization of development and genetic assimilation of acquired characters. *Nature.* 1959;183(4676):1654–1655.
5. Jenuwein T, Allis CD. Translating the histone code. *Science.* 2001;293(5532):1074–1080.
6. Jaenisch R, Bird A. Epigenetic regulation of gene expression: how the genome integrates intrinsic and environmental signals. *Nat Genet.* 2003;33(3s):245–254.
7. Esteller M. Non-coding RNAs in human disease. *Nat Rev Genet.* 2011;12(12):861–874.
8. Jonson-Reid M, Kohl PL, Drake B. Child and adult outcomes of chronic child maltreatment. *Pediatrics.* 2012;129(5):839–845.
9. Russo SJ, Nestler EJ. The brain reward circuitry in mood disorders. *Nat Rev Neurosci.* 2013;14(9):609–625.
10. Annunziato L, Di Renzo GF, Lombardi G, Preziosi P, Scapagnini U. Proceedings: catecholaminergic control of thyroid stimulating hormone (TSH) and adrenocroticotrophic hormone (ACTH) secretion. *Br J Pharmacol.* 1974;52(3):442P–443P.
11. Bird AP, Wolffe AP. Methylation-induced repression: belts, braces, and chromatin. *Cell.* 1999;99(5):451–454.
12. Razin A. CpG methylation, chromatin structure and gene silencing—a three-way connection. *EMBO J.* 1998;17(17):4905–4908.
13. Fuks F. DNA methylation and histone modifications: teaming up to silence genes. *Curr Opin Genet Dev.* 2005;15(5):490–495.
14. Lister R, Mukamel EA, Nery JR, Urich M, Puddifoot CA, Johnson ND, et al. Global epigenomic reconfiguration during mammalian brain development. *Science.* 2013;341 (6146):1237905–1237915.
15. Guo JU, Ma DK, Mo H, Ball MP, Jang M-H, Bonaguidi MA, et al. Neuronal activity modifies the DNA methylation landscape in the adult brain. *Nat Neurosci.* 2011;14 (10):1345–1351.
16. Szulwach KE, Li X, Li Y, Song C-X, Wu H, Dai Q, et al. 5-hmC–mediated epigenetic dynamics during postnatal neurodevelopment and aging. *Nat Neurosci.* 2011;14 (12):1607–1616.
17. Khorasanizadeh S. The nucleosome: from genomic organization to genomic regulation. *Cell.* 2004;116(2):259–272.

18. Ha M, Ng DW., Li W-H, Chen ZJ. Coordinated histone modifications are associated with gene expression variation within and between species. *Genome Res.* 2011;21(4):590–598.
19. Sun HH, Kennedy PJP, Nestler EJE. Epigenetics of the depressed brain: role of histone acetylation and methylation. *Neuropsychopharmacology.* 2013;38(1):124–137.
20. Ma L, Bajic VB, Zhang Z. On the classification of long non-coding RNAs. *RNA Biol.* 2013;10(6):925–933.
21. Huang X, Luo Y-L, Mao Y-S, Ji J-L. The link between long noncoding RNAs and depression. *Prog Neuropsychopharmacol Biol Psychiatry.* 2017;73:73–78.
22. Mitchell AC, Javidfar B, Bicks LK, Neve R, Garbett K, Lander SS, et al. Longitudinal assessment of neuronal 3D genomes in mouse prefrontal cortex. *Nat Commun.* 2016;7:12743.
23. Golden SA, Covington HE, Berton O, Russo SJ. A standardized protocol for repeated social defeat stress in mice. *Nat Protoc.* 2011;6(8):1183–1191.
24. Warren BL, Vialou VF, Iñiguez SD, Alcantara LF, Wright KN, Feng J, et al. Neurobiological sequelae of witnessing stressful events in adult mice. *Biol Psychiatry.* 2013;73(1):7–14.
25. Krishnan V, Han M-H, Graham DL, Berton O, Renthal W, Russo SJ, et al. Molecular adaptations underlying susceptibility and resistance to social defeat in brain reward regions. *Cell.* 2007;131(2):391–404.
26. Takahashi A, Chung J-R, Zhang S, Zhang H, Grossman Y, Aleysin H, Flanigan ME, Pfau ML, Menard C, Dumitriu D, Hodes GE, McEwen BS, Nestler EJ, Han M-H, Russo SJ. Establishment of a repeated social defeat stress model in female mice. *Sci Rep.* 2017;7:12838.
27. Berton O, Mcclung CA, Dileone RJ, Krishnan V, Renthal W, Russo SJ, et al. Essential role of BDNF in the mesolimbic dopamine pathway in social defeat stress. *Science.* 2006;311(5762):864–868.
28. Hodes GE, Pfau ML, Purushothaman I, Ahn HF, Golden SA, Christoffel DJ, et al. Sex differences in nucleus accumbens transcriptome profiles associated with susceptibility versus resilience to subchronic variable stress. *J Neurosci.* 2015;35(50):16362–16376.
29. LaPlant QQ, Vialou VV, Covington HEH, Dumitriu DD, Feng JJ, Warren BLB, et al. Dnmt3a regulates emotional behavior and spine plasticity in the nucleus accumbens. *Nat Neurosci.* 2010;13(9):1137–1143.
30. Poulter MO, Du L, Weaver ICG, Palkovits M, Faludi G, Merali Z, et al. GABAA receptor promoter hypermethylation in suicide brain: implications for the involvement of epigenetic processes. *Biol Psychiatry.* 2008;64(8):645–652.
31. Feng J, Peña CJ, Purushothaman I, Engmann O, Walker D, Brown AN, et al. Tet1 in nucleus accumbens opposes depression- and anxiety-like behaviors. *Neuropsychopharmacology.* 2017;42(8):1657–1669.
32. Gross JA, Pacis A, Chen GG, Drupals M, Lutz P-E, Barreiro LB, et al. Gene-body 5-hydroxymethylation is associated with gene expression changes in the prefrontal cortex of depressed individuals. *Transl Psychiatry.* 2017;7(5):e1119.
33. Córdova-Palomera A, Fatjó-Vilas M, Gastó C, Navarro V, Krebs M-O, Fañanás L. Genome-wide methylation study on depression: differential methylation and variable methylation in monozygotic twins. *Transl Psychiatry.* 2015;5(4):e557.
34. Murphy TM, Crawford B, Dempster EL, Hannon E, Burrage J, Turecki G, et al. Methylomic profiling of cortex samples from completed suicide cases implicates a role for PSORS1C3 in major depression and suicide. *Transl Psychiatry.* 2017;7(1):e989.
35. Chen D, Meng L, Pei F, Zheng Y, Leng J. A review of DNA methylation in depression. *J Clin Neurosci.* 2017;43:39–46.
36. Tsankova NMN, Berton OO, Renthal WW, Kumar AA, Neve RLR, Nestler EJE. Sustained hippocampal chromatin regulation in a mouse model of depression and antidepressant action. *Nat Neurosci.* 2006;9(4):519–525.

37. Uchida S, Hara K, Kobayashi A, Otsuki K, Yamagata H, Hobara T, et al. Epigenetic status of Gdnf in the ventral striatum determines susceptibility and adaptation to daily stressful events. *Neuron.* 2011;69(2):14–24.
38. Castrén E, Kojima M. Brain-derived neurotrophic factor in mood disorders and antidepressant treatments. *Neurobiol Dis.* 2017;97(Pt. B):119–126.
39. Lopez JP, Lim R, Cruceanu C, Crapper L, Fasano C, Labonté B, et al. miR-1202 is a primate-specific and brain-enriched microRNA involved in major depression and antidepressant treatment. *Nat Med.* 2014;20(7):764–768.
40. Chen ESE, Ernst CC, Turecki GG. The epigenetic effects of antidepressant treatment on human prefrontal cortex BDNF expression. *Int J Neuropsychopharmacol.* 2011;14 (3):427–429.
41. Lopez JP, Mamdani F, Labonte B, Beaulieu M-M, Yang JP, Berlim MT, et al. Epigenetic regulation of BDNF expression according to antidepressant response. *Mol Psychiatry.* 2012;18(4):398–399.
42. Ernst C, Chen ES, Turecki G. Histone methylation and decreased expression of TrkB.T1 in orbital frontal cortex of suicide completers. *Mol Psychiatry.* 2009;14(9):830–832.
43. Ernst C, Deleva V, Deng X, Sequeira A, Pomarenski A, Klempan T, et al. Alternative splicing, methylation state, and expression profile of tropomyosin-related kinase B in the frontal cortex of suicide completers. *Arch Gen Psychiatry.* 2009;66(1):22–32.
44. Philibert R, Madan A, Andersen A, Cadoret R, Packer H, Sandhu H. Serotonin transporter mRNA levels are associated with the methylation of an upstream CpG island. *Am J Med Genet B Neuropsychiatr Genet.* 2007;144B(1):101–105.
45. Philibert RA, Sandhu H, Hollenbeck N, Gunter T, Adams W, Madan A. The relationship of 5HTT (SLC6A4) methylation and genotype on mRNA expression and liability to major depression and alcohol dependence in subjects from the Iowa Adoption Studies. *Am J Med Genet B Neuropsychiatr Genet.* 2008;147B(5):543–549.
46. Ismaylova E, Di Sante J, Szyf M, Nemoda Z, Yu W-J, Pomares FB, et al. Serotonin transporter gene promoter methylation in peripheral cells in healthy adults: neural correlates and tissue specificity. *Eur Neuropsychopharm.* 2017;27(10):1032–1041.
47. Olsson CA, Foley DL, Parkinson-Bates M, Byrnes G, McKenzie M, Patton GC, et al. Prospects for epigenetic research within cohort studies of psychological disorder: a pilot investigation of a peripheral cell marker of epigenetic risk for depression. *Biol Psychol.* 2010;83(2):159–165.
48. Elliott EE, Ezra-Nevo GG, Regev LL, Neufeld-Cohen AA, Chen AA. Resilience to social stress coincides with functional DNA methylation of the Crf gene in adult mice. *Nat Neurosci.* 2010;13(11):1351–1353.
49. Sterrenburg L, Gaszner B, Boerrigter J, Santbergen L, Bramini M, Elliott E, et al. Chronic stress induces sex-specific alterations in methylation and expression of corticotropin-releasing factor gene in the rat. *PLoS One.* 2011;6(11):e28128.
50. Mcgowan PO, Sasaki A, D'Alessio AC, Dymov S, Labonté B, Szyf M, et al. Epigenetic regulation of the glucocorticoid receptor in human brain associates with childhood abuse. *Nat Neurosci.* 2009;12(3):342–348.
51. Weaver ICG, Cervoni N, Champagne FA, D'Alessio AC, Sharma S, Seckl JR, et al. Epigenetic programming by maternal behavior. *Nat Neurosci.* 2004;7(8):847–854.
52. Schroeder FA, Lin CL, Crusio WE, Akbarian S. Antidepressant-like effects of the histone deacetylase inhibitor, sodium butyrate, in the mouse. *BPS.* 2007;62(1):55–64.
53. Covington HE, Maze I, LaPlant QC, Vialou VF, Ohnishi YN, Berton O, et al. Antidepressant actions of histone deacetylase inhibitors. *J Neurosci.* 2009;29 (37):11451–11460.
54. Covington HE, Vialou VF, Laplant Q, Ohnishi YN, Nestler EJ. Hippocampal-dependent antidepressant-like activity of histone deacetylase inhibition. *Neurosci Lett.* 2011;493 (3):122–126.

55. Renthal W, Maze I, Krishnan V, Covington HE, Xiao G, Kumar A, et al. Histone deacetylase 5 epigenetically controls behavioral adaptations to chronic emotional stimuli. *Neuron.* 2007;56(3):13–23.
56. Covington III HE, Maze I, Vialou V, Nestler EJ. Antidepressant action of HDAC inhibition in the prefrontal cortex. *Neuroscience.* 2015;298:329–335.
57. Hinwood M, Tynan RJ, Day TA, Walker FR. Repeated social defeat selectively increases δFosB expression and histone H3 acetylation in the infralimbic medial prefrontal cortex. *Cerebral Cortex.* 2011;21(2):262–271.
58. Schmauss C. An HDAC-dependent epigenetic mechanism that enhances the efficacy of the antidepressant drug fluoxetine. *Sci Rep.* 2015;5(1):8.
59. Fuchikami M, Yamamoto S, Morinobu S, Okada S, Yamawaki Y, Yamawaki S. The potential use of histone deacetylase inhibitors in the treatment of depression. *Prog Neuropsychopharmacol Biol Psychiatry.* 2016;64:320–324.
60. Christoffel DJ, Golden SA, Dumitriu D, Robison AJ, Janssen WG, Ahn HF, et al. IκB kinase regulates social defeat stress-induced synaptic and behavioral plasticity. *J Neurosci.* 2011;31(1):314–321.
61. Bagot RC, Parise EM, Peña CJ, Zhang H-X, Maze I, Chaudhury D, et al. Ventral hippocampal afferents to the nucleus accumbens regulate susceptibility to depression. *Nat Commun.* 2015;6:7062–7072.
62. Covington HE, Lobo MK, Maze I, Vialou V, Hyman JM, Zaman S, et al. Antidepressant effect of optogenetic stimulation of the medial prefrontal cortex. *J Neurosci.* 2010;30(48):16082–16090.
63. Hollis F, Duclot F, Gunjan A, Kabbaj M. Individual differences in the effect of social defeat on anhedonia and histone acetylation in the rat hippocampus. *HorBehav.* 2011;59(3):331–337.
64. O'Dushlaine C, Rossin L, Lee PH, Duncan L, Parikshak NN, Newhouse S, et al. Psychiatric genome-wide association study analyses implicate neuronal, immune and histone pathways. *Nat Neurosci.* 2015;18(2):199–209.
65. Covington HE, Maze I, Sun H, Bomze HM, DeMaio KD, Wu EY, et al. A role for repressive histone methylation in cocaine-induced vulnerability to stress. *Neuron.* 2011;71(4):656–670.
66. Robison AJ, Vialou V, Sun H-S, Labonté B, A Golden S, Dias C, et al. Fluoxetine epigenetically alters the CaMKIIα promoter in nucleus accumbens to regulate ΔFosB binding and antidepressant effects. *Neuropsychopharmacology.* 2014;39(5):1178–1186.
67. Hunter RG, McCarthy KJ, Milne TA, Pfaff DW, Mcewen BS. Regulation of hippocampal H3 histone methylation by acute and chronic stress. *Proc Natl Acad Sci USA.* 2009;106(49):20912–20917.
68. Hunter RG, Murakami G, Dewell S, Seligsohn M, Baker MER, Datson NA, et al. Acute stress and hippocampal histone H3 lysine 9 trimethylation, a retrotransposon silencing response. *Proc Natl Acad Sci USA.* 2012;109(43):17657–17662.
69. Jiang Y, Matevossian A, Huang H-S, Straubhaar J, Akbarian S. Isolation of neuronal chromatin from brain tissue. *BMC Neurosci.* 2008;9:42–52.
70. Wilkinson MB, Xiao G, Kumar A, Laplant Q, Renthal W, Sikder D, et al. Imipramine treatment and resiliency exhibit similar chromatin regulation in the mouse nucleus accumbens in depression models. *J Neurosci.* 2009;29(24):7820–7832.
71. Golden SA, Christoffel DJ, Heshmati M, Hodes GE, Magida J, Davis K, et al. Epigenetic regulation of RAC1 induces synaptic remodeling in stress disorders and depression. *Nat Med.* 2013;19(3):337–344.
72. Cruceanu C, Alda M, Nagy C, Freemantle E, Rouleau GA, Turecki G. H3K4 trimethylation in synapsin genes leads to different expression patterns in bipolar disorder and major depression. *Int J Neuropsychopharmacol.* 2013;16(2):289–299.

73. Vogel-Ciernia A, Wood MA. Neuron-specific chromatin remodeling: a missing link in epigenetic mechanisms underlying synaptic plasticity, memory, and intellectual disability disorders. *Neuropharmacology.* 2014;80:18–27.
74. Sun H, Damez-Werno DM, Scobie KN, Shao NY, Dias C, Rabkin J, et al. ACF chromatin-remodeling complex mediates stress-induced depressive-like behavior. *Nat Med.* 2015;21(10):1146–1153.
75. Kraushaar DC, Jin W, Maunakea A, Abraham B, Ha M, Zhao K. Genome-wide incorporation dynamics reveal distinct categories of turnover for the histone variant H3.3. *Genome Biol.* 2013;14(10):R121.
76. Lepack AE, Bagot RC, Peña CJ, Loh Y-HE. Farrelly LA, Lu Y, et al. Aberrant H3.3 dynamics in NAc promote vulnerability to depressive-like behavior. *Proc Natl Acad Sci USA.* 2016;113(44):12562–12567.
77. Maze I, Wenderski W, Noh K-M, Bagot RC, Tzavaras N, Purushothaman I, et al. Critical role of histone turnover in neuronal transcription and plasticity. *Neuron.* 2015;87(1):77–94.
78. Peña CJ, Kronman HG, Walker DM, Cates HM, Bagot RC, Purushothaman I, et al. Early life stress confers lifelong stress susceptibility in mice via ventral tegmental area OTX2. *Science.* 2017;356(6343):1185–1188.
79. Jiang Y, Loh Y-HE. Rajarajan P, Hirayama T, Liao W, Kassim BS, et al. The methyltransferase SETDB1 regulates a large neuron-specific topological chromatin domain. *Nat Genet.* 2017;49(8):1239–1250.
80. Issler O, Chen A. Determining the role of microRNAs in psychiatric disorders. *Nat Rev Neurosci.* 2015;16(4):201–212.
81. O'Connor RM, Dinan TG, Cryan JF. Little things on which happiness depends: microRNAs as novel therapeutic targets for the treatment of anxiety and depression. *Mol Psychiatry.* 2012;17(4):359–376.
82. Lopez JP, Pereira F, Richard-Devantoy S, Berlim M, Chachamovich E, Fiori LM, et al. Co-variation of peripheral levels of miR-1202 and brain activity and connectivity during antidepressant treatment. *Neuropsychopharmacology.* 2017;42(10):2043–2051.
83. Maussion G, Yang J, Yerko V, Barker P, Mechawar N, Ernst C, et al. Regulation of a truncated form of tropomyosin-related kinase B (TrkB) by Hsa-miR-185* in frontal cortex of suicide completers. *PLoS One.* 2012;7(6):e39301–e39311.
84. Smalheiser NR, Lugli G, Rizavi HS, Torvik VI, Turecki G, Dwivedi Y. MicroRNA expression is down-regulated and reorganized in prefrontal cortex of depressed suicide subjects. *PLoS One.* 2012;7(3):e33201.
85. Torres-Berrío A, Lopez JP, Bagot RC, Nouel D, Dal Bo G, Cuesta S, et al. DCC confers susceptibility to depression-like behaviors in humans and mice and is regulated by miR-218. *Biol Psychiatry.* 2017;81(4):306–315.
86. Rinaldi A, Vincenti S, De Vito F, Bozzoni I, Oliverio A, Presutti C, et al. Stress induces region specific alterations in microRNAs expression in mice. *Behav Brain Res.* 2010;208(1):265–269.
87. Meerson A, Cacheaux L, Goosens KA, Sapolsky RM, Soreq H, Kaufer D. Changes in brain MicroRNAs contribute to cholinergic stress reactions. *J Mol Neurosci.* 2010;40(1–2):47–55.
88. Uchida S, Nishida A, Hara K, Kamemoto T, Suetsugi M, Fujimoto M, et al. Characterization of the vulnerability to repeated stress in Fischer 344 rats: possible involvement of microRNA-mediated down-regulation of the glucocorticoid receptor. *Eur J Neurosci.* 2008;27(9):2250–2261.
89. Haramati S, Navon I, Issler O, Ezra-Nevo G, Gil S, Zwang R, et al. MicroRNA as repressors of stress-induced anxiety: the case of amygdalar miR-34. *J Neurosci.* 2011;31(40):14191–14203.

90. Mannironi C, Camon J, De Vito F, Biundo A, De Stefano ME, Persiconi I, et al. Acute stress alters amygdala microRNA miR-135a and miR-124 expression: inferences for corticosteroid dependent stress response. *PLoS One.* 2013;8(9):e73385.
91. Smalheiser NR, Lugli G, Rizavi HS, Zhang H, Torvik VI, Pandey GN, et al. MicroRNA expression in rat brain exposed to repeated inescapable shock: differential alterations in learned helplessness vs. non-learned helplessness. *Int J Neuropsychopharmacol.* 2011;14 (10):1315–1325.
92. Kohen R, Kirov S, Navaja GP, Happe HK, Hamblin MW, Snoddy JR, et al. Gene expression profiling in the hippocampus of learned helpless and nonhelpless rats. *Pharmacogenomics J.* 2005;5(5):278–291.
93. Baudry A, Mouillet-Richard S, Schneider B, Launay J-M, Kellermann O. miR-16 targets the serotonin transporter: a new facet for adaptive responses to antidepressants. *Science.* 2010;329(5998):1537–1541.
94. Dias C, Feng J, Sun H, Shao NY, Mazei-Robison MS, Damez-Werno D, et al. β-Catenin mediates stress resilience through Dicer1/microRNA regulation. *Nature.* 2014;516 (7529):51–55.
95. Pfau ML, Purushothaman I, Feng J, Golden SA, Aleyasin H, Lorsch ZS, et al. Integrative analysis of sex-specific microRNA networks following stress in mouse nucleus accumbens. *Front Mol Neurosci.* 2016;9:144.
96. Heller EA, Cates HM, Peña CJ, Sun H, Shao N, Feng J, et al. Locus-specific epigenetic remodeling controls addiction- and depression-related behaviors. *Nat Neurosci.* 2014;17 (12):1720–1727.
97. Anton T, Bultmann S. Site-specific recruitment of epigenetic factors with a modular CRISPR/Cas system. *Nucleus.* 2017;8(3):1–8.
98. Brocken DJW, Tark-Dame M, Dame RT. dCas9: a versatile tool for epigenome editing. *Curr Issues Mol Biol.* 2018;26:15–32.

CHAPTER FOUR

Epigenetics of Circadian Rhythms in Imprinted Neurodevelopmental Disorders

Rochelle L. Coulson, Janine M. LaSalle[1]
Genome Center, MIND Institute, UC Davis, CA, United States
[1] Corresponding author. E-mail address: jmlasalle@ucdavis.edu

Contents

Abstract

DNA sequence information alone cannot account for the immense variability between chromosomal alleles within diverse cell types in the brain, whether these differences are observed across time, cell type, or parental origin. The complex control and maintenance of gene expression and modulation are regulated by a multitude of molecular and cellular mechanisms that layer on top of the genetic code. The integration of genetic and environmental signals required for regulating brain development and function is achieved in part by a dynamic epigenetic landscape that includes DNA methylation, histone modifications, and noncoding RNAs. These epigenetic mechanisms establish and maintain core biological processes, including genomic imprinting and entrainment of circadian rhythms. This chapter will focus on how the epigenetic layers of DNA methylation and long, noncoding RNAs interact with circadian rhythms at specific imprinted chromosomal loci associated with the human neurodevelopmental disorders Prader–Willi, Angelman, Kagami–Ogata, and Temple syndromes.

Progress in Molecular Biology and Translational Science, volume 157
ISSN 1877-1173
http://dx.doi.org/10.1016/bs.pmbts.2017.11.023

1. DNA METHYLATION

The addition of a methyl group to the fifth carbon of the cytosine ring (5mC) is the most widespread form of DNA methylation in vertebrates, and occurs most frequently at CpG dinucleotides within mammals.[1] Saturation of 5mC methylation has increased with evolutionary time. While the human genome is the most highly methylated at CpG sites, the frequency of CpG dinucleotides is only about one-fifth of the expected frequency based on genomic GC content, due to the susceptibility of methylated cytosines to be deaminated and converted to thymine.[2] The deamination of methylated CpGs is predicted to have led to an overall de-enrichment of CpGs compared to all possible dinucleotides in mammals. In evolutionary time, the loss of methylated CpGs corresponds to the occurrence of "CpG islands" which are predicted to be protected from deamination reactions.

CpG islands are characterized as regions which span more than 200 bp with a greater than 50% CpG content.[3] They often localize at transcription start sites and are generally protected from DNA methylation. In combination with histone acetylation, an unmethylated CpG island corresponds with an accessible, transcriptionally permissive chromatin environment. However, depending on the genomic context and developmental timing of modification, DNA methylation can be a mark of either repression or activation.[4–6] In addition, DNA methylation interacts with histone modifications and variants to regulate gene expression and pre-mRNA splicing, though the nature of this relationship cannot be universally defined, and may be cooperative or antagonistic depending on the particular interaction examined.[7–9]

The addition of methylation to DNA occurs through the activity of DNA methyltransferases (DNMTs). DNA methylation must first be added de novo, then maintained despite constant dilution of signal through nascent DNA strands added during replication. Several DNMTs serve to catalyze different parts of these processes. DNMT3A and DNMT3B catalyze the addition of de novo methylation, assisted by DNMT3L, and play an important role in the establishment of cell and tissue specific DNA methylation patterns in mammalian development.[10,11] Although DNMT3A and DNMT3B have been shown to compensate for each other, the expression of DNMT3A and DNMT3B does not completely overlap spatiotemporally during development.[11,12] *Dnmt3b* is predicted to play an earlier role in DNA methylation in preimplantation embryos and the epiblast, with a switch in the major de novo methyltransferase expression from *Dnmt3b* to *Dnmt3a* occurring around E9.5-E10.5.[12]

DNMT3A and DNMT3B have been shown to physically and functionally interact to stimulate de novo methylation at *Oct4* and *Nanog* in mouse embryonic stem cells, and *Rhox* in E8.5-9.5 embryos, around the window of transition.[13,14] In addition, knockout of *Dnmt3a* only partially reduces DNA methylation at *Brdt*, *Dpep3*, *Cytip*, and *Crygd*, indicating that methylation of these genes requires the cooperation of DNMT3A and DNMT3B.[15] DNMT3A and DNMT3B activity is stimulated by physical interaction of the DNMT3L C-terminal domain with their respective catalytic domains.[16–20]

DNMT1 is responsible for the maintenance of DNA methylation. Although specific activity of DNMT1 for unmethylated DNA has been shown, it catalyzes the methylation of hemimethylated DNA at a rate that is 7–21-fold greater than unmethylated DNA.[21,22] Interestingly, postmitotic neurons maintain high levels of both DNMT3A and DNMT1, indicating a functional role in the adult nervous system. Conditional knockouts of de novo and maintenance DNA methyltransferase activity reveal their roles in lineage specification of neural progenitor cells (NPCs), neuronal maturation, and memory formation.[23–28] Although DNMT1 in postmitotic neurons is not necessary to prevent passive demethylation during cell division, dynamic regulation of DNA methylation is observed in neurological processes, including circadian rhythm, activity-induced neuronal plasticity, and learning and memory, necessitating the persistent availability of methyltransferase activity.[29–34]

Although it is clear that the rapid addition and removal of DNA methylation is required in many neurological processes, the mechanisms of DNA methylation plasticity in the brain are not completely understood. Active DNA demethylation by the ten-eleven translocation (TET) enzymes hydroxylates 5mC to produce 5-hydoxymethyl cytosine (5hmC).[35] While 5mC levels are comparable between tissues, 5hmC is much more variable between tissues, and although it is correlated with cell proliferation, it is observed at high levels in the postnatal brain.[36,37] Enrichment within the brain and its role as an intermediate in TET-mediated demethylation suggests 5hmC as a potential substrate involved in the rapid and reversible changes observed in DNA methylation with neurological activity.

Most standard methods of DNA methylation analysis do not distinguish between 5mC and 5hmC due to the resistance of both bases to deamination by bisulfite treatment; however, the pairing of these protocols with oxidative bisulfite (oxBS) sequencing or 450k array allows for the distinction between 5mC and 5hmC.[38–41] Increasing evidence suggests a functional role for the stable maintenance of 5hmC in the brain beyond its occurrence as a transient intermediate of demethylation. For instance, a recent study demonstrated that 5hmC levels correspond to the depletion of methyl CpG binding 2 (MeCP2) levels over active euchromatic genes in cerebellar granular neurons, suggesting a "functional demethylation" role of 5hmC in gene regulation in postmitotic neurons.[42]

2. LONG, NONCODING RNAs

Genome-wide sequencing of the transcriptome has clearly revealed that not all of the transcribed genome encodes for proteins. Long, noncoding RNAs (lncRNAs) are one of the largest and most functionally diverse classes of noncoding RNAs in the genome. Unlike mRNAs or other functional forms on noncoding RNAs, lncRNAs are frequently retained in the nucleus, often at their chromosomal site of transcription, such as the lncRNA *Xist* which colocalizes with the inactive X chromosome. lncRNAs act at all levels of gene regulation, and can be classified into several broad mechanisms of action.[43] lncRNAs are often expressed, or alternatively spliced, in a developmental or tissue specific manner. lncRNAs can also be highly responsive to exogenous stimuli, such as in an innate immune response, demonstrated by lipopolysaccharide (LPS) treatment of bone derived macrophages.[44] This highly compartmentalized localization of lncRNAs reflects their important roles in the spatiotemporal regulation of gene expression. In particular, lncRNAs are especially abundant in brain, often localized to specific regions, cell-types, or sub-cellular compartments.[45]

Unlike the evolutionary conservation of protein coding genes, lncRNAs exhibit a unique pattern of functional conservation, often with specific sequence conservation over proximal promoter regions and discrete functional domains.[46,47] This is likely due to different selection pressures, mutagenesis rates, and evolutionary patterns compared to protein-coding genes.[47,48] Lack of sequence conservation is not indicative of nonfunctionality, as several well-characterized lncRNAs such as *Air* and *Xist* are poorly conserved.[49] This likely reflects the contribution of characteristics other than underlying sequence, such as folding-energy and secondary structure in the functionality of lncRNAs. lncRNAs are predicted to arise through a variety of evolutionary pathways, including the disruption of protein-coding gene function, chromosomal rearrangement, duplication, and insertions giving rise to functional genetic units.[48]

Some lncRNAs are highly conserved between species, exhibiting similar expression and localization within the brain during early development, suggesting an evolutionarily conserved role of some lncRNAs in neurodevelopment.[46] Others have evolved rapidly, particularly within the human lineage and shed light on the development of human-specific brain complexity.[50] Some lncRNAs mediate transcriptional activity by recruiting or stabilizing the assembly of protein complexes and transcription factors, which may then guide histone modifications, initiate chromatin remodeling, or activate transcription.[43,51,52] These lncRNAs may associate preferentially with specific complexes, loci, or chromosomal domains, such as telomeres or centromeres.[43,51–56] The binding of regulatory factors by lncRNAs, however, does not always facilitate activity. Some lncRNAs act as molecular decoys, binding and sequestering transcription factors, chromatin remodelers, splicing factors, and even microRNAs (miRNAs), thereby controlling their abundance and localization within the nucleus.[43,51,57–60]

lncRNAs also function as signals, with their expression status providing cellular cues regarding time, development, and tissue identity to coordinate appropriate contextual regulatory pathways.[43] The chromosomal organization of protein coding genes and lncRNAs is complex, with the two often overlapping. lncRNAs can be transcribed in both the sense and antisense orientations with regard to the protein coding gene in the same chromosomal locus. They can be associated with nearly every part of the transcript, as promoter-associated, intronically encoded, or 3′ untranslated region (UTR)-associated transcripts.[61,62] These include antisense RNAs, which often arise either from read-through of oppositely oriented nearby genes or from bidirectional promoters.[63] lncRNAs also originate from intergenic enhancers and telomere repeats.[61,62]

The stability of lncRNAs, although highly variable, is comparable to mRNAs. However, several factors, including genomic organization, post-transcriptional localization, composition, and processing may influence the half-life of specific lncRNAs.[64] Intergenic, *cis*-antisense, tail-to-tail bidirectional, and 3′ UTR-associated transcripts are the most stable classes of lncRNAs. Regardless of genomic origin, nuclear-retained transcripts have shorter half-lives than cytoplasmic transcripts, potentially reflecting highly dynamic regulatory functions and/or the lack of mRNA processing steps. Features such as polyadenylation, high conservation, high GC content, and splicing are positively correlated with increased lncRNA stability. Stability of specific lncRNAs is a generally conserved feature between mouse and human, though significant divergence in the stability of some lncRNAs may reflect variation in their posttranscriptional regulation.

While the functional roles that lncRNAs play in the brain are not fully understood, current evidence implicates lncRNAs in neurodevelopment, synaptic plasticity, cognitive function, and memory.[65] Functions in the regulation of pluripotency, cell-fate determination, and differentiation have been attributed to a subset of lncRNAs, including *Sox2OT*, *Nkx2.2AS*, *Rncr2*, and many others.[65] Other lncRNAs, such as *Evf2*, *Malat1*, and *BC200* contribute to proper neurodevelopment through their support of processes involved in synaptic plasticity and learning and memory.[59,66] This includes support of GABAergic inhibitory interneuron development and the regulation of localized protein synthesis at postsynaptic dendritic compartments.[65] Several molecular genetic defects that contribute to the etiology of neurological diseases involve noncoding RNAs (ncRNAs), such as splicing, transcriptional and translational dysregulation, and impaired imprinting. Dysregulation of lncRNAs has been described in an array of neurologic etiologies, including neurodevelopmental disorders (discussed in the forthcoming paragraphs), neurodegenerative disorders (Alzheimer's, Parkinson's, and Huntington's diseases), neuroimmunological disorders (multiple sclerosis), neuro-oncological disease, psychiatric disorders (major depressive disorder (MDD)), and cerebral injury (stroke).[65,67,68]

3. NEURODEVELOPMENTAL DISORDERS INVOLVING IMPRINTING AND CIRCADIAN RHYTHMS

3.1 Prader–Willi and Angelman Syndromes

Prader–Willi syndrome (PWS) and Angelman syndrome (AS) are reciprocally imprinted neurodevelopmental disorders resulting from paternal or maternal deficiencies of the 15q11-13 locus, respectively. PWS and AS are characterized by distinct phenotypes and they present opposite sleep patterns. Patients with PWS experience shorter sleep duration at night and excessive daytime sleepiness, while AS patients experience frequent night time awakenings and reduced overall sleep time.[68–73]

Other features of PWS include hypotonia and failure to thrive at infancy followed by the onset of hyperphagia at about two years of age, as well as intellectual impairment, hypogonadism, and obsessive compulsive tendencies.[68] PWS is caused by the paternal deficiency of 15q11-q13, specifically at *SNORD116*, through three main mechanisms. About 20%–30% of PWS cases result from a maternal uniparental disomy (UPD(15)mat), 65%–75% from a large 15q11-q13 deletion of the paternal allele, and 1%–3% from an imprinting defect, in which the paternal allele carries the maternal imprinting status, sometimes caused by a small deletion.[74] All of these cases exhibit a lack of paternally expressed transcripts from the 15q11-q13 locus as a result of either deletion or silencing. Due to the imprinted status of this locus, the maternal allele cannot compensate for this paternal deficiency, resulting in PWS. While multiple paternally expressed transcripts reside within the 15q11-q13 locus, PWS patients with rare microdeletions have demonstrated that paternal deficiency of *SNORD116* is sufficient to cause PWS.[75–78] Interestingly, no protein coding genes are contained within the *SNORD116* PWS critical region, only a repetitive cluster of small nucleolar RNAs (snoRNAs) and a lncRNA host gene, *116HG*.

AS is characterized by severe intellectual impairment, seizures, hypotonia, and motor delay.[72] AS is caused by maternal deficiency of *UBE3A* through four main mechanisms. About 75% of AS patients carry a large 5–7 Mb de novo deletion, 1%–2% of cases are the result of paternal uniparental disomy (UPD(15)pat), 1%–3% of cases are caused by an imprinting defect, and 5%–10% of patients carry a mutation within *UBE3A*.[72,79] UBE3A encodes an E3 ubiquitin ligase that plays diverse roles in neurodevelopment, including targeting proteins for proteosomal degradation, functioning as a transcriptional coactivator of steroid hormone receptors, and potentially regulating experience-dependent synaptic plasticity through its localization at pre- and post-synaptic compartments.[80] *UBE3A* is imprinted exclusively in neurons, where paternal expression of the locus extends through the oppositely oriented *UBE3A*, producing the antisense transcript, *UBE3A-ATS*, that represses the expression of paternal *UBE3A* in the sense orientation.[81–83]

The 15q11-q13 locus contains one of the only two large clusters of snoRNAs in the mammalian genome, with the second at 14q32.2[83–86] (Fig. 1). These C/D box snoRNAs are embedded within the introns of *SNORD116* host gene (*116HG*) and are uniquely expressed in the brain in mice, with expression observed in several tissues in humans.[83,86,87] The snoRNAs encoded within *SNORD116* currently have no known target sequence; however, they localize to the nucleolus in mature neurons, coinciding with an increase in nucleolar size, and chromatin decondensation of the paternal *SNRPN-UBE3A* allele.[88,89] Processing of the primary transcript

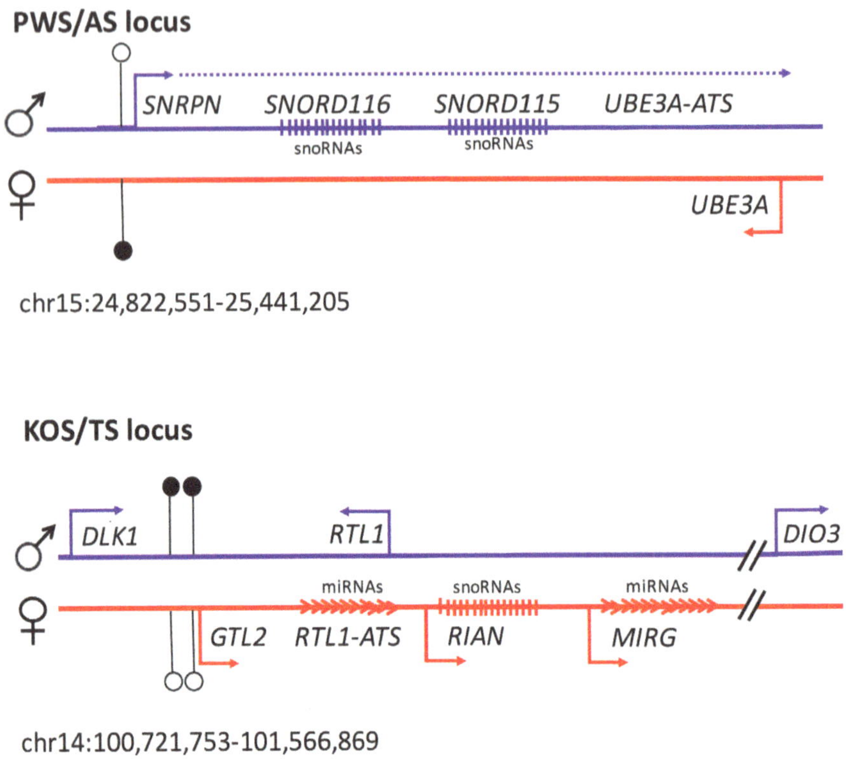

Fig. 1 Chromosomal maps of the *SNRPN-UBE3A* (PWS/AS) and *DLK1-DIO3* (KOS/TS) imprinted loci. Paternally expressed genes in shown in blue and maternally expressed genes shown in red. Imprinted differentially methylated regions shown as filled circles (methylated) or open circles (unmethylated). Coordinates given for human loci (not drawn to scale).

releases these snoRNAs, and the spliced host gene exons of *116HG* are retained within the nucleus, forming an RNA "cloud", which localizes to its site of transcription and is dynamic in size throughout the diurnal cycle.[89,90]

Evidence for diurnal or circadian disruptions in the pathogenic mechanisms in both PWS and AS has been emerging. The *116HG* RNA cloud associates with genes associated with circadian, metabolic, and epigenetic functions. Loss of paternal expression in a *Snord116*$^{+/-}$ mouse model of PWS, which carries a ~150 kb deletion of paternal *Snord116*, results in a light-phase (sleep)-specific dysregulation of these transcripts.[90,91] This includes *Ube3a*, which is expressed at aberrantly high levels during light hours.[90] One of the circadian targets of UBE3A ubiquitin ligase activity is

BMAL1, a circadian clock gene,[92,93] discussed further in detail in the forthcoming paragraphs. In addition, *Ube3a* escapes imprinting in the suprachiasmatic nucleus (SCN) of the hypothalamus which regulates the circadian clock and diurnal cycles in other tissues of the body.[94] Loss of maternal *Ube3a* is associated with disrupted sleep; however, whether this is due to a disruption of sleep homeostasis or circadian rhythms is currently unclear.[92,95] The role of *Ube3a* in maintaining circadian rhythmicity suggests that the dysregulation of *Ube3a* with the loss of *Snord116* likely contributes to the sleep phenotypes associated with PWS.[90]

3.2 Kagami–Ogata and Temple Syndromes

Kagami–Ogata syndrome (KOS) and Temple syndrome (TS) are reciprocally imprinted disorders caused by epimutations, microdeletions, or uniparental disomy (UPD(14)pat and UPD(14)mat, respectively) of 14q32.2.[96,97] Similar to the PWS/AS locus, the KOS/TS locus is organized into a ~1 Mb domain containing both protein coding (*Begain*, *Dlk1*, *Rtl1*, and *Dio3*) and noncoding (*Meg3*/*Gtl2*, *Rtl1as*, *Rian*, and *Mirg*) genes (Fig. 1).[67] *Rian*, also known as *Meg8*, is the noncoding host gene of a repeated cluster of C/D-box snoRNAs (*Snord113* and *Snord114*), the only snoRNA cluster besides the *Snord116* and *Snord115* arrays at the PWS locus.[85] Similar to the PWS snoRNAs, the targets of *Rian* snoRNAs are unknown. *Mirg* contains one of the largest clusters of miRNAs in the mammalian genome.[98] Interestingly, these miRNAs target and suppress the PI3K-mTOR pathway at nearly every level, regulating mitrochondiral biogenesis and metabolic activity in fetal liver.[99] *Rtl1* is regulated by miRNAs processed from its anti-sense transcript *Rtl1as*, maintaining a strict dosage during placental development.[100] KOS-locus noncoding RNAs are maternally expressed and the entire locus undergoes allele-specific chromatin decondensation in neurons, where these noncoding RNAs are highly expressed.[88] TS, in which loss of the paternal allele results in the upregulation of maternally expressed noncoding RNAs, phenocopies PWS, in which paternal expression of *SNORD116* is lost.

Interestingly, chromatin isolation by immunopurification (ChIRP) analysis to investigate the genomic binding sites of *116HG* revealed binding and modulation of KOS/TS locus transcripts in mouse cortex. Loss of *116HG* resulted in light-phase-specific upregulation of maternal noncoding RNAs at the KOS/TS locus.[90] In addition, the PWS locus has been implicated in the repression of the KOS/TS locus though the recruitment of G9a by *IPW* and subsequent chromatin remodeling at the imprinted intergenic differentially methylated region (IG-DMR) in human induced pluripotent stem cells.[101] Several genes at both loci exhibit circadian oscillations in expression, illuminating the diurnal phase specificity of dysregulation observed between two loci.[90,102,103] The regulatory mechanism and the effects of this diurnal regulation on the cross-regulation between the loci, however, have not been elucidated.

4. MECHANISMS OF IMPRINTING IN NEURODEVELOPMENT

Noncoding RNAs are commonly expressed from imprinted gene clusters in the mammalian genome. Imprinting is an epigenetic pattern in which the allelic expression of a locus is dictated by its parental origin. Approximately 1% of all genes, including both protein coding and noncoding, are imprinted in humans. Imprinting is a phenomenon that evolved convergently in mammals and seed plants, with recent evidence suggesting a potential role for allelic bias in social insects.[104–107]

Imprinted genes tend to be clustered within the genome, allowing for their coregulation by imprinting control regions (ICRs), in which parental-specific differential methylation is typically established in the germline and maintained during postfertilization stages of methylation erasure. Once the imprint through the ICR is established, the resultant gene expression pattern is not necessarily static. For instance, genes within imprinted loci may be imprinted in all tissues or only in specific cell types. Furthermore, the imbalance of parental expression of imprinted genes may be either absolute or merely skewed. Imprinted genes often exhibit specific spatiotemporal expression patterns, especially in the developing brain.[108] While differential DNA methylation patterns between parental alleles may be stably maintained throughout development after their establishment, other imprinted clusters acquire patterns specific to different tissues or developmental stages.[109,110]

Differential DNA methylation is the most extensively understood mechanism of genomic imprinting; however, other epigenetic features are known to play a role in the establishment and maintenance of some imprinted loci. Allele-specific histone modifications have been observed at several imprinted clusters, including the *Igf2r* and *Kcnq1* loci, which carry opposing histone marks on their active and inactive alleles.[111–114] Recently, 28 noncanonical imprinted genes were shown to exhibit H3K27me3-dependent transient imprinting in preimplantation embryos independent of DNA methylation.[115]

Transcription itself plays a unique role in the establishment of proper imprints. In some cases, transcription of an imprinted gene is required for the establishment of allele-specific DNA methylation in oocytes, though the relationship between the two is not completely understood.[116] In oocytes, preimplantation embryos, and placenta, global methylation levels are low and active genes are marked by higher relative levels of methylation, providing a potential mechanism of the association of maternal methylation marks with prior transcription in oocytes.[117] Transcription may simply act to mediate chromatin accessibility for DNMT activity. Imprinted loci often exert *cis* regulatory mechanisms that serve to maintain monoallelic expression patterns of other genes within the cluster, particularly though the activity of ncRNAs. The act of transcription alone sometimes provides sufficient repression, as in the case of an antisense transcript, or the ncRNA may function to repress transcription autonomously or through recruitment of chromatin remodelers.[82,118–121] *cis* regulation is observed at several loci, including *UBE3A-ATS*, *RTL1-ATS*, and *GTL2* (Fig. 1). Transcription of some imprinted loci gives rise to a unique class of noncoding antisense transcripts, which may act to silence the expression of genes from the opposite allele.

The paternally expressed *SNRPN-UBE3A* locus produces a particularly large polycistronic transcript, spanning 1 Mb of chromosome 15 and extending through the antisense to the maternally expressed *UBE3A* (*UBE3A-ATS*).[83] Although *SNRPN* is expressed in many tissues, transcription through the *SNORD116*, *SNORD115*, and *UBE3A-ATS*, as well as paternal imprinting of *UBE3A*, is unique to neurons. In neurons and oocytes, expression is initiated at a series of *SNRPN* upstream exons (U exons), producing a polycistronic transcript which is distinct from the *SNURF/SNRPN* transcript which terminates before transcribing the noncoding RNA clusters.[122–125] U exons are preferentially expressed in oocytes, but not in sperm in mammals, providing a potential mechanism for the maternal-specific methylation of the downstream ICR.[126] When transcription of this long paternal transcript is blocked before reaching *UBE3A-ATS*, either by truncation or topotecan-mediated transcriptional stalling, expression of *UBE3A* becomes biallelic.[82,127,128]

Alternatively, some imprinted loci exhibit repression or activation *intrans* to coordinate expression between clusters. Examination of *trans* interactions between imprinted loci has revealed an interconnected regulatory network that acts to coordinate many biological processes underlying complex phenotypes often shared between distinct imprinting disorders.[129–131] These gene networks involve not only imprinted genes, but also biallelically expressed genes to regulate the complex molecular mechanisms involved in growth and development, metabolism, and sleep.[132,133]

Although the function of DNA methylation in imprinting has long been appreciated, an understanding of the role of hydroxymethylation in genomic imprinting has only begun. A recent study revealed the presence of 5hmC on the methylated silent allele at imprinted loci; however, a unique distribution was observed at the *SNRPN-UBE3A* and *DLK1-DIO3* imprinted noncoding RNA clusters, in which 5hmC was enriched outside of known imprinted differentially methylated regions (DMRs) and exclusively on the unmethylated active allele in brain.[134] Allele-specific DNA methylation of ICRs is considered the main mechanism of parental imprinting; however, many epigenetic layers, including histone modifications and noncoding RNAs contribute to the establishment and maintenance of this strict parent-of-origin inheritance. The cooperation of several epigenetic signals ensures the persistence of proper imprinting; however, there is variability in the exact mechanism of imprinting between individual loci. Interestingly, DNMT1 is sufficient to maintain allelic DNA methylation of many imprinted DMRs in preimplantation embryos and embryonic stem cells.[135]

By examining the imprinted status of genes in divergent species, an evolutionary timeline of imprinting can be ascertained. Phylogenetic analysis suggests that imprinting originated after the evolution of viviparity; however, disparate timelines among imprinted clusters suggests that some domains acquired imprinted status much more recently.[136] In particular, imprinting of the PWS locus is restricted to the eutherian (placental) lineage, suggesting it evolved sometime between their divergence from marsupials 180 million years ago (MYA) and mammalian radiation 105 MYA.[136] The rearrangement of the ancestral genomic organization of the genes within the *SNRPN-UBE3A* cluster appears to have been required for the acquisition of an ICR and the establishment of parent-of-origin specific expression at this locus. Interestingly, the ancestral eutherian mammal tenrec (*Echinops telfairi*) lacks the *Snord116* and *Snord115* genes and *Snrpn* and *Ube3a* are on separate chromosomes.[88,109,110] Potentially relevant for the PWS phenotype, tenrecs have a unique metabolic and sleep structure among mammals adapted to long periods of reduced activity and body temperature called torpor.[111–113] Nonrapid eye movement (NREM) sleep and periods of torpor are thought to be ancestrally adaptive to conserve energy and reduce predation.

Eutherian mammals have distinct adaptations for sleep and diurnal activity patterns based on diet, body size, and brain size.[32,114]

In contrast, the chromosomal arrangement of the KOS/TS locus, which bears striking resemblance in organization to the PWS locus in mammals (Fig. 1), is conserved between marsupials and eutherians. Interestingly, despite the earlier evolutionary origin of this locus, imprinting of the PWS and TS loci evolved concurrently within the eutherian lineage.[136,137] This coevolution, in conjunction with evidence of molecular interactions between noncoding elements of these loci, provides strong evidence for a cross-regulatory network involving these two loci and illuminate the phenocopies observed in PWS and TS.

5. CIRCADIAN RHYTHMS AND IMPRINTING IN NEURODEVELOPMENT

From bacteria to plants to mammals, the circadian clock serves as an autonomous mechanism to maintain an internal oscillatory rhythm of physiological processes, with the capacity to adapt in response to changes in the external environment. The discovery of circadian rhythms in cyanobacteria suggests that the circadian clock is an ancient ancestral system.[138] While the presence of a circadian pacemaker is conserved across species, the core elements of regulation are distinct in the different branches of life. A common delayed transcription–translation negative feedback loop suggests that the central circadian system arose through convergent evolution, likely due to improved adaptive fitness.[139,140]

In mammals, the core molecular mechanism of the circadian clock is maintained by the negative feedback loop between the Clock (and its paralog, Npas2) and Bmal1 (Arntl) transcriptional activators and the Per and Cry repressors (Fig. 2). The light-specific accumulation of Clock-Bmal1 activates *Period* (*Per1*, *Per2*, and *Per3*) and *Cryptochrome* (*Cry1* and *Cry2*) genes, leading to the peak of Per and Cry repressors at the end of the light period. Interaction of these repressors with the Clock-Bmal1 complex inhibits their transcription, until proteosomal degradation releases this repression, permitting a new cycle of Clock-Bmal1 accumulation. This feedback loop is supported by an integrated loop of retinoic acid receptor activators and REV-ERB repressors, which modulate the circadian expression of *Bmal1*.[141]

In mammals, the master pacemaker of circadian time resides within the SCN of the hypothalamus; however, nearly all tissues of the body exhibit

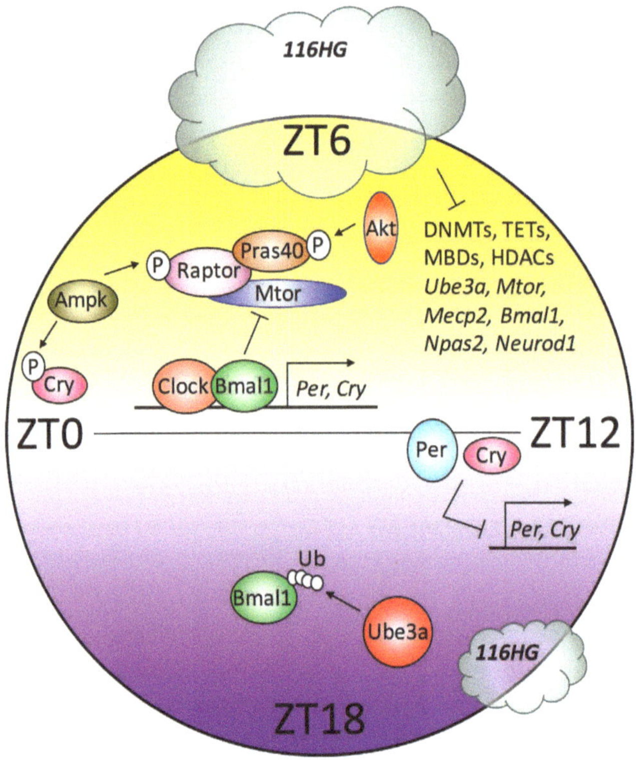

Fig. 2 Molecular mechanisms of the circadian clock. The circadian clock divides biological processes into two phases, light hours (ZT0–ZT12) and dark hours (ZT12–ZT0). The circadian clock shares a cross regulatory network with metabolic pathways, epigenetic factors, and imprinted loci.

environmentally responsive oscillatory rhythms.[142,143] Synchronization of the core circadian SCN rhythms with peripheral cycles integrates the two cycles while adapting to fluctuations in environmental stimuli. Light, temperature, and feeding are some of the environmental cues that help to regulate the changing metabolic and energy requirements across diurnal and seasonal patterns. Coordination of biological and environmental signals requires epigenetic, transcriptional, and translational regulation, though the mechanisms behind these processes are not equally understood.[141–145]

Most extensively investigated are transcriptional rhythms, with 43% of protein coding genes and 32% of conserved ncRNAs oscillating over diurnal time in one or more tissues.[143] Disruption of circadian gene expression is observed in several brain regions of MDD and Alzheimer's (AD) neocortex.[30,146] Rhythmic oscillations in gene expression are a result of many levels of circadian control, including the transcriptional, posttranscriptional, translational, and posttranslational stages. These include chromatin remodeling and RNA polymerase II recruitment, splicing, mRNA and protein stability, and interactions with noncoding RNAs.[147–152] This is exemplified by the observation that rhythmic transcription (nascent RNA) and rhythmic mRNA levels are highly discordant. Rhythmic transcription does not always result in rhythmic mRNA levels and in many cases it is not required to establish rhythmic mRNA oscillations.[153]

In mouse liver, rhythms of RNA polymerase II (RNAPII) recruitment and initiation, as well as several histone modifications (H3K4me3, H3K9ac, and H3K27ac), are observed at thousands of both rhythmically and statically expressed genes.[144,152] Specific antisense inactivation in the SCN and complete ablation of miRNAs in the liver and mouse embryonic fibroblasts significantly alter period length, indicating their role in the posttranscriptional modulation of mRNA rhythms.[151,154,155] Interestingly, alternative splicing, as well as ribosome accumulation and translational efficiency in the liver, is influenced by feeding/fasting and light/dark conditions, demonstrating the responsiveness of circadian cycles of RNA molecules to physiological cues.[147,156]

Although epigenetic mechanisms in circadian rhythms are much less well characterized, evidence of the diurnal plasticity of DNA methylation and its role in circadian entrainment is increasing.[29–31,157] Shift work and sleep deprivation correlate with the epigenetic and transcriptional dysregulation of core circadian genes and are characterized by the aberrant upregulation of *Dnmt3a*.[158–160] Sleep loss correlates with reduced binding of core clock genes Bmal1, Clock, and Npas2 to their targets, which were shown to exhibit phase-specific binding in mouse liver.[161,162] A recent study of 750 human postmortem dorsolateral prefrontal cortex samples revealed striking diurnal and seasonal rhythms in the transcriptome, DNA methylome, and H3K9 acetylome in healthy individuals, which were disrupted in AD patients.[30] Circadian and metabolic rhythms are intrinsically linked through a network of shared pathways, including Mtor/Akt and Ampk signaling.[90,163] Insufficient sleep leads to an imbalance between energy expenditure and intake, resulting in weight gain in healthy individuals, despite normal feeding-induced regulation of leptin, peptide YY, and ghrelin levels.[164] Disruption of the circadian clock entrainment through environmental cues and epigenetic mechanisms therefore poses great risk to human health and is associated with cancer, diabetes, and obesity.

Prader–Willi syndrome offers a unique neurodevelopmental perspective of obesity risk and provides a genetic basis for the study of one of the most prevalent health issues in the United States. Loss of the *116HG* RNA cloud results in the upregulation of many genes specifically during light hours at Zeitgeber 6 (ZT6), including *Ube3a* and *Mtor*. Bmal1 is targeted for ubiquitination and proteasomal degradation by UBE3A to maintain a strict oscillatory pattern and timely activation of *Per* and *Cry* genes within the circadian feedback loop. Precise timing of clock protein turnover is critical for the maintenance of robust circadian rhythmicity. Over-expression and knockdown of UBE3A result in the disruption of this precise feedback loop and circadian rhythmicity in cell culture; however, whether these phenotypes in mice are due to dysregulation of the circadian clock or sleep homeostasis is still unclear.[92,93,95] Interestingly, reciprocal perturbations of *UBE3A* are observed in PWS, in which it is overexpressed, and AS, in which expression is lost. In accord with these molecular changes, PWS and AS present opposing sleep phenotypes of excessive daytime sleepiness and shorter sleep duration, respectively.[69,70,165] Bmal1 inhibits *Mtor* expression and activity, likely acting to suppress excess energy expenditure and nutrient usage during sleep, when Bmal1 accumulation and activity peaks (ZT6 in mice).[166] *Mtor* is bound by *116HG*, and the loss of *Snord116* results in ZT6 upregulation of *Mtor* expression and increased activity, which is typically at its minimum during subjective night.[90,167] This corresponds with light-specific dysregulation of energy expenditure and respiratory exchange ratio in *Snord116*$^{+/-}$ mice.[90] The light-specific dysregulation of *Ube3a* and *Mtor* in *Snord116*$^{+/-}$ mice provides a link between the metabolic and diurnal phenotypes of PWS which may hold potential in understanding how these rhythms cooperate in the development of obesity in the general population.

6. CHRONOTHERAPEUTICS

Chronotherapeutics is the administration of treatment or intervention with respect to circadian rhythms in order to maximize efficacy and minimize toxicity and adverse effects.[168] Nearly half of all genes exhibit circadian oscillations in transcription in one or more tissues in the body, highlighting the importance of chronopharmacodynamics (chronoPD), which specifically considers how drug pathways will be affected by the rhythms of its targets as well as the rhythms of physiological processes modulating the extracellular environment.[143,168] In addition, in order to identify an optimal drug–target interaction, chronopharmacokinetics (chronoPK) explores how physiological rhythms affect the absorption, distribution, metabolism, and elimination of drugs, contributing to changes in potential toxicity effects throughout the circadian cycle.[168] When these factors are considered in conjunction with the half-life of a particular drug, an optimal therapeutic window can be identified within the circadian cycle.

The application of chronotherapy represents a potentially significant benefit for aggressive treatment approaches such as those for cancer. Over 40 cancer therapy drugs have been demonstrated to show circadian variation in tolerability, toxicity, and/or anti-tumor efficacy in rodent studies.[169] These drugs are often administered at doses near the extreme of tolerance; therefore the beneficial potential of chronotherapy to minimize adverse effects and alleviate even a small amount of patient distress is immense. The significance of chronotherapy in various treatment applications has been validated by several studies, including a recent evaluation of 27 studies of the top 30 prescribed drugs in Australia, in which 56% of studies revealed time-dependent variability of drug efficacy.[170] Not all applications of chronotherapy involve drug treatment. Chronotherapy through the manipulation of sleep-wake rhythm, such as sleep deprivation, has been applied in mood disorders such as depression and bipolar disorder.[171,172] The sleep dysregulation observed in many neurodevelopmental disorders, along with the coregulation of sleep and metabolic rhythms and the evolutionary relationship between imprinted loci and sleep, suggests a potential application of chronotherapy in the treatment of a wide variety of neurodevelopmental, neuropsychiatric, metabolic, sleep, and imprinted disorders, including PWS and AS.

ACKNOWLEDGEMENTS

The authors thank Dr. Weston Powell for intellectual input and the financial support of ongoing research support in this field of research by the NIH NINDS (1R01NS076263 and R56NS076263) and the Foundation for Prader-Willi research.

REFERENCES

1. Zemach A, McDaniel IE, Silva P, Zilberman D. Genome-wide evolutionary analysis of eukaryotic DNA methylation. *Science*. 2010;328:916–919.
2. Nabel CS, Manning SA, Kohli RM. The curious chemical biology of cytosine: deamination, methylation, and oxidation as modulators of genomic potential. *ACS Chem Biol*. 2012;7:20–30.
3. International Human Genome Sequencing Consortium.Initial sequencing and analysis of the human genome. *Nature*. 2001;409:860–921.

4. Zhang X, et al. Resource mapping and functional analysis of DNA methylation in *Arabidopsis*. *Cell*. 2006;126:1189–1201.
5. Hellman A, Chess A. Gene body – specific methylation on the active X chromosome. *Science*. 2007;315:1141–1144.
6. Wu H, et al. Dnmt3a-dependent nonpromoter DNA methylation facilitates transcription of neurogenic genes. *Science*. 2010;329:444–448.
7. Rose NR, Klose RJ. Understanding the relationship between DNA methylation and histone lysine methylation. *BBA – Gene Regul Mech*. 2014;1839:1362–1372. 10.1016/j.bbagrm.2014.02.007.
8. Zilberman D, Coleman-Derr D, Ballinger T, Henikoff S. Histone H2A.Z and DNA methylation are mutually antagonistic chromatin marks. *Nature*. 2008;456:125–129.
9. Brown SJ, Stoilov P, Xing Y. Chromatin and epigenetic regulation of pre-mRNA processing. *Hum Mol Genet*. 2012;21:90–96.
10. Lei H, et al. De novo DNA cytosine methyltransferase activities in mouse embryonic stem cells. *Development*. 1996;122:3195–3205.
11. Okano M, Bell DW, Haber DA, Li E. DNA methyltransferases Dnmt3a and Dnmt3b are essential for de novo methylation and mammalian development. *Cell*. 1999;99: 247–257.
12. Watanabe D, Suetake I, Tada T, Tajima S. Stage- and cell-specific expression of Dnmt3a and Dnmt3b during embryogenesis. *Mech Dev*. 2002;118:187–190.
13. Li J, et al. Synergistic function of DNA methyltransferases Dnmt3a and Dnmt3b in the methylation of Oct4 and Nanog. *Mol Cell Biol*. 2007;27:8748–8759.
14. Oda M, et al. DNA methylation regulates long-range gene silencing of an X-linked homeobox gene cluster in a lineage-specific manner. *Genes Dev*. 2006;20:3382–3394.
15. Borgel J, et al. Targets and dynamics of promoter DNA methylation during early mouse development. *Nat Genet*. 2010;42:1093–1100.
16. Gowher H, Liebert K, Hermann A, Xu G, Jeltsch A. Mechanism of stimulation of catalytic activity of Dnmt3A and Dnmt3B DNA-(cytosine-C5)-methyltransferases by Dnmt3L. *J Biol Chem*. 2005;280:13341–13348.
17. Kareta MS, Botello ZM, Ennis JJ, Chou C. Reconstitution and mechanism of the stimulation of de novo methylation by human DNMT3L. *J Biol Chem*. 2006;281:25893–25902.
18. Chedin F, Lieber MR, Hsieh C. The DNA methyltransferase-like protein DNMT3L stimulates de novo methylation by Dnmt3a. *Proc Natl Acad Sci U S A*. 2002;99: 16916–16921.
19. Suetake I, Shinozaki F, Miyagawa J, Takeshima H, Tajima S. DNMT3L stimulates the DNA methylation activity of Dnmt3a and Dnmt3b through a direct interaction. *J Biol Chem*. 2004;279:27816–27823.
20. Chen Z, et al. Physical and functional interactions between the human DNMT3L protein and members of the de novo methyltransferase family. *J Cell Biochem*. 2005; 917:902–917.
21. Bestor T, Laudano A, Mattaliano R, Ingram V. Cloning and sequencing of a cDNA encoding DNA methyltransferase of mouse cells. *J Mol Biol*. 1988;203:971–983.
22. Pradhan S, Bacolla A, Wells RD, Roberts RJ. Recombinant human DNA (cytosine-5) methyltransferase. *J Biol Chem*. 1999;274:33002–33010.
23. Fan G, et al. DNA methylation controls the timing of astrogliogenesis through regulation of JAK-STAT signaling. *Development*. 2002;132:3345–3356.
24. Mohn F, et al. Lineage-specific polycomb targets and de novo DNA methylation define restriction and potential of neuronal progenitors. *Mol Cell*. 2008;30:755–766.
25. Nguyen S, Meletis K, Fu D, Jhaveri S, Jaenisch R. Ablation of de novo DNA methyltransferase Dnmt3a in the nervous system leads to neuromuscular defects and shortened lifespan. *Development*. 2007;263:1663–1676.

26. Golshani P, Hutnick L, Schweizer F, Fan G. Conditional Dnmt1 deletion in dorsal forebrain disrupts development of somatosensory barrel cortex and thalamocortical long-term potentiation. *Thalamus Relat Syst.* 2005;3:227–233.
27. Hutnick LK, et al. DNA hypomethylation restricted to the murine forebrain induces cortical degeneration and impairs postnatal neuronal maturation. *Hum Mol Genet.* 2009;18:2875–2888.
28. Feng J, et al. Dnmt1 and Dnmt3a maintain DNA methylation and regulate synaptic function in adult forebrain neurons. *Nat Neurosci.* 2010;13:423–430.
29. Azzi A, et al. Circadian behavior is light-reprogrammed by plastic DNA methylation. *Nat Neurosci.* 2014;17:377–382.
30. Lim ASP, et al. Diurnal and seasonal molecular rhythms in human neocortex and their relation to Alzheimer's disease. *Nat Commun.* 2017;8:14931. 10.1038/ncomms14931.
31. Stevenson TJ, Prendergast BJ, Mcewen BS. Reversible DNA methylation regulates seasonal photoperiodic time measurement. *Proc Natl ?A3B2 twbch?Acad Sci U S A.* 2013;110:16651–16656.
32. Guo JU, et al. Neuronal activity modifies the DNA methylation landscape in the adult brain. *Nat Neurosci.* 2011;14:1345–1351.
33. Day JJ, Sweatt JD. DNA methylation and memory formation. *Nat Neurosci.* 2011;13: 1319–1323.
34. Levenson JM, et al. Evidence that DNA (cytosine-5) methyltransferase regulates synaptic plasticity in the hippocampus. *J Biol Chem.* 2006;281:15763–15773.
35. Tahiliani M, et al. Conversion of 5-methylcytosine to 5-hydroxymethylcytosine in mammalian DNA by MLL partner TET1. *Science.* 2009;324:930–935.
36. Kriaucionis S, Heintz N. The nuclear DNA base, 5-hydroxymethylcytosine is present in brain and enriched in Purkinje neurons. *Science.* 2009;324:929–930.
37. Bachman M, Uribe-lewis S, Yang X, Williams M. 5-Hydroxymethylcytosine is a predominantly stable DNA modification. *Nat Chem Biol.* 2015;6:1049–1055.
38. Huang Y, et al. The behaviour of 5-hydroxymethylcytosine in bisulfite sequencing. *PLoS ONE.* 2010;5:e8888.
39. Booth MJ, et al. Quantitative sequencing of 5-methylcytosine and 5-hydroxymethylcytosine at single-base resolution. *Science.* 2012;336:934–938.
40. Booth MJ, et al. Oxidative bisulfite sequencing of 5-methylcytosine and 5-hydroxymethylcytosine. *Nat Protoc.* 2013;8:1841–1851.
41. Stewart SK, et al. oxBS-450K: a method for analysing hydroxymethylation using 450K BeadChips. *Methods.* 2015;72:9–15.
42. Mellen M, Ayata P, Heintz N. 5-Hydroxymethylcytosine accumulation in postmitotic neurons results in functional demethylation of expressed genes. *Proc Natl Acad Sci U S A.* 2017;114:E7812–E7821.
43. Wang KC, Chang HY. Molecular mechanisms of long noncoding RNAs. *Mol Cell.* 2011;43:904–914.
44. Ravasi T, et al. Experimental validation of the regulated expression of large numbers of non-coding RNAs from the mouse genome. *Genome Res.* 2006;4:11–19.
45. Mercer TR, Dinger ME, Sunkin SM, Mehler MF, Mattick JS. Specific expression of long noncoding RNAs in the mouse brain. *Proc Natl Acad Sci U S A.* 2008;105:716–721.
46. Chodroff RA, et al. Long noncoding RNA genes: conservation of sequence and brain expression among diverse amniotes. *Genome Biol.* 2010;11:R72.
47. Guttman M, Rinn JL. Modular regulatory principles of large non-coding RNAs. *Nature.* 2012;482:339–346.
48. Ponting CP, Oliver PL, Reik W. Evolution and functions of long noncoding RNAs. *Cell.* 2009;136:629–641.
49. Pang KC, Frith MC, Mattick JS. Rapid evolution of noncoding RNAs: lack of conservation does not mean lack of function. *Trends Genet.* 2006;22:1–5.

50. Pollard KS, et al. An RNA gene expressed during cortical development evolved rapidly in humans. *Nature.* 2006;443:167–172.
51. Wapinski O, Chang HY. Long noncoding RNAs and human disease. *Trends Cell Biol.* 2011;21:354–361.
52. Lee JT. Epigenetic regulation by long, noncoding RNAs. *Science.* 2012;338: 1435–1439.
53. Maison C, et al. SUMOylation promotes de novo targeting of HP1α to pericentric heterochromatin. *Nat Genet.* 2011;43:220–227.
54. Yap KL, et al. Molecular interplay of the noncoding RNA ANRIL and methylated histone H3 lysine 27 by polycomb CBX7 in transcriptional silencing of INK4a. *Mol Cell.* 2010;38:662–674.
55. Gupta RA, et al. Long non-coding RNA HOTAIR reprograms chromatin state to promote cancer metastasis. *Nature.* 2010;464:1071–1076.
56. Lee JT. Lessons from X-chromosome inactivation: long ncRNA as guides and tethers to the epigenome. *Genes Dev.* 2009;23:1831–1842.
57. Kino T, Hurt DE, Ichijo T, Nader N, Chrousos GP. Noncoding RNA gas5 is a growth arrest and starvation-associated repressor of the glucocorticoid receptor. *Sci Signal.* 2010;3:ra8.
58. Tripathi V, et al. The nuclear-retained noncoding RNA MALAT1 regulates alternative splicing by modulating SR splicing factor phosphorylation. *Mol Cell.* 2014;39:925–938.
59. Bernard D, et al. A long nuclear-retained non-coding RNA regulates synaptogenesis by modulating gene expression. *EMBO J.* 2010;29:3082–3093.
60. Poliseno L, et al. A coding-independent function of gene and pseudogene mRNAs regulates tumour biology. *Nature.* 2010;465:1033–1038.
61. Qureshi IA, Mehler MF. Emerging roles of non-coding RNAs in brain evolution, development, plasticity and disease. *Nat Rev Neurosci.* 2012;13:528–541.
62. Mercer TR, Dinger ME, Mattick JS. Long non-coding RNAs: insights into functions. *Nat Rev Genet.* 2009;10:155–159.
63. Villegas VE, Zaphiropoulos PG. Neighboring gene regulation by antisense long non-coding RNAs. *Int J Mol Sci.* 2015;16:3251–3266.
64. Clark MB, et al. Genome-wide analysis of long noncoding RNA stability. *Genome Res.* 2012;22:885–898.
65. Chen Y, Zhou J. LncRNAs: macromolecules with big roles in neurobiology and neurological diseases. *Metab Brain Dis.* 2017;32:281–291.
66. Eom T, Berardi V, Zhong J, Risuleo G, Tiedge H. Dual nature of translational control by regulatory BC RNAs. *Mol Cell Biol.* 2011;31:4538–4549.
67. Howard M, Charalambous M. Molecular basis of imprinting disorders affecting chromosome 14: lessons from murine models. *Reproduction.* 2015;149:R237–R249.
68. Cassidy SB, Schwartz S, Miller JL, Driscoll DJ. Prader–Willi syndrome. *Genet Med.* 2012;14:10–26.
69. Lassi G, et al. Deletion of the Snord116/SNORD116 alters sleep in mice and patients with Prader–Willi syndrome. *Sleep.* 2016;39:637–644.
70. Gibbs S, Wiltshire E, Elder D. Nocturnal sleep measured by actigraphy in children with Prader–Willi syndrome. *J Pediatr.* 2013;162:765–769.
71. Butler JV, et al. Prevalence of, and risk factors for, physical ill-health in people with Prader–Willi syndrome: a population-based study. *Dev Med Child Neurol.* 2002;44: 248–255.
72. Buiting K, Williams C, Horsthemke B. Angelman syndrome — insights into a rare neurogenetic disorder. *Nat Rev Neurol.* 2016;12:584–593.
73. Colas D, Wagstaff J, Fort P, Salvert D, Sarda N. Sleep disturbances in Ube3a maternal-deficient mice modeling Angelman syndrome. *Neurobiol Dis.* 2005;20:471–478.

74. Ohta T, et al. Imprinting-mutation mechanisms in Prader–Willi syndrome. *Am J Hum Genet.* 1999;64:397–413.
75. Sahoo T, et al. Prader–Willi phenotype caused by paternal deficiency for the HBII-85 C/D box small nucleolar RNA cluster. *Nat Genet.* 2008;40:719–721.
76. Duker AL, et al. Paternally inherited microdeletion at 15q11.2 confirms a significant role for the SNORD116 C/D box snoRNA cluster in Prader–Willi syndrome. *Eur J Hum Genet.* 2010;18:1196–1201.
77. de Smith AJ, et al. A deletion of the HBII-85 class of small nucleolar RNAs (snoRNAs) is associated with hyperphagia, obesity and hypogonadism. *Hum Mol Genet.* 2009;18:3257–3265.
78. Bieth E, et al. Highly restricted deletion of the SNORD116 region is implicated in Prader–Willi syndrome. *Eur J Hum Genet.* 2014;23:252–255. 10.1038/ejhg.2014.103.
79. Kishino T, Lalande M, Wagstaff J. UBE3A/E6-AP mutations cause Angelman syndrome. *Nat Genet.* 1997;15:70–73.
80. Lasalle JM, Reiter T, Stormy J. Epigenetic regulation of UBE3A and roles in human neurodevelopmental disorders. *Epigenomics.* 2015;7:1213–1228.
81. Rougeulle C, Cardoso C, Fontes M, Colleaux L, Lalande M. An imprinted antisense RNA overlaps UBE3A and a second maternally expressed transcript. *Nat Genet.* 1998;19:15–16.
82. Meng L, Person RE, Beaudet AL. Ube3a-ATS is an atypical RNA polymerase II transcript that represses the paternal expression of Ube3a. *Hum Mol Genet.* 2012;21:3001–3012.
83. Runte M, et al. The IC-SNURF–SNRPN transcript serves as a host for multiple small nucleolar RNA species and as an antisense RNA for UBE3A. *Hum Mol Genet.* 2001;10:2687–2700.
84. Bortolin-Cavaillé ML, Cavaillé J. The SNORD115 (H/MBII-52) and SNORD116 (H/MBII-85) gene clusters at the imprinted Prader–Willi locus generate canonical box C/D snoRNAs. *Nucleic Acids Res.* 2012;40:6800–6807.
85. Cavaille J, Zeitz H, Paulsen M, Ferguson-Smith AC, Bachellerie J-P. Identification of tandemly-repeated C/D snoRNA genes at the imprinted human 14q32 domain reminiscent of those at the Prader–Willi/Angelman syndrome region. *Hum Mol Genet.* 2002;11:1527–1538.
86. Cavaillé J, et al. Identification of brain-specific and imprinted small nucleolar RNA genes exhibiting an unusual genomic organization. *Proc Natl Acad Sci U S A.* 2000;97:14311–14316.
87. Galiveti CR, Raabe CA, Konthur Z, Rozhdestvensky TS. Differential regulation of non-protein coding RNAs from Prader–Willi syndrome locus. *Sci Rep.* 2014;4:.
88. Leung KN, Vallero RO, Dubose AJ, Resnick JL, Lasalle JM. Imprinting regulates mammalian snoRNA-encoding chromatin decondensation and neuronal nucleolar size. *Hum Mol Genet.* 2009;18:4227–4238.
89. Vitali P, Royo H, Marty V, Bortolin-Cavaille M-L, Cavaille J. Long nuclear-retained non-coding RNAs and allele-specific higher-order chromatin organization at imprinted snoRNA gene arrays. *J Cell Sci.* 2010;123:70–83.
90. Powell WT, et al. A Prader–Willi locus lncRNA cloud modulates diurnal genes and energy expenditure. *Hum Mol Genet.* 2013;22:4218–4328.
91. Ding F, et al. SnoRNA Snord116 (Pwcr1/MBll-85) deletion causes growth deficiency and hyperphagia in mice. *PLoS ONE.* 2008. 10.1371/journal.pone.0001709.
92. Shi SQ, Bichell TJ, Ihrie RA, Johnson CH. Ube3a imprinting impairs circadian robustness in Angelman syndrome models. *Curr Biol.* 2015;5:537–545. 10.1016/j.cub.2014.12.047.

93. Gossan NC, et al. The E3 ubiquitin ligase UBE3A is an integral component of the molecular circadian clock through regulating the BMAL1 transcription factor. *Nucleic Acids Res.* 2014;42:5765–5775. 10.1093/nar/gku225.
94. Jones KA, Han JE, DeBruyne JP, Philpot BD. Persistent neuronal Ube3a expression in the suprachiasmatic nucleus of Angelman syndrome model mice. *Sci Rep.* 2016;6:28238.
95. Ehlen JC, et al. Maternal Ube3a loss disrupts sleep homeostasis but leaves circadian rhythmicity largely intact. *J Neurosci.* 2015;35:13587–13598.
96. Ogata T, Kagami M. Kagami–Ogata syndrome: a clinically recognizable upd(14)pat and related disorder affecting the chromosome 14q32.2 imprinted region. *J Hum Genet.* 2015;61:87–94.
97. Ioannides Y, Lokulo-Sodipe K, Mackay DJG, Davies JH, Temple IK. Temple syndrome: improving the recognition of an underdiagnosed chromosome 14 imprinting disorder: an analysis of 51 published cases. *J Med Genet.* 2014;51:495–501.
98. Seitz H, et al. A large imprinted microRNA gene cluster at the mouse Dlk1-Gtl2 domain. *Genome Res.* 2004;14:1741–1748.
99. Qian P, et al. The Dlk1-Gtl2 locus preserves LT-HSC function by inhibiting the PI3K-mTOR pathway to restrict mitochondrial metabolism. *Cell Stem Cell.* 2016;18:214–228. 10.1016/j.stem.2015.11.001.
100. Ito M, et al. A trans-homologue interaction between reciprocally imprinted miR-127 and Rtl1 regulates placenta development. *Development.* 2015;142:2425–2430.
101. Stelzer Y, Sagi I, Yanuka O, Eiges R, Benvenisty N. The noncoding RNA IPW regulates the imprinted DLK1-DIO3 locus in an induced pluripotent stem cell model of Prader–Willi syndrome. *Nat Genet.* 2014;46:551–557.
102. Labialle S, et al. Novel imprinted transcripts from the Dlk1-Gtl2 intergenic region, Mico1 and Mico1os, show circadian oscillations. *Epigenetics.* 2008;3:322–329.
103. Labialle S, et al. Coordinated diurnal regulation of genes from the Dlk1-Dio3 imprinted domain: implications for regulation of clusters of non-paralogous genes. *Hum Mol Genet.* 2008;17:15–26.
104. Barton SC, Surani MAH, Norris ML. Role of paternal and maternal genomes in mouse development. *Nature.* 1984;311:374–376.
105. Mcgrath J, Solter D. Completion of mouse embryogenesis requires both the maternal and paternal genomes. *Cell.* 1984;37:179–183.
106. Scott RJ, Spielman M, Bailey J, Dickinson HG. Parent-of-origin effects on seed development in *Arabidopsis thaliana. Development.* 1998;125:3329–3341.
107. Kocher SD, et al. A search for parent-of-origin effects on honey bee gene expression. *Genes Genomes Genet.* 2015;5:1657–1662.
108. Davies W, Isles AR, Wilkinson LS. Imprinted gene expression in the brain. *Neurosci Biobehav Rev.* 2005;29:421–430.
109. Tremblay KD, Duran KL, Bartolomei MS. A 5′ 2-kilobase-pair region of the imprinted mouse H19 gene exhibits exclusive paternal methylation throughout development. *Mol Cell Biol.* 1997;17:4322–4329.
110. Feil R, Walter J, Allen ND, Reik W. Developmental control of allelic methylation in the imprinted mouse Igf2 and H19 genes. *Development.* 1994;120:2933–2943.
111. Lewis A, et al. Imprinting on distal chromosome 7 in the placenta involves repressive histone methylation independent of DNA methylation. *Nat Genet.* 2004;36:1291–1295.
112. Lewis A, et al. Epigenetic dynamics of the Kcnq1 imprinted domain in the early embryo. *Development.* 2006;133:4203–4210.
113. Yang Y, et al. The histone code regulating expression of the imprinted mouse Igf2r gene. *Endocrinology.* 2003;144:5658–5670.

114. Dindot SV, et al. Epigenetic profiling at mouse imprinted gene clusters reveals novel epigenetic and genetic features at differentially methylated regions. *Genome Res.* 2009;19:1374–1383.
115. Inoue A, Jiang L, Lu F, Suzuki T, Zhang Y. Maternal H3K27me3 controls DNA methylation-independent imprinting. *Nature.* 2017;547:419–424.
116. Chotalia M, et al. Transcription is required for establishment of germline methylation marks at imprinted genes. *Genes Dev.* 2009;23:105–117.
117. Schroeder DI, et al. Early developmental and evolutionary origins of gene body DNA methylation patterns in mammalian placentas. *PLoS Genet.* 2015;11:e1005442.
118. Nagano T, et al. The air noncoding RNA epigenetically silences transcription by targeting G9a to chromatin. *Science.* 2008;322:1717–1721.
119. Edwards CA, Ferguson-smith AC. Mechanisms regulating imprinted genes in clusters. *Curr Opin Cell Biol.* 2007;19:281–289.
120. Latos PA, et al. Airn transcriptional overlap, but not its lncRNA products, induces imprinted Igf2r silencing. *Science.* 2012;338:1469–1472.
121. Pandey RR, et al. Kcnq1ot1 antisense noncoding RNA mediates lineage-specific transcriptional silencing through chromatin-level regulation. *Mol Cell.* 2008;32:232–246. 10.1016/j.molcel.2008.08.022.
122. Bressler J, et al. The SNRPN promoter is not required for genomic imprinting of the Prader–Willi/Angelman domain in mice. *Nat Genet.* 2001;28:232–240.
123. Landers M, et al. Regulation of the large (~1000 kb) imprinted murine Ube3a antisense transcript by alternative exons upstream of Snurf/Snrpn. *Nucleic Acids Res.* 2004;32:3480–3492.
124. Smith EY, Futtner CR, Chamberlain SJ, Johnstone KA, Resnick JL. Transcription is required to establish maternal imprinting at the Prader–Willi syndrome and Angelman syndrome locus. *PLoS Genet.* 2011;7:e1002422. 10.1371/journal.pgen.1002422.
125. Mapendano CK, et al. Expression of the Snurf-Snrpn IC transcript in the oocyte and its putative role in the imprinting establishment of the mouse 7C imprinting domain. *J Hum Genet.* 2006;51:236–243. 10.1007/s10038-005-0351-8.
126. Lewis MW, et al. Angelman syndrome imprinting center encodes a transcriptional promoter. *Proc Natl Acad Sci USA.* 2014;112:6871–6875. 10.1073/pnas.1411261111.
127. Meng L, et al. Truncation of Ube3a-ATS unsilences paternal Ube3a and ameliorates behavioral defects in the Angelman syndrome mouse model. *PLoS Genet.* 2013;9: e1004039.
128. Powell WT, et al. R-loop formation at Snord116 mediates topotecan inhibition of Ube3a-antisense and allele-specific chromatin decondensation. *Proc Natl Acad Sci USA.* 2013;110:13938–13943.
129. Al Adhami H, et al. A systems-level approach to parental genomic imprinting: the imprinted gene network includes extracellular matrix genes and regulates cell cycle exit and differentiation. *Genome Res.* 2015;25:353–367.
130. Cattanach BM, Beechey CV, Peters J. Interactions between imprinting effects in the mouse. *Genetics.* 2004;168:397–413.
131. Eggermann T, et al. Imprinting disorders: a group of congenital disorders with overlapping patterns of molecular changes affecting imprinted loci. *ClinEpigenet.* 2015;7:123.
132. Tinarelli F, Garcia-garcia C, Nicassio F, Tucci V. Parent-of-origin genetic background affects the transcriptional levels of circadian and neuronal plasticity genes following sleep loss. *Philos Trans R Soc B.* 2014;369:20120471.
133. Tucci V. Genomic imprinting: a new epigenetic perspective of sleep regulation. *PLoS Genet.* 2016;12:e1006004. 10.1371/journal.pgen.1006004.
134. Ramon Hernandez Mora J, et al. Profiling of oxBS-450K 5-hydroxymethylcytosine in human placenta and brain reveals enrichment at imprinted loci. *Epigenetics.* 2017;2294:.

135. Hirasawa R, et al. Maternal and zygotic Dnmt1 are necessary and sufficient for the maintenance of DNA methylation imprints during preimplantation development. *Genes Dev.* 2008;22:1607–1616.
136. Rapkins RW, et al. Recent assembly of an imprinted domain from non-imprinted components. *PLoS Genet.* 2006;2:1666–1675.
137. Edwards CA, et al. The evolution of the DLK1-DIO3 imprinted domain in mammals. *PLoS Biol.* 2008;6:e135.
138. Kondo T, et al. Circadian rhythms in prokaryotes: luciferase as a reporter of circadian gene expression in cyanobacteria. *Proc Natl Acad Sci U S A.* 1993;90:5672–5676.
139. Dunlap JC. Molecular bases for circadian clocks. *Cell.* 1999;96:271–290.
140. Harmer SL, Panda S, Kay SA. Molecular bases of circadian rhythms. *Annu Rev Cell Dev Biol.* 2001;17:215–253.
141. Takahashi JS. Transcriptional architecture of the mammalian circadian clock. *Nat Rev Genet.* 2016;18:164–179.
142. Yan J, Wang H, Liu Y, Shao C. Analysis of gene regulatory networks in the mammalian circadian rhythm. *PLoS Comput Biol.* 2008;4:e1000193.
143. Zhang R, et al. A circadian gene expression atlas in mammals: implications for biology and medicine. *Proc Natl Acad Sci U S A.* 2014;111:16219–16224.
144. Koike N, et al. Transcriptional architecture and chromatin landscape of the core circadian clock in mammals. *Science.* 2013;338:349–354.
145. Papazyan R, Zhang Y, Lazar MA. Genetic and epigenomic mechanisms of mammalian circadian transcription. *Nat Struct Mol Biol.* 2016;23:1045–1052.
146. Li JZ, et al. Circadian patterns of gene expression in the human brain and disruption in major depressive disorder. *Proc Natl Acad Sci U S A.* 2013;110:9950–9955.
147. McGlincy NJ, Valomon A, Chesham JE, Maywood ES, Hastings MH. Regulation of alternative splicing by the circadian clock and food related cues. *Genome Biol.* 2012;13: R54.
148. Hughes ME, Grant GR, Paquin C, Qian J, Nitabach MN. Deep sequencing the circadian and diurnal transcriptome of *Drosophila* brain. *Genome Res.* 2012;22: 1266–1281.
149. Lim C, Allada R. Emerging roles for post-transcriptional regulation in circadian clocks. *Nat Neurosci.* 2013;16:1544–1550.
150. Partch CL, Green CB, Takahashi JS. Molecular architecture of the mammalian circadian clock. *Trends Cell Biol.* 2015;24:90–99.
151. Du N, Arpat AB, De Matos M, Gatfield D. MicroRNAs shape circadian hepatic gene expression on a transcriptome-wide scale. *Elife.* 2014;1–29. 10.7554/eLife.02510.
152. Le Martelot G, et al. Genome-wide RNA polymerase II profiles and RNA accumulation reveal kinetics of transcription and associated epigenetic changes during diurnal cycles. *PLoS Biol.* 2012;10:e1001442.
153. Menet JS, Rodriguez J, Abruzzi KC, Rosbash M. Nascent-Seq reveals novel features of mouse circadian transcriptional regulation. *Elife.* 2012;1–25. 10.7554/eLife.00011.
154. Cheng H-Y, et al. microRNA modulation of circadian-clock period and entrainment. *Neuron.* 2007;54:813–829.
155. Chen R, D'Alessandro M, Lee C. miRNAs are required for generating a time delay critical for the circadian oscillator. *Curr Biol.* 2013;23:1959–1968.
156. Sinturel F, et al. Diurnal oscillations in liver mass and cell size accompany ribosome assembly cycles. *Cell.* 2017;169:651–63.e14.
157. Lim ASP, et al. 24-Hour rhythms of DNA methylation and their relation with rhythms of RNA expression in the human dorsolateral prefrontal cortex. *PLoS Genet.* 2014;10: e1004792.
158. Bhatti P, et al. Nightshift work and genome-wide DNA methylation. *Chronobiol Int.* 2015;32:103–112.

159. Massart R, et al. The genome-wide landscape of DNA methylation and hydroxymethylation in response to sleep deprivation impacts on synaptic plasticity genes. *Transl Psychiatry.* 2014;4:e347.
160. Cedernaes J, et al. Acute sleep loss induces tissue-specific epigenetic and transcriptional alterations to circadian clock genes in men. *J Clin Endocrinol Metab.* 2015;100: E1255–E1261.
161. Mongrain V, La Spada F, Curie T, Franken P. Sleep loss reduces the DNA-binding of BMAL1, CLOCK, and NPAS2 to specific clock genes in the mouse cerebral cortex. *PLoS ONE.* 2011;6:e26622.
162. Rey G, Rougemont J, Reinke H, Brunner M. Genome-wide and phase-specific DNA-binding rhythms of BMAL1 control circadian output functions in mouse liver. *PLoS Biol.* 2011;9:e1000595.
163. Jordan SD, Lamia KA. AMPK at the crossroads of circadian clocks and metabolism. *Mol Cell Endrocrinol.* 2013;366:163–169.
164. Markwald RR, et al. Impact of insufficient sleep on total daily energy expenditure, food intake, and weight gain. *Proc Natl Acad Sci U S A.* 2013;110:5695–5700. 10.1073/pnas.1216951110.
165. Thibert RL, Larson AM, Hsieh DT, Raby AR, Thiele EA. Neurologic manifestations of Angelman syndrome. *Pediatr Neurol.* 2013;48:271–279.
166. Khapre RV, et al. BMAL1-dependent regulation of the mTOR signaling pathway delays aging. *Aging (Albany NY).* 2014;6:48–57.
167. Cao R, Anderson FE, Jung Y-J, Dziema H, Obrietan K. Circadian regulation of mTOR signaling in the mouse SCN. *Neuroscience.* 2011;181:79–88. 10.1016/j.neuroscience.2011.03.005.Circadian.
168. Ballesta A, Innominato PF, Dallmann R, Rand DA, Lévi FA. Systems chronotherapeutics. *Phamacol Rev.* 2017;69:161–199.
169. Levi F, Okyar A, Dulong S, Innominato PF, Clairambault J. Circadian timing in cancer treatments. *Annu Rev Pharmacol Toxicol.* 2010;50:377–421.
170. Kaur G, Phillips CL, Wong K, Mclachlan AJ. Timing of administration: for commonly-prescribed medicines in Australia. *Pharmaceutics.* 2016;8:.
171. Benedetti F. Antidepressant chronotherapeutics for bipolar depression. *Dialogues Clin Neurosci.* 2012;14:401–411.
172. Dallaspezia S, Benedetti F. Sleep deprivation therapy for depression. *Curr Top Behav Neurosci.* 2015;25:483–502.

CHAPTER FIVE

The Role of DNA Methylation in Drug Addiction: Implications for Diagnostic and Therapeutics

Elad Lax, Moshe Szyf[1]

Department of Pharmacology and Therapeutics, McGill University Medical School, Montreal, QC, Canada

[1] Corresponding author. E-mail address: moshe.szyf@mcgill.ca

Contents

Abstract

Drug addiction is a devastating health problem that is a very heavy burden on the individual affected and the society in general. Recent research defines addiction as a neurobehavioral disorder. Underpinning biological mechanisms of drug addiction are abnormal neuronal and brain activity following acute and repeated drug exposure. Abnormal gene expression is found in reward and decision-making brain regions of addicts and in animal models and is possibly responsible for changes in brain function. DNA methylation is an epigenetic modification that regulates gene expression. Global and site-specific changes in DNA methylation are observed in addiction. Here, we discuss recent findings on the involvement of DNA methylation in drug addiction from animal and human studies. We also propose future directions for utilizing DNA methylation-based approaches for diagnosis, therapeutics, and evaluation of response to therapy in drug addiction.

Progress in Molecular Biology and Translational Science, volume 157
ISSN 1877-1173
http://dx.doi.org/10.1016/bs.pmbts.2018.01.003

1. INTRODUCTION

Drug abuse is defined as a chronic brain disorder of compulsory and repetitive drug-intake despite negative consequences.[1] Drug addiction is becoming an alarming medical and social problem in modern society with an increasing prevalence and overdose rates.

In recent years, it has become evident that addiction is a neuropsychiatric disorder that involves neuronal circuits and morphological and molecular brain abnormalities.[2] The basis for these abnormalities is altered gene expression in brain regions related to reward, motivation, and decision-making which lead to compromised neuronal functionality.[3] The main challenges in treating drug addiction are the mental and physiological hardships that accompany drug abstinence and mostly the high rates of relapse to drug-intake even after long abstinence periods.[4,5]

2. ANIMAL MODELS OF DRUG ADDICTION

To decipher the biological mechanisms that govern drug-intake and abuse and to test different treatments for drug addiction, several animal models were developed, mostly in rodents, that recapitulate different aspects of human addiction.

The simplest model is drug-induced sensitization in which repeated administration of a drug to the animal results in increased behavioral effects due to changes in the mesolimbic dopamine system.[6,7]

Another model is the conditioned-place preference (CPP); in this model, animals are injected with drug or vehicle and placed in one or other compartment of an apparatus. This way the animal learns to associate and prefer one specific environment with the rewarding effect of the drug. After animals are trained under this scheme, they can freely explore the apparatus and exhibit high preference to the drug-associated compartment. This Pavlovian-conditioned preference is translated as a drug-seeking behavior. This model revealed the involvement of brain memory formation and retrieval mechanisms in the evolution of drug addiction disorders.[8–10]

Both sensitization and CPP share a common drawback as models for addiction. Although a human addict actively decides when and how much drug to consume, animals used in these models are passively exposed to fixed drug dosages on a fixed schedule set by the experimenter. Therefore, it is hard to directly draw conclusions from these models to human subjects. To answer the need for a better model, a system which allows animals to control drug-intake was developed. The self-administration model allows animals to freely administer drug in an operant conditioning chamber (also known as Skinner box).[11] The chamber has two levers, a press on one of the levers results in drug administration (usually intravenously) and a simultaneous associated cue (light and/or sound). Presses of the other lever will not yield any reward. The animals learn over time to press the rewarding lever. An increase in the number of presses over time is an indicator of learned drug-intake. Drug-seeking and craving-like behaviors can then be measured by removing the drug from the administration system and counting the number of pressings on the lever. Translated as measuring the effort of the animal to find and consume the drug, high levels of lever pressings mean higher drug seeking. A more recent development of this method is the "incubation of craving."[12] In this model, longer abstinence periods resulted in a robust increase of seeking behavior when animals were reintroduced to the chamber. This incubation of craving response over time mimics the human phenomenon of increased craving after prolonged abstinence and therefore this version of the self-administration model is considered the most translational one.[13,14]

3. DNA METHYLATION IN ANIMAL MODELS OF DRUG ADDICTION

DNA methylation is the enzymatic addition of a methyl group onto cytosine bases in the DNA converting cytosine to 5-methylcytosine. This happens mostly, though not exclusively, on cytosines which reside next to guanine (a CpG dinucleotide).[15,16] The addition of the methyl group on cytosine is catalyzed by DNA methyltransferases (DNMT1, DNMT3a, and DNMT3b), whereas demethylation could be promoted by ten-eleven translocation methylcytosine dioxygenases (Tet1–3) which further convert the methyl moiety to hydroxymethyl.[17] DNA methylation in gene promoters is associated with gene silencing through several mechanisms. First, it blocks binding of transcription factors.[18] Furthermore, a group of proteins preferentially binds methylated DNA to regulate gene expression.[19] This group includes the methyl-CpG-binding protein 2 (Mecp2) and the methyl-CpG-binding domain (MBD) family (MBD1-4).[20] The group of proteins translate DNA methylation status to gene regulation decisions through attracting other transcriptional regulatory complexes to promoters and enhancers. The 5-hydroxymethylcytosine moiety might play a role as an epigenetic mark that is different from 5-methylcytosine.[21,22]

A long line of evidence suggests that abnormal patterns of gene expression occur in brain regions related to drug addiction such as the nucleus accumbens, prefrontal cortex, amygdala, and the ventral tegmental area.[23–25]

As DNA methylation is a fundamental mechanism of gene regulation, it is reasonable to believe that it also regulates gene expression in the brain in general and particularly in addiction. Indeed, several studies in animal models and human subjects found substantial differences in DNA methylation following drug experience.[26,27] Altered levels of DNA methylation are found in different experimental conditions and in response to different classes of drugs of abuse such as methamphetamine,[28–30] alcohol,[31–33] opioids,[34] cannabis,[35] and cocaine.[36–38] This illustrates the general and widespread involvement of DNA methylation in drug addiction.

Dietary supplementation or injection of methyl donors like methionine and *S*-adenosyl methionine (SAM) provides methyl group substrates to DNMT enzymes which could result in increased DNA methylation.[39,40] Hypomethylation can be obtained by treatment with DNMT inhibitors like 5-azacytidine[41] or RG108[42]; however, the latter should be directly infused to the brain because it does not cross the blood–brain barrier.

Barbier *et al.*[43] found increased protein levels of neuronal DNMT1 and hypermethylation in the medial prefrontal cortex of rats exposed to alcohol; and blockade of DNMT by the inhibitor RG108 attenuated alcohol-intake.

Supplementation of methyl donors was suggested as a potential treatment for addiction. However, attempts of systemic methyl supplementation to reduce drug-seeking behaviors exhibit mixed results. A study found reduced DNA methylation in the prefrontal cortex of mice trained in the CPP paradigm with cocaine, but not with morphine or food. Supplementation of methionine prevented the establishment of cocaine preference but not that of morphine or food.[44] Others found that methyl supplementation by chronic l-methionine (MET) blocked sensitization to cocaine and attenuated drug-induced reinstatement in a self-administration paradigm.[38] Furthermore, DNMT3a expression in the nucleus accumbens is dynamically regulated by cocaine. Pharmacological inhibition of DNMT activity by RG108 directly in the nucleus accumbens reduces DNA methylation and increases CPP and locomotor sensitization, whereas DNMT3a overexpression reduces CPP.[36] Others showed that cocaine has a bidirectional, gene-specific effect on DNA methylation in the nucleus accumbens.

Cocaine causes hypermethylation and increases Mecp2 binding to the promoter of the gene *PP1c* (protein phosphatase-1 catalytic subunit) while promoting hypomethylation and decreases Mecp2 binding to the promoter of *FosB*.[45] Also, systemic SAM treatment potentiates cocaine-induced locomotor sensitization in mice.[46]

Two limitations are common among the abovementioned studies. First, none of these studies used the "incubation of craving" model to explore the long-term effects of abstinence on methylation. The second limitation is the lack of genome-wide analysis of methylation. Only global levels of methylation or the methylation status of a few genes were tested. These issues were addressed in a recent study by Massart *etal.*[47] Using the "incubation of craving" model in rats trained to self-administer cocaine, and treated with either SAM or RG108, the genome-wide DNA methylation and gene expression landscape in the nucleus accumbens after short (1 day) and long (30 days) abstinence periods and the effects of epigenetic treatments were delineated. The main findings are that a long incubation period results in robust changes in methylation and that direct accumbal infusion of SAM that is paired with a "cue" after long incubation times increases drug-seeking behavior, whereas a single treatment with RG108 decreases this behavior (Fig. 1). Importantly, the effects of these single administrations of a DNA methylation inhibitor remain stable for 30 more days. These data suggest that DNA methylation might be mediating the impact of "incubation" on the craving phenotype and that this phenotype could be reprogrammed by a DNA demethylation agent. These studies point to the possibility that DNA methylation inhibitors could be used for the treatment of addiction in humans, a treatment that might target the epigenetic underpinnings of this condition rather than the symptoms and proximal biochemical and physiological alterations.

4. ALTERATIONS IN DNA METHYLATION ASSOCIATED WITH DRUG ADDICTION IN HUMAN

Studies of brain tissue from postmortem drug addicts found altered DNA methylation in the prefrontal cortex of alcohol[48] and heroin users.[49] Interestingly, gene expression shows a complex relationship to DNA methylation in the alcohol study, suggesting that some of the effects of DNA methylation on gene expression are indirect. It is very probable that DNA methylation controls upstream regulators and thereby altering many downstream genes as was shown previously in a mechanistic study.[50]

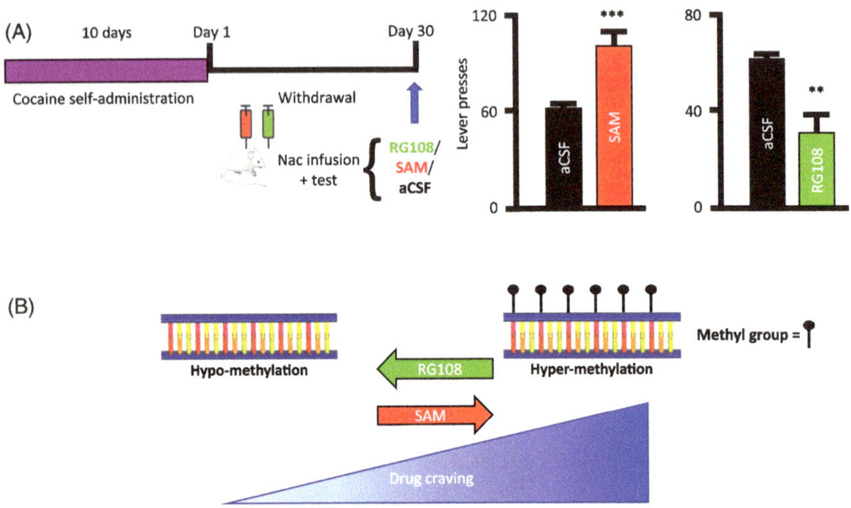

Fig. 1 ***Nucleus accumbens infusions of DNA methylation modulators manipulate drug craving after prolonged withdrawal (incubation of craving).*** (A) Infusion of RG108, a DNA methyltransferase inhibitor, following a 30-day withdrawal period, reduces drug-seeking behavior. To the contrary, infusion of the methyl donor *S*-adenosyl methionine (SAM) increased drug-seeking behavior. (B) A model depicting the effects of DNA methylation modulators on drug craving. RG108 promoted hypomethylation and reduced drug craving, whereas SAM caused hypermethylation and increased drug craving. *aCSF*, Artificial cerebrospinal fluid; *NAc*, nucleus accumbens.[47]

DNA methylation levels cannot be assessed in the brains of living humans. Therefore, human studies in the field of epigenetics and addiction use either postmortem brain samples or peripheral tissues such as whole-blood, isolated types of white blood cells such as T-cells or saliva. The relevance of DNA methylation status in peripheral tissues to neuropsychiatric diseases is not trivial and is hotly debated in the field because DNA methylation profiles are "cell-type specific"[51,52] and the brain is the tissue of interest and not blood or saliva for obvious reasons. Nevertheless, an accumulating line of evidence shows good correlational relationships between DNA methylation levels in periphery and the brain. These studies show strong correlations in rodent models with chronic pain[53] and stress in monkeys.[54] There are also several examples for association (correlations) between blood and brain-associated disorders in human studies, for example, in maternal depression,[55] and even across species.[56] Methylation levels of Illumina 450k methylation array probes were compared across brain regions and peripheral tissues in dozens of postmortem human subjects.[57,58] These studies found many CpGs whose state of methylation strongly correlates within subjects across brain and peripheral tissues. Taken together, these findings support measuring DNA methylation levels in peripheral tissues of drug addicts.

Considering the apparent relevance of peripheral DNA methylation to neuropsychiatric disorders, a growing body of studies associate altered DNA methylation with addiction. A study in adolescence showed that hypermethylation of the promoter of catechol-*O*-methyltransferase (COMT) is associated with reduced risk of cannabis use.[59] In a few studies, promoter hypermethylation of the *Oprm1* (opioid receptor mu 1) gene across peripheral tissues (whole blood, lymphocytes, and sperm) is associated with opioid addiction.[60–62] The *Oprm1* genes encodes the mu-opioid receptor. This receptor is the site of action of opioids; hypermethylation of this gene can result in reduced expression of the receptor. As this receptor is involved in mediating the rewarding effects of opioids and other drugs,[63,64] its altered expression can increase craving and drug misuse.

To the best of our knowledge, most human studies were done on alcohol abusers.

A cross-tissue study in postmortem brains and buccal cells of alcohol users found more than 400 differentially methylated CpGs which also correlate across tissues.[65]

Higher methylation levels in few genes were found in alcohol-dependent people and the findings were ethnically specific. African-Americans had higher methylation levels of CpGs in the *GABRB3* (gamma-aminobutyric acid type A receptor beta3 subunit) and *POMC* (proopiomelanocortin) genes, whereas European-Americans had higher methylation levels of CpGs in the genes *HTR3A* (5-hydroxytryptamine receptor 3A), *NCAM1* (neural cell adhesion molecule 1), *DRD4* (dopamine receptor D4), *MBD3* (methyl-CpG-binding domain protein 3), *HTR2B* (5-hydroxytryptamine receptor 2B), and *GRIN1* (glutamate ionotropic receptor NMDA-type subunit 1).[66] Altered methylation levels of the *POMC* gene were also found in alcohol-dependent patients in a different study.[67] Interestingly, DNA methylation is dynamically altered in alcohol-dependent patients, whereas the promoter of the *BDNF* (brain-derived neurotrophic factor) gene is hypermethylated in these patients; it underwent a significant hypomethylation during alcohol withdrawal (up to 14 days).[68] The gene *LEP* (leptin), encoding the leptin gene, also shows reduced methylation levels during alcohol withdrawal which correlates with increased expression of the gene and increased alcohol craving.[69]

Few large studies were done at a genome-wide scale. In one such study,[70] a genome-wide analysis in a discovery cohort of monozygotic twin pairs discordant for alcohol abuse found that the top hypermethylated site was located in the body of the *PPM1G* gene (protein phosphatase, Mg2+/Mn2+-dependent 1G). Next, in a large replication study the methylation status of this gene associates with alcohol use, behavioral traits, and imaging. Overall, *PPM1G* hypermethylation is associated with alcohol abuse and increased impulsiveness.

A different study[71] found hypomethylation of the promoter of the *PCSK9* gene (proprotein convertase subtilisin/kexin type 9) in three brain discovery cohorts and in blood and liver tissues in a replication cohort. When the researchers exposed rodents to alcohol, they observed reduced mRNA and protein levels of *PCSK9*, confirming the relevance of methylation of this promoter to expression.

5. DNA METHYLATION AS A BIOMARKER

Because peripheral DNA methylation measures are readily feasible, some studies aim to test whether DNA methylation biomarkers might be used as diagnostic tools for drug abusive behaviors and for drug addiction status. For example, the gene *GDAP1* (ganglioside-induced differentiation-associated protein 1) is hypomethylated in alcohol abusers.[72] These findings were confirmed by a successful replication study.[73] The replication study also shows correlation with years of alcohol dependence as well as reduced hypomethylation after a short-term treatment program, suggesting that *GDAP1* methylation levels might serve as a biomarker for alcohol dependence severity and treatment outcomes.

In an epigenomic-wide association study (EWAS), Liu *et al.*[74] performed a meta-analysis of population-based cohorts which included over 13,000 participants and found hundreds of differentially methylated sites between alcohol consumers and controls. A set of 144 CpGs is highly predictive in discriminating between heavy drinkers and nondrinkers, suggesting an epigenetic "fingerprint" for heavy alcohol use.

In a study done on two cohorts of opioid addicts stabilized on low and high levels of methadone,[75] a small subset of 13 CpGs was consistently differentially methylated and associated with the dose of methadone. These data suggest that efficacy and adherence to methadone maintenance can be evaluated by measuring methylation levels on a small set of CpGs.

6. FUTURE DIRECTIONS

Although there is an increasing line of evidence from preclinical models of addiction, there are only a few human studies that systematically assessed DNA methylation in addiction. Most of the studies were done on small cohorts and focused on one or a few candidate genes, except in the case of alcohol use where larger studies have been carried out.[70,71] Therefore, there is a need for more human studies, mostly to explore whether epigenetic treatments such as DNA methylation inhibitors might be beneficial in treating drug addicts and reducing drug relapse. It might be worthwhile to examine the possibility that DNA methylation markers might be used to screen people who are at high risk for developing drug addiction which could be combined with new strategies for epigenetically guided early interventions.

ACKNOWLEDGMENTS

This work was supported by the Canadian Institute of Health Research Grant MOP-42411 (MS), and the Ministere du Developpement Economique de l'Innovation et de l'Exportation program of the government of Quebec (MS). EL was supported by the Richard and Edith Strauss Canada Fund Post-Doctoral Fellowship in Medicine.

REFERENCES

1. American Psychiatric Association. *Substance-Related and Addictive Disorders, Diagnostic and Statistical Manual of Mental Disorders*. 5th ed Arlington, VA: American Psychiatric Publishing; 2013.
2. Kalivas PW, Volkow ND. The neural basis of addiction: a pathology of motivation and choice. *Am J Psychiatry*. 2005;162:1403–1413.
3. Nestler EJ. Molecular basis of long-term plasticity underlying addiction. *Nat Rev Neurosci*. 2001;2:119–128.
4. Bossert JM, Ghitza UE, Lu L, Epstein DH, Shaham Y. Neurobiology of relapse to heroin and cocaine seeking: an update and clinical implications. *Eur J Pharmacol*. 2005;526:36–50.
5. Mendelson JH, Mello NK. Management of cocaine abuse and dependence. *N Engl J Med*. 1996;334:965–972.
6. Kalivas PW, Pierce RC, Cornish J, Sorg BA. A role for sensitization in craving and relapse in cocaine addiction. *J Psychopharmacol*. 1998;12:49–53.
7. Robinson TE, Berridge KC. The neural basis of drug craving: an incentive-sensitization theory of addiction. *Brain Res Brain Res Rev*. 1993;18:247–291.
8. Nader MA. Animal models for addiction medicine: from vulnerable phenotypes to addicted individuals. *Prog Brain Res*. 2016;224:3–24.
9. Sartor GC, Aston-Jones G. Post-retrieval extinction attenuates cocaine memories. *Neuropsychopharmacology*. 2014;39:1059–1065.
10. Tzschentke TM. Measuring reward with the conditioned place preference (CPP) paradigm: update of the last decade. *Addict Biol*. 2007;12:227–462.

11. Skinner BF. *The Behavior of Organisms: An Experimental Analysis*. New York: Appleton-Century-Crofts; 1938.
12. Grimm JW, Hope BT, Wise RA, Shaham Y. Neuroadaptation. Incubation of cocaine craving after withdrawal. *Nature*. 2001;412:141–142.
13. Li X, Caprioli D, Marchant NJ. Recent updates on incubation of drug craving: a mini-review. *Addict Biol*. 2015;20:872–876.
14. Venniro M, Caprioli D, Shaham Y. Animal models of drug relapse and craving: from drug priming-induced reinstatement to incubation of craving after voluntary abstinence. *Prog Brain Res*. 2016;224:25–52.
15. Cedar H, Razin A. DNA methylation and development. *Biochim Biophys Acta*. 1990;1049:1–8.
16. Razin A, Cedar H. DNA methylation and genomic imprinting. *Cell*. 1994;77:473–476.
17. Day JJ, Kennedy AJ, Sweatt JD. DNA methylation and its implications and accessibility for neuropsychiatric therapeutics. *Annu Rev Pharmacol Toxicol*. 2015;55:591–611.
18. Moore LD, Le T, Fan G. DNA methylation and its basic function. *Neuropsychopharmacology*. 2013;38:23–38.
19. Nan X, Cross S, Bird A. Gene silencing by methyl-CpG-binding proteins. *Novartis Found Symp*. 1998;214:6–16.
20. Ludwig AK, Zhang P, Cardoso MC. Modifiers and readers of DNA modifications and their impact on genome structure, expression, and stability in disease. *Front Genet*. 2016;7:115.
21. Massart R, et al. Hydroxymethylation and DNA methylation profiles in the prefrontal cortex of the non-human primate rhesus macaque and the impact of maternal deprivation on hydroxymethylation. *Neuroscience*. 2014;268:139–148.
22. Szyf M. The elusive role of 5'-hydroxymethylcytosine. *Epigenomics*. 2016;8:1539–1551.
23. Feng J, Nestler EJ. Epigenetic mechanisms of drug addiction. *Curr Opin Neurobiol*. 2013;23:521–528.
24. Nestler EJ. Epigenetic mechanisms of drug addiction. *Neuropharmacology*. 2014;76(pt B):259–268.
25. Robison AJ, Nestler EJ. Transcriptional and epigenetic mechanisms of addiction. *Nat Rev Neurosci*. 2011;12:623–637.
26. Alaghband Y, Bredy TW, Wood MA. The role of active DNA demethylation and Tet enzyme function in memory formation and cocaine action. *Neurosci Lett*. 2016;625:40–46.
27. Tuesta LM, Zhang Y. Mechanisms of epigenetic memory and addiction. *EMBO J*. 2014;33:1091–1103.
28. Cheng MC, Hsu SH, Chen CH. Chronic methamphetamine treatment reduces the expression of synaptic plasticity genes and changes their DNA methylation status in the mouse brain. *Brain Res*. 2015;1629:126–134.
29. Itzhak Y, Ergui I, Young JI. Long-term parental methamphetamine exposure of mice influences behavior and hippocampal DNA methylation of the offspring. *Mol Psychiatry*. 2015;20:232–239.
30. Jayanthi S, et al. Methamphetamine induces TET1- and TET3-dependent DNA hydroxymethylation of Crh and Avp genes in the rat nucleus accumbens. *Mol Neurobiol*. 2017. 10.1007/s12035-017-0750-9.
31. Bleich S, et al. Epigenetic DNA hypermethylation of the HERP gene promoter induces down-regulation of its mRNA expression in patients with alcohol dependence. *Alcohol Clin Exp Res*. 2006;30:587–591.
32. Bonsch D, Lenz B, Reulbach U, Kornhuber J, Bleich S. Homocysteine associated genomic DNA hypermethylation in patients with chronic alcoholism. *J Neural Transm (Vienna)*. 2004;111:1611–1616.

33. Ponomarev I, Wang S, Zhang L, Harris RA, Mayfield RD. Gene coexpression networks in human brain identify epigenetic modifications in alcohol dependence. *J Neurosci.* 2012;32:1884–1897.
34. Zhu J, et al. Methylation of glucocorticoid receptor gene promoter modulates morphine dependence and accompanied hypothalamus-pituitary-adrenal axis dysfunction. *J Neurosci Res.* 2017;95:1459–1473.
35. Watson CT, et al. Genome-wide DNA methylation profiling reveals epigenetic changes in the rat nucleus accumbens associated with cross-generational effects of adolescent THC exposure. *Neuropsychopharmacology.* 2015;40:2993–3005.
36. LaPlant Q, et al. Dnmt3a regulates emotional behavior and spine plasticity in the nucleus accumbens. *Nat Neurosci.* 2010;13:1137–1143.
37. Tian W, et al. Demethylation of c-MYB binding site mediates upregulation of Bdnf IV in cocaine-conditioned place preference. *Sci Rep.* 2016;6:22087.
38. Wright KN, et al. Methyl supplementation attenuates cocaine-seeking behaviors and cocaine-induced c-Fos activation in a DNA methylation-dependent manner. *J Neurosci.* 2015;35:8948–8958.
39. Lan J, Hua S, He X, Zhang Y. DNA methyltransferases and methyl-binding proteins of mammals. *Acta Biochim Biophys Sin (Shanghai).* 2010;42:243–252.
40. MacDonald JL, Roskams AJ. Epigenetic regulation of nervous system development by DNA methylation and histone deacetylation. *Prog Neurobiol.* 2009;88:170–183.
41. Stresemann C, Lyko F. Modes of action of the DNA methyltransferase inhibitors azacytidine and decitabine. *Int J Cancer.* 2008;123:8–13.
42. Brueckner B, et al. Epigenetic reactivation of tumor suppressor genes by a novel small-molecule inhibitor of human DNA methyltransferases. *Cancer Res.* 2005;65:6305–6311.
43. Barbier E, et al. DNA methylation in the medial prefrontal cortex regulates alcohol-induced behavior and plasticity. *J Neurosci.* 2015;35:6153–6164.
44. Tian W, et al. Reversal of cocaine-conditioned place preference through methyl supplementation in mice: altering global DNA methylation in the prefrontal cortex. *PLoS One.* 2012;7:e33435.
45. Anier K, Malinovskaja K, Aonurm-Helm A, Zharkovsky A, Kalda A. DNA methylation regulates cocaine-induced behavioral sensitization in mice. *Neuropsychopharmacology.* 2010;35:2450–2461.
46. Anier K, Zharkovsky A, Kalda A. *S*-Adenosylmethionine modifies cocaine-induced DNA methylation and increases locomotor sensitization in mice. *Int J Neuropsychopharmacol.* 2013;16:2053–2066.
47. Massart R, et al. Role of DNA methylation in the nucleus accumbens in incubation of cocaine craving. *J Neurosci.* 2015;35:8042–8058.
48. Wang F, Xu H, Zhao H, Gelernter J, Zhang H. DNA co-methylation modules in postmortem prefrontal cortex tissues of European Australians with alcohol use disorders. *Sci Rep.* 2016;6:19430.
49. Kozlenkov A, et al. DNA methylation profiling of human prefrontal cortex neurons in heroin users shows significant difference between genomic contexts of hyper- and hypomethylation and a younger epigenetic age. *Genes (Basel).* 2017;8:E152.
50. Cheishvili D, et al. DNA methylation controls unmethylated transcription start sites in the genome in trans. *Epigenomics.* 2017;9:611–633.
51. Bernstein BE, et al. The NIH roadmap epigenomics mapping consortium. *Nat Biotechnol.* 2010;28:1045–1048.
52. Roost MS, et al. DNA methylation and transcriptional trajectories during human development and reprogramming of isogenic pluripotent stem cells. *Nat Commun.* 2017;8:908.
53. Massart R, et al. Overlapping signatures of chronic pain in the DNA methylation landscape of prefrontal cortex and peripheral T cells. *Sci Rep.* 2016;6:19615.

54. Provencal N, et al. The signature of maternal rearing in the methylome in rhesus macaque prefrontal cortex and T cells. *J Neurosci.* 2012;32:15626–15642.
55. Nemoda Z, et al. Maternal depression is associated with DNA methylation changes in cord blood T lymphocytes and adult hippocampi. *Transl Psychiatry.* 2015;5:e545.
56. Nieratschker V, et al. MORC1 exhibits cross-species differential methylation in association with early life stress as well as genome-wide association with MDD. *Transl Psychiatry.* 2014;4:e429.
57. Edgar RD, Jones MJ, Meaney MJ, Turecki G, Kobor MS. BECon: a tool for interpreting DNA methylation findings from blood in the context of brain. *Transl Psychiatry.* 2017;7:e1187.
58. Hannon E, Lunnon K, Schalkwyk L, Mill J. Interindividual methylomic variation across blood, cortex, and cerebellum: implications for epigenetic studies of neurological and neuropsychiatric phenotypes. *Epigenetics.* 2015;10:1024–1032.
59. van der Knaap LJ, et al. Catechol-O-methyltransferase gene methylation and substance use in adolescents: the TRAILS study. *Genes Brain Behav.* 2014;13:618–625.
60. Chorbov VM, Todorov AA, Lynskey MT, Cicero TJ. Elevated levels of DNA methylation at the OPRM1 promoter in blood and sperm from male opioid addicts. *J Opioid Manag.* 2011;7:258–264.
61. Ebrahimi G, et al. Elevated levels of DNA methylation at the OPRM1 promoter region in men with opioid use disorder. *Am J Drug Alcohol Abuse.* 2017;43:1–7.
62. Nielsen DA, et al. Increased OPRM1 DNA methylation in lymphocytes of methadone-maintained former heroin addicts. *Neuropsychopharmacology.* 2009;34:867–873.
63. Kreek MJ, LaForge KS, Butelman E. Pharmacotherapy of addictions. *Nat Rev Drug Discov.* 2002;1:710–726.
64. Matthes HW, et al. Loss of morphine-induced analgesia, reward effect and withdrawal symptoms in mice lacking the mu-opioid-receptor gene. *Nature.* 1996;383:819–823.
65. Hagerty SL, Bidwell LC, Harlaar N, Hutchison KE. An exploratory association study of alcohol use disorder and DNA methylation. *Alcohol Clin Exp Res.* 2016;40:1633–1640.
66. Zhang H, et al. Array-based profiling of DNA methylation changes associated with alcohol dependence. *Alcohol Clin Exp Res.* 2013;37(suppl 1):E108–E115.
67. Muschler MA, et al. DNA methylation of the POMC gene promoter is associated with craving in alcohol dependence. *J Neural Transm (Vienna).* 2010;117:513–519.
68. Heberlein A, et al. Do changes in the BDNF promoter methylation indicate the risk of alcohol relapse? *Eur Neuropsychopharmacol.* 2015;25:1892–1897.
69. Hillemacher T, et al. DNA methylation of the LEP gene is associated with craving during alcohol withdrawal. *Psychoneuroendocrinology.* 2015;51:371–377.
70. Ruggeri B, et al. Association of protein phosphatase PPM1G with alcohol use disorder and brain activity during behavioral control in a genome-wide methylation analysis. *Am J Psychiatry.* 2015;172:543–552.
71. Lohoff FW, et al. Methylomic profiling and replication implicates deregulation of PCSK9 in alcohol use disorder. *Mol Psychiatry.* 2017. 10.1038/mp.2017.168.
72. Philibert RA, et al. A pilot examination of the genome-wide DNA methylation signatures of subjects entering and exiting short-term alcohol dependence treatment programs. *Epigenetics.* 2014;9:1212–1219.
73. Bruckmann C, Di Santo A, Karle KN, Batra A, Nieratschker V. Validation of differential GDAP1 DNA methylation in alcohol dependence and its potential function as a biomarker for disease severity and therapy outcome. *Epigenetics.* 2016;11:456–463.
74. Liu C, et al. A DNA methylation biomarker of alcohol consumption. *Mol Psychiatry.* 2016. 10.1038/mp.2016.192.
75. Marie-Claire C, et al. Variability of response to methadone: genome-wide DNA methylation analysis in two independent cohorts. *Epigenomics.* 2016;8:181–195.

CHAPTER SIX

DNA Methylation in Animal Models of Psychosis

Dennis R. Grayson[1], Alessandro Guidotti
Center for Alcohol Research in Epigenetics, The Psychiatric Institute, Department of Psychiatry, University of Illinois, Chicago, IL, United States
[1] Corresponding author. E-mail address: dgrayson@uic.edu

Contents

Abstract

Schizophrenia (SZ) is a debilitating disease that impacts 1% of the population worldwide. Association studies have shown that inherited genetic mutations account for a portion of disease risk. However, environmental factors play an important role in the pathophysiology of the disease by altering cellular epigenetic marks at the level of chromatin. Postmortem brain studies of SZ subjects suggest that the dynamic equilibrium between DNA methylation and demethylation network components is disrupted at the level of individual SZ target genes. Herein, we review the role of DNA methylation and demethylation in the context of what is currently known regarding SZ. Furthermore, we describe the deficits that accompany two mouse models of SZ. The chronic methionine mouse model of SZ is predicated on the administration of methionine to SZ patients and controls in the context of clinical studies that were carried out during the 1960s and 1970s. The prenatal restraint stress model of SZ is based on a prolonged stress paradigm administered to pregnant dams during gestation days 7–21. The adult offspring of these dams show various behavioral and

Progress in Molecular Biology and Translational Science, volume 157
ISSN 1877-1173
http://dx.doi.org/10.1016/bs.pmbts.2017.12.012

biochemical deficits in adulthood. Both models are epigenetic in origin and mimic the positive and negative symptoms, as well as the cognitive endophenotypes commonly observed in SZ patients. We also discuss the utility of typical and atypical antipsychotic drugs in alleviating these symptoms in each model.

1. THE EPIGENETIC HYPOTHESIS OF PSYCHIATRIC DISEASE

Current thinking suggests that both genetic predisposition and environmental factors have interacting roles in the pathophysiology of the more common psychiatric disorders, such as schizophrenia (SZ), bipolar illness (BP), and major depression. Familial studies support a strong hereditary component, whereas twin studies implicate the role of additional environmental variables, such as prenatal and early life stressors. There is evidence that environmental stressors impact the development of key brain structures and circuitries, and can lead to the development of psychiatric disease. Genetic studies have implicated some 108 distinct genomic loci associated with the development of SZ.[1] However, association is not the same as causation and each genome variant is likely present in only a small number of individuals (1–2%) that are diagnosed with the disorder. He gene mutations that have been identified are intimately involved in the biology and pathogenesis of the disorder but are not, by themselves, sufficient to cause the disease. As described in the sections that follow, environmental factors interact with the expression of multiple genes that show overlap with these 108 loci impacting the levels of expression of these genes in the brain. This suggests that fundamental epigenetic mechanisms are operative in regulating finite gene networks that, when dysregulated, lead to the development of psychiatric disorders. The epigenetic hypothesis of SZ postulates that interactions between environmental factors and the genetic landscape of affected individuals are ultimately responsible for the pathogenesis of the disorder.[2] The aforementioned concepts will be elaborated upon in subsequent sections.

1.1 Neurodevelopmental Models—Multiple Hit Hypothesis

Most psychiatric disorders exhibit complex patterns of inheritance. That is, their inheritance is non-Mendelian indicating that it likely involves the interaction of one or more genes with environmental factors. For purposes of this article, we will focus primarily on the effects of epigenetics in the context of SZ, which is a devastating and chronic neuropsychiatric disorder. Moreover, it is triggered by an array of interacting genetic, developmental, and environmental factors that interfere with normal brain development. SZ has a worldwide prevalence of 1% and is a major socioeconomic burden. With some exceptions, it is usually diagnosed in young adults following the first episode of psychosis. Symptoms of SZ are classified as positive, negative, or cognitive. Positive symptoms include hallucinations that can be visual and/or auditory. Delusions are beliefs that may arise from the hallucinations that are also part of the spectrum of positive symptoms. People with SZ have difficulty in concentrating and often have trouble keeping track of their thoughts. Negative symptoms of SZ often appear temporally before onset of the disorder (during the so-called prodromal period). People experiencing negative symptoms may lose interest and motivation in life and activities, including relationships and sex. They become asocial and often feel uncomfortable with people, avoiding conversations, and so on. Patients with SZ over longer periods of time experience declines in cognitive capacity and executive function. Because of the complexities of neuropsychiatric disorders like SZ, modeling these diseases in animals is not an exact science. Nevertheless, there are behaviors that can be used to assess various aspects of social activity in rodents that mimic behaviors reminiscent of those observed in patients diagnosed with SZ. For example, deficits in social interaction can be assessed by examining the amount of active interaction time a mouse spends with a novel or nonfamiliar mouse. These will be discussed in more detail as the various animal models are described.

In the multihit hypothesis of SZ, the initial hit is presumed to be genetic. Environmental factors (stressors) that are triggered by detrimental events act progressively and the effects of these accumulate during the pre- and postadolescent lifespan of the individual (Fig. 1).[3] How these stressors impact brain development is currently under intense scrutiny in a large number of laboratories worldwide. Both the intensity of the stressor and the developmental timing of when the stress occurs are important factors that determine the ultimate consequences to the mature brain. Prenatal stresses often include maternal stress during pregnancy (physical or psychological abuse, trauma), maternal infections during pregnancy, obstetrical complications both before and during delivery, exposure to drugs of abuse in utero, maternal and paternal age at conception, and malnutrition and exposure to environmental pollutants in utero. Early postnatal stresses, which occur following birth, directly affect the developing brain and some of the major factors include exposure to physical and psychological abuse and trauma, malnutrition, and

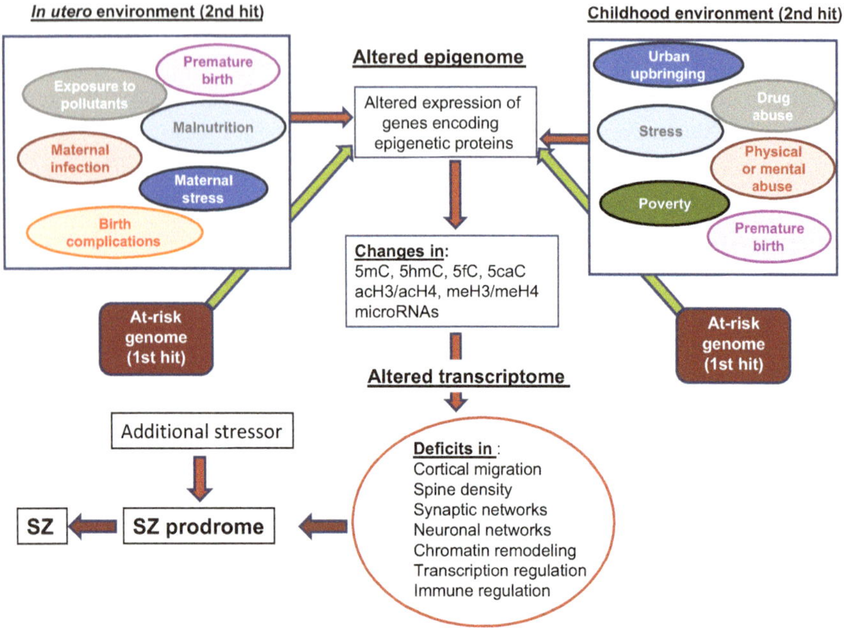

Fig. 1 Epigenetic impact of the pre- and postnatal environment on brain development. In the two-hit hypothesis of SZ, a predisposing genome would be considered a first hit. Environmentally induced epigenetic mechanisms acting on the genome during prenatal and postnatal development facilitate genome-wide changes in the epigenome that alter cell type-specific transcriptional profiles at distinct developmental time points. These environmental insults that then act on the predisposing genome at distinct development times represent a second hit. The subsequent mRNA alterations, in turn, produce altered protein profiles that cause deficits in multiple processes including cortical migration, chromatin remodeling, transcriptional regulation, immune-response regulation, and the formation of synaptic connections. Moreover, changes at the cellular level can translate into altered neuronal network connections that ultimately impact the phenotype of the individual. Genome-wide association and whole exome sequencing studies have identified a large number of genes associated with SZ that encode proteins that function in synapse formation, transcriptional regulation, and chromatin-remodeling pathways.

poverty. In addition to physical and psychological abuse, poverty, urban city living, later postnatal environmental influences also include abnormal hormonal influences, exposure to drugs of abuse such as cannabis or psychostimulants and bullying.[4] It is widely thought that any one environmental factor is not a necessary and sufficient cause of any mental disorder. Instead, it seems that cumulative exposure to multiple environmental influences at different stages in the lifespan is more likely to be predictive of psychiatric disorder risk. However, it should be pointed out that research into how environmental stressors impact the risk of psychiatric disease is at an early stage. Moreover, the list of impactful environmental factors is likely to grow considerably before we clearly understand how relevant these factors may be.[4] It seems clear that one of the mechanisms by which stressors impact individual cells during development is by altering the methylation status of vulnerable genes in the genome.

1.2 DNA Methylation/Demethylation Pathway in the Brain

1.2.1 DNA Methyltransferases

DNA methylation is a postreplicative epigenetic modification of DNA that occurs at the 5 position of cytosine and is used by cells as one mechanism to control patterns of gene expression.[2,5,6] The DNA methylation/demethylation pathway is schematically shown in Fig. 2. In general, DNA methylation negatively impacts transcription through an interaction with various methyl CpG-binding proteins and other repressor proteins that tend to regionally condense chromatin in the vicinity of transcriptional start sites of targeted genes. DNA methylation is carried out enzymatically by members of the DNA methyltransferase (DNMT) enzyme family which consists of DNMT1, DNMT3a, DNMT3b, and DNMT3l.[6,7] The DNMTs catalyze the transfer of a methyl group from S-adenosylmethionine (SAM) to the 5 position of cytosine forming 5-methylcytosine (5mC) and S-adenosylhomocysteine (SAH). For many years, it was thought that only the cytosine in the dinucleotide pair CpG is methylated. However, we now know that non-CpG methylation is also abundant in the brain and other tissues and is present throughout the genome.[8]

The DNMT proteins have a similar domain structure, although there are considerable differences between DNMT1, the maintenance methyl transferase, and the de novo methyltransferases (DNMTs 3a and 3b).[10] DNMT1 differs significantly at the amino terminus where it has numerous protein domains that allow it to interact with multiple additional proteins that are necessary for it to target replication foci while carrying out its function during DNA replication. That is, DNMT1 serves as a maintenance methyltransferase that methylates hemimethylated DNA by adding a methyl group to the complementary strand of DNA. In this way, methylation patterns are faithfully maintained following DNA replication. It is present at high levels in mature, postmitotic GABAergic neurons of the motor and piriform cortices, striatum, CA1 region of the hippocampus, dentate gyrus, and basolateral amygdala of mouse brains.[11] The DNMT3 proteins share sequence and structural similarity and are responsible for adding new (de novo) methylation marks to previously nonmethylated DNA. DNMT3a is predominantly present in postmitotic neurons and oligodendrocytes, whereas DNMT3b is expressed during early stages of neurogenesis.[12] That is, both DNMT1 and DNMT3a are present in mature differentiated neurons, whereas DNMT3b expression is either low or is not detectable in these cells. Mice with conditional knockouts of both DNMT1 and DNMT3a in

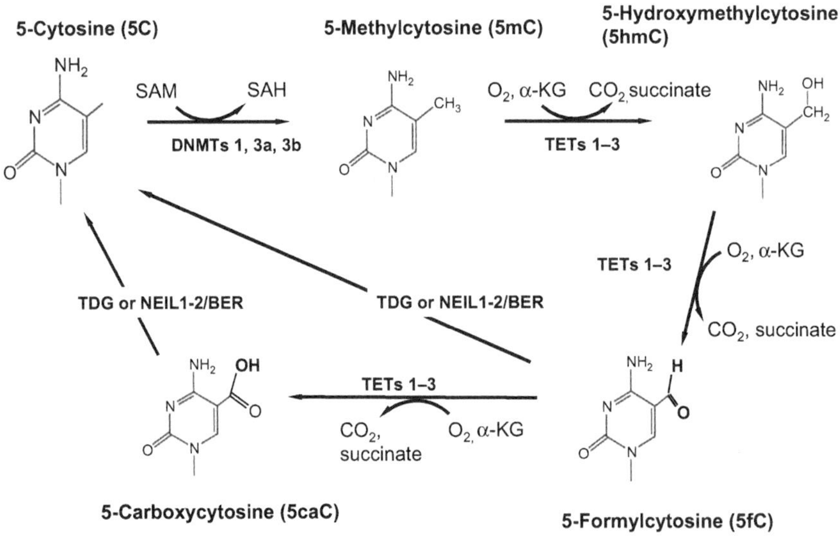

Fig. 2 The DNA demethylation pathway. The DNA methyltransferase enzymes (DNMTs 1, 3a, and 3b) that catalyze DNA methylation utilize the universal methyl donor SAM (S-adenosylmethionine) to transfer a methyl group to the 5 position of the heterocyclic aromatic ring of cytosine (5C). Methylcytosine (5mC) is a repressive chromatin mark that negatively correlates with gene expression. Cytosine demethylation, which often reactivates transcription, occurs through several steps, the first of which involves hydroxylation of the methylated cytosine. DNA hydroxymethylation is catalyzed by members of the TET (1–3) family of heme-containing Fe(II)/α-ketoglutarate-dependent dioxygenases. TET proteins further oxidize the hydroxymethyl group to form 5-formylcytosine (5fC) and 5-carboxylcytosine (5caC). Both 5fC and 5caC can be excised by thymine deglycosylase generating an abasic site which is subsequently replaced by base excision repair (BER) enzymes which regenerate the unmodified cytosine base. Unmodified cytosines (5C) are recognized and bound by proteins such as DNMT1 and TET1 that contain a CXXC motif in their zinc finger DNA-binding domain.[9] Methyl-binding domain proteins, like MeCP2 and MBD2, bind to 5mC with high affinity.

excitatory forebrain neurons show abnormal long-term potentiation, as well as learning and memory deficits.[13] While no neuronal loss was detected in double DNMT1/3a knockouts, these neurons showed substantially deregulated patterns of gene expression.[13] DNMT3l is most similar to the DNMT3 proteins although it is catalytically inactive.[5,6,14] There is evidence showing that DNMT3l binds to the tails of histone H3 and recruits DNMT3a and 3b to methylate DNA.[15] However, while DNMT3l is expressed in the developing brain, it is downregulated during neuronal differentiation and is not abundantly expressed postnatally.[14,16,17]

It was initially reported that DNMT1 mRNA is more highly expressed in GABAergic interneurons of the prefrontal cortex of postmortem SZ cortex (BA 10) than in nonpsychiatric subjects.[17–21] This finding was accompanied by the observation that Reelin (RELN) and glutamic acid decarboxylase 1 (GAD1) mRNAs are downregulated in these same postmortem brain samples. In situ hybridization studies show that these changes occur in the same cortical GABAergic neurons, suggesting a causal link between DNA methylation and the decreased expression of RELN and GAD1. Additional evidence for this regulation was provided by a study using primary neuronal cultures in vitro,[22] in which a DNMT1 antisense oligonucleotide was used to show that the downregulation of DNMT1 protein is coincident with the increased expression of RELN and GAD1 mRNAs. In addition to the increased expression of DNMT1 in the brain, it was also observed in lymphocytes of patients diagnosed with SZ as compared with nonpsychiatric control lymphocytes.[23] Furthermore, using laser-capture microdissection, it was shown that DNMT1 is overexpressed in layer 1 GABAergic neurons of the PFC of SZ subjects.[21] Collectively, these data form the basis for subsequent studies that demonstrate that DNA methylation is increased at selected promoters including RELN in SZ PFC.[24,25]

DNMT1 is the maintenance methyltransferase and it adds a methyl group to hemimethylated DNA to make a double-stranded copy of the methylated residues in the daughter strands. While DNMT1 has been shown to exhibit limited de novo activity, it favors hemimethylated DNA as a substrate over nonmethylated DNA. That is, the intrinsic preference for hemimethylated DNA as a substrate is estimated at 30- to 40-fold.[26] However, siRNA knockdown of DNMT1 do not alter maintenance methylation patterns.[27,28] Moreover, DNMT1 possesses a cysteine-rich CXXC protein domain which allows it to bind to unmethylated C residues with higher affinity than either hemimethylated or methylated DNA.[29,30] The CXXC domain is also required for DNMT enzymatic activity.[29] It seems likely that DNMT1 acts both to methylate DNA and as a general repressor protein repressing transcription in the region to where it is bound. While DNMT1 is usually downregulated in nonproliferating cells, significant expression occurs in the brain where the majority of cells are postmitotic.[31] In the brain, DNMT1 is predominantly expressed in neurons[11] and immunohistochemistry shows it to localize to the cytoplasmic compartment.[31] This observation is particularly curious, as most of the known functions of DNMT1 reside in the nucleus. A clue to this immunohistochemical staining pattern was recently revealed when it was discovered that DNMT1 is encoded by multiple isoforms. That is, upstream and in-frame with the highly conserved amino acid sequence of DNMT1, there is the equivalent of 101 codons downstream of two upstream ATG start codons.[32] These additional start codons are functional producing two isoforms in addition to the highly conserved ATG translational start codon specifying isoform 1.[33] More recently, it was discovered that isoform 3 of DNMT1 is targeted to the mitochondria where it modulates mitochondrial function by methylating the mitochondrial genome.[34] This alternatively spliced isoform lacks the typical nuclear localization signal normally present on mature DNMT1 protein isoforms, which likely accounts, in part, for much of the cytoplasmic DNMT1 antibody staining. In addition to the nuclear localization signal, there are numerous additional protein domains that allow it to interact with histone tails and other proteins in the formation of repressor complexes.[35,36]

1.2.2 DNA Methylation in Schizophrenia

DNA methylation plays an important role during development by epigenetically altering cell-specific patterns of gene expression across distinct cells of the brain. Epigenetic regulation is particularly critical during development because gene expression is dynamic during fetal and infant life and becomes more stable during later periods of the lifespan. It has long been thought that the dysregulation of epigenetic mechanisms plays a vital role in the pathogenesis of psychiatric disorders such as SZ.[6] It is known that external or environmental factors facilitate changes in DNA methylation at both specific regulatory loci and also globally.[37] In addition, a large body of research supports the concept that the development of SZ is dependent on environmental variables especially during fetal and perinatal development (Fig. 1).

Studies of DNA methylation in the brain show that the dorsolateral prefrontal cortex exhibits dramatic changes in DNA methylation during fetal compared with postnatal development.[38,39] These methylation changes are likely due to alterations in neuronal composition that occurs rapidly during early fetal development.[38] That is, the large gene expression changes that accompany early fetal development are also characterized by a decrease in progenitor-like cells and an increase in differentiated adult neurons and non-neuronal cells. These cellular composition profiles explain some of the variability observed in methylation at individual CpGs especially during the transition from prenatal to postnatal life.[37,38] In SZ patients, methylated CpGs associated with this disorder strongly correlate with changes related to the prenatal to postnatal transition.[37] Using the Infinium HumanMethylation450 array, the methylation status of a large number of postmortem SZ subjects and nonpsychiatric controls was evaluated.[40] They identified 920 CpGs that were found to be differentially methylated at a false discovery rate of <0.2, 625 of which changed in the same direction.[40] These findings are complicated by the observation that the majority of SZ subjects were on antipsychotic medications at their time of death. Antipsychotics, such as clozapine, may alter the methylation profile of brain CpGs.[41] Interestingly, nearly a quarter of the significantly differentially methylated CpGs observed in SZ subjects included loci identified from genome-wide association studies.[40] That is, the developmentally associated changes in DNA methylation were significantly enriched in genomic regions that confer inherited risk for SZ.[40] These data, which were derived from 335 SZ patients and 108 normal subjects, support an important role for DNA methylation in the developmental origins of SZ. In addition to genome-wide studies of DNA methylation, there are numerous candidate gene reports showing DNA hyper- and hypomethylation at various selected candidate genes (recently reviewed).[42]

A key question in contemporary psychiatry concerns the extent to which methylation signatures observed in accessible tissues, such as blood, are also present in brain. This question is driven due to the lack of brain tissue availability from recently diagnosed patients. One study that was done to compare the methylation profiles between brain and blood showed that a large proportion of tissue-specific, differentially methylated regions were located near genes involved in functional pathways related to neurodevelopment and neuronal differentiation.[43] They also showed high correlations (~70%) between blood and brain tissues for sites that demonstrate significance between individual variations in DNA methylation.[43] Another study utilized methyl-binding domain enrichment sequencing to compare the profiles of haloperidol treated and untreated mice.[44] They reported that haloperidol treatment leaves biomarker signatures in blood and that the methylation status of many sites in the blood mirrors those in brain.[44] These authors conclude that findings in the blood have value as a proxy for brain tissue. In contrast, another group evaluated DNA methylation from the blood and brain of the same individuals suffering from chronic bouts of epilepsy.[45] These investigators observed that only 7.9% of CpG sites showed a statistically significant, large correlation between blood and brain.[45] A methylation-wide association study (MWAS) of over 700 SZ subjects and 700 nonpsychiatric subjects identified the top MWAS finding to be located near *FAM63B*, a participant in networks regulated by microRNAs linked to neuronal differentiation and dopaminergic gene expression.[46] The authors report that their association studies suggest testable hypotheses relevant to disease mechanisms and yield biomarkers that could prove useful to improve disease management.[46] This report was followed by a second MWAS assessing the methylation status of all common CpG SNPs using brain and blood by methyl domain-binding protein sequencing.[47] Their results indicate that >90% of CpG SNPs that are methylated in brain are also methylated in blood.[47] Collectively, it seems clear that the amount and distribution of 5mC reflect the effects of environmental stressors on the brain. Moreover, this is also reflected in the blood of affected individuals.

1.2.3 Ten-Eleven Translocation Methyldioxygenases

In addition to DNA methylation, DNA hydroxymethylation is a common DNA modification in mammalian cells.[48] In fact, 5hmC represents between 0.3% and 0.7% of total nucleotides with highest levels present in the central nervous system.[49,50] DNA hydroxymethylation is catalyzed by one of three members of the Ten-Eleven Translocation (TET) family of methyl dioxygenases. These enzymes (TETs 1, 2, and 3) are Fe(II)-dependent and use molecular oxygen and α-ketoglutarate to transfer a hydroxyl group to 5mC producing 5-hydroxymethylcytosine (5hmC) (Fig. 2). The expression of TET proteins is cell-type specific, with TET1 being abundant in embryonic stem cells and TETs 2 and 3 showing comparable expression profiles in other tissues.[51,52] Both TET1 and TET3 have cysteine-rich CXXC domains in their structure which are associated with DNA binding and targeting.[30,53]

For many years, 5mC was thought to be thermodynamically stable and the demethylation reaction was thought to be an unfavorable reaction. However, we now know that DNA demethylation is viable and that hydroxymethylation of 5mC is the first step in the pathway. In addition to 5hmC, TET proteins can further oxidize 5hmC forming 5-formylcytosine (5fC) and also 5-carboxylcytosine (5caC).[54] As shown in Fig. 2, these groups are further processed and removed by thymidine DNA deglycosylase (TDG). Depletion of TDG in murine embryonic stem cells leads to the accumulation of detectable levels of 5caC.[55] 5hmC, 5fC, and 5caC are each stable epigenetic marks that accumulate to different extents in the brain. 5hmC is particularly prominent in Purkinje cells of the cerebellum.[29,49] At the same time, 5hmC, 5fC, and 5caC show different abundances and tissue specificities. 5hmC is from 10- to 100-fold more abundant than 5fC/5caC and is enriched in neurons.[54] This is likely due to the preferences of each of the TET proteins for particular substrates and their differential distribution in distinct cell types. Both TET1 and TET2 show a preference for 5mC as substrate as compared with either 5fC or 5caC.[56] Similar to DNMT1, TET3 has a cysteine-rich CXXC domain and it has a high affinity for 5caC to which it binds during 5caC excision by base repair enzymes.[54,57] The observation that these additional epigenetic marks are stable and accumulate to differing extents supports the view that 5hmC, 5fC, and 5caC each play distinct roles in the epigenome.

Thymine deglycosylase is a DNA repair enzyme that removes 5fC and 5caC through the base excision repair (BER) pathway. The iterative oxidation of 5mC forming 5hmC, 5fC, and 5caC is the primary means by which active DNA demethylation of 5mC begins in mammalian cells (Fig. 2). The terminal base is removed by either TDG or members of the prototypical endonuclease VIII (Nei) or Nei-like (NEIL-1-3) proteins. NEIL1 and NEIL2 promote TDG-mediated excision of 5fC and 5caC.[58] While it has been suggested that 5hmC is instead deaminated by activation-induced deaminase (AID) forming 5hmU which is subsequently removed by BER enzymes,[59] no detectable deamination of 5hmC has been found even for cells overexpressing AID/APOBEC.[60] A role for AID/APOBEC enzymes in demethylating 5mC or its oxidized substrates is still unclear. It has recently been shown that while APOBEC3A readily deaminates 5mC, it still has a higher catalytic rate for unmodified cytosine in the identical sequence context. It also has a vastly reduced affinity for 5hmC, 5fC, or 5caC.[61] This group developed sensitive assays for detecting deamination and they showed that APOBEC3A deaminates 5hmC at a 5600-fold lower rate than 5mC.[61] They note that 5fC was modified at an even lower rate while there was no detectable activity for 5caC.[61] Overall, their data suggest that deamination of oxidized 5hmC and other oxidation products is highly unfavorable. It seems instead that the main pathway for DNA demethylation is the iterative oxidation of 5mC to 5fC and 5caC which is then excised by TDG or related enzymes (e.g., NEIL1-3) and replaced by BER enzymes.[54]

1.2.4 Distribution of 5hmC in the Genome

In the previous section, we noted that 5mC is further oxidized by members of the TETenzyme family. 5hmC is a stable epigenetic mark and accumulates to measurable levels in the genome. In the adult human brain, 13.4% of all CpGs are highly hydroxymethylated, which strongly implicates a regulatory role in the modification.[39] While 5mC is symmetrically situated on both strands of DNA in CpG pairs, 5hmC is usually present on only one of two strands. That is, while one strand is hydroxymethylated, the opposing CpG site usually remains methylated indicating that 5hmC/5mCpG is one of the major states for 5hmC in the mammalian genome.[39,62] Interestingly, while 5hmC is generally associated with gene transcription, it is found at high levels in the gene bodies of active genes. However, similar to what has been observed with DNA methylation and transcriptional repression, the relationship between hydroxymethylation and transcription is complex. As one example, genes that are actively transcribed have low levels of 5hmC, whereas genes that are expressed at low levels have high amounts of 5hmC at their promoters. At the same time, the distribution of 5hmC is very different between neural stem cells and embryonic stem cells. It is interesting to note that enrichment of 5hmC at promoters and proximal to transcription start sites most often associates with TET1 and 5hmC located in gene bodies and at exon boundaries correlates with TET2 activity.[63] It appears that many of the studies examining 5hmC abundance and gene expression are not consistent.[54] Moreover, it seems like the effect of 5hmC on gene transcription is both cell-type dependent and context dependent. It has also been shown that there is a transcription correlated bias of 5hmC on the sense strand of the DNA while there is a 5mC bias on the antisense strand.[39]

Whole genome mapping of 5hmC has shown that 5hmC is present in promoters, enhancers, gene bodies, repetitive DNA elements, and intergenically.[39] 5hmC levels are 10-fold higher in the adult brain as compared with the fetal brain. Analysis of the genomic distribution of 5hmC during developmental periods indicates that there are still a large number of CpGs that have higher 5hmC levels in the fetal brain as compared with the same sites in the adult brain. The distribution of these fetal 5hmC sites shows that they are highly enriched in enhancers and not at other genomic features consistent with a role for 5hmC at regulatory domains in the fetal brain.[39] Using TET-assisted bisulfite sequencing, the distribution of 5hmC in the genome was mapped. They showed that there were some 28.4 million 5hmCs in the adult human prefrontal cortex.[39] In general, poised enhancers, which are marked by H3K4me1, have the highest levels of 5hmC, followed by active enhancers (marked by H3K4me1 and H3K27ac), introns, exons, and intergenic regions.[39] Prominent 5hmC peaks are associated with exon-intron boundaries where they mark 5' splice sites.[39,64] Collectively, the results imply novel roles for 5hmC in regulating mRNA splicing and gene regulation. Highly hydroxymethylated CpGs are strongly enriched at multiple genomic features where they comprise ~13% of all CpGs in the genome.

While a considerable amount of information is known regarding hydroxymethylation in the brain, considerably less information is available regarding 5hmC in the context of psychiatric disorders. There is considerable interest in the role of 5hmC in SZ, especially since the report that showed elevated levels of TET1 in the inferior parietal lobule of a group of 9 BP with psychosis and 10 SZ subjects. At the same time, a group of 11 nonpsychiatric controls and 12 major depressed subjects failed to share this increase in TET1 mRNA levels.[65] Interestingly, the CXXC domain containing protein DNMT1 shows increased binding to CpG rich GABAergic promoters in the prefrontal cortex of a cohort of SZ and BP patients obtained from the Harvard Brain Tissue Resource Center.[66] Studies carried out thus far on 5hmC and SZ have been primarily focused on candidate genes associated with the disease. For example, significant increases in 5hmC levels were reported in this same postmortem SZ brain cohort at the GAD1 promoter. In another study, investigators observed increased amounts of both 5mC and 5hmC at BDNF promoter IXabcd in parietal cortical samples of psychotic patients from the Stanley Foundation Neuropathology Consortium.[67] These increases were accompanied by decreased expression of the BDNF IXabcd mRNA levels. While studies thus far implicate a role for aberrant hydroxymethylation in the context of SZ, this is an understudied area of research.

2. METHIONINE MOUSE MODEL OF PSYCHOSIS

2.1 Role of Methionine in Schizophrenia

The idea for the methionine (MET) mouse model of SZ originated with clinical studies that were carried out in the 1960s and 1970s in which SZ subjects were administered large amounts of the amino acid MET along with a monoamine oxidase inhibitor. MET is an essential amino acid and is a methyl group donor which, when administered exogenously, increases the levels of SAM in the brain. The transmethylation hypothesis of SZ was first proposed in 1952,[68] in which it was hypothesized that increases in the abnormal metabolism of dopamine, noradrenaline, and serotonin might be responsible for the psychotic symptoms of the disease.[68] In other words, the disease might be the consequence of the production of psychotomimetic methylated derivatives of catecholamines or indolealkylamines in the brain.[69] Excess MET was expected to increase the activity of enzymes that transmethylate biogenic amine precursors through *O*-methylation.[70] The intention of the 10 clinical trials that were carried out was to alleviate the positive symptoms of patients by decreasing the levels of psychomimetic biogenic amines. However, what happened, instead, was that 64% of participating SZ subjects exhibited a worsening of their symptoms.[70] MET had a profound pharmacologic effect on mood, perception, and function in patients diagnosed with SZ. The response of patients varied in severity and included an increase in psychosis, auditory and visual hallucinations, confusion and disorientation, autonomic responses, and psychomotor retardation. The recrudescence of psychotic symptoms varied in severity and in the numbers of patients responding with adverse symptoms between clinical trials. Of the SZ patients administered MET, two groups were identified, reactors and nonresponders. Reactors responded with an exacerbation of their symptoms, whereas nonresponders were relatively unaffected by the treatments. Interestingly, there were no diagnostic differences between these groups that might help to explain the differential responses.[71]

To study the effects of exogenous METon epigenetic marks in neurons in vitro, we administered a large dose of MET (2 mM) to primary cultures of cortical neurons prepared from embryonic day 16 rat pups.[22] These cultures are mixed, with primarily GABAergic neurons based on the extent of GAD1 immunoreactivity and some glial cells. The addition of MET caused a downregulation of GAD1 and RELN mRNAs with no change in the amount of neuron-specific enolase mRNA.[22] This result was accompanied by an increase in the levels of DNA methylation at the RELN promoter suggesting a mechanism for the reduced expression of the corresponding mRNA. That is, it seemed likely that the downregulation of the RELN mRNA was caused by increased promoter methylation mediated by the action of one of the DNMTs. To test this hypothesis directly, we used an antisense knockdown strategy. To determine whether the increased expression of DNMT1 was responsible for the reduced expression of RELN mRNA, we designed an antisense oligo that was previously used to knockdown DNMT1 mRNA in various cancers in mice ex vivo.[72] The antisense was introduced into primary cortical neurons and the primary cultures were treated with MET. The downregulation of DNMT1 mRNA and protein was confirmed by qRT-PCR and western blotting, respectively. The reduced expression of DNMT1 was accompanied by an increase in both RELN and GAD1 mRNAs in the MET-treated cultures. These data were used to provide a mechanistic hypothesis regarding the recrudescence of psychotic symptoms in the MET-treated SZ subjects. We proposed the hypothesis that MET treatment of SZ subjects facilitates changes in the genome-wide methylation profiles of genes actively participating in GABAergic and glutamatergic neurons of the brain such that the expression of genes involved in neuronal function is downregulated in response to the amino acid.[73]

2.2 Methionine Mouse Model of Schizophrenia

It has been known for some time that both dietary and subcutaneously administered MET is converted to SAM by the enzyme MET adenosyltransferase 2A (MAT2A). SAM is the universal methyl donor in virtually all living cells.[73] Our group had also recently shown that SAM levels in the prefrontal cortex (BA9) of patients with SZ and BP are increased nearly twofold when compared with the levels in nonpsychiatric controls.[74] The subcutaneous injection of a large dose of MET (5.2 mmol/kg, twice daily for 5 days) in mice facilitates a more than twofold increase in both SAM and SAH levels.[75] Moreover, the increased levels of SAM correlate with an increase in DNMT1 mRNA levels in GABAergic neurons of the frontal cortex.[75] Preliminary experiments were used to establish the MET dose needed to yield the maximal response in terms of elevated levels of SAM and SAH for the model which was 6.6 mmol/kg (twice daily) of MET. The brain SAM and SAH levels increased with increasing MET dose and time after subcutaneous injection.[75]

2.2.1 Behavioral Phenotype

We showed that prepulse inhibition (PPI) of startle, which represents a measure of sensory gating in both humans and rodents, decreased at a faster rate as the prepulse/startle interval increased in MET-treated mice compared with saline-treated mice. In a follow-up to the aforementioned study, the behaviors associated with chronic MET treatment in mice were expanded to include measures of social interaction. The chronic administration of MET to mice not only impairs sensory gating (PPI), but also induces a deficit in social interaction time in treated animals.[76] The deficit in social interaction is reminiscent of the negative symptoms of SZ.[77] In addition, chronic injection of MET into mice prevents the development of aggressive behavior toward intruder mice following a 2–3-week period of social isolation.[76] In this latter study, VPA when simultaneously administered with MET was shown to prevent both the PPI and social interaction deficits.

More recently, the MET-induced mouse model of SZ was revisited by investigators that showed several additional behaviors associated with the model.[77] First, they not only replicated the deficit in sensory gating (PPI) and the social interaction deficit observed by others, but also showed that these mice do not exhibit any pathological effects due to the treatment. They also showed that chronically treated mice show greater levels of stereotyped behaviors. In the forced swim test, MET-treated mice display more immobility time than saline-treated animals indicating a lower level of motivated affection. Using the novel object recognition test, MET-treated mice exhibit impaired recognition. Finally, in the inhibitory avoidance assay, while there were no differences in latency to enter the dark chamber, following 48 h, MET-treated mice displayed an increase in the latency time to enter the dark chamber and a decreased number of animals avoided entering the dark chamber.[77] The latter two tests are indicative of impaired cognition. The chronic MET treatment had no effect on body weight and did not alter their nociceptive response as measured by the tail flick assay.[77] Comparison of both behavioral and biochemical deficits in the MET mouse model of SZ with SZ-like behaviors are listed in Table 1.

Table 1 Comparison of Molecular and Behavioral Abnormalities in SZ and BP Disorder Patients With the Mouse MET and PRS Models of Psychiatric Disease.

	SZ + BP Patients	MET Mice	PRS Mice
Molecular changes			
DNMT1, DNMT3a, TET expression	↑	↑	↑
DNMT1 or MECP2 binding to GAD1, RELN, BDNF promoters	↑	↑	↑
5mC, 5hmC enrichment at RELN or GAD1	↑	↑	↑
Candidate gene (GAD1, RELN, BDNF) expression	↓	↓	↓
Behavioral changes			
Positive symptoms (stereotyped behavior, sensitivity to NMDA receptor antagonist)	↑	↑	↑
Negative symptoms (social interaction deficits)	↑	↑	↑
Cognitive, information processing deficit (PPI, fear conditioning)	↑	↑	↑

2.2.2 Biochemical Phenotype

A number of different assays were used to examine aspects of chromatin architecture in the vicinity of candidate genes relevant to SZ pathophysiology. One of the first experiments performed was to determine whether RELN and GAD1 mRNAs were reduced in MET-treated mice. Both RELN and GAD1 are reduced in this model. Another study showed that the downregulation of RELN and GAD1 was accompanied by a loss of layer III pyramidal cell dendritic spin density in the frontal cortex of treated animals.[78] Using bisulfite genomic sequencing, it was next shown that the region upstream of the mouse RELN promoter was more heavily methylated, by nearly twofold, in MET-treated mice as compared with saline-treated mice.[75] No change was observed in RELN promoter methylation status following a comparable injection of the amino acid glycine. Finally, coinjection of the mood stabilizer and histone deacetylase (HDAC) inhibitor, VPA, reverted the MET-induced downregulation of RELN and GAD1 through an epigenetic action involving HDACs.[75] The same dose of VPA increased H3 histone acetylation in mouse brain nearly fourfold.[75] The latter experiment suggested the possibility that the inhibition of HDACs might represent a suitable strategy for alleviating the epigenetic risk of individuals genetically susceptible to SZ.

Some of the aforementioned data were replicated in a separate study that showed that in addition to hypermethylation of the RELN and GAD1 promoters, MET treatment facilitates a recruitment of MeCP2 to the corresponding promoter regions in the frontal cortex of treated mice.[79] No change in promoter 5mC levels at the GAD2 promoter was detected indicating a degree of gene specificity to the response. The MET-induced RELN and GAD1 mRNA downregulation is reversed after 7 days of withdrawal. Moreover, the recruitment of MeCP2 to the RELN and GAD1 promoter is time dependent and is reversed nearly completely after 6–9 days following MET withdrawal.[80] Similar to previous studies, VPA and the chemically unrelated benzamide HDAC inhibitor, MS275,[81] reverses the RELN and GAD1 mRNA downregulation and promoter hypermethylation.[80] VPA and MS275 have overlapping but distinct inhibitory profiles and different IC50s at individual Class I HDACs. Finally, the DNMT1 inhibitor, procainamide,[82] when administered simultaneously with MET, blocks the accumulation of MeCP2 present at the RELN and GAD1 promoters. These data expand the existing knowledge of MET-induced epigenetic changes at the RELN and GAD1 promoters by providing information regarding the reversibility of these effects. In addition, the studies with inhibitors now include the HDAC inhibitor MS275 and the DNMT1 inhibitor procainamide.

2.2.3 Responsiveness of the Methionine Model to Medications

To test whether VPA used in combination with typical or atypical antipsychotics might reverse the RELN and GAD1 mRNA decrease and promoter hypermethylation, clozapine (CLZ), sulpiride (SULP), haloperidol (HAL), and olanzapine (OLZ) were tested in MET-treated mice.[83] CLZ, SULP, HAL, and OLZ were administered to mice following 7 days of MET treatment. Results showed that doses of CLZ and SULP, but not HAL or OLZ, facilitated a demethylation of the MET-mediated promoter hypermethylation at the RELN and GAD1 promoters in the frontal cortex and striatum.[83] Interestingly, the action of these atypical antipsychotics was potentiated by the coadministration of a relevant dose of VPA.[83] Neither HAL nor OLZ alone or in combination with VPA were able to elicit demethylation at these two promoters or to alter the behavioral deficits. This contrasts with another study that showed that a single injection of HAL and CLZ reverses the locomotor hyperactivity and stereotyped behavior induced by chronic MET administration.[77] While these differences could be due to the doses used in the respective studies, only the more recent study examined sterotyped behaviors. Nevertheless, the chronic MET mouse model of SZ has face and predictive validity and may be useful in the preclinical search for new neuroleptic drugs.

3. PRENATAL RESTRAINT STRESS MODEL OF SCHIZOPHRENIA

As indicated earlier, SZ is characterized by an abnormal genetic load and an environmental insult that is necessary for onset of the disease. Stressors associated with gestation include malnutrition, premature birth, complications during delivery, and so on (Fig. 1). Animal studies confirm the sensitivity of the developing brain to environmental disturbances. Exposure of rats and mice to stresses, immune challenges, infection, and malnutrition during pregnancy leads to a disruption of various behavioral and neurochemical parameters in adult offspring.[84] The effects of chronic hyperactivation of the hypothalamus–pituitary–adrenal axis in fetal rodents have been examined for a number of years now.[85] Prenatal restraint stress in mice has been shown to produce delays in the tangential migration of inhibitory neuron progenitor cells of the developing neocortex.[86] In addition, prenatal stress has been shown to impair the emotional responses to subsequent stress implicating a reduced ability to adapt to new stresses.[87] This impaired adaptive response is linked to a disruption of the development of serotonergic neurons in the embryonic hindbrain.[87] Moreover, chronic prenatal stress also affects lymphoid cells and alters neurotrophin and cytokine expression in lymphocytes.[88] Collectively, prenatal stress affects multiple systems in the developing animal including long-lasting neurobiological and behavioral changes that are linked to neuroplasticity. Because different strains of mice respond to prenatal stress differently, it is important to test behavior in each strain at the start of experimentation.

Chronic overactivity of the HPA axis produces effects that include the repression of reproduction, growth, as well as thyroid and immune functions, which have been shown to be associated with pathological states.[85,89] In addition, exposure of the brain to prolonged periods of stress in utero likely results in the hyperactivity of the stress system, defective negative feedback by glucocorticoids, alterations in cognition, novelty seeking, increased vulnerability to addictive behaviors, and mood disorders.[85] Moreover, mice that have been subjected to chronic prenatal stress show evidence of behaviors reminiscent of the positive and negative symptoms of SZ.[90]

3.1 Behavioral Phenotype of PRS Mice

The adult offspring of dams exposed to repeated episodes of restraint stress during gestation (termed PRS mice) develop a SZ-like endophenotype. While at birth the PRS mice exhibit 10% reduced body weight and are born from 12 to 18 h earlier than nonstressed (NS) mice, there are no differences in body weight, auditory sensitivity, pain sensitivity, or motor coordination at adulthood.[84] Swiss-albino ND4 mice, prenatally stressed from gestation days 7 to 21, show a robust and persistent hyperlocomotor activity as measured by horizontal activity.[84] The increase observed in locomotor activity was recently replicated in another study.[91,92] In addition, PRS mice exhibited a decrease in social interactions with an intruder mouse.[84] More recently, the social interaction deficit was also replicated in a similar study.[91,92] Social withdrawal is a negative symptom of SZ. PRS mice also show deficits in PPI, which reflects sensory information processing or sensory gating.[84] PRS mice demonstrate a significant deficit in fear conditioning response in contextual fear conditioning tests.[84] Moreover, PRS mice exhibit increased stereotyped behaviors which is sensitive to the administration of the NMDA subtype of glutamate receptor.[84] Collectively, the behavioral deficits in PRS mice are representative of behaviors observed in patients diagnosed with SZ and BP (Table 1).

3.2 Biochemical Phenotype of PRS Mice

To better understand the underlying mechanism associated with the altered epigenetic framework in adult PRS mice, experiments were performed early in development to examine the expression of DNMT1 and DNMT3a during development following prenatal stress.[84] Examination of the postnatal mRNA profile of DNMT1 and 3a in the frontal cortex shows that both mRNAs are highest at postnatal day 1 (PD1). Both mRNAs show a robust decrease by PD7 which continues until PD14. There is a gradual decline thereafter until adulthood (PD60) at which point, expression of both mRNAs is lowest.[84] NS mice show a similar temporal decline in mRNAs but at each time point, mRNA levels in NS mice are about 50% of that in the PRS mice. The levels of both mRNAs were also about half of that observed in the PRS mouse hippocampus.[84] These results suggest that both DNMTs are overexpressed in the frontal cortex and hippocampus, and this suggests that the levels of DNA methylation are temporally increased in PRS mice as well.

Epigenetic studies of adult PRS mice at the biochemical level show that subtypes of metabotropic glutamate receptors (mGluR) are downregulated in PRS adult offspring. In one study, it was shown that the downregulation of mGluR2 and mGluR3 in the frontal cortex of PRS mice is linked with an increased expression of DNMT1.[93] There was increased DNMT1 mRNA and protein in the frontal cortex and increased DNMT1 binding to the mGluR2 and mGluR3 promoters in the frontal cortex of PRS mice.[93] In addition, MeCP2 binding to these promoters increased in PRS mice suggesting that the increased DNMT1 binding was accompanied by increased levels of DNA methylation.[93] It was further established that the GABAergic neuron-specific GAD1 and the neurotrophin BDNF-IX mRNAs were downregulated.[93] It was also shown that there was increased binding of MeCP2 to the corresponding GAD1 and BDNF-IX promoter regions.[93]

In a follow-up study, these investigators extended their observations to include RELN and GAD1 expression.[84] Similar to what was observed with the mGluR mRNAs, RELN and GAD1 mRNAs were downregulated and there was increased binding of DNMT1 and MeCP2 to the corresponding promoter regions. The levels of 5mC and 5hmC at these two promoters were elevated in the PRS frontal cortex relative to NS mice.[84] In addition to elevated DNMT1 mRNA, a time-dependent increase in the levels of TET1 mRNA in the frontal cortex and hippocampus of PRS mice was noted.[91] This was accompanied by decreased amounts of BDNF I–IV, VI, and IX in the frontal cortex and BDNF III, IV, VI, and IX mRNA transcripts in the hippocampus of PRS mice.[91] The levels of 5mC and 5hmC were similarly increased at each of the BDNF promoter regions. In addition, the decrease in corresponding BDNF transcript levels was accompanied by an enrichment of 5mC and 5hmC at the respective promoters.[91] Interestingly, the expression of BDNF transcripts IV and IX positively correlates with the social interaction deficits observed.[91] Lastly, they observed that there was increased binding of DNMT1 to BDNF promoters III, IV, VI, and IX. These studies provide a compelling set of data for comparison with the epigenetic profile derived from postmortem human SZ brain (Table 1).

3.3 Responsiveness of the PRS Mouse to Medications

Initial studies of the PRS model were designed to test the efficacy of the preclinical compound LY379268, which was structurally related to the compound Eli Lilly was testing in phase III clinical trials for treatment of SZ, LY2140023. This potent and selective mGlu2/3 receptor antagonist, LY379268, was originally synthesized by Jim Monn and Darryl Schoep at Eli Lilly and Company.[94] Systemic administration of LY379268 corrected all of the behavioral and biochemical deficits exhibited by PRS mice. LY2140023 was recently tested in a three-armed clinical trial against placebo and Zyprexa. A large body of preclinical data support a role for mGluR agonists as new antipsychotic drugs for the symptomatic relief from SZ.[95] Unfortunately, the mGluR agonist failed to outperform the placebo, which actually outperformed Zyprexa, Lilly's currently marketed SZ drug.

In addition to mGluR agonists, various epigenetic drugs have been used to evaluate their utility in treating the PRS mouse. For example, VPA was shown to correct the stereotyped behavior in PRS mice.[84] This result was anticipated as VPA has the same effect in the MET-treated mouse model of SZ. While CLZ corrects the behavioral phenotype of PRS mice, HAL fails to do so. Moreover, CLZ normalizes the biochemical phenotype in PRS mice, including the BDNF IV, VI and IX, RELN, and GAD1 mRNA downregulation, the elevated 5mC and 5hmC levels, % MeCP2 binding, and DNMT1 binding at corresponding promoters.[92] In addition, CLZ also normalizes the elevated DNMT1 and TET1 mRNA levels in PRS mice, whereas HAL fails to do the same. Collectively, CLZ, the gold standard in antipsychotic therapy,[96] corrects both the behavioral and epigenetic alterations associated with PRS mice. CLZ is the only antipsychotic medication capable of attenuating the epigenetic components of psychiatric disorders.

The PRS model in rats, and presumably also in mice, shows sexual dimorphism in terms of the outcome of the offspring. Most studies that investigate PRS mice generally use male mice in their studies. Males exhibit a pronounced response, whereas females often show opposing responses.[97] Where male rats show increased anxiety, decreased neurogenesis in the hippocampal dentate gyrus, reduced activity of mGlu1/5 activity, and increased BDNF in the hippocampus, females display reduced anxiety, improved learning in the Morris water maze, increased activity of mGlu1/5 receptor activity in the hippocampus, and no changes in hippocampal neurogenesis or BDNF levels.[97] These data strongly suggest that epigenetic changes in the hippocampus of PRS rats induced by early life environmental perturbations are dependent on sex and that the behavioral effects diverge between males and females. Moreover, the effects of gender in epigenetic models of psychiatric disorders need to be examined in more detail in future studies.

4. CONCLUSIONS

SZ is a debilitating disorder that has both genetic and epigenetic components that have a role in its pathophysiology. Both the MET mouse and PRS mouse mimic the positive, negative, and cognitive aspects of SZ. The data showing that CLZ is able to reverse both the behavioral and biochemical deficits associated with these models indicate that this drug may serve as a model compound in the development of new drugs that can normalize the epigenetic aspects of these disorders. Having said this, we recognize that the mechanistic details of CLZ action in reversing these endophenotypes is still incompletely understood.

ACKNOWLEDGMENT

This work is supported in part by grant 5P50 AA022538 to DRG and AG.

REFERENCES

1. Schizophrenia Working Group of the Psychiatric Genomics Consortium.Biological insights from 108 schizophrenia-associated genetic loci. *Nature*. 2014;511:421–427.
2. Grayson DR. Schizophrenia and the epigenetic hypothesis. *Epigenomics*. 2010;2: 341–344.

3. Millan MJ. An epigenetic framework for neurodevelopmental disorders: from pathogenesis to potential therapy. *Neuropharmacology*. 2013;68:2–82.
4. Uher R, Zwicker A. Etiology in psychiatry: embracing the reality of poly-gene-environmental causation of mental illness. *World Psychiatry*. 2017;16:121–129.
5. Jurkowska RZ, Jurkowski TP, Jeltsch A. Structure and function of mammalian DNA methyltransferases. *Chembiochem*. 2011;12:206–222.
6. Grayson DR, Guidotti A. The dynamics of DNA methylation in schizophrenia and related psychiatric disorders. *Neuropsychopharmacology*. 2013;38:138–166.
7. Bayraktar G, Kreutz MR. Neuronal DNA methyltransferases: epigenetic mediators between synaptic activity and gene expression? *Neuroscientist*. 2017. [Epub ahead of print].
8. Jang HS, Shin WJ, Lee JE, Do JT. CpG and non-CpG methylation in epigenetic gene regulation and brain function. *Genes (Basel)*. 2017;8(6):148.
9. Long HK, Blackledge NP, Klose RJ. ZF-CxxC domain-containing proteins, CpG islands and the chromatin connection. *Biochem Soc Trans*. 2013;41:727–740.
10. Edwards JR, Yarychkivska O, Boulard M, Bestor TH. DNA methylation and DNA methyltransferases. *Epigenetics Chromatin*. 2017;10:23.
11. Kadriu B, Guidotti A, Chen Y, Grayson DR. DNA methyltransferases1 (DNMT1) and 3a (DNMT3a) colocalize with GAD67-positive neurons in the GAD67-GFP mouse brain. *J Comp Neurol*. 2012;520:1951–1964.
12. Watanabe D, Uchiyama K, Hanaoka K. Transition of mouse de novo methyltransferases expression from Dnmt3b to Dnmt3a during neural progenitor cell development. *Neuroscience*. 2006;142:727–737.
13. Feng J, Zhou Y, Campbell SL, et al. Dnmt1 and Dnmt3a maintain DNA methylation and regulate synaptic function in adult forebrain neurons. *Nat Neurosci*. 2010;13:423–430.
14. Moore LD, Thuc L, Fan G. DNA methylation and its basic function. *Neuropsychopharmacology*. 2013;38:23–38.
15. Ooi SK1. Qiu C, Bernstein E, et al. DNMT3L connects unmethylated lysine 4 of histone H3 to de novo methylation of DNA. *Nature*. 2007;448:714–717.
16. Lee MS, Jun DH, Hwang CI, et al. Selection of neural differentiation-specific genes by comparing profiles of random differentiation. *Stem Cells*. 2006;24:1946–1955.
17. Kovacheva VP, Mellott TJ, Davison JM, et al. Gestational choline deficiency causes global and Igf2 gene DNA hypermethylation by up-regulation of Dnmt1 expression. *J Biol Chem*. 2007;282:31777–31788.
18. Veldic M, Caruncho HJ, Liu WS, et al. DNA-methyltransferase 1 mRNA is selectively overexpressed in telencephalic GABAergic interneurons of schizophrenia brains. *Proc Natl Acad Sci USA*. 2004;101:348–353.
19. Veldic M, Guidotti A, Maloku E, Davis JM, Costa E. In psychosis, cortical interneurons overexpress DNA-methyltransferase 1. *Proc Natl Acad Sci USA*. 2005;102:2152–2157.
20. Veldic M, Kadriu B, Maloku E, et al. Epigenetic mechanisms expressed in basal ganglia GABAergic neurons differentiate schizophrenia from bipolar disorder. *Schizophr Res*. 2007;91:51–61.
21. Ruzicka WB, Zhubi A, Veldic M, Grayson DR, Costa E, Guidotti A. Selective epigenetic alteration of layer I GABAergic neurons isolated from prefrontal cortex of schizophrenia patients using laser-assisted microdissection. *Mol Psychiatry*. 2007;12:385–397.
22. Noh JS, Sharma RP, Veldic M, et al. DNA methyltransferase 1 regulates reelin mRNA expression in mouse primary cortical cultures. *Proc Natl Acad Sci USA*. 2005;102:1749–1754.
23. Zhubi A, Veldic M, Puri NV, et al. An upregulation of DNA-methyltransferase 1 and 3a expressed in telencephalic GABAergic neurons of schizophrenia patients is also detected in peripheral blood lymphocytes. *Schizophr Res*. 2009;111:115–122.
24. Grayson DR, Jia X, Chen Y, et al. Reelin promoter hypermethylation in schizophrenia. *Proc Natl Acad Sci USA*. 2005;102:9341–9346.

25. Abdolmaleky HM, Cheng KH, Russo. et al. Hypermethylation of the reelin (RELN) promoter in the brain of schizophrenic patients: a preliminary report. *Am J Med Genet B Neuropsychiatr Genet.* 2005;134B:60–66.
26. Jeltsch A. On the enzymatic properties of Dnmt1: specificity, processivity, mechanism of linear diffusion and allosteric regulation of the enzyme. *Epigenetics.* 2006;1:63–66.
27. Ting AH, Jair KW, Schuebel KE, Baylin SB. Differential requirement for DNA methyltransferase 1 in maintaining human cancer cell gene promoter hypermethylation. *Cancer Res.* 2006;66:729–735.
28. Ting AH, Jair KW, Suzuki H, Yen RW, Baylin SB, Schuebel KE. CpG island hypermethylation is maintained in human colorectal cancer cells after RNAimediated depletion of DNMT1. *Nat Genet.* 2004;36:582–584.
29. Pradhan M, Estève PO, Chin HG, Samaranayke M, Kim GD, Pradhan S. CXXC domain of human DNMT1 is essential for enzymatic activity. *Biochemistry.* 2008;47: 10000–10009.
30. Frauer C, Rottach A, Meilinger D, et al. Different binding properties and function of CXXC zinc finger domains in Dnmt1 and Tet1. *PLoS One.* 2011;6:e16627.
31. Shock LS, Thakkar PV, Peterson EJ, Moran RG, Taylor SM. DNA methyltransferase 1, cytosine methylation, and cytosine hydroxymethylation in mammalian mitochondria. *Proc Natl Acad Sci USA.* 2011;108:3630–3635.
32. Inano K, Suetake I, Ueda T, et al. Maintenance-type DNA methyltransferase is highly expressed in post-mitotic neurons and localized in the cytoplasmic compartment. *J Biochem.* 2000;128:315–321.
33. Yoder JA, Yen RW, Vertino PM, Bestor TH, Baylin SB. New 5' regions of the murine and human genes for DNA (cytosine-5)-methyltransferase. *J Biol Chem.* 1996;271:31092–31097.
34. Saini SK, Mangalhara KC, Prakasam G, Bamezai RNK. DNA Methyltransferase1 (DNMT1) Isoform3 methylates mitochondrial genome and modulates its biology. *Sci Rep.* 2017;7:1525.
35. Tajima S, Suetake I, Takeshita K, Nakagawa A, Kimura H. Domain structure of the Dnmt1, Dnmt3a, and Dnmt3b DNA methyltransferases. *Adv Exp Med Biol.* 2016;945: 63–86.
36. Kar S, Deb M, Sengupta D, et al. An insight into the various regulatory mechanisms modulating human DNA methyltransferase 1 stability and function. *Epigenetics.* 2012;7:994–1007.
37. Jaffe AE, Gao Y, Deep-Soboslay A, et al. Mapping DNA methylation across development, genotype and schizophrenia in the human frontal cortex. *Nat Neurosci.* 2016;19:40–47.
38. Jaffe AE, Shin J, Collado-Torres L, et al. Developmental regulation of human cortex transcription and its clinical relevance at single base resolution. *Nat Neurosci.* 2015;18:154–161.
39. Wen L, Li X, Yan L, Tan Y, et al. Whole-genome analysis of 5-hydroxymethylcytosine and 5-methylcytosine at base resolution in the human brain. *Genome Biol.* 2014;15:R49.
40. Montano C, Taub MA, Jaffe A, et al. Association of DNA methylation differences with schizophrenia in an epigenome-wide association study. *JAMA Psychiatry.* 2016;73:506–514.
41. Guidotti A, Dong E, Grayson DR. Epigenetic basis of clozapine action. *J Drug Des Res.* 2017;4:1055.
42. Pries LK, Gülöksüz S, Kenis G. DNA methylation in schizophrenia. *Adv Exp Med Biol.* 2017;978:211–236.
43. Davies MN, Volta M, Pidsley R, et al. Functional annotation of the human brain methylome identifies tissue-specific epigenetic variation across brain and blood. *Genome Biol.* 2012;13:R43.

44. Aberg KA, Xie LY, McClay JL, et al. Testing two models describing how methylome-wide studies in blood are informative for psychiatric conditions. *Epigenomics.* 2013;5:367–377.
45. Walton E, Hass J, Liu J, et al. Correspondence of DNA methylation between blood and brain tissue and its application to schizophrenia research. *Schizophr Bull.* 2016;42: 406–414.
46. Aberg KA, McClay JL, Nerella S, et al. Methylome-wide association study of schizophrenia: identifying blood biomarker signatures of environmental insults. *JAMA Psychiatry.* 2014;71:255–264.
47. van den Oord EJ, Clark SL, Xie LY, et al. A whole methylome CpG-SNP association study of psychosis in blood and brain tissue. *Schizophr Bull.* 2016;42:1018–1026.
48. Williams K, Christensen J, Helin K. DNA methylation: TET proteins-guardians of CpG islands? *EMBO Rep.* 2011;13:28–35.
49. Kriaucionis S, Heintz N. The nuclear DNA base 5-hydroxymethylcytosine is present in Purkinje neurons and the brain. *Science.* 2009;324:929–930.
50. Globisch D, Münzel M, Müller M, et al. Tissue distribution of 5-hydroxymethylcytosine and search for active demethylation intermediates. *PLoS One.* 2010;5:e15367.
51. Ito S, Shen L, Dai Q, et al. Tet proteins can convert 5-methylcytosine to 5-formylcytosine and 5-carboxylcytosine. *Science.* 2011;333:1300–1303.
52. Nestor CE, Ottaviano R, Reddington J, et al. Tissue type is a major modifier of the 5-hydroxymethylcytosine content of human genes. *Genome Res.* 2012;22:467–477.
53. Xu Y, Xu C, Kato A, et al. Tet3 CXXC domain and dioxygenase activity cooperatively regulate key genes for Xenopus eye and neural development. *Cell.* 2012;151:1200–1213.
54. Shi DQ, Ali I, Tang J, Yang WC. New insights into 5hmC DNA modification: generation, distribution and function. *Front Genet.* 2017;8:100.
55. He YF, Li BZ, Li Z, et al. Tet-mediated formation of 5-carboxylcytosine and its excision by TDG in mammalian DNA. *Science.* 2011;333:1303–1307.
56. Hu L, Lu J, Cheng J, et al. Structural insight into substrate preference for TET-mediated oxidation. *Nature.* 2015;527:118–122.
57. Jin SG, Zhang ZM, Dunwell TL, et al. Tet3 reads 5-carboxylcytosine through its CXXC domain and is a potential guardian against neurodegeneration. *Cell Rep.* 2016;14:493–505.
58. Spruijt CG, Gnerlich F, Smits AH, et al. Dynamic readers for 5-(hydroxy)methylcytosine and its oxidized derivatives. *Cell.* 2013;152:1146–1159.
59. Hendrich B, Hardeland U, Ng HH, Jiricny J, Bird A. The thymine glycosylase MBD4 can bind to the product of deamination at methylated CpG sites. *Nature.* 1999;401: 301–304.
60. Nabel CS, Jia H, Ye Y, et al. AID/APOBEC deaminases disfavor modified cytosines implicated in DNA demethylation. *Nat Chem Biol.* 2012;8:751–758.
61. Schutsky EK, Nabel CS, Davis AKF, DeNizio JE, Kohli RM. APOBEC3A efficiently deaminates methylated, but not TET-oxidized, cytosine bases in DNA. *Nucleic Acids Res.* 2017;45:7655–7665.
62. Song CX, Diao J, Brunger AT, Quake SR. Simultaneous single-molecule epigenetic imaging of DNA methylation and hydroxymethylation. *Proc Natl Acad Sci USA.* 2016;113:4338–4343.
63. Huang Y, Chavez L, Chang X, et al. Distinct roles of the methylcytosine oxidases Tet1 and Tet2 in mouse embryonic stem cells. *Proc Natl Acad Sci USA.* 2014;111:1361–1366.
64. Khare T, Pai S, Koncevicius K, et al. 5-hmC in the brain is abundant in synaptic genes and shows differences at the exon-intron boundary. *Nat Struct Mol Biol.* 2012;19:1037–1043.
65. Dong E, Gavin DP, Chen Y, Davis J. Upregulation of TET1 and downregulation of APOBEC3A and APOBEC3C in the parietal cortex of psychotic patients. *Transl Psychiatry.* 2012;2:e159.

66. Dong E, Ruzicka WB, Grayson DR, Guidotti A. DNA-methyltransferase1 (DNMT1) binding to CpG rich GABAergic and BDNF promoters is increased in the brain of schizophrenia and bipolar disorder patients. *Schizophr Res.* 2015;167:35–41.
67. Gavin DP, Sharma RP, Chase KA, Matrisciano F, Dong E, Guidotti A. Growth arrest and DNA-damage-inducible, beta (GADD45b)-mediated DNA demethylation in major psychosis. *Neuropsychopharmacology.* 2012;37:531–542.
68. Osmond H, Smythies J. Schizophrenia: a new approach. *J Ment Sci.* 1952;98:309–315.
69. Smythies JR. The transmethylation hypotheses of schizophrenia re-evaluated. *Trends Neurosci.* 1984;7:45–47.
70. Grayson DR, Chen Y, Dong E, Kundakovic M, Guidotti A. From trans-methylation to cytosine methylation: evolution of the methylation hypothesis of schizophrenia. *Epigenetics.* 2009;4:144–149.
71. Antun FJ, Burnett GB, Cooper AJ, Daly RJ, Smythies JR, Zealley AK. The effect of l-methionine (without MAOI) in schizophrenia. *J Psychiatr Res.* 1971;8:63–71.
72. Szyf M. Utilization of antisense oligonucleotides to study the role of 5-cytosine DNA methyltransferase in cellular transformation and oncogenesis. *Methods.* 2002;27: 184–191.
73. Costa E, Chen Y, Dong E, et al. GABAergic promoter hypermethylation as a model to study the neurochemistry of schizophrenia vulnerability. *Expert Rev Neurother.* 2009;9:87–98.
74. Guidotti A, Ruzicka W, Grayson DR, et al. S-adenosyl methionine and DNA methyltransferase-1 mRNA overexpression in psychosis. *Neuroreport.* 2007;18:57–60.
75. Tremolizzo L, Carboni G, Ruzicka WB, et al. An epigenetic mouse model for molecular and behavioral neuropathologies related to schizophrenia vulnerability. *Proc Natl Acad Sci USA.* 2002;99:17095–17100.
76. Tremolizzo L, Doueiri MS, Dong E, et al. Valproate corrects the schizophrenia-like epigenetic behavioral modifications induced by methionine in mice. *Biol Psychiatry.* 2005;57:500–509.
77. Wang L, Alachkar A, Sanathara N, Belluzzi JD, Wang Z, Civelli O. A methionine-induced animal model of schizophrenia: face and predictive validity. *Int J Neuropsychopharmacol.* 2015;18:pyv054.
78. Tueting P, Davis JM, Veldic M, et al. L-Methionine decreases dendritic spine density in mouse frontal cortex. *Neuroreport.* 2010;21:543–548.
79. Dong E, Agis-Balboa RC, Simonini MV, Grayson DR, Costa E, Guidotti A. Reelin and glutamic acid decarboxylase67 promoter remodeling in an epigenetic methionine-induced mouse model of schizophrenia. *Proc Natl Acad Sci USA.* 2005;102:12578–12583.
80. Dong E, Guidotti A, Grayson DR, Costa E. Histone hyperacetylation induces demethylation of reelin and 67-kDa glutamic acid decarboxylase promoters. *Proc Natl Acad Sci USA.* 2007;104:4676–4681.
81. Kundakovic M, Chen Y, Guidotti A, Grayson DR. The reelin and GAD67 promoters are activated by epigenetic drugs that facilitate the disruption of local repressor complexes. *Mol Pharmacol.* 2009;75:342–354.
82. Lee BH, Yegnasubramanian S, Lin X, Nelson WG. Procainamide is a specific inhibitor of DNA methyltransferase 1. *J Biol Chem.* 2005;280:40749–40756.
83. Dong E, Nelson M, Grayson DR, Costa E, Guidotti A. Clozapine and sulpiride but not haloperidol or olanzapine activate brain DNA demethylation. *Proc Natl Acad Sci USA.* 2008;105:13614–13619.
84. Matrisciano F, Tueting P, Dalal I, et al. Epigenetic modifications of GABAergic interneurons are associated with the schizophrenia-like phenotype induced by prenatal stress in mice. *Neuropharmacology.* 2013;68:184–194.
85. Maccari S, Morley-Fletcher S. Effects of prenatal restraint stress on the hypothalamus-pituitary-adrenal axis and related behavioural and neurobiological alterations. *Psychoneuroendocrinology.* 2007;32(Suppl. 1):S10–S15.

86. Stevens HE, Su T, Yanagawa Y, Vaccarino FM. Prenatal stress delays inhibitory neuron progenitor migration in the developing neocortex. *Psychoneuroendocrinology.* 2013;38: 509–521.
87. Miyagawa K, Tsuji M, Ishii D, Takeda K, Takeda H. Prenatal stress induces vulnerability to stress together with the disruption of central serotonin neurons in mice. *Behav Brain Res.* 2015;277:228–236.
88. Pascuan CG, Di Rosso ME, Pivoz-Avedikian JE, Wald MR, Zorrilla Zubilete MA, Genaro AM. Alteration of neurotrophin and cytokine expression in lymphocytes as novel peripheral markers of spatial memory deficits induced by prenatal stress. *Physiol Behav.* 2017;173:144–155.
89. Weinstock M. Prenatal stressors in rodents: effects on behavior. *Neurobiol Stress.* 2016;6: 3–13.
90. Guidotti A, Dong E, Tueting P, Grayson DR. Modeling the molecular epigenetic profile of psychosis in prenatally stressed mice. *Prog Mol Biol Transl Sci.* 2014;128:89–101.
91. Dong E, Dzitoyeva SG, Matrisciano F, Tueting P, Grayson DR, Guidotti A. Brain-derived neurotrophic factor epigenetic modifications associated with schizophrenia-like phenotype induced by prenatal stress in mice. *Biol Psychiatry.* 2015;77:589–596.
92. Dong E, Tueting P, Matrisciano F, Grayson DR, Guidotti A. Behavioral and molecular neuroepigenetic alterations in prenatally stressed mice: relevance for the study of chromatin remodeling properties of antipsychotic drugs. *Transl Psychiatry.* 2016;6:e711.
93. Matrisciano F, Tueting P, Maccari S, Nicoletti F, Guidotti A. Pharmacological activation of group-II metabotropic glutamate receptors corrects a schizophrenia-like phenotype induced by prenatal stress in mice. *Neuropsychopharmacology.* 2012;37:929–938.
94. Monn JA, Valli MJ, Massey SM, et al. Synthesis, pharmacological characterization, and molecular modeling of heterobicyclic amino acids related to (+)-2-aminobicyclo[3.1.0] hexane-2,6-dicarboxylic acid (LY354740): identification of two new potent, selective, and systemically active agonists for group II metabotropic glutamate receptors. *J Med Chem.* 1999;42:1027–1040.
95. Wierońska JM, Zorn SH, Doller D, Pilc A. Metabotropic glutamate receptors as targets for new antipsychotic drugs: historical perspective and critical comparative assessment. *Pharmacol Ther.* 2016;157:10–27.
96. Meltzer HY. Update on typical and atypical antipsychotic drugs. *Annu Rev Med.* 2013;64:393–406.
97. Zuena AR, Mairesse J, Casolini P, et al. Prenatal restraint stress generates two distinct behavioral and neurochemical profiles in male and female rats. *PLoS One.* 2008;3:e2170.

CHAPTER SEVEN

Epigenetic Programming by Early-Life Stress

Reema Abdulrahman S. Alyamani, Chris Murgatroyd[1]
School of Healthcare Science, Manchester Metropolitan University, Manchester, United Kingdom
[1] Corresponding author. E-mail address: c.murgatroyd@mmu.ac.uk

Contents

Abstract

Studies show that adverse conditions during early life can increase risks of developing mood disorders later in life. It is currently hypothesized that levels of environmental adversity in this early developmental period are able to shape the experience-dependent maturation of stress-regulating pathways leading to long-lasting alterations in stress responsivity during adulthood; a phenomenon often referred to as "early-life programming." Research is addressing the molecular mechanisms underlying this programming by which gene–environment interactions can predispose individuals toward psychopathology. Here we review key findings from animal and clinical studies examining the effects of prenatal and postnatal environment in shaping development of the neuroendocrine regulation of stress and the role of epigenetic mechanisms in translating early-life conditions into long-lasting gene expression changes underpinning stress-related behaviors. We also review latest ideas regarding interactions between environments at different developmental stages and evidence for

Progress in Molecular Biology and Translational Science, volume 157
ISSN 1877-1173
http://dx.doi.org/10.1016/bs.pmbts.2018.01.004

transgenerational effects. Understanding how prenatal and postnatal experiences can give rise to lasting epigenetic marks conferring increased vulnerability to mental disorders is a major focus of molecular psychiatry and should pave new guidelines for therapeutic interventions.

1. INTRODUCTION

The close relationship between the quality of early life and mental health in later life is well described. Many human and rodent studies illustrate that aspects of the early environment can lead to dramatic changes in physical and mental development compromising altered cognition, mood, and behavior. In particular, stress and adverse conditions during early life periods of development can shape individual differences in vulnerability to stress-related disorders throughout life.[1] This raises the question of how environmental experiences become incorporated at the cellular and molecular level leading to long-term alterations in various neuronal functions, including stress responses and risk to mental disease.

To understand gene–environment interactions and to elucidate pathways through which early life experiences are able to program long-term responses, researchers are using animal models, in which the early environment can be manipulated in a controlled fashion, and clinical studies. This has revealed the importance of epigenetic mechanisms of gene regulation in explaining how early life experiences could leave lasting chemical marks on the brain and influence both physical and mental health later in life even when the initial trigger is long gone.

In this review, we highlight important animal and human studies addressing epigenetic regulation of gene expression in sustaining the effects of early life experiences. We focus on studies that have investigated how adversity is able to shape biological stress systems, particularly the hypothalamic-pituitary-adrenal (HPA) system—an integral component of the body's stress response—leading to lasting alterations in stress responsivity.[2] We discuss what we know about the function of epigenetic systems and their roles in programming brain development and disease in response to prenatal and postnatal environmental stressors and research demonstrating interactions between prenatal and postnatal periods. We then review research-investigating mechanisms underlying possible transgenerational effects of stress. Finally, we highlight work linking animal findings to human studies and effectiveness of epigenetic markers in blood in relation to mood disorders and environmental impacts in clinical studies.

2. EARLY LIFE ADVERSITY SHAPES LATER LIFE STRESS RESPONSIVITY

Following exposure to a stressor, our autonomic nervous system initiates a quick "fight-or-flight" response and a slower more protracted HPA axis response. Tight regulation of the HPA axis is core to the long-term control of systems governing stress responsivity. Following a stressor, the paraventricular nucleus (PVN) of the hypothalamus releases the neuropeptides, corticotropin-releasing hormone (CRH), and arginine vasopressin (AVP). This stimulates the release of adrenocorticotrophic hormone (ACTH) from the anterior pituitary, which, in turn, acts on the adrenal cortex to release glucocorticoid hormones, that is, cortisol (humans) or corticosterone (animals). A negative feedback loop, through glucocorticoid receptors (GR) ensures efficient return to a homeostatic balance when the stress system is no longer challenged (Fig. 1). It has been shown that this negative feedback can become dysregulated, particularly following periods of chronic stress. This may critically impact the development of affective disorders; altered HPA activity is one of the most commonly observed neuroendocrine symptoms in major depressive disorder.[3] Childhood stress is a strong predictor of impaired inhibitory feedback regulation of the HPA axis.[4] Studies in rodent models further support the concept that exposure to a chronic stressor can lead to long-term changes in HPA regulation and behavior stemming from changes in neuropeptide regulation.[5]

3. EPIGENETIC MECHANISMS

"Epigenetics" is a term used to describe the study of stable alterations in gene expression potential that arise during development and differentiation, and is under the influence of the environment.[6] Genomes can therefore be considered to contain two layers of information: first, there is the DNA sequence that is conserved throughout life and is mostly identical in all cells of a body. Second, there are the epigenetic marks that are cell-specific and can be dynamic.

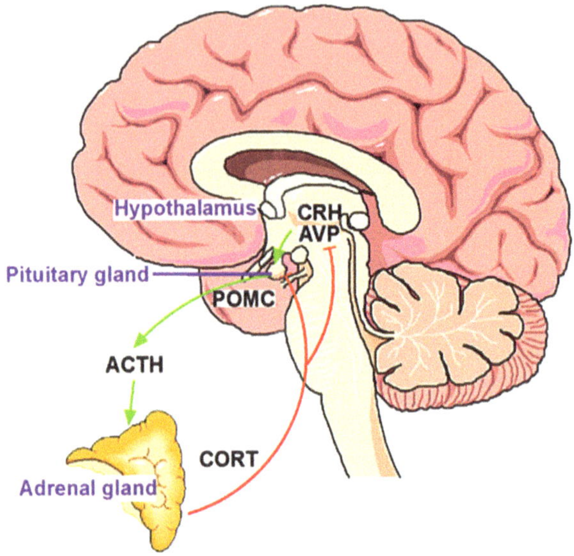

Fig. 1 ***The HPA stress axis.*** The neuropeptides corticotrophin-releasing hormone (CRH) and arginine vasopressin (AVP) are expressed in the parvocelluar neurons of the paraventricular nucleus. The joint release of CRH and AVP into the portal blood vessels leads to potent stimulation of anterior pituitary adrenocorticotrophic hormone (ACTH) secretion and, in turn, of corticosterone from the adrenal glands. The activational effects of the hypothalamic-pituitary-adrenal (HPA) axis are counteracted by the inhibitory effects of glucocorticoid receptors expressed in the hippocampus, hypothalamus, and anterior pituitary.

Epigenetic mechanisms enable regulation of a gene, independent of DNA sequence. DNA methylation and modification of core histones that package the DNA into chromatin represent the best-understood epigenetic marks. These marks are able to govern accessibility of the DNA to the machinery driving gene expression; inaccessible genes become silenced, whereas accessible genes are actively transcribed.

These epigenetic processes enable a cell and an organism to integrate intrinsic and environmental signals into the genome resulting in regulatory control of gene expression and thus facilitating adaptation. In this way, epigenetic mechanisms could be thought of as conferring plasticity to the hard-coded genome. In the context of the early life environment, epigenetic changes offer a plausible mechanism by which early experiences could be integrated into the genome as a kind of memory to program adult hormonal and behavioral responses.

3.1 DNA Methylation

DNA methylation, describes the addition of a methyl group to DNA at the cytosine side-chain in cytosine–guanine (CpG) dinucleotides, known as a CpG. The conventional view is that DNA methylation is a silencing mark and that, in general, the endpoint of DNA methylation is either long-term silencing or fine-tuning of gene expression potential. Hence, promoters and enhancer elements of transcriptionally active genes are usually not methylated but may become silenced once targeted by DNA methylation. The insertion of methyl groups changes the appearance and structure of DNA, which may either: (1) directly block DNA binding of transcription factors or (2) attract factors that preferentially bind to methylated or unmethylated DNA to interfere with transcription factor accessibility (Fig. 2A).

The vast majority (70%–80%) of all CpGs are methylated. Around 85% of these are located in repetitive sequences, such as transposons that constitute around half of the human genome while the other 15% typically cluster within GC-rich regions known as "CpG islands." CpG islands are defined as regions of greater than 500 bp that have cytosine/guanine content of greater

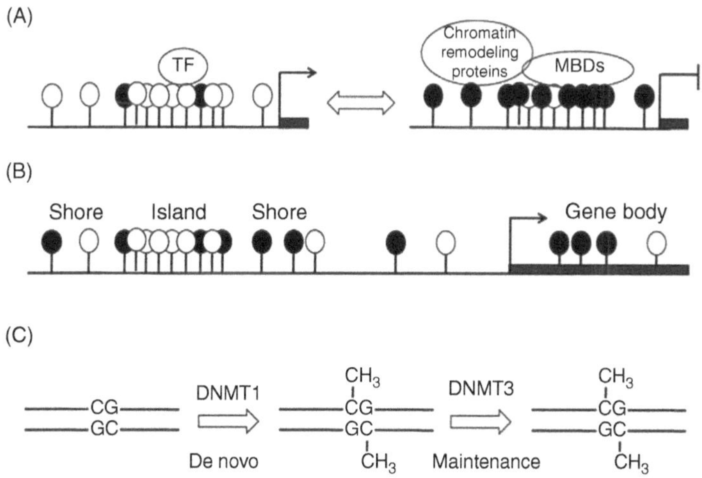

Fig. 2 ***DNA methylation.*** (A) DNA methylation *(black circle)*, can lead to gene repression through inhibiting transcription factor (TF) binding and recruitment of methylated DNA binding proteins (MBDs) and chromatin remodeling factors. (B) GC-rich regions are known as CpG islands and areas around 2 kb away, within sequences of intermediate CpG density, are referred to as CpG shores. (C) Dnmt1 is considered the primary maintenance methyltransferase, whereas Dnmt3a and Dnmt3b are considered de novo methyltransferases important in establishing methylation.

than 55%. Up to 60% of CpG islands are in the 5′ regulatory (promoter) regions of genes. Although CpG islands, promoters and enhancer regions appear to attract most attention in regard to DNA methylation and gene transcription, it is becoming increasingly clear that methylation at other gene elements play important regulatory roles. Indeed genome-wide sequencing of DNA methylation patterns among different tissues are revealing the presence and locations of tissue-specific differentially methylated regions. Strikingly, many of these do not seem to occur in CpG islands themselves but tend to locate in regions next to CpG islands around 2 kb away within sequences of intermediate CpG density. These are now referred to as "CpG shores."[7] Further regions can be divided up as "CpG shelves" as 2 kb regions extending from the shores while CpGs located yet further away from CpG islands are defined as being in "open sea" (Fig. 2B). In contrast to repression-associated DNA methylation in promoters and enhancers, higher CpG methylation within gene bodies is being frequently observed in transcriptionally active genes perhaps suggesting a further functional role of DNA methylation in transcriptional regulation.

The process of DNA methylation depends on a family of enzymes known as DNA methyltransferases (DNMTs). DNMT1 recognizes hemimethylated DNA and methylates appropriate cytosines in newly synthesized daughter strands formed during replication, acting as the maintenance DNMT. DNMT3A and DNMT3B can methylate unmethylated DNA supportive of their roles as de novo methylases (Fig. 2C). DNA methylation can also be reversed. For example, during development there are global and gene-specific increases and decreases in levels of 5mC. The TET family of enzymes can modify 5mC to 5-hydroxymethylcytosine (5hmC) as part of the processes of active demethylation. This modification appears to be particularly abundant in brain and might allow dynamic regulation of DNA methylation. For an overview of methods used to analyze DNA methylation, see Ref. 8.

3.2 Chromatin

Chromatin is a complex of DNA wrapped around histone proteins. Histones can be modified at their N-terminal tails by methylation, phosphorylation, acetylation, and ubiquitination as part of a histone signature serving to define accessibility to the DNA, that is, densely packaged "closed" chromatin (heterochromatin) in contrast to accessible "open" chromatin (euchromatin). Histone acetylation is known to be a predominant signal for active chromatin configurations while some specific histone methylation

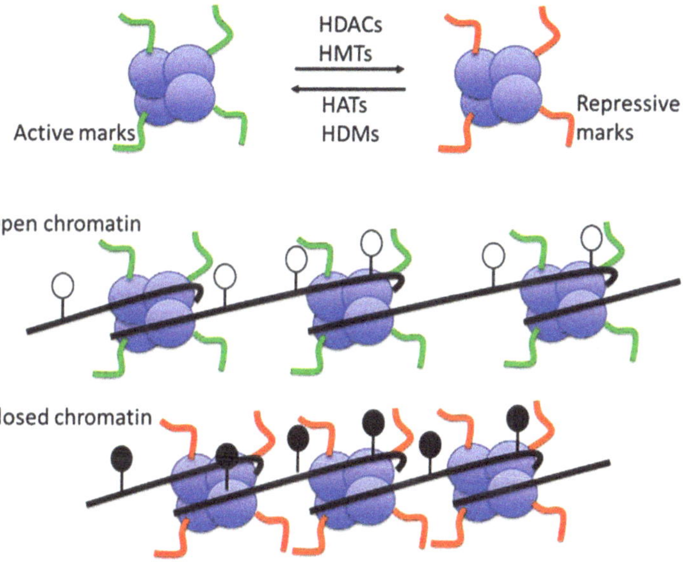

Fig. 3 ***Chromatin.*** DNA wraps around histones to form a complex referred to as chromatin. The N-terminal tails of histones serve as sites for epigenetic marking by acetylation and methylation via chromatin remodeling enzymes (e.g., HDACs, HATs, HMTs, and HDMs). DNA methylation refers to the chemical transfer of methyl groups to CpG acceptor sites through a class of enzymes known DNA methyltransferases (DNMTs). Active marks (e.g., histone acetylation and DNA hypomethylation) characterize "open" chromatin, while repressive marks (e.g., histone methylation and DNA hypermethylation) occur at "closed" chromatin.

reactions are associated with either gene silencing or activation. These are regulated through an array of histone modifying enzymes that catalyze the addition or removal of an array of covalent modifications. This "histone code" can be "read" by specific proteins, which have particular domains that recognize certain histone modifications to produce an effect on gene expression. Chromatin remodeling plays a central role in gene regulation by providing the transcription machinery with dynamic access to an otherwise tightly packaged genome. For an overview of methods used to analyze chromatin and histone modifications, see Ref. 8 (Fig. 3).

3.3 Noncoding RNA

The vast majority of the human genome is transcribed, but only around 2% of these transcripts are translated into protein. However, many of these are nonetheless functional: known as noncoding RNA (ncRNA) they can function to regulate gene expression at the transcriptional and posttranscriptional level. These can be divided into two main groups: the short ncRNAs (<30 nts) and the long ncRNAs (>200 nts). The three major classes of short noncoding RNAs are microRNAs (miRNAs), short interfering RNAs (siRNAs), and piwi-interacting RNAs (piRNAs). Both major groups are shown to play a role in heterochromatin formation, histone modification, DNA methylation targeting, and gene silencing.

4. EPIGENETIC PROGRAMMING BY PRENATAL STRESS

Aside from controlling constitutive gene expression, epigenetic mechanisms can also serve to fine-tune gene expression potential in response to environmental cues. Hence, it has been proposed that conditions of prenatal and postnatal environment can evoke changes in DNA methylation facilitating epigenetic programming of critical genes involved in regulating stress responsivity that may in turn manifest with neuroendocrine and behavioral symptoms in adulthood (Fig. 4).

4.1 Prenatal Stress

In animal models, prenatal stress has been shown to lead to various anxiety and depressive-like phenotypes.[9] It has been demonstrated that the long-term effects of early-life stress are mediated through alterations in the maternal and fetal HPA axes leading to exposure of the fetus to excess maternal glucocorticoids[10] that may pass the placental barrier and disrupt fetal brain development.

An important factor in regulating the supply of maternal glucocorticoids to the developing fetus is the placental enzyme, 11b-hydroxysteroid dehydrogenase 2 (11b-HSD-2), that deactivates cortisol.[11] It is important to control this placental diffusion as maternal glucocorticoid levels are up to 10-fold higher than in the fetus.[12] Regulation of placental 11b-HSD-2 is altered in response to a number of maternal components, including maternal undernutrition and protein deficiency.[13] Rodent studies have further shown it is upregulated following acute stress in pregnancy[14] and that chronic stress and prenatal maternal anxiety causes a reduction of HSD11B2 placental expression that is associated with increased CpG methylation of the *HSD11B2* gene in the placenta.[15]

The human *HSD11B2* gene contains four CpG islands and studies have shown that its activity is regulated by methylation.[16] In agreement with

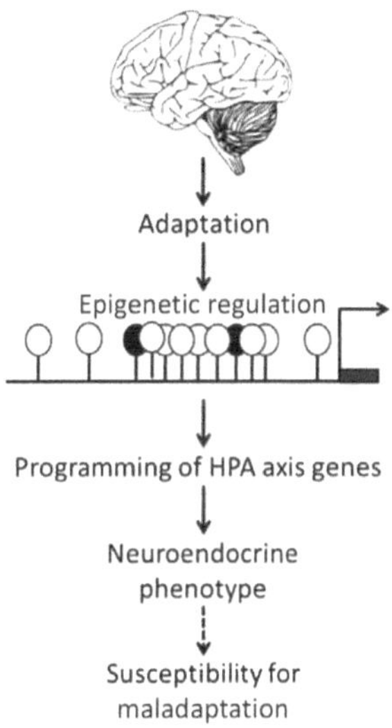

Fig. 4 ***Model for epigenetic programming by early-life stress.*** Early life experience can persistently alter expression levels of key genes through epigenetic marking, which can underpin changes in behavior, neuroendocrine, and stress responsivity throughout later life. Collectively, this process is referred to as epigenetic programming. The nature of the environment throughout later life, in addition to the impact of biological processes associated with aging and genetic sex, may exacerbate the effects of programming established during early life resulting in increased vulnerability to mood disorders. *HPA*, Hypothalamic-pituitary-adrenal.

findings from the aforementioned animal models, a study by O'Donnell *et al.*[17] reveals that prenatal trait anxiety negatively correlates with placental *11β-HSD2* mRNA expression. When testing infants, a study by Conradt *et al.*[18] revealed that intrauterine exposure to maternal depression negatively affects infant behavior and this is associated with increased methylation of the placental *11b-HSD-2* gene and the GR gene, *NR3C1*, while Appleton *et al.*[19] also found that infants whose mothers experienced the greatest levels of socioeconomic adversity during pregnancy had the lowest extent of methylation of the placental *11*b-*HSD2* gene. This suggests that epigenetic regulation of *11b-HSD2* links environmental cues transmitted from the mother to the foetus during pregnancy that could program the response to adverse postnatal environment via less exposure to glucocorticoids during development. Further studies have also identified differential epigenetic regulation of glucorticoids. For example, a study by Oberlander *et al.* reveals that prenatal exposure to maternal depressed/anxious mood during the third trimester of gestational development results in increased methylation of a CpG-rich region in the promoter and exon1F of the *NR3C1* gene in the cord blood of newborns, which correlates with salivary cortisol levels in infants at 3 months of age.[20]

4.2 Epigenetic Programming by Postnatal Stress

Early postnatal life is another sensitive time period for stress programming effects. Numerous animal studies of impaired mother-infant interactions (such as maternal stress, separation, or deprivation) during the early postnatal period have been shown to disrupt behaviors, such as memory, learning, and anxiety- and depressive-like behaviors together with neuroendocrine dysregulation of the HPA axis.[5]

A study of early-life adversity through the use of maternal separation in mice led to a sustained increase in expression of hypothalamic *Avp* in the paraventricular nucleus (PVN) that underlines elevated corticosterone secretion, heightens endocrine responsiveness to subsequent stressors, and alters HPA axis feedback inhibition. Importantly, this altered expression associated with reduced levels of DNA methylation in the PVN at particular CpG dinucleotides within an enhancer region important for *Avp* gene activity.[21] Mechanistic investigations reveal that hypomethylation at this region reduces the ability of MeCP2 to bind and recruit further epigenetic machinery, such as HDACs and DNMTs, supporting previous evidence for a role of MeCP2 as an epigenetic platform upon which histone and DNA modifications are carried out to confer transcriptional repression. Importantly, this hypomethylation was induced by neuronal depolarization triggering Ca^{2+}-dependent phosphorylation of MeCP2 causing MeCP2 to dissociate from the *Avp* enhancer resulting in a loss of both its repressing activity and its organization of DNA methylation and histone marks.[22]

Human studies reveal compelling evidence that adversity during early childhood may cause a range of physical and mental problems in adulthood, including depressive-related disorders and alterations in HPA axis activity.[1] Importantly, a number of key studies demonstrate that the epigenetic programming seen in early-life stress animal models may extrapolate to human studies. For example, McGowan *et al.*[23] found hypermethylation of *NR3C1* promoter among suicide victims with a history of abuse in childhood, but not among controls or suicide victims who did not suffer such early life stress. These data appear consistent with a seminal study (described later) at the same homologous region of the rat *Nr3c1* promoter.[24] Numerous further studies support *NR3C1* to be susceptible to alterations in DNA methylation, in response to early environmental stress.[25]

In a genome-wide study of promoter methylation in individuals with history of severe abuse during childhood, Labonté *et al.*[26] identified 362 differentially methylated promoters in hippocampal neurons from postmortem brain tissues of individuals with a history of severe childhood abuse compared to control subjects. Among these genes, those involved in neuronal plasticity are the most differentially methylated. More recently, Suderman *et al.*[27] demonstrated that 997 gene promoters in whole blood DNA of adult subjects are differentially methylated in association with childhood abuse. Most of these genes are known to be involved in key cell signaling pathways linked to development and regulation of transcription. Provencal *et al.*,[28] by analyzing the genome-wide promoter DNA methylation profiles in T cells from adult men, revealed that 448 distinct gene promoters are differentially methylated in persons with a history of parental physical aggression from 6 to 15 years of age compared to a control group. Functionally, most of these genes were previously demonstrated to play a substantial role in aggression and are enriched in biological pathways that are affected by behavior.

Numerous studies have demonstrated that childhood trauma increases risk for developing depression in adulthood, particularly in response to additional stress.[29] The hypothesis is that neuroendocrine, particularly HPA activity, resultant from early-life stress could increase risk to develop depression in response to stress later in life. A study by Klengel *et al.*[30] showed that an increased risk of developing adult stress-related psychiatric disorders associated with allele-specific, childhood trauma-dependent DNA demethylation at glucocorticoid response DNA elements of the FK506 binding protein 5 (*FKBP5*) gene, which encodes a cochaperone that regulates GR sensitivity. The demethylation of *FKBP5* related to an increased stress-dependent gene transcription followed by a long-term dysregulation of the stress hormone system and a global effect on immune functions and brain areas linked to stress regulation.

Together this suggests that epigenetic processes in the brain could mediate the effects of the early environment on gene expression, and that stable epigenetic marks, such as DNA methylation might then persist into adulthood to influence HPA axis activity.

4.3 Interactions of Prenatal and Postnatal Stress

The fetal origin hypothesis of human disease proposes that in utero environmental exposures lead to modifications during fetal development, which are adaptive (or maladaptive) dependent whether the subsequent postnatal environment is matches. Discontinuities between prenatal and postnatal environments create vulnerability. This effect is best exemplified in the associations of low fetal growth with diabetes and hypertension over several decades, that are thought to arise from fetal adaptations that confer advantage in food-scarce environments but create risk in western food-rich environments.[31,32] Low birth weight is also associated with adolescent depression in the presence of childhood adversities, consistent with the hypothesis.[33]

A study by Murgatroyd *et al.* found interactive effects of prenatal and postnatal maternal depression on *NR3C11F* promoter methylation, in young children; specifically, infants of mothers with low prenatal depression were vulnerable to the effects of postnatal depression, at the level of DNA methylation: consistent with an interplay between prenatal- and postnatal environments.[34] Further work would be important to address mechanisms, especially those genes impacted by prenatal stress that might confer resistance to later effects of postnatal stress.

5. EPIGENETIC PROGRAMMING BY EARLY-LIFE CARE

Variations in parental investment in mammals commonly involve the quantity and quality of behavioral interactions between parent and offspring. Numerous studies highlight the importance of maternal influences on subsequent performance both in terms of physiological functions and behavior of the offspring.[35] Importantly, the nature of mother-offspring interaction influences gene expression and the development of behavioral responses in the offspring, which remain stable from early development to adulthood.

One seminal animal study investigated the epigenetic effects of early life environment on stress programming through variations in the quality of early postnatal maternal care, as measured by levels of licking and grooming. Rats who received high levels of maternal care during early life developed sustained elevations in *Nr3c1* expression within the hippocampus and reduced HPA axis responses to stress. Meaney and coworkers further found an important role for epigenetic regulation revealing that the enhanced *Nr3c1* expression associated with a persistent DNA hypomethylation at specific CpG dinucleotides within the hippocampal *Nr3c1* exon 17 promoter and increased histone acetylation. The lower CpG methylation facilitated binding of the transcriptional activator nerve growth factor-inducible protein A (NGF1a) to this region.[24]

Two recent studies have reported moderation of prenatal stress effects by early human tactile stimulation (assessed by how often mothers reported stroking their babies) over the first weeks of life on behavioral and physiological stress reactivity in infancy[36,37] demonstrating that the effect of maternal prenatal depression on infant reactivity would be modified by tactile stimulation. This was followed by another study showing a reduction of *NR3C1* gene methylation associated with maternal stroking of these children, supporting the role of epigenetic mechanisms linking early life stress with long-term effects of maternal care.[34]

Recent research is demonstrating that the rewarding sensation of touch in affiliative interactions could be underpinned by a specialized system of nerve fibers called C-Tactile afferents (CTs), which respond optimally to slowly moving, gentle touch, typical of a caress.[38,39] Interestingly, studies in rodents have shown that tactile stimuli activate hypothalamic oxytocin neurons.[40] This, together with other recent research, suggests C-tactile afferent stimulation may mediate oxytocin release during affiliative tactile interactions.[41]

6. TRANSGENERATIONAL EFFECTS OF STRESS

Numerous studies show that children of depressed parents are at risk for developing depressive disorders themselves and other internalizing and externalizing disorders,[42,43] whether this be through prenatal or postnatal exposure to maternal stress. However, whether these risks are transmitted beyond 2 generations is less well known with only a few published 3-generation studies of major depression.[44,45] One study of 800 depressed and never-depressed women together with information from their children (at 15 years) and grandmothers detected an intergenerational transmission of depression: specifically, the grandmother's depression affected the mother's depression and her own stressful life context, and maternal and grandmother depression affected youth depression. This was mediated by interpersonal stress processes, which in turn affect parenting and children's social functioning.[46] A further study on 251 sets of grandchildren, parents and grandparents found those with 2 previous generations affected with major depression were at highest risk for major depression; though the study was too small to test for sex effects.[47]

Factors contributing to transgenerational transmission of depression have been difficult to identify: most studies based on clinical samples either represent unusually adverse contexts that are difficult to generalize and, for the most part, have not been large enough, with multiple collected environmental and biological variables, ideally longitudinally. A recent study by Murgatroyd and Nephew investigated the effects of chronic social stress during lactation in F0 mothers[48] across 3 generations and found that maternal care behavior in the F2 dams was more severely impaired than in the F0 and F1 dams and the expression of maternal anxiety is augmented in F2 dams. This occurred together with reduced basal cortisol (in contrast to an increase in F1 dams), a lack of changes in neuroendocrine gene expression, and reduced serum ICAM-1 (intercellular adhesion molecule-1) levels—a marker for inflammation and blood–brain barrier integrity.[49] Furthermore, treatment of F1 dams with OXT prevented some of the above effects on anxiety-related behavior and induced levels of interferon-γ (IFNγ) in the F2 offspring, while treatment with AVP prevented some of the impairments in social contact.[50] Together this highlights the possible role of maternal neuropeptides in regulating offspring behavior, which may form part of the mechanisms underlying effects of transgenerational stress.

7. BLOOD EPIGENETIC MARKERS FOR EARLY-LIFE STRESS

Access to brain tissues is obviously limited in human subjects. As such, many studies on specific epigenetic pathways in developmental programming of adult physical and mental problems by early-life stressful conditions have been performed on peripheral tissue samples. Studies using peripheral DNA, from blood or saliva of infants, adolescents or adults, have shown increased levels of *NR3C1* methylation in response to perinatal stress[20,34,51] and abuse or neglect during childhood.[52,53] Most studies thus far reported on DNA methylation after enduring stress or traumatic events, such as abuse or neglect.[52–55] Interestingly, many of these studies have looked at different regions and CpG residues of the *NR3C1* 1F promoter regions, though several recent clinical studies examining leukocytes have reported elevated methylation of the homologous human *NR3C1* 1F promoter (homologous to the rat 1–7 promoter) at a specific CpG associated with prenatal maternal depression[18,20,56] and childhood stress.[53,57] This suggests perhaps, different CpG residues might be sensitive to specific environmental stressors across this region.[25]

This raises the question of whether epigenetic changes in blood provide a clinically valuable surrogate to what is happening in the brain. Initial studies in rhesus macaques by Provencal *et al.*[58] examining DNA methylation in the prefrontal cortex and peripheral T cells in parallel in response to differential rearing do reveal differences in common between both tissues. Though results, as a whole, tend to suggest the T cells were not a direct surrogate markers of the brain tissue, but probably reflect the response of the immune system to early life stress. However, the idea that the epigenomic response to early life stress is not limited to the brain and could be studied in peripheral lymphocytes is a major area of study.

8. PERSPECTIVES

There is increasing evidence that exposure to stressful events during prenatal and postnatal periods associate with long-lasting mental health problems. While the molecular basis for such long-term effects may not be entirely clear, much evidence supports a role for the involvement of epigenetic regulation. The idea is that epigenetic tuning in response to the early life environment may prepare genes for responses to future environment, that is, enabling the organism to adapt to changing environmental conditions. These adaptive processes can, however, enhance the risk of pathology in the later life especially if there is mismatch between early and later life environments. Thereby, epigenetic research promises to improve the understanding of the genes and pathways by which organisms show widely diverse responses to stressful events in early life. Several specific genes linked to GR regulation, such as *NR3C1* are proving to be important markers.

With the advent of new high-throughput sequencing technologies, research is now scaling up to epigenome-wide analyses of all epigenetic marks throughout a genome in a specific tissue. Given that epigenetic modifications are sensitive to changes in the environment, it might be anticipated that these efforts will identify epigenomic signatures for mental disorders and molecular dysregulations resulting from early life stress. While such a strategy appears highly attractive in the field of cancer research addressing clonally expanded cell populations, rather than heterogeneous tissues, the field of epigenetic association studies in neuroendocrine disorders and psychiatry might be more challenging. Epigenetic modifications reported to date in animal models and postmortem brain tissues are seldom "all or none" but gradual, and seem to occur in a highly cell-type-specific manner. In this view, the extraordinary complexity and heterogeneity of neural tissues poses a major hurdle to derive epigenetic biomarkers in psychiatric disease.

In sum, understanding how early life experiences can give rise to lasting epigenetic memories conferring increased risk for mental disorders is emerging at the epicenter of modern psychiatry. Whether suitable social or pharmacological interventions could reverse deleterious epigenetic programming triggered by adverse conditions during early life, should receive highest priority on future research agendas. Progress in this field will further garner public interest, a general understanding and appreciation of the consequences of childhood abuse and neglect for victims in later life.

REFERENCES

1. Murgatroyd C, Spengler D. Epigenetics of early child development. *Front Psychiatry.* 2011;2:16.
2. Murgatroyd C, Spengler D. Epigenetic programming of the HPA axis: early life decides. *Stress.* 2011;14:581–589.
3. Holsboer F. The corticosteroid receptor hypothesis of depression. *Neuropsychopharmacology.* 2000;23:477–501.
4. Gonzalez A. The impact of childhood maltreatment on biological systems: implications for clinical interventions. *Paediatr Child Health.* 2013;18:415–418.
5. Murgatroyd C, Bradburn S. Epigenetics, the Environment, and Children's Health Across Lifespans. In: Switzerland: Springer International Publishing: 2016. 207–229.
6. Murgatroyd CA. Molecular Neuroendocrinololgy From Genome to Physiology. In: West Sussex, UK: John Wiley & Sons, Ltd; 2015.
7. Gomez S, et al. Comparative analysis of iron homeostasis in Sub-Saharan African children with sickle cell disease and their unaffected siblings. *Front Pediatr.* 2016;4:8.
8. Murgatroyd CA. *Epigenetics in Psychiatry.* Amsterdam: Elsevier/AP; 2014.
9. Babenko O, Kovalchuk I, Metz GAS. Stress-induced perinatal and transgenerational epigenetic programming of brain development and mental health. *Neurosci Biobehav Rev.* 2015;48:70–91.
10. Reynolds RM, et al. Transmitting biological effects of stress in utero: implications for mother and offspring. *Psychoneuroendocrinology.* 2013;38:1843–1849.
11. Stewart PM, Rogerson FM, Mason JI. Type 2 11 beta-hydroxysteroid dehydrogenase messenger ribonucleic acid and activity in human placenta and fetal membranes: its relationship to birth weight and putative role in fetal adrenal steroidogenesis. *J Clin Endocrinol Metab.* 1995;80:885–890.
12. Seckl JR. Glucocorticoid programming of the fetus; adult phenotypes and molecular mechanisms. *Mol Cell Endocrinol.* 2001;185:61–71.
13. Bertram C, Trowern AR, Copin N, Jackson AA, Whorwood CB. The maternal diet during pregnancy programs altered expression of the glucocorticoid receptor and type 2 11β-hydroxysteroid dehydrogenase: potential molecular mechanisms underlying the programming of hypertension in utero. *Endocrinology.* 2001;142:2841–2853.
14. Welberg LAM. Chronic maternal stress inhibits the capacity to up-regulate placental 11-hydroxysteroid dehydrogenase type 2 activity. *J Endocrinol.* 2005;186:R7–R12.
15. Jensen Peña C, Monk C, Champagne FA. Epigenetic effects of prenatal stress on 11β-hydroxysteroid dehydrogenase-2 in the placenta and fetal brain. *PLoS One.* 2012;7:e39791.
16. Togher KL, et al. Epigenetic regulation of the placental HSD11B2 barrier and its role as a critical regulator of fetal development. *Epigenetics.* 2014;9:816–822.
17. O'Donnell KJ, et al. Maternal prenatal anxiety and downregulation of placental 11β-HSD2. *Psychoneuroendocrinology.* 2012;37:818–826.

18. Conradt E, Lester BM, Appleton AA, Armstrong DA, Marsit CJ. The roles of DNA methylation of *NR3C1* and *11β-HSD2* and exposure to maternal mood disorder in utero on newborn neurobehavior. *Epigenetics.* 2013;8:1321–1329.
19. Appleton AA, Lester BM, Armstrong DA, Lesseur C, Marsit CJ. Examining the joint contribution of placental NR3C1 and HSD11B2 methylation for infant neurobehavior. *Psychoneuroendocrinology.* 2015;52:32–42.
20. Oberlander TF, et al. Prenatal exposure to maternal depression, neonatal methylation of human glucocorticoid receptor gene (NR3C1) and infant cortisol stress responses. *Epigenetics.* 2008;3:97–106.
21. Murgatroyd C, et al. Dynamic DNA methylation programs persistent adverse effects of early-life stress. *Nat Neurosci.* 2009;12:1559–U108.
22. Murgatroyd C, Spengler D. Polycomb binding precedes early-life stress responsive DNA methylation at the Avp enhancer. *PLoS One.* 2014;9:e90277.
23. McGowan PO, et al. Promoter-wide hypermethylation of the ribosomal RNA gene promoter in the suicide brain. *PLoS One.* 2008;3:e2085.
24. Weaver ICG, et al. Epigenetic programming by maternal behavior. *Nat Neurosci.* 2004;7:847–854.
25. Smart C, Strathdee G, Watson S, Murgatroyd C, McAllister-Williams RH. Early life trauma, depression and the glucocorticoid receptor gene: an epigenetic perspective. *Psychol Med.* 2015;45:3393–3410.
26. Labonté B, et al. Genome-wide epigenetic regulation by early-life trauma. *Arch Gen Psychiatry.* 2012;69:722–731.
27. Suderman M, et al. Childhood abuse is associated with methylation of multiple loci in adult DNA. *BMC Med Genomics.* 2014;7:13.
28. Provençal N, et al. Association of childhood chronic physical aggression with a DNA methylation signature in adult human T cells. *PLoS One.* 2014;9:e89839.
29. Heim C, Binder EB. Current research trends in early life stress and depression: review of human studies on sensitive periods, gene–environment interactions, and epigenetics. *Exp Neurol.* 2012;233:102–111.
30. Klengel T, et al. Allele-specific FKBP5 DNA demethylation mediates gene–childhood trauma interactions. *Nat Neurosci.* 2012;16:33–41.
31. Barker DJ. Fetal origins of coronary heart disease. *BMJ.* 1995;311:171–174.
32. Bateson P, et al. Developmental plasticity and human health. *Nature.* 2004;430:419–421.
33. Costello EJ, Worthman C, Erkanli A, Angold A. Prediction from low birth weight to female adolescent depression: a test of competing hypotheses. *Arch Gen Psychiatry.* 2007;64:338–344.
34. Murgatroyd C, Quinn JP, Sharp HM, Pickles A, Hill J. Effects of prenatal and postnatal depression, and maternal stroking, at the glucocorticoid receptor gene. *Transl Psychiatry.* 2015;5:e560.
35. Babb JA, Deligiannidis KM, Murgatroyd CA, Nephew BC. Peripartum depression and anxiety as an integrative cross domain target for psychiatric preventative measures. *Behav Brain Res.* 2015;276:32–44.
36. Pickles A, Sharp H, Hellier J, Hill J. Prenatal anxiety, maternal stroking in infancy, and symptoms of emotional and behavioral disorders at 3.5 years. *Eur Child Adolesc Psychiatry.* 2017;26:325–334.
37. Sharp H, Hill J, Hellier J, Pickles A. Maternal antenatal anxiety, postnatal stroking and emotional problems in children: outcomes predicted from pre- and postnatal programming hypotheses. *Psychol Med.* 2014;28:1–15.
38. Pawling R, Cannon PR, McGlone FP, Walker SC. C-tactile afferent stimulating touch carries a positive affective value. *PLoS One.* 2017;12:e0173457.
39. McGlone F, Wessberg J, Olausson H. Discriminative and affective touch: sensing and feeling. *Neuron.* 2014;82:737–755.

40. Okabe S, Yoshida M, Takayanagi Y, Onaka T. Activation of hypothalamic oxytocin neurons following tactile stimuli in rats. *Neurosci Lett.* 2015;600:22–27.
41. Walker SC, Trotter PD, Swaney WT, Marshall A, Mcglone FP. C-tactile afferents: cutaneous mediators of oxytocin release during affiliative tactile interactions? *Neuropeptides.* 2017;64:27–38.
42. National Research Council (US) and Institute of Medicine (US) Committee on Depression, Parenting Practices, and the Healthy Development of Children. In: England MJ, Sim LJ, eds. In: *Depression in Parents, Parenting, and Children. Opportunities to Improve Identification, Treatment, and Prevention. Washington, DC: National Academies Press; 2009.*
43. Lyons-Ruth K, Easterbrooks MA, Cibelli CD. Infant attachment strategies, infant mental lag, and maternal depressive symptoms: predictors of internalizing and externalizing problems at age 7. *Dev Psychol.* 1997;33:681–692.
44. Pettit JW, Olino TM, Roberts RE, Seeley JR, Lewinsohn PM. Intergenerational transmission of internalizing problems: effects of parental and grandparental major depressive disorder on child behavior. *J Clin Child Adolesc Psychol.* 2008;37:640–650.
45. Olino TM, et al. Influence of parental and grandparental major depressive disorder on behavior problems in early childhood: a three-generation study. *J Am Acad Child Adolesc Psychiatry.* 2008;47:53–60.
46. Hammen C, Shih JH, Brennan PA. Intergenerational transmission of depression: test of an interpersonal stress model in a community sample. *J Consult Clin Psychol.* 2004;72:511–522.
47. Weissman MM, et al. A 30-year study of 3 generations at high risk and low risk for depression. *JAMA Psychiatry.* 2016;73:970.
48. Murgatroyd CA, Nephew BC. Effects of early life social stress on maternal behavior and neuroendocrinology. *Psychoneuroendocrinology.* 2013;38:219–228.
49. Murgatroyd CA, et al. Transgenerational social stress, immune factors, hormones, and social behavior. *Front Ecol Evol.* 2016;3:149.
50. Murgatroyd CA, et al. Effects of chronic social stress and maternal intranasal oxytocin and vasopressin on offspring interferon-γ and behavior. *Front Endocrinol (Lausanne).* 2016;7:155.
51. Mulligan CJ, D'Errico NC, Stees J, Hughes DA. Methylation changes at NR3C1 in newborns associate with maternal prenatal stress exposure and newborn birth weight. *Epigenetics.* 2012;7:853–857.
52. Perroud N, et al. Increased methylation of glucocorticoid receptor gene (NR3C1) in adults with a history of childhood maltreatment: a link with the severity and type of trauma. *Transl Psychiatry.* 2011;1:e59.
53. Tyrka AR, Price LH, Marsit C, Walters OC, Carpenter LL. Childhood adversity and epigenetic modulation of the leukocyte glucocorticoid receptor: preliminary findings in healthy adults. *PLoS One.* 2012;7:e30148.
54. McGowan PO, et al. Epigenetic regulation of the glucocorticoid receptor in human brain associates with childhood abuse. *Nat Neurosci.* 2009;12:342–348.
55. Perroud N, et al. Childhood maltreatment and methylation of the glucocorticoid receptor gene NR3C1 in bipolar disorder. *Br J Psychiatry.* 2013;204:30–35.
56. Hompes T, et al. Investigating the influence of maternal cortisol and emotional state during pregnancy on the DNA methylation status of the glucocorticoid receptor gene (NR3C1) promoter region in cord blood. *J Psychiatr Res.* 2013;47:880–891.
57. Melas PA, et al. Genetic and epigenetic associations of MAOA and NR3C1 with depression and childhood adversities. *Int J Neuropsychopharmacol.* 2013;1:1–16.
58. Provencal N, et al. The signature of maternal rearing in the methylome in rhesus macaque prefrontal cortex and T cells. *J Neurosci.* 2012;32:15626–15642.

CHAPTER EIGHT

Epigenetic Drugs for Mood Disorders

Jacob Peedicayil[1], Aniket Kumar
Christian Medical College, Vellore, India
[1] Corresponding author. E-mail address: jpeedi@cmcvellore.ac.in

Contents

Abstract

There is increasing evidence that changes in epigenetic mechanisms of gene expression are involved in the pathogenesis of mood disorders. Such evidence stems from studies conducted on postmortem brain tissues and peripheral cells or tissues of patients with mood disorders. This article describes and discusses the epigenetic changes in the mood disorders (major depressive disorder and bipolar disorder) found to date. The article also describes and discusses preclinical drug trials of epigenetic drugs for treating mood disorders. In addition, nonrandomized and randomized controlled trials of nutritional drugs with effects on epigenetic mechanisms of gene expression in patients with major depressive disorder and bipolar disorder are discussed. Trials of epigenetic drugs and nutritional drugs with epigenetic effects are showing promising results for the treatment of mood disorders. Thus, epigenetic drugs and nutritional drugs with epigenetic effects could be useful in the treatment of patients with these disorders.

Progress in Molecular Biology and Translational Science, volume 157
ISSN 1877-1173
http://dx.doi.org/10.1016/bs.pmbts.2018.01.005

1. INTRODUCTION

According to the International Classification of Mental and Behavioral Disorders, Version 10,[1] mood disorders, which are sometimes also called affective disorders, are classified into the following categories: manic episode, bipolar disorder (BD), depressive episode, recurrent depressive disorder, persistent mood disorders, other mood disorders, and unspecified mood disorders. Major depressive disorder (MDD) includes depressive episodes, recurrent depressive disorders, and persistent mood disorder. Mood disorders comprise an important and common category of psychiatric disorders, and are associated with impairments in interpersonal, social, and occupational functioning.[2] They are known to cause considerable health, social, economic, and human costs to society.[3] Once thought to be mainly adult disorders, mood disorders are increasingly being diagnosed in children and adolescents.[4] Mood disorders are associated with high rates of suicide.[2]

Until recently, the monoaminergic neurotransmitters, norepinephrine, dopamine, and serotonin were the main focus of theoretical and experimental research on the pathogenesis of these disorders. However, a progressive shift has occurred from studying single neurotransmitter systems to studying neurobehavioral systems, neural circuits, and more intricate neuroregulatory mechanisms.[2] Hence, the monoaminergic systems are at present viewed as more broad neuromodulatory systems, and any abnormalities involving them in mood disorders could be secondary or epiphenomenal effects.[2] Psychosocial factors also underlie mood disorders. For example, childhood adversity can cause exaggerated responses to stress later in life. These responses are associated with abnormalities in serotonergic activity and in the hypothalamic–pituitary–adrenal axis.[5,6]

From family, twin, and adoption studies, it has long been known that mood disorders have a hereditary basis. Since finding the genes predisposing to these disorders could shed light on the underlying pathophysiology of these disorders, and lead to better treatments, a lot of effort has been put into finding such genes by genetic mapping studies. However, to date no genetic polymorphism or mutation underlying these disorders has been definitively identified.[7,8] Instead, there is increasing evidence that abnormalities of epigenetic mechanisms of gene expression underlie these disorders.[9] Among the various mood disorders, so far the role of epigenetics has been studied in MDD and BD. This article discusses the role of epigenetic abnormalities in the pathogenesis of MDD and BD, and the possible reversal of these abnormalities with epigenetic drugs and nutritional drugs that modify epigenetic mechanisms of gene expression.

1.1 Major Depressive Disorder

The lifetime prevalence of MDD varies from 5% to 17% with an average of 12%,[2] making it the commonest of all psychiatric disorders.[3] The diagnosis of MDD requires a distinct change of mood, characterized by sadness or irritability and accompanied by at least several psychophysiological changes such as disturbances in sleep, appetite, or sexual desire; constipation; loss of the ability to experience pleasure at work or with friends; crying; suicidal thoughts; and slowness of speech and action. These changes should last for at least 2 weeks and interfere with work and family relations.[10] MDD is known to be 2 times more prevalent in women than in men. Patients with MDD are at an increased risk for one or more comorbid disorders such as alcohol abuse or dependence, panic disorder, obsessive–compulsive disorder, and social anxiety disorder.[2] Cognitive impairment, manifesting as impaired ability to think, concentrate, or make decisions, occurs in about two-thirds of patients with MDD.[11] Cognitive impairment in patients with MDD is associated with a poor response to treatment.[11]

The average age of onset of MDD is about 40 years, with half of all patients having an onset between 20 and 50 years of age.[2] MDD can also begin during childhood and old age.[2] It occurs most commonly in individuals without close interpersonal relationships and in divorced or separated individuals.[2] MDD is commonly present in patients presenting in primary care and general medical practice.[12,13] This disorder is also known to be under-recognized, underdiagnosed, and undertreated by doctors.[14] Some of the clinical features of MDD are listed in Table 1.

Table 1 Clinical Features of Major Depressive Disorder and Bipolar Disorder.

	MDD	BD
Average age at onset, year	40	30
Life time prevalence, %	12	About 5
Male:female ratio	1:2	1:1
Variation with race/ethnicity	Nil	Nil

BD, Bipolar disorder; *MDD*, major depressive disorder.

1.1.1 *Current Pharmacotherapy of Major Depressive Disorder*

Current pharmacotherapy of MDD comprises several classes of drugs such as tricyclic antidepressants, selective serotonin-reuptake inhibitors, and monoamine oxidase inhibitors.[15–17] In addition to adult patients with MDD, these drugs are also used to treat MDD in children and adolescents.[18] These drugs act primarily by facilitating neurotransmission mediated by monoamines (norepinephrine, dopamine, and serotonin) in the brain.[16,17] This primary effect can lead to long-term secondary effects involving adaptive or regulatory mechanisms that enhance the effectiveness of therapy.[17] Such secondary effects include the expression of gene products such as brain-derived neurotrophic factor (BDNF).[17] New antidepressants like ketamine, an antagonist at the *N*-methyl-D-aspartate (NMDA) receptor, one of the receptors for the excitatory neurotransmitter glutamic acid, have shown beneficial effects in MDD[19]; but the fleeting nature of ketamine's therapeutic benefit and its potential for abuse and neurotoxicity suggest that its clinical use requires caution. Another new antidepressant is agomelatine, a melatonin receptor agonist and serotonin 5-HT_{2C} receptor antagonist.[20]

The currently used drugs are imperfect. Many patients have significant residual symptoms despite drug therapy.[21] About 30%–40% of depressed patients are thought to not show clinical improvement despite therapy with currently available drugs.[22] The current drugs also have drawbacks like delayed onset of action (about 2–3 weeks) and adverse effects.[17] Moreover, a large percentage of patients do not respond adequately, and are said to have treatment-resistant depression.[23]

1.2 Bipolar Disorder

BD is characterized by a clinical course of one or more manic episodes and sometimes major depressive episodes. In the former there is a distinct period of an abnormally and persistently elevated, expansive, or irritable mood lasting for at least 1 week, or less if the patient has to be admitted into a hospital.[2] In mania there can also be euphoria, overactivity and a lack of necessity for sleep, and increased optimism, which can lead to impaired judgement.[24] A less severe form of a manic episode is termed a hypomanic episode, which lasts for at least 4 days and which is not sufficiently severe to affect social or occupational functioning, and where there are no psychotic features.[2] In the clinical course of bipolar I disorder, there are one or more manic episodes and sometimes major depressive episodes. There is a variant of BD called bipolar II disorder where there are major depressive episodes and hypomania and not mania. The lifetime prevalence of BD was thought to be about 1% previously, but is presently thought to be higher, at about 5%.[4] BD is frequently associated with other coexisting conditions especially anxiety disorder and substance abuse disorder.[25] Although the severity of mania is evident in BD, most disability is thought to be due to the depressive phases of the disorder.[22] BD is known to be a chronic, lifelong illness.[26] In addition to changes in mood, patients with BD can have cognitive impairment, presenting as impairments of executive function, verbal memory, psychomotor speed, and sustained attention.[27] Another complication that can present in BD patients is rapid cycling BD, where patients have four or more episodes of mania and/or major depression in 1 year.[28] Some of the clinical features of BD are listed in Table 1.

1.2.1 Current Pharmacotherapy of Bipolar Disorder

The drugs used to manage BD comprise the mood stabilizers, augmented by antidepressants, antipsychotics, and benzodiazepines.[2,26] Mood stabilizers, drugs that prevent and treat manic and depressive episodes in patients with BD,[29] include lithium and the anticonvulsants valproic acid, lamotrigine, carbamazepine, and oxcarbazepine.[30] Among these drugs, lithium is considered the first-line and most effective drug because it is effective in treating acute manic and depressive episodes, and in preventing manic and depressive episodes.[29]

As in the case of the drug therapy of MDD, there are shortcomings with the current drug therapy of BD.[31] Despite treatment, about 37% of patients with BD relapse into depression or mania within 12 months, and 60% within 2 years.[26] For many patients, despite treatment, manic and depressive episodes dominate their lives. Such patients are said to have treatment-resistant BD.[32] Usually, treatment resistance is said to occur if patients do not respond adequately after treatment for at least 6 weeks in mania, 12 weeks in BD depression, and 12 months or longer during long-term maintenance treatment.[33] Among currently used drugs for managing BD, the mood stabilizers control mood swings, but do not adequately improve cognitive decline, and indeed could aggravate cognitive decline.[34]

2. EPIGENETIC CHANGES IN MOOD DISORDERS

2.1 Epigenetic Changes in Major Depressive Disorder

Epigenetic changes in the postmortem brain in patients with MDD are listed in Table 2. Several changes have been found including those involving

Table 2 Epigenetic Changes in Postmortem Brain Tissues in Major Depressive Disorder.

Tissue	Epigenetic Change	References
PFC	Expression of 21 miRNAs is downregulated	35
FC	Differentially methylated DNA of many genes underlying neuronal development	39
PFC	Several miRNAs associated with polyamines are upregulated	36
PFC	Increased H3K4 trimethylation in three synapsin genes	52
Hippocampus, PFC	Aberrant NMDA receptor gene DNA methylation	38
ACG, DLPFC DRN, NLC	Changes in DNA methylation of genes encoding galanin neuropeptides	53
PFC	DNA hypomethylation of synapsin II CpG islands	40
BA11, BA25	Hypomethylation of DMR upstream of PSORS1C3 gene	41
Hippocampus	DNA methylation of DLG4 and GJA-1 genes is unchanged	42
PFC	Dysregulation of expression and activity of miRNA-124-3p	37
PFC	Gene body 5-hydroxymethylation associated with gene expression changes	54

ACG, Anterior cingulate gyrus; *BA*, Brodmann area of cerebral cortex; *DLPFC*, dorsolateral prefrontal cortex; *DMR*, differentially methylated region; *DRN*, dorsal raphe nucleus; *FC*, frontal cortex; *miRNAs*, microRNAs; *NLC*, noradrenergic locus coeruleus; *PFC*, prefrontal cortex.

microRNAs (miRNAs)[35–37] and abnormalities in DNA methylation.[38] Regarding DNA methylation, genes have shown hypomethylation, hypermethylation, or no changes in methylation.[38–42] As is evident from the available data, the study of epigenetic changes in the postmortem brain in MDD is in its early stages, and more work needs to be done in this area. Table 3 lists epigenetic changes in peripheral tissues/cells in patients with MDD. Among the several changes noted, the most consistent appears to be hypermethylation of the promoter of the gene encoding BDNF.[43–46] BDNF is one of four members of the neurotrophin family which are important for the development of the brain and for neuronal plasticity.[47] Of the four members, BDNF has the most abundant and widespread function in the mammalian nervous system. BDNF is released from neurons both presynaptically and postsynaptically. Hypermethylation of the BDNF promoter

Table 3 Epigenetic Changes in Peripheral Cells/Tissues in Major Depressive Disorder.

Tissue	Epigenetic Change	References
Leukocytes	Hypermethylation of promoter of ACE gene	48
Peripheral blood	Hypermethylation of BDNF gene promoter	43
Peripheral blood	Hypermethylation of BDNF gene promoter	44
PBMC	Hypermethylation of BDNF gene promoter	45
PBMC	Hypermethylation of BDNF gene promoter	46
Leukocytes	Age-associated decrease in global DNA methylation	49
Peripheral blood	Circulating lncRNAs are associated with MDD	55
PFC, leukocytes, sperm	Epigenetic changes in all tissues studied	56
Leukocytes	Genome-wide DNA methylation differences in monozygotic twins discordant for MDD	57
Leukocytes	Reduced DNA methylation in glucocorticoid receptor gene	50
Whole blood DNA	Inverse correlation between DNA methylation of SERT gene and white matter integrity of corpus callosum	51
Peripheral blood	Changes in DNA methylation in oxytocin receptor gene	58
PBMC	Promoters of many stress-related genes are hypermethylated	59
Whole blood	miR4646-associated methylation locus is hypomethylated	60

ACE, Angiotensin converting enzyme; *BDNF*, brain-derived neurotrophic factor; *FC*, frontal cortex; *lncRNAs*, long noncoding RNAs; *MDD*, major depressive disorder; *PBMC*, peripheral blood mononuclear cells; *PFC*, prefrontal cortex.

would lead to decreased expression of BDNF and decreased levels of BDNF. These epigenetic data support the biochemical data regarding lower levels of BDNF in patients with MDD.[47] Disturbances in BDNF signaling in the brain, mainly due to a decrease in expression or release, has been linked in recent years to several psychiatric and neurological disorders.[47] As in the case of DNA methylation changes in postmortem brain, even in peripheral tissues/cells, there is both hypermethylation and hypomethylation of genes.[48–50] Other epigenetic changes in peripheral tissues/cells in patients with MDD include DNA methylation changes in the gene encoding the serotonin transporter (SERT).[51] This transporter is involved in the reuptake of serotonin from the synaptic cleft into the presynaptic neuron, and is inhibited by the commonly used antidepressants, selective serotonin-reuptake inhibitors.[17]

2.2 Epigenetic Changes in Bipolar Disorder

Epigenetic changes found in the postmortem brain in patients with BD are listed in Table 4. The changes found include changes in DNA methylation of several genes and changes in the expression of miRNAs. A finding that appears to be well established is that in the genes encoding reelin and glutamic acid decarboxylase-67 (GAD_{67}) in the prefrontal cortex of

Table 4 Epigenetic Changes in Postmortem Brain Tissues in Bipolar Disorder.

Tissue	Epigenetic Change	References
PFC	Overexpression of DNMT1 in GABAergic neurons	61
Frontal lobe	DNA hypomethylation of gene encoding MB-COMT	67
PFC	Overexpression of DNMT1 in GABAergic neurons	62
Frontal cortex	Widespread changes in DNA methylation	68
PFC	Twenty two miRNAs are differentially expressed	69
CC, OC, PFC, PC, WBC, sperm	Hypomethylation of HLA complex group 9 gene	70
Frontal cortex	Global DNA hypermethylation and histone H3 phosphorylation	71
DLPFC	Five differentially expressed genes	72
Parietal cortex	Upregulation of TET1; downregulation of APOBEC3A and APOBEC3C	64
PFC	Differential expression of exosomal miRNAs	73
PFC	Hypomethylation of exon 11 of KCNQ3 gene	74
Hippocampus	DNA methylation changes at GABA-related genes	75
PFC	CpG and CpH changes in HLA complex group 9 compared to controls	76

CC, Corpus callosum; *DNMT1*, DNA methyltransferase 1; *DLPFC*, dorsolateral prefrontal cortex; *MB-COMT*, membrane-bound catechol-*O*-methyl transferase; *miRNAs*, microRNAs; *OC*, occipital cortex; *PC*, parietal cortex; *PFC*, prefrontal cortex; *WBC*, white blood cells.

postmortem brains of patients with BD, there is overactivity of DNA methyltransferase-1 (DNMT-1).[61,62] This would lead to increased DNA methylation of these genes and decreased expression of reelin and GAD_{67}. Of these two gene products, reelin is a glycoprotein essential for the development of the brain, and GAD_{67} is the enzyme which catalyzes the conversion of glutamic acid to γ-amino butyric acid (GABA). These epigenetic data support the biochemical data that in the prefrontal cortex of postmortem brains of BD patients there is decreased expression of reelin and GAD_{67} proteins and mRNAs.[63] In addition to the study of methylation of DNA, the study of DNA demethylation has also been conducted. Thus, one study[64] found that there is upregulation of ten-eleven translocation 1 (TET1), an enzyme involved in demethylation of DNA. Table 5 lists epigenetic changes in peripheral tissues/cells in patients with BD. Such studies have mainly focused on DNA methylation. It is interesting that like in the peripheral cells/tissues of patients with MDD, hypermethylation of the promoter of the gene encoding BDNF has been found in patients in peripheral blood mononuclear cells of patients with BD.[65] Changes in DNA methylation of the promoter of the gene encoding SERT have also been observed.[66]

Table 5 Epigenetic Changes in Peripheral Cells/Tissues in Bipolar Disorder.

Tissue	Epigenetic Change	References
Lymphoblastoid cells	Alteration of DNA methylation of PPIEL gene	77
Lymphoblastoid cell lines	Alteration of expression of 7 of 13 miRNAs	78
Leukocytes	Global DNA methylation is not altered	79
Lymphoblastoid cells	No alteration in acetylated histone H3 in ADRBK2 gene promoter	80
Saliva	Hypomethylation of HTR2A gene promoter	81
Leukocytes	Hypermethylation of 5HTR1A gene promoter	82
Lymphoblastoid cell lines	Hypermethylation of SERT gene promoter	66
Saliva	Hypomethylation of MB-COMT promoter	83
PBMC	Hypermethylation of BDNF gene promoter	65
Buccal cells	Many disease-associated loci show DNA methylation changes	84
Leukocytes	Several miRNAs and their targets underlie BD	85

BDNF, Brain-derived neurotrophic factor; *DMR*, differentially methylated region; *HTR2A*, serotonin receptor type 2A; *miRNAs*, microRNAs, *MB-COMT*, membrane-bound catechol-*O*-methyl transferase; *PBMC*, peripheral blood mononuclear cells; *PPIEL*, peptidylprolyl isomerase E-like; *SERT*, serotonin transporter.

3. EPIGENETIC THERAPY OF MOOD DISORDERS

Epigenetic therapy, the use of drugs that modify epigenetic mechanisms of gene expression to correct disease states, is a new and expanding area of research in pharmacology. Although several classes of epigenetic drugs are being investigated, at present most research focuses on two classes of drugs: drugs that inhibit DNA methyltransferase (DNMTi) and drugs that inhibit histone deacetylase (HDACi). Indeed, it is well established that the clinically used mood stabilizer valproic acid acts as a HDACi, and that this action could contribute to its mood stabilizing properties. However, valproic acid, in addition to inhibiting HDACs, also has nonepigenetic mechanisms of action on neuronal conduction, and both these mechanisms of action could contribute to the mood stabilizing effects of this drug. At therapeutic concentrations, it relieves histone deacetylase (HDAC)-dependent transcriptional repression and causes hyperacetylation in cells and in vivo.[86] It also inhibits HDAC in vitro, most probably by binding to the catalytic center of HDACs, and also indirectly via transcriptional activation of many promoters.[86] In addition, it is known that valproic acid and other HDACi can induce DNA demethylation by activating DNA demethylating mechanisms.[87] The present state of research on the use of epigenetic therapy in mood disorders is given below.

3.1 Epigenetic Drugs for Mood Disorders

At present, the study of the use of epigenetic drugs for mood disorders is in the preclinical stages, except for valproic acid which has been in clinical use as a mood stabilizer for several years. The findings of preclinical trials of epigenetic drugs for the treatment of MDD are shown in Table 6. Such studies have so far used DNMTi and HDACi, although the majority of studies have investigated the use of HDACi. One study[88] investigated the use of the newer antipsychotic quetiapine which was found to reverse depressive-like behavior in rats, but which was also found to inhibit DNMT-1 in the hippocampus of the rats. As shown, both DNMTi and HDACi have demonstrated beneficial results in preclinical trials of MDD. However, HDACi appear to have more promising effects than DNMTi as antidepressant drugs because in one study DNMTi did not show antidepressant effects.[89] Table 7 lists the preclinical trials of epigenetic drugs for the treatment of BD. All such studies have used HDACi as the investigational

Table 6 Preclinical Trials of Epigenetic Drugs for Major Depressive Disorder.

Drug	Class	Model	Finding	References
Azacytidine, Decitabine	DNMTi	Rats	Has antidepressant effects	90
Cpd-60	HDAC 1/2 inhibitor	Mice	Has antidepressant effects	91
Azacytidine	DNMTi	Rats	Does not have antidepressant effects	89
Azacytidine	DNMTi	Rats	Potentiaties effects of antidepressants	92
SB	HDACi	FSL rats	Has antidepressant effects	93
Knockdown of HDAC5	Transgenic	Mice	Has antidepressant-like effects	94
SAHA	HDACi	CRTC1-deficient mice	Has antidepressant-like effects	95
SB, VA	HDACi	Mice	Has antidepressant-like effects	96
33i	HDACi	Mice	Has antidepressant-like effects	97
Quetiapine	Antipsychotic	Rats	Has DNMTi-like effects; decreases depression	88

CRTC1, CREB-regulated transcription coactivator 1; *DNMTi*, DNA methyltransferase inhibitor; *FSL*, Flinders sensitive line; *HDACi*, histone deacetylase inhibitor; *SAHA*, suberoylanilide hydroxamic acid; *SB*, sodium butyrate; *TSA*, trichostatin A; *VA*, valproic acid.

drug. As shown, HDACi have so far been found to show favorable results for treating BD in preclinical drug trials.

3.2 Nutritional Drugs Having Epigenetic Effects in Mood Disorders

An individual's nutrition affects his or her epigenome, resulting in changes in gene expression.[110] These effects can occur prenatally as well as postnatally, and also throughout the lifespan. Nutritional epigenetics is an active area of epigenetic research with most work being done on the effects of the hemopoietic drugs folic acid and vitamin B_{12} (cyanocobalamin) during fetal and early postnatal life. These drugs are required for the conversion of

Table 7 Preclinical Trials of Epigenetic Drugs for Bipolar Disorder.

Drug	Class	Model	Finding	References
VPA, TSA	HDACi	Human lymphocytes	Increases acetylated histone3 (acH3)	98
VPA	HDACi	Human lymphocytes	Increases GAD_{67} expression	99
VPA	HDACi	Rat cortical neurons	Activates promoter IV of BDNF gene	100
SB	HDACi	Rats	Decreases mania-like symptoms	101
VA	HDACi	CRFtg mice	Improves prepulse inhibition deficits	102
VA	HDACi	Human astrocytes	Regulates glutamate carboxypeptidase II stability	103
VA	HDACi	Adult zebrafish tectum	Upregulates Notch signaling	104
SB	HDACi	Rat	Blocks oubain-induced hyperlocomotion	105
VA	HDACi	Serotonergic cell line	Increases histone acetylation	106
VA	HDACi	Mice	Increases hippocampal p21 expression	107
VA	HDACi	Developing rat	Stimulates proliferation of glial precursors	108
SB	HDACi	Rats	Reduces oubain-induced hyperactivity	109

CRFtg, Transgenic mice overexpressing corticotrophin-releasing factor; *HDACi*, histone deacetylase inhibitor; *SB*, sodium butyrate; *TSA*, trichostatin A; *VA*, valproic acid.

homocysteine to methionine. Methionine is metabolized to *S*-adenosyl methionine (SAMe), the main methyl donor for most methylation reactions in the body, including methylation of DNA (Figs. 1 and 2). For the use of nutritional drugs in the management of psychiatric disorders like the mood disorders, instead of cyanocobalamin and folic acid, SAMe and L-methylfolate (LMF) are recommended. SAMe and LMF, unlike cyanocobalamin and folic acid, act as methyl donors in the body.[110]

The results of nonrandomized controlled trials of nutritional drugs modifying epigenetic patterns are given in Table 8. All such studies have so far made use of SAMe, except for one study which used LMF.[112] For all the

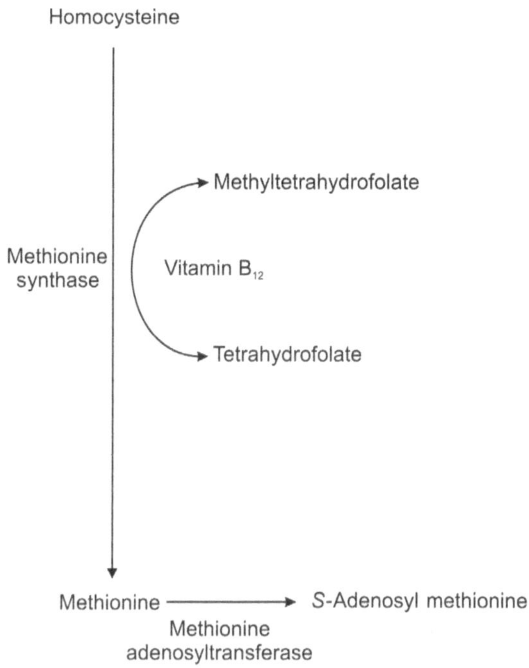

Fig. 1 Schematic representation of conversion of homocysteine to methionine, and of methionine to *S*-adenosyl methionine. *Reproduced with permission from Peedicayil J. Role of epigenetics in pharmacotherapy, psychotherapy and nutritional management of mental disorders. J Clin Pharm Ther. 2012;37:499–501.*[110]

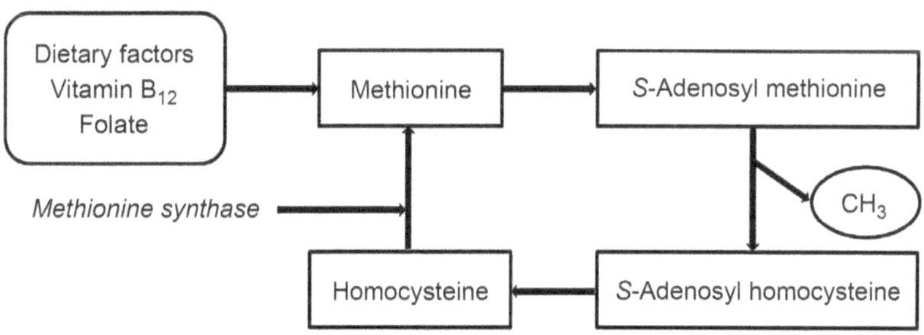

Fig. 2 Schematic representation of the role of the dietary factors vitamin B_{12} and folic acid in the formation of *S*-adenosyl methionine from methionine. *Reproduced with permission from Matt S, Roth ED, Roth TL. Role of epigenetics in the brain. In: Peedicayil J, Grayson DR, Avramopoulos D, eds. Epigenetics in Psychiatry. Waltham, MA: Elsevier; 2014:79–99.*[111]

Table 8 Nutritional Drugs With Epigenetic Effects in Mood Disorders: Nonrandomized Controlled Trials.

Drug	Mood Disorder	Effect	References
SAMe	MDD	Clinically improves seven of nine treated patients	116
SAMe	MDD	Improves symptoms	117
SAMe	MDD	Significantly improves symptoms	118
SAMe	MDD	Rapidly improves symptoms	119
SAMe	MDD	Improves symptoms	120
SAMe	MDD	Augments effects of standard antidepressants	121
SAMe	MDD	Rapidly decreases depression	122
SAMe	MDD	Augments SSRI effects; improves cognitive symptoms	123
LMF	BD	Augments effects of standard antidepressants	112

BD, Bipolar disorder; *MDD*, major depressive disorder; *LMF*, L-methylfolate; *SAMe*, *S*-adenosyl methionine; *SSRI*, selective serotonin-reuptake inhibitor.

studies the diagnosis of the subjects was MDD, except for one study which was conducted on patients with BD.[112] As shown, all nonrandomized controlled trials conducted so far investigating nutritional drugs acting epigenetically have been found to show beneficial effects. The results of randomized controlled trials (RCTs) investigating nutritional drugs acting epigenetically are given in Table 9. Most studies have investigated the use of SAMe, although three studies have investigated the use of LMF, and one investigated the use of folic acid. Except for one study which was conducted on patients with BD,[113] and another on patients with postmenopausal depression,[114] the studies were conducted on patients with MDD. As shown, most, but not all of the RCT using SAMe so far conducted, have found beneficial effects of SAMe in treated patients. It is interesting that all RCT that used LMF demonstrated significant clinical improvement, whereas the RCT that investigated the use of folic acid did not find significant clinical improvement.[115] As alluded to earlier, folic acid unlike LMF does not act as a methyl donor in the body.

4. CONCLUDING REMARKS

There is experimental evidence that changes in epigenetic mechanisms of gene expression are involved in the pathogenesis of the mood

Table 9 Nutritional Drugs With Epigenetic Effects in Mood Disorders: Randomized Controlled Trials.

Drug	Mood Disorder	Effect	References
LMF	MDD	Causes significant clinical and social improvement	124
SAMe	PMD	Has rapid antidepressant effect	114
SAMe	MDD	Causes significant clinical improvement	125
SAMe	MDD	Causes significant clinical improvement	126
SAMe	MDD	The effect of SAMe is similar to that of imipramine	127
SAMe	MDD	The effect of SAMe is similar to that of imipramine	128
SAMe	MDD	SAMe is effective for nonresponders to SSRIs	129
LMF	MDD	Causes significant clinical improvement	130
SAMe	MDD	SAMe is an effective antidepressant	131
Folic acid	MDD	Does not improve clinical symptoms	115
SAMe	BD	Does not improve clinical symptoms	113
SAMe	MDD	SAMe is not superior to placebo	132
LMF	MDD	Significantly improves depressive symptoms	133

BD, Bipolar disorder; *LMF*, L-methylfolate; *MDD*, major depressive disorder; *PMD*, postmenopausal depression; *SAMe*, *S*-adenosyl methionine; *SSRIs*, selective serotonin-reuptake inhibitors.

disorders (MDD and BD). So far the best established finding appears to be hypermethylation of the promoter of the BDNF gene in peripheral cells and tissues of patients with MDD. There also appears to be abnormalities in the expression and activity of miRNAs underlying mood disorders. However, much more work needs to be done to elucidate the precise miRNAs and the abnormalities involved. Epigenetic drugs, both DNMTi and HDACi, are showing favorable results in preclinical trials for the treatment of MDD. HDACi appear to show more favorable results than DNMTi in this respect. Preclinical trials of epigenetic drugs for BD have so far only investigated the use of HDACi, which are showing favorable results. Among the 18 HDAC isozymes, HDAC2 is thought to play a major role in memory and cognition, and is thought to negatively regulate these brain functions.[134,135] Hence, inhibition of HDAC2 by HDAC2-specific inhibitors that cross the blood–brain barrier could be useful clinically in improving cognition in patients with mood disorders. Indeed, such drugs are undergoing preclinical trials and are showing favorable results in this regard.[136,137] Efforts are being made to develop HDAC2-specific inhibitors.[138,139] Nutritional drugs with epigenetic effects are showing promising results in nonrandomized controlled trials and RCT of these drugs for treating mood disorders. So far most studies have tried nutritional drugs with epigenetic effects for the treatment of MDD, and only a few have been conducted on patients with BD.

In addition to DNMTi and HDACi, several other classes of epigenetic drugs are undergoing preclinical and clinical trials, at present mainly for use as anticancer drugs. Such drugs could also be investigated for use in the treatment of mood disorders. Such drugs include inhibitors of the bromodomain and extra-terminal domain (BET) family of proteins.[140,141] They could be useful epigenetic drugs for targeting the BET family of proteins, known to be involved in controlling epigenetic mechanisms of gene expression in the brain.[142] Another potential epigenetic target for drugs to treat mood disorders is histone lysine demethylase, the enzyme that catalyzes the demethylation of lysine amino acid residues on histones.[143]

REFERENCES

1. World Health Organization. The ICD-10 Classification of Mental and Behavioural Disorders. Diagnostic Criteria for Research. Geneva: World Health Organization; 1993
2. Sadock BA, Sadock VA, Ruiz P. *Synopsis of Psychiatry*. 11th ed. Philadelphia, PA: Wolters Kluwer; 2015.
3. Baingana F, al'Absi M, Becker AE, et al. Global research challenges and opportunities for mental health and substance-use disorders. *Nature*. 2015;527:S172–S177.
4. Akiskal HS. Mood disorders: historical introduction and conceptual overview. In: Sadock BJ, Sadock VA, Ruiz P, eds. In: *Comprehensive Textbook of Psychiatry. Philadelphia, PA: Wolters Kluwer; 2017. 13.1.*
5. aan het Rot M, Mathew SJ, Charney DS. Neurobiological mechanisms in major depressive disorder. *CMAJ*. 2009;180:305–313.
6. Pemberton R, Tyszkiewicz F. Factors contributing to depressive mood states in everyday life: a systematic review. *J Affect Disord*. 2016;200:103–110.
7. Gebicke-Haerter PJ. Systems psychopharmacology: a network approach to developing novel therapies. *World J Psychiatry*. 2016;6:66–83.
8. Callaway E. Genome studies attract criticism. *Nature*. 2017;546:463.
9. Lee R, Avramopoulos D. Introduction to epigenetics in psychiatry. In: Peedicayil J, Grayson DR, Avramopoulos D, eds. In: *Epigenetics in Psychiatry. Waltham, MA: Elsevier; 2014:3–25.*
10. Belmaker RH, Agam G. Major depressive disorder. *N Engl J Med*. 2008;358:55–68.
11. Rock PL, Roiser JP, Riedel WJ, et al. Cognitive impairment in depression: a systematic review and meta-analysis. *Psychol Med*. 2014;44:2029–2040.
12. Blacker CV, Clare AW. Depressive disorder in primary care. *Br J Psychiatry*. 1987;150:737–751.
13. Katon W, Schulberg H. Epidemiology of depression in primary care. *Gen Hosp Psychiatry*. 1992;14:237–247.

14. Davidson JR, Meltzer-Brody SE. The underrecognition and undertreatment of depression: what is the breadth of the problem? *J Clin Psychiatry.* 1999;60(suppl 7):4–9.
15. Nierenberg AA. Current perspectives on the diagnosis and treatment of major depressive disorder. *Am J Manag Care.* 2001;7:S353–S366.
16. Mann JJ. The medical management of depression. *N Engl J Med.* 2005;353:1819–1834.
17. O'Donnell JM, Shelton RC. Drug therapy of depression and anxiety disorders. In: Brunton LL, Chabner BA, Knollmann BC, eds. In: *The Pharmacological Basis of Therapeutics. New York: McGraw-Hill; 2011:397–415.*
18. Cipriani A, Zhou X, Del Giovane C, et al. Comparative efficacy and tolerability of antidepressants for major depressive disorder in children and adolescents: a network meta-analysis. *Lancet.* 2016;388:881–890.
19. Newport DJ, Carpenter LL, McDonald WM, et al. Ketamine and other NMDA antagonists: early clinical trials and possible mechanisms in depression. *Am J Psychiatry.* 2015;172:950–966.
20. Ghosh A, Hellewell JS. A review of the efficacy and tolerability of agomelatine in the treatment of major depression. *Expert Opin Investig Drugs.* 2007;16:1999–2004.
21. Kupfer DJ, Frank E, Phillips ML. Major depressive disorder: new clinical, neurobiological, and treatment perspectives. *Lancet.* 2012;379:1045–1055.
22. Rang HP, Ritter JM, Flower RJ, Henderson G. *Rang & Dales Pharmacology.* 8th ed. London: Elsevier; 2016.
23. Jenkins E, Goldner EM. Approaches to understanding and addressing treatment-resistant depression: a scoping review. *Depression Res Treat.* 2012;2012:469680.
24. Belmaker RH. Bipolar disorder. *N Engl J Med.* 2004;351:476–486.
25. Frye MA. Bipolar disorder—a focus on depression. *N Engl J Med.* 2011;364:51–59.
26. Grande I, Berk M, Birmaher B, Vieta E. Bipolar disorder. *Lancet.* 2016;387:1561–1571.
27. Latalova K, Prasko J, Diveky T, et al. Cognitive impairment in bipolar disorder. *Biomed Pap Med Fac Univ Palacky lomouc Czech Repub.* 2011;155:19–26.
28. Barrios C, Chaudhry TA, Goodnick PJ. Rapid cycling bipolar disorder. *Expert Opin Pharmacother.* 2001;2:1963–1973.
29. Bauer MS, Mitchner L. What is a "mood stabilizer"? An evidence-based response. *Am J Psychiatry.* 2004;161:3–18.
30. Grande I, Berk M, Birmaher B, et al. Bipolar disorder. *Lancet.* 2016;387:1561–1572.
31. Peedicayil J. Epigenetic approaches for bipolar disorder drug discovery. *Expert Opin Drug Discov.* 2014;9:917–930.
32. Gitlin M. Treatment-resistant bipolar disorder. *Mol Psychiatry.* 2006;11:227–240.
33. Poon SH, Sim K, Sum MY, et al. Evidence-based options for treatment-resistant adult bipolar disorder. *Bipolar Disord.* 2012;14:573–584.
34. MacQueen GM, Memedovich KA. Cognitive dysfunction in major depression and bipolar disorder: assessment and treatment options. *Psychiatry Clin Neurosci.* 2017;71: 18–27.
35. Smalheiser NR, Lugli G, Rizavi HS, et al. MicroRNA expression is down-regulated and reorganized in prefrontal cortex of depressed suicide subjects. *PLoS One.* 2012;7: e33201.
36. Lopez JP, Fiori LM, Gross JA, et al. Regulatory role of miRNAs in polyamine gene expression in the prefrontal cortex of depressed suicide completers. *Int J Neuropsychopharmacol.* 2014;23–32.
37. Roy B, Dunbar M, Shelton RC, et al. Identification of microRNA-124-3p as a putative epigenetic signature of major depressive disorder. *Neuropsychopharmacology.* 2017;42: 864–875.
38. Kaut O, Schmitt I, Hofmann A, et al. Aberrant NMDA receptor DNA methylation detected by epigenome-wide analysis of hippocampus and prefrontal cortex in major depression. *Eur Arch Psychiatry Clin Neurosci.* 2015;265:331–341.

39. Sabunciyan S, Aryee MJ, Irizarry RA, et al. Genome-wide DNA methylation scan in major depressive disorder. *PLoS One*. 2012;7:e34451.
40. Cruceanu C, Kutsarova E, Chen ES, et al. DNA hypomethylation of synapsin II CpG islands associates with increased gene expression in bipolar disorder and major depression. *BMC Psychiatry*. 2016;16:286.
41. Murphy TM, Crawford B, Dempster EL, et al. Methylomic profiling of cortex samples from completed suicide cases implicates a role for PSORS1C3 in major depression and suicide. *Transl Psychiatry*. 2017;7:e989.
42. Kaut O, Sharma A, Schmitt I, et al. DNA methylation of DLG4 and GJA-1 of human hippocampus and prefrontal cortex in major depression is unchanged in major depression is unchanged in comparison to healthy individuals. *J Clin Neurosci*. 2017;43:261–263.
43. Kang H-J, Kim J-M, Lee J-Y, et al. BDNF promoter methylation and suicidal behavior in depressive patients. *J Affect Disord*. 2013;151:679–685.
44. Kim J-M, Stewart R, Kang H-J, et al. A longitudinal study of BDNF promoter methylation and genotype in poststroke depression. *J Affect Disord*. 2013;149:93–99.
45. D'Addario C, Dell'Osso B, Galimberti D, et al. Epigenetic modulation of BDNF gene in patients with major depressive disorder. *Biol Psychiatry*. 2013;73:e6–e7.
46. Lopez JP, Mamdani F, Labonte B, et al. Epigenetic regulation of BDNF expression according to antidepressant response. *Mol Psychiatry*. 2013;18:398–489.
47. Mitchelmore C, Gede L. Brain derived neurotrophic factor: epigenetic regulation in psychiatric disorders. *Brain Res*. 2014;1586:162–172.
48. Zill P, Baghai TC, Schüle C, et al. DNA methylation analysis of the angiotensin converting enzyme (ACE) gene in major depression. *PLoS One*. 2012;7:e40479.
49. Tseng P-T, Lin P-Y, Lee Y, et al. Age-associated decrease in global DNA methylation in patients with major depression. *Neuropsychiatr Dis Treat*. 2014;10:2105–2114.
50. Tyrka AR, Parade SH, Welch ES, et al. Methylation of the leukocyte glucocorticoid receptor gene promoter in adults: associations with early adversity and depressive, anxiety and substance-use disorders. *Transl Psychiatry*. 2016;6:e848.
51. Won E, Choi S, Kang J, et al. Association between reduced white matter integrity in the corpus callosum and serotonin transporter gene DNA methylation in medication-naive patients with major depressive disorder. *Transl Psychiatry*. 2016;6:e866.
52. Cruceanu C, Alda M, Nagy C, et al. H3K4 tri-methylation in synapsin genes leads to different expression patterns in bipolar disorder and major depression. *Int J Neurosychopharmacol*. 2013;16:289–299.
53. Barde S, Rüegg J, Prud'homme J, et al. Alterations in the neuropeptide galanin system in major depressive disorder involve levels of transcripts, methylation, and peptide. *Proc Natl Acad Sci USA*. 2016;113:E8472–E8481.
54. Gross JA, Pacis A, Chen GG, et al. Gene-body 5-hydroxymethylation is associated with gene expression changes in the prefrontal cortex of depressed individuals. *Transl Psychiatry*. 2017;7:e1119.
55. Liu Z, Li X, Sun N. Microarray profiling and co-expression network analysis of circulating lncRNAs and mRNAs associated with major depressive disorder. *PLoS One*. 2014;9:e93388.
56. Oh G, Wang S-C, Pal M, et al. DNA modification study of major depressive disorder: beyond locus-by-locus comparisons. *Biol Psychiatry*. 2015;77:246–255.
57. Malki K, Koritskaya E, Harris F, et al. Epigenetic differences in monozygotic twins discordant for major depressive disorder. *Transl Psychiatry*. 2016;6:e839.
58. Kimmel M, Clive M, Gispen F, et al. Oxytocin receptor DNA methylation in postpartum depression. *Psychoneuroendocrinology*. 2016;69:150–160.
59. Roy B, Shelton RC, Dwivedi Y. DNA methylation and expression of stress related genes in PBMC of MDD patients with and without serious suicidal ideation. *J Psychiatr Res*. 2017;89:115–124.

60. Boström AE, Ciuculete DM, Attwood M, et al. A MIR4646 associated methylation locus is hypomethylated in adolescent depression. *J Affect Dis.* 2017;220:117–128.
61. Veldic M, Guidotti A, Maloku E, et al. In psychosis, cortical interneurons overexpress DNA-methyltransferase 1. *Proc Natl Acad Sci USA.* 2005;102:2152–2157.
62. Guidotti A, Ruzicka W, Grayson DR, et al. *S*-Adenosyl methionine and DNA methyltransferase-1 mRNA overexpression in psychosis. *Neuroreport.* 2007;18:57–60.
63. Guidotti A, Auta J, Davis JM, et al. Decrease in reelin and glutamic acid decarboxylase$_{67}$ (GAD_{67}) expression in schizophrenia and bipolar disorder: a post-mortem brain study. *Arch Gen Psychiatry.* 2000;57:1061–1069.
64. Dong E, Gavin DP, Chen Y, et al. Upregulation of TET1 and downregulation of APOBEC3A in the parietal cortex of psychotic patients. *Transl Psychiatry.* 2012;2: e159.
65. D'Addario C, Dell'Osso B, Palazzo MC, et al. Selective DNA methylation of BDNF promoter in bipolar disorder: differences among patients with BDI and BDII. *Neuropsychopharmacology.* 2012;37:1647–1655.
66. Sugawara H, Iwamoto K, Bundo M, et al. Hypermethylation of serotonin transporter gene in bipolar disorder detected by epigenome analysis of discordant monozygotic twins. *Transl Psychiatry.* 2011;1:e24.
67. Abdolmaleky HM, Cheng K-h. Faraone SV, et al. Hypomethylation of *MB-COMT* promoter is a major risk factor for schizophrenia and bipolar disorder. *Hum Mol Genet.* 2006;15:3132–3145.
68. Mill J, Tang T, Kaminsky Z, et al. Epigenomic profiling reveals DNA methylation changes associated with major psychosis. *Am J Hum Genet.* 2008;82:696–711.
69. Kim AH, Reimers M, Maher B, et al. microRNA expression profiling in the prefrontal cortex of individuals affected with schizophrenia and bipolar disorders. *Schizophr Res.* 2010;124:183–191.
70. Kaminsky Z, Tochigi M, Jia P, et al. A multi-tissue analysis identifies HLA complex group 9 gene methylation differences in bipolar disorder. *Mol Psychiatry.* 2012;17: 728–740.
71. Rao JS, Keleshian VL, Klein S, et al. Epigenetic modifications in frontal cortex from Alzheimer's disease and bipolar disorder patients. *Trans Psychiatry.* 2012;e132.
72. Akula N, Barb J, Jiang X. RNA-sequencing of the brain transcriptome implicates dysregulation of neuroplasticity, circadian rhythms and GTPase binding in bipolar disorder. *Mol Psychiatry.* 2014;19:1179–1185.
73. Banigan MG, Kao PF, Kozubek JA, et al. Differential expression of exosomal microRNAs in prefrontal cortices of schizophrenia and bipolar disorder patients. *PLoS One.* 2013;8:e48814.
74. Kaminsky Z, Jones I, Verma R, et al. DNA methylation and expression of KCNQ3 in bipolar disorder. *Bipolar Disord.* 2015;17:150–159.
75. Ruzicka WB, Subburaju S, Benes FM. Circuit- and diagnosis-specific DNA methylation changes at γ-aminobutyric acid-related genes in postmortem human hippocampus in schizophrenia and bipolar disorder. *JAMA Psychiatry.* 2015;72:541–551.
76. Pal M, Ebrahimi S, Oh G, et al. High precision DNA modification analysis of HCG9 in major psychosis. *Schizophr Bull.* 2016;42:170–177.
77. Kuratomi G, Iwamoto K, Bundo M, et al. Aberrant DNA methylation associated with bipolar disorder identified from discordant monozygotic twins. *Mol Psychiatry.* 2008;13:429–441.
78. Chen H, Wang N, Burmeister M, et al. microRNA expression changes in lymphoblastoid cell lines in response to lithium treatment. *Int J Neuropsychopharmacol.* 2009;12:975–981.
79. Bromberg A, Bersudsky Y, Levine J, et al. Global leukocyte DNA methylation is not altered in euthymic bipolar patients. *J Affect Disord.* 2009;118:234–239.

80. McCarthy MJ, Barrett TB, Nissen S, et al. Allele specific analysis of the *ADRBK2* gene in lymphoblastoid cells from bipolar disorder patients. *J Psychiatr Res.* 2010;44: 201–208.
81. Ghadirivasfi M, Nohesara S, Ahmadkhaniha H-R, et al. Hypomethylation of the serotonin receptor type-2A gene (HTR2A) at T102C polymorphic site in DNA derived from the saliva of patients with schizophrenia and bipolar disorder. *Am J Med Genet B.* 2011;156:536–545.
82. Carrard A, Salzmann A, Malafosse A, et al. Increased DNA methylation status of the serotonin receptor *5HTR1A* gene promoter in schizophrenia and bipolar disorder. *J Affect Dis.* 2011;132:450–453.
83. Nohesara S, Ghadirivasfi M, Mostafavi S, et al. DNA hypomethylation of MB-COMT promoter in the DNA derived from saliva in schizophrenia and bipolar disorder. *J Psychiatr Res.* 2011;45:1432–1438.
84. Dempster EL, Wong CC, Lester KJ, et al. Genome-wide methylomic analysis of monozygotic twins discordant for adolescent depression. *Biol Psychiatry.* 2014;76:977–983.
85. Forstner AJ, Hofman A, Maaser A, et al. Genome-wide analysis implicates microRNAs and their target genes in the development of bipolar disorder. *Transl Psychiatry.* 2015;5: e678.
86. Machado-Vieira R, Ibrahim L, Zarate CA. Histone deacetylases and mood disorders: epigenetic programming in gene-environment interactions. *CNS Neurosci Ther.* 2011;17:699–704.
87. Guidotti A, Grayson DR. DNA methylation and demethylation as targets for antipsychotic therapy. *Dialogues Clin Neurosci.* 2014;16:419–429.
88. Ignácio ZM, Réus GZ, Abelaira HM, et al. Quetiapine treatment reverses depressive-like behavior and reduces DNA methyltransferase activity induced by maternal deprivation. *Behav Brain Res.* 2017;320:225–232.
89. Xing B, Liu P, Xu F-y. et al. Effect of microinjecting 5-aza-2-deoxycytidine into ventrolateral orbital cortex on depressive-like behavior in rats. *Neurosci Lett.* 2014;574:11–14.
90. Sales AJ, Biojone C, Terceti MS, et al. Antidepressant-like effect induced by systemic and intra-hippocampal administration of DNA methylation inhibitors. *Br J Pharmacol.* 2011;164:1711–1721.
91. Schroeder FA, Lewis MC, Fass DM, et al. A selective HDAC 1/2 inhibitor modulates chromatin and gene expression in brain and alters mouse behavior in two mood-related tests. *PLoS One.* 2013;8:e71323.
92. Sales AJ, Joca SRL. Effects of DNA methylation inhibitors and conventional antidepressants on mice behaviour and brain DNA methylation levels. *Acta Neuropsychiatr.* 2016;28:11–22.
93. Wei YB, Melas PA, Wegener G, et al. Antidepressant-like effect of sodium butyrate is associated with an increase in TET1 and in 5-hydroxymethylation levels in the Bdnf gene. *Int J Neuropsychopharmacol.* 2014;18(2).
94. Su C-L, Su C-W, Hsiao Y-H, et al. Epigenetic regulation of BDNF in the learned helplessness-induced model of depression. *J Psychiatr Res.* 2016;76:101–110.
95. Meylan EM, Halfon O, Magistretti PJ, et al. The HDAC inhibitor SAHA improves depressive-like behavior of CRTC1-deficient mice: possible relevance for treatment-resistant depression. *Neuropharmacology.* 2016;107:111–121.
96. Hayase T. Putative epigenetic involvement of the endocannabinoid system in anxiety- and depression-related behaviors caused by nicotine as a stressor. *PLoS One.* 2016;11: e0158950.
97. Muñoz-Cobo I, Belloch FB, Diaz-Perdigón T, et al. SIRT2 inhibition reverses anhedonia in the VGLUT1+/- depression model. *Behav Brain Res.* 2017;335:128–131.

98. Sharma RP, Rosen C, Kartan S, et al. Valproic acid and chromatin remodeling in schizophrenia and bipolar disorder: preliminary results from a clinical population. *Schizophr Res.* 2006;88:227–231.
99. Gavin DP, Kartan S, Chase K, et al. Histone deacetylase inhibitors and candidate gene expression: an in vivo and in vitro approach to studying chromatin remodeling in a clinical population. *J Psychiatr Res.* 2009;43:870–876.
100. Yasuda S, Liang MH, Marinova Z, et al. The mood stabilizers lithium and valproate selectively activate the promoter IV of brain-derived neurotrophic factor in neurons. *Mol Psychiatry.* 2009;14:51–59.
101. Moretti M, Valvassori SS, Varela RB, et al. Behavioral and neurochemical effects of sodium butyrate in an animal model of mania. *Behav Pharmacol.* 2011;22:766–772.
102. Douma TN, Millan MJ, Verdouw PM, et al. Valproate improves prepulse inhibition deficits induced by corticotrophin-releasing factor independent of GABA(A) and GABA(B) receptor activation. *Neuropharmacology.* 2014;79:66–74.
103. Choi JY, Kim JH, Jo SA. Acetylation regulates the stability of glutamate carboxypeptidase II protein in human astrocytes. *Biochem Biophys Res Commun.* 2014;450:372–377.
104. Dozawa M, Kono H, Sato Y, et al. Valproic acid, a histone deacetylase inhibitor, regulates cell proliferation in the adult zebrafish optic tectum. *Dev Dyn.* 2014;243: 1401–1415.
105. Varela RB, Valvassori SS, Lopes-Borges J, et al. Sodium butyrate and mood stabilizers block oubain-induced hyperlocomotion and increase BDNF, NGF and GDNF levels in brain of Wistar rats. *J Psychiatr Res.* 2015;61:114–121.
106. Balasubramanian D, Deng AX, Doudney K, et al. Valproic acid exposure leads to upregulation and increased promoter histone acetylation of sepiapterin reductase in a serotonergic cell line. *Neuropharmacology.* 2015;99:79–88.
107. Aizawa S, Yamamuro Y. Valproate administration to mice increases hippocampal p21 expression by altering genomic DNA methylation. *Neuroreport.* 2015;26:915–920.
108. Lee HJ, Dreyfus C, DiCicco-Bloom E. Valproic acid stimulates proliferation of glial precursors during cortical gliogenesis in developing rat. *Dev Neurobiol.* 2016;76: 780–798.
109. Valvassori SS, Dal-Pont GC, Steckert AV, et al. Sodium butyrate has an antimanic effect and protects the brain against oxidative stress in an animal model of mania induced by oubain. *Psychiatr Res.* 2016;235:154–159.
110. Peedicayil J. Role of epigenetics in pharmacotherapy, psychotherapy and nutritional management of mental disorders. *J Clin Pharm Ther.* 2012;37:499–501.
111. Matt S, Roth ED, Roth TL. Role of epigenetics in the brain. In: Peedicayil J, Grayson DR, Avramopoulos D, eds. In: *Epigenetics in Psychiatry. Waltham, MA: Elsevier; 2014:79–99.*
112. Nierenberg AA, Montana R, Kinrys G, et al. L-Methylfolate for bipolar I depressive episodes: an open trial proof-of-concept registry. *J Affect Disord.* 2017;207:429–433.
113. Murphy BL, Babb SM, Ravichandran C, et al. Oral SAMe in persistent treatment-refractory bipolar depression: a double-blind, randomized clinical trial. *J Clin Psychopharmacol.* 2014;34:413–416.
114. Kagan BL, Sultzer DL, Rosenlicht N, et al. Oral *S*-adenosylmethionine in depression: a randomized, double-blind, placebo-controlled trial. *Am J Psychiatry.* 1990;147:591–595.
115. Bedson E, Bell D, Carr D, et al. Folate augmentation of treatment-evaluation for depression (FolATED): randomized trial and economic evaluation. *Health Technol Assess.* 2014;18(48).
116. Lipinski JF, Cohen BM, Frankenburg F, et al. Open trial of *S*-adenosylmethionine for treatment of depression. *Am J Psychiatry.* 1984;141:448–450.

117. Rosenbaum JF, Fava M, Falk WE, et al. An open-label pilot study of oral S-adenosyl-L-methionine in major depression: interim results. *Psychopharmacol Bull.* 1988;24:189–194.
118. Rosenbaum JF, Fava M, Falk WE, et al. The antidepressant potential of oral S-adenosyl-L-methionine. *Acta Psychiatr Scand.* 1990;81:432–436.
119. Fava M, Giannelli A, Rapisarda V, et al. Rapidity of onset of the antidepressant effect of parenteral S-adenosyl-L-methionine. *Psychiatry Res.* 1995;56:295–297.
120. Di Rocco A, Rogers JD, Brown R, et al. S-Adenosyl-methionine improves depression in patients with Parkinson's disease in an open-label clinical trial. *Movement Disord.* 2000;15:1225–1229.
121. Alpert JE, Papakostas G, Mischoulon D, et al. S-Adensyl-L-methionine (SAMe) as an adjunct for resistant major depressive disorder. *J Clin Psychopharmacol.* 2004;24:661–664.
122. Shippy RA, Mendez D, Jones K, et al. S-Adenosylmethionine (SAM-e) for the treatment of depression in people living with HIV/AIDS. *BMC Psychiatry.* 2004;4:38.
123. Levkovitz Y, Alpert JE, Brintz CE, et al. Effects of S-adenosylmethionine augmentation of serotonin-reuptake inhibitor antidepressants on cognitive symptoms of major depressive disorder. *J Affect Disord.* 2012;136:1174–1178.
124. Godfrey PSA, Toone BK, Carney MWP, et al. Enhancement of recovery from psychiatric illness by methylfolate. *Lancet.* 1990;336:392–395.
125. Salmaggi P, Bressa GM, Nicchia G, et al. Double-blind, placebo-controlled study of S-adenosyl-L-methionine in depressed postmenopausal women. *Psychother Psychosom.* 1993;59:34–40.
126. Bell KM, Potkin SG, Carreon D, et al. S-Adenosylmethionine blood levels in major depression: changes with drug treatment. *Acta Neurol Scand Suppl.* 1994;154:15–18.
127. Pancheri P, Scapicchio P, Chiaie RD. A double-blind, randomized parallel-group, efficacy and safety study of intramuscular S-adenosyl-L-methionine 1,4-butanedisulphonate (SAMe) versus imipramine in patients with major depressive disorder. *Int J Neuropsychopharmacol.* 2002;5:287–294.
128. Delle Chiaie R, Pancheri P, Scapicchio P. Efficacy and tolerability of oral and intramuscular S-adenosyl-L-methionine 1,4-butanedisulfonate (SAMe) in the treatment of major depression: comparison with imipramine in 2 multicenter studies. *Am J Clin Nutr.* 2002;76:1172S–1176S.
129. Papakostas GI, Mischoulon D, Shyu I, et al. S-Adenosyl methionine (SAMe) augmentation of serotonin reuptake inhibitors for antidepressant nonresponders with major depressive disorder: a double-blind, randomized clinical trial. *Am J Psychiatry.* 2010;167:942–948.
130. Papakostas GI, Shelton RC, Zajecka JM, et al. Effect of adjunctive L-methylfolate 15 mg among inadequate responders to SSRIs in depressed patients who were stratified by biomarker levels and genotype: results from a randomized clinical trial. *J Clin Psychiatry.* 2014;75:855–863.
131. Sarris J, Papakostas GI, Vitolo O, et al. S-Adenosyl methionine (SAMe) versus escitalopram and placebo in major depression RCT: efficacy and effects of histamine and carnitine as moderators of response. *J Affect Disord.* 2014;164:76–81.
132. Mischoulon D, Price LH, Carpenter LL, et al. A double-blind, randomized, placebo-controlled clinical trial of S-adenosyl-L-methionine (SAMe) versus escitalopram in major depressive disorder. *J Clin Psychiatry.* 2014;75:370–376.
133. Shelton RC, Pencina MJ, Barrentine LW, et al. Association of obesity and inflammatory marker levels on treatment outcome: results from a double-blind, randomized study of adjunctive L-methylfolate calcium in patients with MDD who are inadequate responders to SSRIs. *J Clin Psychiatry.* 2015;76:1635–1641.
134. Gräff J, Tsai LH. Histone acetylation: molecular mnemonics on the chromatin. *Nat Rev Neurosci.* 2013;14:97–111.

135. Penney J, Tsai LH. Histone deacetylases in memory and cognition. *Sci Signal.* 2014;7: re12.
136. Wagner FF, Weiver M, Steinbacher S, et al. Kinetic and structural insights into the binding of histone deacetylase 1 and 2 (HDAC 1,2) inhibitors. *Bioorg Med Chem.* 2016;24:4008–4015.
137. Wagner FF, Zhang Y-L, Fass DM, et al. Kinetically selective inhibitors of histone deactylase 2 (HDAC2) as cognition enhancers. *Chem Sci.* 2015;6:804–815.
138. Zhou J, Li M, Chen N, et al. Computational design of a time-dependent histone deacetylase 2 selective inhibitor. *ACS Chem Biol.* 2015;10:687–692.
139. Basant N, Lin X, Reid TE, et al. Discovery of a novel HDAC2 inhibitor by a scaffold-merging hybrid query. *Comb Chem High Throughput Screen.* 2015;18:693–700.
140. Noguchi-Yachide T. BET bromodomain as a target of epigenetic therapy. *Chem Pharm Bull (Tokyo).* 2016;64:540–547.
141. Ferri E, Petosa C, McKenna CE. Bromodomains: structure, function and pharmacology of inhibition. *Biochem Pharmacol.* 2016;106:1–18.
142. Wahlestedt C. Emerging epigenetic therapies in neuroscience: focus on bromodomain-containing drug targets. *Neuropsychopharmacology.* 2017;42:374–375.
143. Peter CJ, Akbarian S. Balancing histone methylation activities in psychiatric disorders. *Trends Mol Med.* 2011;17:372–379.

CHAPTER NINE

DNA Methylation and Psychiatric Disorders

Chunyu Liu*[†,1], Chuan Jiao[†], Kangli Wang[†], Ning Yuan[‡]**
* University of Illinois, Chicago, IL, United States
† School of Life Science, Central South University, Changsha, China
‡ Hunan Brain Hospital, Changsha, China
1 Corresponding author. E-mail address: liuch@upstate.edu

Contents

Abstract

DNA methylation has been an important area of research in the study of molecular mechanism to psychiatric disorders. Recent evidence has suggested that abnormalities in global methylation, methylation of genes, and pathways could play a role in the etiology of many forms of mental illness. In this article, we review the mechanisms of

Progress in Molecular Biology and Translational Science, volume 157
ISSN 1877-1173
http://dx.doi.org/10.1016/bs.pmbts.2018.01.006

DNA methylation, including the genetic and environmental factors affecting methylation changes. We report and discuss major findings regarding DNA methylation in psychiatric patients, both within the context of global methylation studies and gene-specific methylation studies. Finally, we discuss issues surrounding data quality improvement, the limitations of current methylation analysis methods, and the possibility of using DNA methylation-based treatment for psychiatric disorders in the future.

1. DNA METHYLATION IN HUMAN GENOME

DNA methylation is the epigenetic process by which methyl groups are added to DNA nucleotides, primarily cytosine and adenine. Methylation is thought to have a significant impact on the structure and functions of DNA through covalent chemical modification. This article focuses on DNA methylation in the human genome, specifically as it relates to psychiatric disorders.

1.1 Cytosine Methylation and Demethylation

In the human genome, DNA methylation predominantly occurs at cytosine nucleotides that are paired with guanine, resulting in what is termed CpG methylation. In addition, the majority of CpG sites (60%–90%) are methylated in mammals.[1] CpG methylation was once considered to be the only form of DNA methylation in the human genome, but recently, examples of non-CpG methylation have been discovered.

There are three major known forms of cytosine—cytosine, methylcytosine, and hydroxymethylcytosine, commonly symbolized as C, 5mC, and 5hmC, respectively, as the methylation and hydroxymethylation occur at the 5th carbon atom of the cytosine ring. 5mC is the most prevalent form of DNA methylation. 5hmC occurs more frequently in brain, heart, and sperm and less frequently in spleen, liver, and cancer cell lines.[2–4] 5hmC constitutes 0.6% of total nucleotides in Purkinje cells and 0.2% of total nucleotides in granule cells. In contrast, 5mC accounts for 1.5% of total nucleotides in Purkinje cells and 2% of granule cell nucleotides. About 0.5% of the total nucleotides in Purkinje and granule cells are native, unmodified C.[3]

5hmC is one of the intermediate products in mammalian DNA demethylation.[5] Ten-eleven translocation cytosine dioxygenases (TETs) oxidize 5mC to produce 5-hydroxymethylcytosine (5hmC), 5-formylcytosine (5fC), and 5-carboxylcytosine (5caC). These intermediate products can then be removed by base excision repair (BER) proteins. These oxidized cytosines (5hmC, 5fC, and 5caC) may possess unique regulatory functions in the mammalian genome.[6] 5fC and 5caC are present at extremely low levels in the genome, and their levels of biological significance remain to be seen.

Both 5hmC and 5mC levels change over development. 5hmC levels were found to increase during neuronal differentiation in embryonic mouse brain[7]; 5hmC levels were also found to increase initially and then slowly decrease afterward in mouse embryonic stem cells. In contrast, 5mC levels were found to increase gradually over time in the same process.[2,3]

1.2 CpG and Non-CpG (CpH) Methylation

Even though CG dinucleotides are the primary sites for DNA methylation in mammals, DNA methylation at non-CpG sites (CpA, CpT, and CpC) have been discovered as well[8] and account for ~25% of DNA methylation in human and mouse brains.[9] Non-CpG methylation (denoted CpH) sites are most likely to be followed by a G; they are commonly seen in stem cells,[10] but have also been discovered in other tissues types, including skeletal muscle cell culture,[11] occyte,[12] and brain.[13] Most methylated CpH methylation sites are near highly methylated CpGs.[9] Neurons appear to have the highest level of CpH methylation.[13] Although CpG methylation typically exhibits a bimodal distribution of methylation levels, CpH shows intermediate methylation levels (mostly ~25%).[9]

1.3 Enzymes Involved in Cytosine Methylation and Demethylation

In mammals, there are three major DNA methyltransferases (DNMTs): DNMT1, DNMT3a, and DNMT3b. Typically, DNMT1 acts as a maintenance DNMT, whereas DNMT3a and 3b are de novo DNMTs that establish new CpG methylation. However, these functions are not entirely exclusive to one type of enzyme.[14,15] For example, DNMT3a has been found to be critical for maintaining CpH in postmitotic neurons after the CpH has been established during neuronal maturation.[9,16] DNA demethylation involves TET and BER.[5]

1.4 Adenine Methylation

Methylation of N^6-adenine (N6mA, or m6A) was reported in mammalian embryonic stem cells.[17] The gene family complex including methyltransferase like 3 (METTL3), methyltransferase like 14 (METTL14), and WTAP[18] has been linked to methylation of N6mA.[19] The gene *Alkbh1* encodes a N6mA demethylase.[20] The METTL3–METTL14 complex methylates both DNA and RNA, with a preference for RNA.[21] Other N6Ma methyltransferases and demethylases might be discovered in the future, as METTL is a big gene family.

N6mA may perform many epigenetic regulatory functions, including suppressing gene expression and L1 element activity,[20] but much remains to be explored. A genome-wide sequencing method, DA-6mA-seq (DpnI-Assisted N^6-methylAdenine sequencing), has been developed to measure m6A in DNA.[22] Hopefully, this technology will lead to more studies of m6A in the human genome. Several reviews on N6mA are available for further reading.[23–26]

1.5 Functionality of DNA Methylation

Functions of CpG methylation: The primary function of DNA methylation has long been thought to be regulation of gene expression, particularly through suppression of gene expression, which is likely accomplished through the changing of transcription factor binding or chromatin organization. Unbiased genome-wide studies have revealed that DNA methylation levels of CpGs near transcription start sites frequently negatively correlate with expression levels of their corresponding genes, and methylation levels of gene body CpGs are often positively correlated with expression of their host genes. Moreover, both positive and negative correlations with gene expression occur at CpG sites across all kinds of genomic regions. Intergenic CpGs appear to correlate with gene expression less frequently. These phenomena have been observed in fibroblast cells and brain,[27,28] indicating that DNA methylation regulates gene expression differently according to its genomic context.

Altering transcription factor binding and changing chromatin organization are the two major ways in which DNA methylation may regulate gene expression.

Transcription factor binding: There is a possibility that some transcription factors are more sensitive to methylation status of DNA than others. For example, a study on CTCF showed that DNA methylation status has limited impact on CTCF binding patterns,[29] whereas a number of TFs, such as the ETS family and KLF4, have been found to be sensitive to methylated DNA.[30,31] EpiSELEX-seq[32] was recently developed to probe the sensitivity of transcription factor binding to DNA modifications in vitro. A catalog of DNA methylation-sensitive transcription factors will likely be needed for a better understanding of the impact of DNA methylation on gene expression.

Chromatin structure: DNA methylation established in early embryogenesis has been found to be critical to setting up chromatin structure, as measured by chromatin immunoprecipitation (ChIP) analysis.[33] More research is needed to resolve the relationship between DNA methylation and chromatin structure, though they may have a chicken–egg relationship: changes in methylation in a genomic region may lead to chromatin folding or unfolding, whereas, conversely, open chromatin could allow for binding of enzymes and changes in DNA methylation.[34]

However, a number of other biological effects, including gene expression, chromatin stability, and related mutation rates, could be products of chromatin changes,[35] genomic imprinting,[36] X chromosome inactivation,[37] and so on. The mutation rate was found to be much higher in medium (20%–60%) methylation level than in low (≤20%) methylation level and high (>80%) methylation level CpG sites within the human genome. Unmethylated CpGs had the lowest mutation rate.[35] X chromosome inactivation is highly correlated with methylation status at CpG islands.[37]

The functions of non-CpG methylation are still largely unknown. A few studies on promoter regions indicate that non-CpG methylation at gene promoters is associated with reduced gene expression, as in the case of the old model for CpG methylation.[38,39] Genome-wide assessment will be needed to get a complete picture of CpH methylation function, as has been done for CpG methylation.

1.6 Mitochondrial DNA Methylation

5mC and 5hmC were detected in mitochondrial DNA, where non-CpG methylation was found to account for ~50% of methylation within the D-loop region of mitochondria. The D-loop region controls mitochondrial DNA replication and transcription.[40] Therefore, it was suggested that all forms of DNA methylation (5mC, 5hmC, and non-CpG) may contribute to regulation of mitochondrial functions.

1.7 Techniques for Measuring DNA Methylation

A number of reviews have compared different methods for studying DNA methylation.[41–43] Global methylation level can be measured using ELISA, mass spectrometry,[44] and HPLC.[45] Measuring DNA methylation at the level of individual nucleotides in the genome requires methods such as microarray and sequencing.

Whole-genome bisulfite sequencing is the gold standard for measuring CpG and non-CpG methylation, but it is currently significantly more expensive than other methods. Choosing an ideal method requires a trade-off between costs and benefits. The best method for each experiment depends on the target methylation type (non-CpG, 5mhC, or 5mC), the coverage of the methylomic region being measured (candidate region vs. genome-wide), desired resolution (regional measure vs. single-base measure), input quantity (amount of DNA in experiment), and required sensitivity (extent of ability to detect small changes). For example, to quantify 5mhC, Tet-assisted bisulfite sequencing (TAB-Seq) is needed.[46] Some methods, such as methylated DNA immunoprecipitation (MeDIP), methyl-CpG-binding domain (MBD), and Cap-seq, only offer methylation measures of blocks of genomic regions, whereas other methods, such as bisulfite sequencing-based methods, have single-base resolution. Beyond this, data from different experimental platforms are associated with different costs and downstream analysis complexity. These factors should be considered when selecting methods for quantifying DNA methylation.

1.8 Materials for Studying DNA Methylation in Psychiatric Disorders

Tissues from live humans, postmortem specimens, and preserved clinical samples can be used to study DNA methylation. To study psychiatric disorders, we choose tissues based on a combination of accessibility and epigenetic similarities to brain. Because epigenetics is strongly tissue specific, we should know the pros and cons of all tissues for potential study. Brain is the most ideal tissue, though live brain tissue is only accessible via brain surgery, which is extremely rare for psychiatric patients today. For accessibility reasons, peripheral blood is the most commonly used tissue; saliva samples can also be used. Olfactory epithelium tissue and olfactory epithelium-derived cell lines have been proposed as promising surrogates for studying neuropsychiatric disorders[47]. However, harvesting olfactory epithelium is still a painful clinical procedure and so is less frequently used.

Brain data come with many confounding factors. What we can measure through postmortem tissue reveals only a snapshot of a human's life. Sex, age, life stress, brain injury, substance use, duration of illness, medication, cause of death, agonal state, and postmortem interval are all factors that could potentially bias findings and that therefore must be controlled for properly. The brain is a complex organ with many different regions composed of many different cell types. Most previous and current studies have used bulk tissue, which measures the average of all cells sampled in each tissue dissection. The random composition of cell types could bias the results. Ideally, cell composition should be controlled using statistical methods. Cell sorting, laser capture microdissection, and microfluidics are experimental methods used to collect relatively homogeneous cells that can be used for either pooled-cell analysis or single-cell analysis.

Peripheral blood is valuable for developing biomarkers, even though it may be difficult to use blood as a proxy for brain to study CNS function. It is important to note that DNA methylation profiling in blood is very different from profiling in brain.[48] When searching for a proxy in blood to represent brain, it is important that the loci of interest have highly correlated methylation levels in blood and brain; the loci should also respond to environmental insult consistently in both tissues.

One DNA methylation study with paired blood and temporal lobe biopsy samples from 12 epilepsy patients detected 7.9% of CpG sites as being significantly correlated between blood and brain tissue.[48] A principal component analysis of DNA methylation in blood and brain from 17 subjects found a principal component associated with age independent of tissue type.[49] Based on this study, a web application, BECon (Blood–Brain Epigenetic Concordance; https://redgar598.shinyapps.io/BECon/), was developed to catalog the blood–brain methylation correlation[50] for CpG sites assayed on the Illumina 450 Methylation chip.

A study of blood, cortex, and hippocampus (HIP) from haloperidol-treated and untreated C57BL/6 mice detected highly correlated (correlation greater than 0.5) changes between blood and brain, showing there is hope for using blood to study haloperidol treatment-induced DNA methylation changes in brain.[51]

These results indicate the subset of CpG sites that could be analyzed from blood data to predict brain changes associated with disorders or treatments. In the future, it is quite possible that blood could regularly be used to study disease progression and treatment responses.

Cultured cells are another important category of biological material that we can use to study biological mechanisms, development, and drug response. Induced pluripotent stem cells (iPSC) carry potential for studying DNA methylation dynamics in early-stage development. iPSC can be induced from patient fibroblast, lymphocyte, or even urine cells for studying DNA methylation and its potential biological impact at early developmental stages.

Lymphoblastoid cell lines (LCLs), transformed from B cells by Epstein–Barr virus, can be preserved nearly immortally and serve as an almost unlimited source of material. LCLs have been widely used for case–-control comparison, mostly as a way of studying biomarkers for drug effects. Due to its relatively homogeneous, controlled cultural environment, the LCL model is less influenced by environmental factors associated with human life, giving it unique power. The major catch of the LCL model is that the DNA methylation pattern in LCLs is significantly different from the DNA methylation pattern in whole blood.[52] Therefore, the results obtained from LCLs should be interpreted with caution.

Animal models will not be discussed in this article even though they are valuable in the study of general biology. DNA methylation in mouse brain cannot be compared with human brain at the single-base level. Besides this, animal behavior arguably differs significantly from human behavior.

2. FACTORS INFLUENCING DNA METHYLATION

DNA methylation is a product of interaction between genetic variants and environmental influence.

2.1 Genetic Factors

In the study of methylation analysis, the genetic factors governing methylation can be treated as quantitative traits. Mapping of methylation quantitative trait loci (mQTL) allows for analysis of association between SNP genotype and DNA methylation level, where a significant correlation indicates that the genomic elements regulate DNA methylation.[53] Like eQTL, genetic variants influence DNA methylation in a tissue-specific manner, making mQTL tissue specific.[54] A number of studies have identified mQTL in human brain.[53,55]

Allele-specific methylation (ASM) is another presentation of mQTL,[56,57] which focuses on *cis*-acting mQTL in relatively small genomic regions. A study of a four-generation family showed that ASM was pervasive at the nonimprinted loci and also present in somatic and germ cells.[58]

SNPs associated with mQTL or ASM may be useful in interpreting biological functions of SNPs and, further, in analyzing results of genetic associations for psychiatric disorders. mQTL SNPs have been found to be enriched in GWAS signals for bipolar disorder[59] and schizophrenia[60] and have also been used to identify disease risk genes.[61]

Paternal age at time of birth has been related to the risk of developing psychiatric disorders, particularly schizophrenia and autism, possibly through increasing de novo DNA mutation rates.[62–64] In a mouse study, a genome-wide DNA methylation comparison revealed a significant loss of methylation in the older mice's sperm in regions regulating transcription.[65] This loss in DNA methylation in older sperm may contribute to de novo mutations, which could, in turn, increase disease risk in offspring.

2.2 Environmental Factors

DNA methylation status is not a constant like DNA sequence. It is dynamic, displaying effects of environmental changes at various times. Environmental factors can affect DNA prenatally or postnatally; some are relevant at a single point in life, whereas others are relevant to multiple stages of life.

2.2.1 Prenatal Exposures

The prenatal period is critical for fetal development and is particularly critical to brain development in the context of psychiatric disorders. Neurogenesis, neuron migration, maturation, and networking, as well as synaptic pruning, all require precise procedures that could be affected by environmental factors.

The fetus is, in general, well protected by the mother's body from biological and physical insults, though viral infection, malnutrition, and maternal substance and medication use pose potential biological threats. Many of these factors have been shown to increase risk of the fetus developing psychiatric disorders, and DNA methylation may play an important role in the process.[66] Even though a causal relationship has not been established,[67] much evidence connecting prenatal factors to DNA methylation changes are present in animal models.[68]

At least two studies have connected maternal malnutrition to the risk of schizophrenia.[69,70] Perinatal malnutrition was found to be linked to changes in DNA methylation, which may further lead to growth and metabolic disease.[71,72] DNA methylation is at its most dynamic period during early embryonic development; maternal intake of methyl-group donors such as folate, betaine, and folic acid was found to be associated with infant buccal cell DNA methylation, though only in the periconception period.[73]

Prenatal alcohol exposure was found to be associated with DNA methylation changes in genes related to protocadherins, glutamatergic synapses, and hippo signaling.[74]

Psychological stress experienced by the mothers may have an impact on the well-being of the fetus as well, possibly by affecting DNA methylation. A small study on pregnancy during the 1998 Quebec ice storm identified altered DNA methylation in blood and saliva of children 8 and 11 years old; the DNA methylation changes were correlated with objective maternal stress.[75] A new study of 121 subjects showed that even grand-maternal exposure to psychosocial stress during pregnancy had an effect on DNA methylation of the grandchildren.[76]

It is unlikely that DNA methylation is the only change maternal malnutrition and other stressors impose on fetal development and consequent risk of psychiatric and other disorders. Fetal neuroendocrine alteration[77] and inflammation reaction[78] could also lead to pathology risk.

2.2.2 Postnatal Exposures

Postnatal development and aging also involve neuronal changes, neuroplasticity, and degeneration. Moreover, biological and physical insult can lead to damage to neurons and their supporting systems.

Early life stress has been shown to have a significant adverse effect on psychiatric disease risk. Childhood abuse, neglect, poverty, and other forms of stressors can increase risk of adulthood psychiatric disorders, primarily mood and disorders such as depression, anxiety, and post-traumatic stress disorder. Mouse models indicate that at least some of these early life stressors could lead to sustained DNA hypomethylation in post-mitotic neurons.[79]

In adulthood, smoking, drinking, and substance use likely change DNA methylation levels. Studies have found that methamphetamine use can change Dnmt1, Dnmt2, and Reelin levels in rat brain[80,81]; nicotine use has been shown to decrease Dnmt1 levels in mouse GABAergic neurons and to increase expression of GAD67[82]; and DNMT-3a and DNMT-3b expression levels are reduced in alcoholic patients.[83] As DNMTs are methyltransferases, which are critical for DNA methylation, it is likely that their expression changes have an impact on DNA methylation. Indeed, alcohol use[84] and smoking[85] have been found to be associated with changes in DNA methylation in humans. Methamphetamine use was found to alter DNA methylation in a rodent model.[86] Studies of environmental factors' effects on DNA methylation changes in human brain or blood biomarkers have produced compelling results, though these require replication in large samples.

3. DNA METHYLATION IN MAJOR PSYCHIATRIC DISORDERS

Major psychiatric disorders such as schizophrenia (SZ), bipolar disorder (BD), major depressive disorder (MDD), autism spectrum disorder (ASD), eating disorders [including anorexia nervosa (AN) and bulimia nervosa (BN)], alcohol use disorder (AUD), and attention deficit hyperactivity disorder (ADHD) are characterized by complex symptoms and development of pathology determined by intricate underlying genetics and environmental factors.[87,88] Although many of these disorders are thought to be highly heritabile,[87] all psychiatric concordance rates are substantially below 100% for monozygotic twins (MZTs),[87] suggesting that environmental factors contribute significantly to the development of pathology. DNA methylation, a product of the interaction between genetics and environment, could play a significant role in the process. Here, we summarize the major studies on DNA methylation in psychiatric disorders within the realm of both global methylation and gene-specific methylation changes in candidate gene and epigenome-wide studies.

3.1 Global DNA Methylation Changes in Major Psychiatric Disorders

Global methylation changes have been studied for diseases and related traits. In the context of psychiatric disorders, global DNA methylation levels can be influenced by genetic variants,[89] specific disease, medication,[90–92] age of disease onset,[91] sex,[90,93,94] smoking status,[92] and brain region(s) being examined.[95]

Through the examination of *peripheral tissue* (blood and buccal cell) samples, many psychiatric conditions have been linked to either low or high global methylation levels. Studies have generally reported high global methylation levels in anxiety[96] and AUD.[97] In contrast, low global methylation levels have been found in smokers (vs. nonsmokers),[92] SZ,[90,91,94,98] BD,[99] MDD,[100] ASD,[101,102] and AN.[103,104] SZ patients on medication have higher global DNA methylation levels—almost as high as those of healthy controls—than drug-free patients.[90,91]

Five studies have investigated methylation level differences in brain. Measurement of methylation level in the frontal cortex (FC) revealed hypermethylation in AUD[97] and hypomethylation in BD,[95] whereas both hypermethylation[105] and hypomethylation[95] were associated with SZ. In anterior cingulate cortex (ACC), hypermethylation was identified for SZ[95,105] and BD.[95] In HIP, hypermethylation was reported for SZ[105] and AUD.[97]

In summary, abnormal global DNA methylation levels have been detected for most psychiatric disorders, with some consistent results. In peripheral tissues and multiple brain regions, AUD patients have been recorded as having higher methylation levels than controls. In SZ and BD, increased methylation has mainly been detected in brain, whereas decreased methylation has mainly been detected in peripheral tissues. Other studies have reported negative results for methylation differences (Table 1).

3.2 Gene-Specific Methylation Studies in Psychiatric Disorders

Global methylation analysis does not capture changes of specific genes, biomarkers, and pathways. Gene-specific methylation is investigated in candidate gene studies and epigenome-wide association studies (EWASs). Neurotransmitter, neurodevelopment,[112,113] and immune system[114–116] have been hypothesized as mechanisms possibly underlying psychiatric disorders.[117,118] Results from candidate gene studies are summarized later.

3.2.1 DNA Methylation of Candidate Genes in Psychiatric Disorders

Current hypotheses in psychiatry pertaining to neurotransmitter systems include the overactivity of the dopaminergic system and hypofunction of the serotonergic system, the γ-aminobutyric acid (GABA)-ergic system and the glutamatergic system, primarily in SZ, BD, and MDD. The methylation changes consistently reported in candidate genes are plotted in the neurotransmitter systems in Fig. 1.

3.2.1.1 DNA Methylation of Dopaminergic System Genes in SZ, BD, MDD, AN, ADHD, and AUD

Dopamine (DA) overactivation has been strongly associated with SZ.[119] Antipsychotics that are antagonists of DRD2 are the most effective drugs for treating schizophrenia.[120] Methylation changes in DA system genes in major psychiatric disorders such as SZ, BD, MDD, AN, ADHD, and AUD have also been reported, in particular for the DA receptors (DRDs), catechol-*O*-methyltransferase (*COMT*), and DA transporter (*DAT*).

Five subtypes of DRD, namely, DRD1-DRD5, have been identified in mammals. Hypomethylation and high-expression of *DRD2*,[121–123] *DRD4*, and *DRD5* were identified in SZ.[123] Hypomethylation and high-expression of *DRD4* were also detected for ADHD.[124] However, in AN, hypermethylation of *DRD2* was reported.[125,126]

Table 1 Global DNA Methylation Studies of Psychiatric Disorders.

Studies	Samples	Tissues	Experimental Platforms	Major Findings
SZ and BD				
Shimabukuro *et al.* (2007)[94]	210 SZ, 237 C, Japanese	LCLs	HPLC	↓Male SZ
Dempster *et al.* (2011)[89]	22 discordant MZT pairs: 11 SZ, 11 BD	WB; B for	replication of TOP	HM27, EpiTYPER (Sequenom)
No systemic changes.				
Melas *et al.* (2012)[91]	177 SZ, 171 C	WB	LUMA; PCR	↓In SZ; ↑haloperidol (control-like); ↓early onset (a putative marker of SZ severity)
Bönsch *et al.* (2012)[90]	20 twin pairs discordant for SZ, 8 twin pairs concordant for SZ, 42 C twin pairs	WB	Methylation-sensitive restriction (HpaII) and PCR	↓In SZ twins compared with healthy C twins (males); ↑patients on medication (control-like) compared with patients not on medication
Nishioka *et al.* (2013)[98]	18 first-episode SZ, 15 C, Japanese	WB	HM27	↓In SZ
Xiao *et al.* (2014)[95]	5 SZ, 7 BD, 6 C	FC, ACC (BA9 and BA24)	MeDIP-Seq	↓SZ, BD in FC; ↑SZ, BD in ACC
Alelú-Paz *et al.* (2016)[105]	6 SZ DLPFC, 6 SZ HIP, 7 SZ ACC, 3 C	DLPFC, ACC, HIP	HM450	↑SZ

(Continued)

Table 1 Global DNA Methylation Studies of Psychiatric Disorders.—cont'd.

Studies	Samples	Tissues	Experimental Platforms	Major Findings
Bromberg *et al.* (2009)[106]	49 BD, 27 C	LCLs	Cytosine-extension assay	×BD
Rao *et al.* (2012)[107]	10 BD. 10 C	FC	ELISA-based assay	↑BD
Soeiro-de-Souza *et al.* (2013)[108]	50 BD, 50 C	WB	ELISA-based assay	↓5hmC levels in BD I patients; ×5mC
Burghardt *et al.* (2015)[92]	115 BD I	LCLs	LUMA	Average global methylation was significantly influenced by insulin resistance (↑), second-generation antipsychotic use (×), and smoking (↓)
Huzayyin *et al.* (2014)[99]	14 BD, 16 un-AR, 14 AR (BD or MDD), 16 C	TL	ELISA-based assay	↓In BD patients and their relatives; ↓in BD patients, with lithium treatment, not in relatives
MDD				
Tseng *et al.* (2014)[100]	49 MDD, 25 C	LCLs	Enzyme-linked immunosorbent assay-based method	Significant↓ in 5hmC and trends of↓ in 5mC, but only in the old group of MDD; 5mC level inversely correlated with disease severity
Byrne *et al.* (2013)[93]	12 MZT pairs discordant for MDD and 12 MZT pairs concordant for no MDD and low neuroticism	WB	HM450	× All; different in females
ASD				
	86 parents of cases, 200 mothers	WB		↓ASD

James *et al.* (2008)[102]			Enzymatic digestion and HPLC	
Melnyk *et al.* (2012)[101]	68 ASD, 54 C	WB	Enzymatic digestion and HPLC	↓ASD
Wong *et al.* (2014)[109]	50 MZT pairs discordant and concordant for ASD/ASD severity	WB	HM27	×ASD
Anxiety				
Murphy *et al.* (2015)[96]	25 anxious, 22 C	WB	RT-PCR, methylflash methylated DNA quantification kit	↑Anxiety
AN				
Frieling *et al.* (2007)[104]	22 AN, 24 BN, 30 C	WB	—	↓AN
Saffrey *et al.* (2013)[110]	10 AN, 10 C	Buccal cells	Sequenom mass array platform	×AN
Tremolizzo *et al.* (2014)[103]	32 AN, 13 C	WB	HpaII	↓AN
AUD				
Manzardo *et al.* (2012)[111]	10 AUD, 10 C	FC	Immunoprecipitation	×AUD
Otero *et al.* (2012)[97]	Rat	HIP, PFC	HpaII	↑AUD

ACC, Anterior cingulate cortex; *ADHD*, attention deficit-hyperactivity disorder; *AN*, anorexia nervosa; *ASD*, autism spectrum disorder; *AUD*, alcohol use disorder; *B*, brain; *BD*, bipolar disorder; *BN*, bulimia nervosa; *C*, controls; *DLPFC*, dorsolateral prefrontal cortex; *DNAm*, DNA methylation; *FC*, frontal cortex; *HIP*, hippocampus; *HM27*, Illumina Infinium HumanMethylation 27K; *HM450*, Illumina Infinium HumanMethylation 450K; *HPLC*, high-performance liquid chromatography; *LCLs*, leukocyte cell lines; *LUMA*, luminometric methylation assay; *MDD*, major depressive disorder; *SZ*, schizophrenia; *TL*, transformed lymphoblasts; *WB*, white blood cells; ↓↑ and × indicate reduced, increased, and unchanged global methylation, respectively.

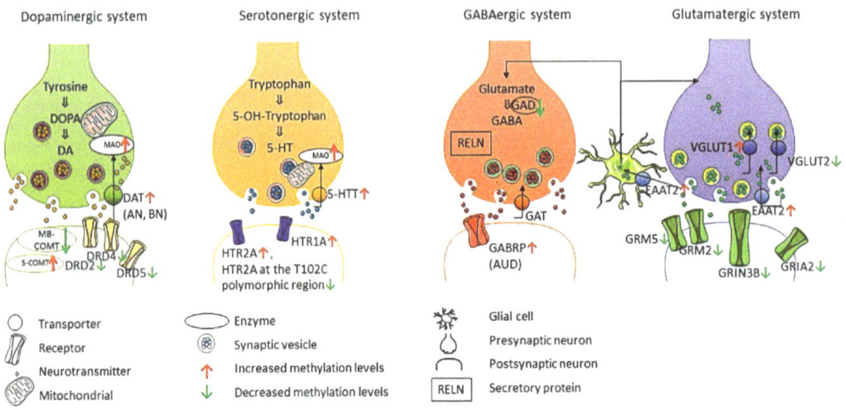

Fig. 1 Methylation changes of candidate genes related to the neurotransmitters in SZ and BD. Note: Two genes reported in non-SZ/BD are also plotted: DAT in Dopaminergic system for eating disorders and GABRP in GABAergic system for AUD because they play important roles in these neurotransmitter systems. Reports regarding RELN have been largely inconsistent.

COMT, one of the DA degradation enzymes, regulates the homeostatic levels of DA in the synapses. This gene is located on chromosome 22q11.21, a region genetically linked to SZ in several studies.[87,127] Two isoforms of *COMT* exist: the soluble short form (S-COMT) and the membrane-bound long form (MB-COMT); each isoform is believed to have distinct cellular distributions and roles. MB-COMT displays some preference for brain tissues, notably in the HIP, and primarily aids in terminating dopaminergic synaptic neurotransmission, whereas S-COMT mainly works to eliminate biologically active or toxic catechol in blood.[128]

No overall *COMT* methylation changes were identified for FC in SZ or BD.[117] Hypomethylation was found for MB-COMT in SZ and BD in both brain and saliva.[122,129,130] In brain, MB-COMT was found to be hypomethylated, especially in the left frontal lobe.[122,129,130] In blood, S-COMT hypermethylation was reported in SZ patients.[91] In the promoter of S-COMT, no methylation changes were found in SZ patients.[131]

Monoamine oxidase A (*MAOA*) encodes an enzyme that degrades DA, norepinephrine, and serotonin. Hypermethylation of *MAOA* at its promoter and exon1 was detected in SZ blood.[132]

DAT (also known as *SLC6A3*) is a DA transporter. Hypomethylation of *DAT* was detected in AUD,[133,134] and hypermethylation of *DAT* was found for AN and BN.[104]

Three main kinds of genes in dopaminergic system have been associated with methylation changes in psychiatric disorders: receptor genes (hypomethylation), transporter genes (hypermethylation), and degradation enzyme genes [both hypomethylation and hypermethylation (Fig. 1)]. These pieces of evidence support the idea that hyperactivity of DA is, to some extent, a product of methylation levels. Interestingly, the methylation changes in DA system genes for eating disorders (AN and BN) seem to be opposite to those in other disorders such as SZ and BD.

3.2.1.2 DNA Methylation of the Serotonergic System in SZ, BD, MDD, and ADHD

Serotonin, or 5-hydroxytryptamine (5-HT), a monoamine inhibitory neurotransmitter, regulates mood, appetite, and sleep. Serotonergic mechanisms play a role in the action of atypical antipsychotic drugs. Several EWASs have discovered signals possibly relevant to psychiatric illnesses in the serotonergic system pathways.[135]

Serotonin has several receptors. HT1A and HT2A, two of these receptors, are relevant to the symptoms of SZ. The *HTR1A* promoter is hypermethylated for SZ and BD in a study using LCLs.[136] In *HTR2A*, hypermethylation and low gene expression in SZ and BD were detected for FC samples,[137,138] but at its *T102C* polymorphic region, hypomethylation was reported in SZ and BD saliva.[139]

The serotonin transporter (*5-HTT*, also known as *SLC6A4*) encodes the sodium-dependent serotonin transporter. 5-HTT hypermethylation was reported in MDD[140,141] and ADHD.[140] No significant difference ($P = 0.15$) was reported between SZ and controls.[91] Hypermethylation of the *5-HTT* promoter is evident in SZ and BD.[142]

So far, serotonergic system gene methylation changes in receptor genes and transporter genes (Fig. 1) have been linked to major psychiatric disorders, suggesting that methylation could contribute to serotonin dysfunction. It is interesting that *5-HTT* was found to be mainly hypermethylated in SZ, BD, MDD, and ADHD.

3.2.1.3 DNA Methylation of Glutamatergic System in SZ and BD

Glutamate (Glu) plays a critical role in synaptic plasticity and is the most abundant fast excitatory neurotransmitter in the mammalian nervous system. Several findings have pointed to the activity in the Glu system as a partial determinant of psychosis.[89,117] Antagonists of NMDA receptors, such as ketamine, can produce a psychotic-like state in healthy subjects.

Chemical receptors of Glu were used to divide glutamate receptors into ionotropic and metabotropic. Ionotropic Glu receptors include AMPA, NMDA, and Kainate receptors. Metabotropic Glu receptors indirectly activate ion channels on the plasma membrane and include mGluR1 through mGluR8, which are encoded by genes *GRM1* through *GRM8*.

Hypomethylation of *GRM2* and *GRM5*, and no methylation irregularities for *GRM3* and *GRM8*, was identified in SZ.[143] Both glutamate ionotropic receptor AMPA-type subunit 2 (*GRIA2*) and glutamate ionotropic receptor NMDA-type subunit 3B (*GRIN3B*) were hypomethylated in male SZ patients.[117]

Glu transporters, which move Glu across the membrane, are composed of two families: the excitatory amino acid transporter (EAAT) family and the vesicular Glu transporter (VGLUT) family. VGLUT1 and VGLUT2 are complementarily expressed in brain. VGLUT1 is mainly concerned with cortico-cortical projections to relay information, with low fidelity, whereas VGLUT2 is concerned with thalamo-cortical projections, with higher fidelity.[144] VGLUT1 (encoded by *SLC17A7*) was found to be hypermethylated and has low expression in female SZ patients. VGLUT2 (encoded by *SLC17A6*), though it serves a similar function as VGLUT1, was found to be hypomethylated and highly expressed in female SZ patients.[117] This may be due to different modes of Glu release in VGLUT1 and VGLUT2. EAAT2 (encoded by *SLC1A2*) is hypermethylated in BD patients.[145] SCG2, a secretory protein located in neuronal vesicles, stimulates the release of Glu. Hypomethylation of SCG2 was detected in female SZ patients.[117]

In brief, Glu-related methylation changes have been found for SZ and BD but no other psychiatric disorders. In SZ and BD, methylation changes have been reported for glutamatergic receptor genes, secretory protein genes (hypomethylation), and transporter genes (Fig. 1).

3.2.1.4 DNA Methylation of GABAergic System in SZ, BD, and AUD

GABA acts as a potent inhibitory neurotransmitter in the adult human brain. Some EWASs have found that GABAergic neurotransmission may be involved in SZ,[117] Hypofunction of GABAergic interneurons appears to be important in the etiology of major psychosis. GABA can influence the development of neural progenitor cells via *brain-derived neurotrophic factor* (*BDNF*) expression.

Two general classes of GABA receptor are known: $GABA_A$ receptors, which are part of a ligand-gated ion channel complex, and $GABA_B$

receptors, which are metabotropic G protein-coupled receptors. *GABRP*, which encodes the $GABA_A$ receptor, was found to be hypermethylated in AUD.[146]

GAD1 (Glu decarboxylase 1, also known as GAD67), is an enzyme that synthesizes GABA from Glu. Brain region-specific results were observed in the differential methylation studies of *GAD1*. In the HIP, hypomethylation and high expression of *GAD1* have been identified in both SZ and BD.[147] In BA39-40 (parietal cortex) and cerebellar cortex (CBL), on the contrary, hypermethylation and low expression of *GAD1* were detected.[148] *GAD1* hypomethylation was identified in AUD in blood.[146]

Reelin (RELN), an extracellular matrix protein expressed by GABAergic interneurons of the cortex and glutamatergic cerebellar neurons,[149] is primarily related to neuronal migration, axonal branching, synaptogenesis, and cell signaling. The findings for methylation changes of *RELN* have been inconsistent. Hypermethylation and low expression were detected for SZ in FC[150,151] and occipital cortex[151] according to some studies, but other studies reported no significant methylation changes in SZ[117,152,153] or BD[117,153] in FC, forebrain,[153] or blood.[90] The brain studies detecting changes used small samples, lowering the reliability of the positive results.

Three genes, *GABRP*, *GAD1*, and *RELN* (Fig. 1), related to the GABAergic system, have been linked to methylation changes in major psychiatric disorders. *RELN* plays an important role in the neuron, but findings regarding the gene have been inconsistent. *GAD1* has shown brain region-specific changes.

3.2.1.5 DNA Methylation of Genes Associated With Immune System and Inflammation in SZ, BD, MDD, and PTSD

Cytokines, which can be proinflammatory or anti-inflammatory, are small secreted proteins important for communication between cells. They also act as important coordinators of immune and inflammatory responses.[154] Cytokines mobilize resources for handling physical and psychological stress by maintaining homeostasis. There has been increasing evidence for the role of inflammation in psychiatric disorders. Altered cytokine levels have been observed in multiple disorders, including SZ, MDD, OCD,[155] BD,[155,156] ASD,[157] PTSD, anxiety and panic disorder,[158,159] and possibly eating disorders.[160] Therefore, it is important to investigate immune-/inflammation-related methylation signatures for involvement in psychiatric disorders.

DNA methylation has been studied in immune-related candidate genes in PTSD, SZ, aggression, and also in treatment response to antidepressants.

Global methylation was found to be increased in PTSD patients. Differentially methylated CpG sites in PTSD mapped to genes that could be involved with or interact with inflammation pathways,[161] even though the methylation changes were not found for these cytokine genes per se. Another candidate gene study of PTSD in military service members detected DNA methylation changes at promoter regions of *H19* and *IL-18* in case–-control and pre- to postdeployment comparison.[162] The HLA complex group *HCG9* gene shows consistently lower DNA methylation in brain, blood, and sperm of BD patients,[163] and also in another study of postmortem brains of SZ.[164] Differential DNA methylation of a number of interleukin genes and their transcription factor regulators are associated with chronic physical aggression.[165] IL-6 and C-reactive protein (*CRP*) were elevated in depressed patients. Furthermore, in depressive patients only, *IL-6* DNA methylation shows an inverse correlation with circulating *IL-6* and *CRP* in blood.[166] DNA methylation of the *IL-11* CpG island correlates with treatment effects of antidepressants, escitalopram, or nortriptyline in MDD patients.[167]

Glucocorticoids are a class of steroid hormones critical to the maintenance of basal and stress-related homeostasis; glucocorticoids also mediate stress and inflammation's relationship with immune response.[168] DNA methylation of the glucocorticoid receptor gene (*NR3C1*) has been frequently studied in relation to epigenetic impact of early life stress and regulation of hypothalamic–pituitary–adrenal (HPA) axis. Increased *NR3C1* promoter methylation was associated with higher cortisol responses to stress.[169]

Maternal stress can have an impact on the fetus in utero. Depressive mood in mothers correlates with *NR3C1* promoter methylation in newborn cord blood mononuclear cells.[169] In another study, *NR3C1* DNA methylation in children was shown to be related to the stress levels of their mothers during pregnancy.[170] Postnatal stress, like childhood sexual abuse, also affects this promoter's methylation level,[171] suggesting childhood maltreatment leads to permanent DNA methylation changes. DNA methylation of the *NR3C1* promoter is also found to be associated with BN,[172] and, along with interferon-gamma (*IFN-gamma*) and *IL*-6, associated with psychological traits.[173] Consequently, *NR3C1* promoter methylation has become one of the most promising epigenetic research targets in the immune–inflammation regulatory system, but whether it is involved in other psychiatric disorders such as SZ and BD remains to be seen.

Because glucocorticoids can modulate cytokine signaling,[174] the interaction between *NR3C1* and cytokines is an important topic for future epigenome-wide studies on psychiatric disorders that possibly involve immune–inflammation changes.

3.2.1.6 DNA Methylation of Other Candidate Genes

3.2.1.6.1 **BDNF in SZ, BD, and BN** *BDNF*, located on chromosome 11p14.1, has been found to be related to many biological functions, including NMDA receptor activity, synapse stability, dopaminergic, cholinergic, serotonergic, and GABAergic signaling, synaptogenesis, and dendritogenesis.

Several studies using blood samples have examined *BDNF* methylation. In *BDNF* promoter I, hypermethylation in SZ[175] is found, but *BDNF* promoter IV yields no significant results in SZ.[175] *BDNF* hypomethylation in drug-free SZ[176] was identified in another study, as was hypermethylation of *BDNF* in BN.[177] *BDNF* is associated with higher levels of methylation in BD II than in BD I.[178,179] Hypermethylation of *BDNF* is also detected in depressed patients compared with manic/mixed episode patients.[179] The respective methylation results for BD I and BD II patients seem congruent with the fact that mania is a clinical feature of BD I, whereas BD II only includes hypomania. In FC of SZ patients, *BDNF* hypomethylation is detected.[180]

3.2.1.6.2 **Candidate Genes From Genetic Studies** A number of candidate genes identified from genetic, linkage, and association studies have been further investigated for DNA methylation changes. These genes include *FOXP2*, *DTNBP1*, *AUTS2*, and *RAI1*.

FOXP2 (*forkhead box P2*), a transcription factor located on 7q31, was first found mutated in a large multigenerational family study, in association with a rare severe speech and language disorder; the gene was later implicated in SZ and MDD[181] by an association study. *FOXP2* is hypermethylated in the brains of SZ patients.[182]

DTNBP1 (*dystrobrevin-binding protein 1*), located on 6p22.3, encodes a neuronal protein that binds to beta-dystrobrevin, may be part of the dystrophin protein complex, and has been reported in genetic studies on SZ,[183–185] BD,[183,186,187] and MDD.[188] *DTNBP1* is hypermethylated in SZ and BD, according to a saliva study[189] and an FC study.[117]

AUTS2 (autism susceptibility candidate 2), located on 7q11, encodes a nuclear protein that is expressed primarily in the developing brain. It is mainly detected in ASD,[190] SZ,[191] and AUD.[192] *AUTS2* is found to be hypermethylated in SZ male FC samples.[117]

RAI1 (*retinoic acid induced 1*), located on 17p11.2, was identified in Smith–Magenis syndrome[193] and SZ[194] through genetic analyses. These genes are all hypermethylated in SZ and BD. RAI1 is located at a region that encodes a polymorphic polyglutamine tract associated with SZ and response to antipsychotic medication; it is found to be hypermethylated in the FC of SZ females.[117]

3.2.1.6.3 **Other Candidate Genes** Five genes located in two chromosomal regions, 1p36.22 and 11p15.5, were reported in several papers.

ANP (atrial natriuretic peptide), located in 1p36.22, is hypermethylated in BN[195]; *MTHFR* (*methylenetetrahydrofolatereductase*), which is also located in 1p36.22, is hypomethylated in ADHD.[124]

11p15.5 is enriched for imprinted genes and includes three relevant candidate genes, *IGF2*, *H19*, and *KCNQ1OT1*. *IGF2* is a paternally inherited member of the insulin family of polypeptide growth factors; *H19* is near *IGF2* and is only expressed from the maternally inherited chromosome; *KCNQ1OT1* is a maternally inherited long noncoding RNA gene located in an intron of *KCNQ1*. Hypomethylation of *IGF2*,[124,196] *H19*, and *KCNQ1OT1* was identified in ADHD.[124]

Table 2 summarizes all the candidate genes we found in the methylation studies associated with psychiatric disorders.

3.2.2 Epigenome-Wide Association Studies

EWASs are currently the best unbiased method for uncovering disease-related changes. Brain tissue is thought to be the tissue most linked to psychiatric disorders and is of great importance in the study of disease-related pathways and mechanisms. In addition, peripheral tissue studies have revealed some important findings, especially ones related to immune-related pathways and biomarkers. EWAS has examined data from both of these tissues; the findings are presented later.

3.2.2.1 DNA Methylation Changes in Brain Highlight the Importance of Early Brain Development

About 16 brain EWASs have been published so far. Alelú-Paz *et al.* found differentially methylated probes (DMPs) related to neurotransmission pathway in DLPFC, antigen processing and presentation in HIP, and protein kinase activity and signal transduction in ACC.[105] Mill *et al.*, using FC samples, found 51 genes in men and 37 genes in women that were differentially methylated in SZ patients in comparison to controls. These genes are enriched in brain development and mitochondrial functions.[117] Pathways

Table 2 DNA Methylation Studies of Candidate Genes for Psychiatric Disorders.

Gene	Locus	Function	Tissues	Major Findings	References
Dopaminergic system					
DRD2	11q23.2	DA receptor	B, WB	↓M and ↑E (SZ), ↑M (AN)	121–123,125,126,197
DRD4	11p15.5	DA receptor	B, WB	↓M and ↑E (SZ, ADHD)	123,124
DRD5	4p16.1	DA receptor	WB	↓M and ↑E (SZ)	123
MB-COMT	22q11.21	Regulates the homeostatic levels of DA in the synapses	B, S	↓M (especially in the left frontal lobe for SZ and BD)	122,129,130
S-COMT	22q11.21	Regulates the homeostatic levels of DA in the synapses	B, WB, S	↑M (SZ)	91,131
SLC6A3 (DAT)	5p15.33	DA transporter	WB	↑M (AN, BN), ↓M (AUD)	125,133,134
MAOA	Xp11.3	Degradation enzyme of DA, serotonin, and NE	WB	↑Male	132
DBH	9q34.2	DA beta-hydroxylase	WB	↓M (AUD)	146
Serotonergic system					
HTR1A	5q12.3	Serotonin receptor	WB	↑M (SZ and BD)	136
HTR2A	13q14.2	Serotonin receptor	B	↑M and ↓E (SZ)	137,138
HTR2A at the T102C	13q14.2	Serotonin receptor	S	↓M (SZ and BD)	139

(Continued)

Table 2 DNA Methylation Studies of Candidate Genes for Psychiatric Disorders.—cont'd.

Gene	Locus	Function	Tissues	Major Findings	References
polymorphic region					
SLC6A4 (5-HTT)	17q11.2	Serotonin transporter	B, WB, S	↑M (SZ and BD (antipsychotic-free, MZT), MDD, ADHD)	91,124,140–142,198
GABAergic system					
GAD1 (GAD67)	2q31.1	Enzyme that synthesizes GABA from Glu	B, WB	↓M (SZ, BD, AUD)	117,146–148
RELN	7q22.1	Cell positioning and neuronal migration during brain development	B, WB	↑M (SZ brain)	90,117,150–153
GABRP		GABA receptor	WB	↑M (AUD)	146
Glutamatergic system					
GRM2	3p21.2	Glu metabotropic receptor	B	↓M (SZ)	117
GRM5	11q14.2-q14.3	Glu metabotropic receptor	B	↓M (SZ)	117
GRIN3B	19p13.3	Located ~10 kb upstream of the NMDA-receptor-subunit	B	↓M (male SZ)	117
GRIA2	4q32.1	In the promoter of the AMPA-receptor-subunit	B	↓M and↑E (male SZ)	117
SCG2	2q36.1	A secretory protein located in neuronal vesicles that is known to stimulate the release of Glu	B	↓M (female SZ)	117
VGLUT1	19q13.33		B		117

		Packs Glu into synaptic vesicles		↑M and↓E (female SZ)	
VGLUT2	11p14.3	Glutamate transporter	B	↓M and↑E (female SZ)	117
SLC1A2 (EAAT2)	11p13	A member of the solute transporter proteins	WB	↑M (BD)	145
BDNF	11p14.1	Brain-derived neurotrophic factor; a principal regulator and developer, involved in function and plasticity of neurons, as well as in progression of pathology	B, WB	↑M (blood in BD II; blood in SZ at promoter I; blood in BN) ↓M (FC, B in drug-free SZ)	175–177,179,180,199
Immune and inflammation					
H19	2q33.2	HLA complex group	WB	↑M (PTSD)	162
IL18	11q23.1	Interleukin 18	WB	↑M (PTSD)	162
IL-6	7p15.3	Interleukin 6	WB	↑M (depression)	166
CRP	1q23.2	C-reactive protein	WB	↑M (depression)	166
IL11	19q13.42	Interleukin 11	WB	↑M (escitalopram), ↓M (nortriptyline)	167
HCG9	6p21.3	HLA complex group 9	B, WB,	sperm	↓M (BD, SZ)
163,164					
NR3C1	5q31.3	Glucocorticoid receptor	WB	↓M (ADHD)	124
NR3C1 promoter 169–172	5q31.3	Glucocorticoid receptor	WB, cord	blood	↑M (BN, stress, PTSD, BN)

(*Continued*)

Table 2 DNA Methylation Studies of Candidate Genes for Psychiatric Disorders.—cont'd.

Gene	Locus	Function	Tissues	Major Findings	References
FKBP5	6p21.31	Immunoregulation	WB	↑M (BD)	200
Candidate genes from genetic studies					
FOXP2	7q31.1	Positive selection in human lineage and genes associated with SZ	B	↑M (SZ)	182
DTNBP1	6p22.3	Prenominated "psychosis-candidate gene"	B, S	↑M (SZ)	117,189
AUTS2	7q11.22	Spans a translocation breakpoint associated with mental retardation and autism	B	↑M (SZ)	117
RAI1	17p11.2	Polymorphic polyglutamine tract associated with SZ and response to antipsychotic medication	B	↑M (SZ)	117
1p36.22					
ANP	1p36.22	Atrial natriuretic peptide	WB	↑M (BN)	195
MTHFR	1p36.22	Methylenetetrahydrofolate reductase	WB	↓M (ADHD)	124
11p15.5					
IGF2DMR (IGF2)	11p15.5	Part of the insulin family of polypeptide GF, paternally inherited	WB, CB	↓M (ADHD)	124,196
H19	11p15.5	Near IGF2, maternally inherited	WB	↓M (ADHD)	124
KCNQ1OT1	11p15.5	LncRNA, maternally inherited	WB	↓M (ADHD)	124
Other candidate genes					
RPP21	6p22.1	A component of rib nuclease	B	↑M (SZ, female BD)	117
KEL	7q34	Part of the Kell blood group, metalloendopeptidase	B	↑M (SZ, female BD)	117
RPL39	Xq24	Ribosomal protein	B	↑M (female BD)	117

MECP2	Xq28	Chromosomal protein that binds to methylated DNA	FC	↑M and ↓E (ASD)	201
SNCA	4q22.1	Alpha synuclein	WB	↑M (AN, AUD) and ↓E	104,202
HERP	16q13	Homocysteine-induced endoplasmic reticulum protein	WB	↑M (AUD) and ↓E	203
PPM1G	2p23.3	Negative regulation of cell stress response	WB	↑M (AUD)	204
OXTR	3p25.3	Oxytocin receptor	BC	↑M (AN)	205
SSTR4	20p11.21	Somatostatin receptor 4	WB	↓M (AUD)	146
ALDH1L2	12q23.3	Aldehyde dehydrogenase 1 family member L2	WB	↓M (AUD)	146
ST6GALNAC1	17q25.1	Related "synthesis of substrates in *N*-glycan biosynthesis"	WB	↓M (BD)	89
KCNQ3	8q24.22	Regulation of neuronal excitability	B	↓M (BD)	206
COX-2	1q31.1	Prostaglandin biosynthesis	B	↓M (BD)	107
FAM63B	15q21.3	Removal of Lys-48-linked conjugated ubiquitin from proteins	WB	↓M (BD)	207

ADHD, Attention deficit-hyperactivity disorder; *AN*, anorexia nervosa; *AUD*, alcohol use disorder; *B*, brain; *BC*, buccal cells; *BD*, bipolar disorder; *BN*, bulimia nervosa; *CB*, cord blood; *DA*, dopamine; *GABA*, gamma-aminobutyric acid; *GF*, growth factors; *Glu*, glutamate; *LCLs*, leukocyte cell lines; *MDD*, major depressive disorder; *S*, saliva; *SZ*, schizophrenia; *TF*, transcription factor; *TL*, transformed lymphoblasts; *WB*, white blood cells.

for neuron development,[208–210] synaptic transmission,[105,209,211] nervous system differentiation,[95,210] embryo development,[210] calcium ion binding,[209] and cell fate commitment[210] (Table 3) are enriched in discovered DMPs and DMRs and are also identified in other studies. The neuron development pathway is frequently identified in brain EWASs, which indicates the importance of further studying the relationship between brain development methylation changes and psychiatric disorders.

Two papers have focused on the relationship between DNA methylation changes in fetal brain development and SZ. Jaffe *et al.*[210] found 2104 DMPs when comparing PFC samples of patients and controls. Furthermore, they found widespread DNA methylation changes when comparing prenatal to postnatal life; these changes were enriched for SZ-GWAS risk loci and related to age of onset.[210] Interestingly, Hannon *et al.*,[60] using fetal PFC, striatum, and CBL, found that, in fetal-specific brain mQTLs, there was an enrichment of SZ-GWAS loci.[60] DMPs at fetal and postnatal stages, especially CpGs with higher methylation in fetal PFC, were more likely to be enriched for GWAS loci than the remaining age-regulated CpGs. Jaffe *et al.*'s and Hannon *et al.*'s papers point to the important influence of prenatal brain development on SZ pathology.

Further EWAS results detected 1812 DMPs for males with AUD and zero DMPs for females with AUD, possibly because of the limited female sample sizes in this study.[208] For ASD, interestingly, decreased methylation probes were enriched in immune function, and increased methylation probes were related to synaptic membrane.[211]

3.2.2.2 DNA Methylation Changes in Peripheral Tissues Highlight the Relevance of Immune Function in Psychiatric Etiology

Cellular heterogeneity in the brain, cause of death, agonal state, postmortem interval, and other factors can contribute to inconsistent findings in postmortem brain studies such as those mentioned earlier. Peripheral tissues are better than postmortem brain tissues for studying the response to the environment because the peripheral tissues can be collected from living patients. Peripheral tissues, especially blood, play important roles in identifying biomarkers.

SZ and BD EWASs utilizing peripheral tissues have detected several DMPs. Of the DMPs, 65.4% were located in CpG islands (CGIs) and the regions flanking CGIs (CGI: 40.6%; CGI shore: 13.3%; CGI shelf: 11.5%). Of the 95 DMPs in the CGIs, most were located in the promoter regions (promoter: 75.8%; gene body: 14.7%; 3′-UTR: 2.1%).[218] Valproic acid and

Table 3 EWAS of Psychiatric Disorders in Brain.

Studies	Sample Sizes	Tissues	Platforms	Main Findings	Pathway Enrichment
SZ and BD					
Mill *et al.* (2008)[117]	35 **SZ**, 35 **BD**, 35 C	FC	Microarray	DMPs (51 genes in men and 37 genes in women)	Brain development, **mitochondrial** functions
Pidsley *et al.* (2014)[209]	43 PFC (20 SZ and 23 C), 44 CBL (21 **SZ** and 23 C)	PFC CBL	HM450	22 DMPs in PFC	Neuron development; synaptic transmission; calcium ion binding
Chen *et al.* (2014)[28]	39 **SZ**, 36 **BD**, 43 C	CBL	HM27	488 DMPs	—
Numata *et al.* (2014)[212]	106 **SZ**, 110 C	DLPFC	HM27	107 DMPs	—
Wockner *et al.* (2014)[213]	24 **SZ**, 24 C	FC	HM450	4641 DMPs	—
Xiao *et al.* (2014)[95]	5 **SZ**, 7 **BD**, 6 C	FC, ACC (BA9, BA24)	MeDIP-Seq	4985 DMRs in BA9; 3867 DMRs in BA24 for SZ, ↑hsa-mir-4266 in SZ	Guidance and signaling, multicellular organismal development and signaling, axon guidance, oligodendrocyte differentiation
Zhao *et al.* (2015)[214]	5 **SZ**, 7 **BD**, 6 C	BA9	MeDIP-Seq		—

(Continued)

Table 3 EWAS of Psychiatric Disorders in Brain.—cont'd.

Studies	Sample Sizes	Tissues	Platforms	Main Findings	Pathway Enrichment
				SZ: 10,961 (7,880↑ and 3,081↓) DMRs; BD: 16,599 (6,836↑ and 9,763↓)	
Ruzicka *et al.* (2015)[147]	8 **SZ**, 8 **BD**, 8 C	HIP	HM450	↑ GAD1, SZ and BD (family)	—
Alelú-Paz *et al.* (2016)[105]	6 **SZ** DLPFC, 6 **SZ** HIP, 7 **SZ** ACC, 3 C (DLPFC, HIP, ACC)	Several brain regions (DLPFC, ACC, HIP)	HM450	139 DMPs; DMPs may be enriched in CpG shelves and gene bodies; DLPFC: 66; HIP: 18; ACC: 55	DLPFC: neurotransmission; HIP: antigen processing and presentation; ACC: protein kinase activity and signal transduction
Jaffe *et al.* (2016)[210]	191 **SZ**, 335 C	PFC (DLPFC, BA46, and BA9)	HM450	2104 DMPs; DNAm differences between fetal and postnatal life were enriched for SZ risk loci and were associated with the age of illness onset	Embryo development, cell fate commitment, **nervous system differentiation**
MDD					
Sabunciyan *et al.* (2012)[215]	39 **MDD**, 26 C	FC	CHARM	224 DMPs	—
ASD					
	9 **ASD**, 9 C	BA19	HM27	×	—

Ginsberg *et al.* (2012)[216]					
Ladd-Acosta *et al.* (2014)[217]	19 **ASD**, 21 C	TC, PFC, CBL	HM450	4 significant DMPs (TC, CBL), 3 replicated	—
Nardone *et al.* (2014)[211]	23 **ASD**, 23 C	BA10, BA24	HM450	>5,000 DMPs in BA10, >10,000 DMPs in BA24	In BA10, ↓DMPs: **immune function**, ↑DMPs: **synaptic membrane**
AUD					
Wang *et al.* (2015)[208]	46 **AUD** PFC (16 male pairs, 7 female pairs), 46 C	PFC	HM450	1812 DMPs in male, 0 in female	**Neural development** and transcriptional regulation

ADHD, Attention deficit-hyperactivity disorder; *AN*, anorexia nervosa; *AR*, affected relatives; *AUD*, alcohol use disorder; *BC*, buccal cells; *BD*, bipolar disorder; *BN*, bulimia nervosa; *C*, control; *CBL*, cerebellum; *DMPs*, differential methylation probes; *DMRs*, differential methylation regions; *DNAm*, DNA methylation; *FC*, frontal cortex; *GABA*, gamma-aminobutyric acid; *HIP*, hippocampus; *HM27*, Illumina Infinium HumanMethylation 27K; *HM450*, Illumina Infinium HumanMethylation 450K; *MDD*, major depressive disorder; *MeDIP-Seq*, methylated DNA immunoprecipitation sequencing; *PFC*, prefrontal cortex; *SZ*, schizophrenia; *TC*, temporal cortex.

The bolded words indicate studies related to immunity and neurodevelopment.

quetiapine are significantly associated with altered methylation signatures after adjustment for drug-related changes of cell-type composition.[219] Interestingly, hypermethylated CGI genes are found to be enriched in neuroactive ligand–receptor interaction pathways and the Jak-STAT signaling pathway in SZ; hypomethylated CGI shores are enriched in infection; meanwhile, genes lacking CGIs are enriched in metabolism, cell adhesion, and axon guidance.[220] For MDD, most of the DMPs (85.7%) are located in CGIs, in the gene promoter regions.[221]

In analyzing the pathways in these studies, we found the DMPs identified from peripheral tissue EWAS are mainly related to stress,[222] immune function,[166,219,220,222–226] infection,[220] organismal injury and abnormalities,[227] neurogenesis,[219] neurodevelopment,[166,228] embryonic development,[219,227] neurological disease,[227] postnatal synaptic transmission,[229] modulation of monoamine and cholinergic neurotransmission,[118,223] mitochondrion organization,[98] calcium ion binding,[228] neuroactive ligand–receptor,[220] cholesterol storage and lipid transport,[118] and lipoprotein.[166] Interestingly, many studies on peripheral tissues were able to detect DNA methylation changes in pathways related to immune function in multiple disorders, including SZ, BD, ADHD, and AUD[166,220,222–226] (Table 4). Lipid-related pathways are detected in studies on AN and depression.[118,166]

In summary, EWASs have revealed that increased DMPs are likely more related to brain development, whereas decreased DMPs appear to be more related to immune function, according to both brain[211] and peripheral samples.[166,220]

3.2.3 Monozygotic Twin Studies on Psychiatric Disorders

MZTs that are discordant for psychiatric disorders bring unique value to the study of epigenetics. There is increasing evidence that epigenetic differences between MZTs are related to the etiology of psychiatric disorders. For example, in MZT studies, hypomethylation of *PPIEL*[233] in BP II and hypomethylation of *RELN* promoter in SZ[90] are detected. Furthermore, hypermethylation of *SOX10*,[90] *SLC6A*,[135] *RORA*,[234] and *ZBTB20*[235] are detected in MZT studies as candidate genes for SZ, BD, ASD, and MDD, respectively (as detailed in Section 3.1).

High correlations of global methylation levels based on blood samples have been calculated within discordant MZT pairs in SZ,[89] BD,[89] ASD,[109] and MDD,[93,236] indicating no systemic changes in genome-wide epigenetic programming. In addition, no differences are detected between discordant

Table 4 EWAS of Psychiatric Disorders in Peripheral Tissues.

Studies	Sample Sizes	Tissues	Platforms	Main Findings	Pathway Enrichments
SZ and BD					
Kinoshita *et al.* (2013)[218]	24 medication-free **SZ**, 23 C, 3 pairs MZT males	LCLs	HM450	234 DMPs	—
Kinoshita *et al.* (2014)[230]	63 **SZ**, 42 C	WB	HM450	2552 DMPs	Transcription-related functions
Liu *et al.* (2014)[226]	98 **SZ**, 108 C	LCLs	HM27	20 DMPs (16 validated, 11 of which were related to reality distortion symptoms)	**Inflammatory response**
van den Oord *et al.* (2016)[225]	712 **SZ**, 696 C	WB	MBD-seq	7 DMPs	**Immunity**
Montano *et al.* (2016)[227]	689 **SZ**, 645 C	WB	HM450	625 (68%) showed association in the same direction and 172 (19%) of these had a replication P value of less than 0.05	Neurological disease, organismal injury and abnormalities, and embryonic development.
Nishioka *et al.* (2013)[98]	18 first-episode **SZ**, 15 C; Japanese	WB	HM27	1429 CpG sites (1352 genes) in males and 173 CpG sites (172 genes) in females had significantly different DNA methylation statuses (compared to controls). Only 32 genes	Mitochondrion organization, the nuclear lumen, transcription factor binding, and nucleotide binding

(Continued)

Table 4 EWAS of Psychiatric Disorders in Peripheral Tissues.—cont'd.

Studies	Sample Sizes	Tissues	Platforms	Main Findings	Pathway Enrichments
				were found in common (between males and females)	
Sabunciyan *et al.* (2015)[224]	20 mania, 20 C	WB	HM450	↑*CYP11A1*	**Inflammatory markers in serum**
Houtepen *et al.* (2016)[219]	172 BD with 6 frequently used psychotropic drugs	WB	HM450 (50), HM27 (122)	DMPs	**Immunity**, neurogenesis, embryonic and regulatory functions
Walker *et al.* (2016)[228]	Scottish family with multiple cases of BD or MDD	WB	HM450	Altered methylation in carriers of a haplotype linked to BD and MDD	Neurodevelopment, calcium ion binding
Li *et al.* (2015)[220]	6 SZ, 3 BD, 1 C	WB	MeDIP-Seq	Thousands of DMPs located preferentially in promoters 3′-UTRs and 5′-UTRs of genes in SZ; 0 in BD	↑CGI genes: neuroactive ligand–receptor interactions and the Jak-STAT signaling pathway in SZ; ↓CGI shores: **infection**; genes lacking CGIs: metabolism, cell adhesion, and axon guidance
MDD					
Uddin *et al.* (2011)[166]	33 MDD, 67 C	WB	HM27	DMPs; IL-6 and CRP levels were elevated among those	↑: Brain development; ↓: **lipoprotein**

				with lifetime depression, and IL-6 methylation held an inverse correlation with circulating IL-6 and CRP among those with depression exclusively.	
Numata *et al.* (2015)[221]	20 medication-free MDD, 19 C; replication: 12 medication-free MDD, 12 C	LCLs	HM450	363 DMPs↓ in the discovery sets; most of these (85.7%) were located in CGIs in the gene promoter regions. Of the top 100 DMPs, 84 were replicated	—
Weder *et al.* (2014)[231]	94 maltreated children (35% with MDD), 96 C	Saliva	HM450	3 genome-wide significant predictors of depression: *ID3*, *GRIN1*, *TPPP*	*ID3*: stress response, *GRIN1*: neural plasticity, *TPPP*: neural circuitry development
ASD					
Berko *et al.* (2014)[229]	47 ASD, 48 C	Ectodermal cells	HM450	DNA methylation patterns are dysregulated in ectodermal cells in these individuals. *OR2L13*↓	Postnatal synaptic transmission
AN					
Booij *et al.* (2015)[118]	29 AN (10 with AN-restrictive type, 19 with AN-binge/purge type) and 15 C	LCLs	HM450	14 ↑DMPs; age of onset (development of the brain and spinal cord); chronicity of illness (synaptogenesis, neurocognitive deficits,	Histone acetylation, RNA modification, cholesterol storage and lipid transport, and dopamine and Glu signaling

(*Continued*)

Table 4 EWAS of Psychiatric Disorders in Peripheral Tissues.—cont'd.

Studies	Sample Sizes	Tissues	Platforms	Main Findings	Pathway Enrichments
				anxiety, altered social functioning, and bowel, kidney, liver, and immune function)	
ADHD					
Walton *et al.* (2017)[232]	1018 mother-offspring pairs (ADHD)	WB	HM450	13 DMPs, TOP: *SKI*, a gene related to transforming growth factor-beta signaling and neural tube development	Peroxisomal processes
Wilmot *et al.* (2015)[223]	46 ADHD, 46 C; Replication: 10 ADHD, 10 C	Saliva	HM450 and BS	*VIPR2* and *MYT1L*	**Inflammatory** processes and modulation of monoamine and cholinergic neurotransmission
AUD					
Zhang *et al.* (2013)[222]	63 AUD, 65 C (Chinese)	WB	HM27	DMPs	Stress, **immune response**, and signal transduction
Zhao *et al.* (2013)[146]	20 AUD (Chinese)	WB	HM450	865 ↓DMPs, 716 ↑DMPs; *SSTR4*↓, *ALDH1L2*↓, *GAD1*↓, *DBH*↓, and *GABRP*↑	AUD

ADHD, Attention deficit-hyperactivity disorder; *AN*, anorexia nervosa; *AUD*, alcohol use disorder; *BC*, buccal cells; *BD*, bipolar disorder; *BN*, bulimia nervosa; *CGI*, CpG islands; *DMPs*, differential methylation probes; *DMRs*, differential methylation regions; *DNAm*, DNA methylation; *GABA*, gamma-aminobutyric acid; *HM27*, Illumina Infinium HumanMethylation 27K; *HM450*, Illumina Infinium HumanMethylation 450K; *LCLs*, leukocyte cell lines; *MBD-seq*, methyl-CpG-binding domain protein-enriched genome sequencing; *MDD*, major depressive disorder; *MeDIP-Seq*, methylated DNA immunoprecipitation sequencing; *MTZ*, monozygotic twin; *SZ*, schizophrenia; *WB*, white blood cells.

The bolded words pertain to immunity and neurodevelopment.

MZT pairs and their matched concordant MZT pairs at any individual probe in a global methylation MDD study[93] using blood samples.

Numerous DMPs were detected for MZT pairs in BD,[89,237] SZ,[238] ASD,[234] and MDD.[93,236,239,240] The genes located via these DMPs are enriched for nervous system development and function,[89,236,237] fetal development,[234] morphogenesis,[234] apoptosis,[234] synaptic regulation,[234] steroid biosynthesis,[234] psychological disorders,[89,234,236,237] rapid glucocorticoid signaling,[239] interleukin-2-mediated signaling events,[239] cell proliferation on forebrain,[240] fat cell differentiation,[240] dopaminergic synapse,[89,237,239] and Glu signaling.[89,237,240] These pathways are similar to those found in EWAS (Section 3.2.2), which may be because the conventional EWAS having almost 1000 samples weakens some confounding influences (like genetics and placenta environment).

Because MZTs share unusually similar genetics, placentas, and (often) environments, identical twin studies, all other things being equal, should allow for more confidence in the true positivity of results when compared to basic case–control studies. However, the validity of twin study findings is frequently limited by small sample sizes and tissue specificity. No significant global methylation difference in psychiatric disorders is found for MZT pair blood samples, as opposed to in the other studies (described earlier), which identify global methylation differences between psychiatric cases and controls. It is possible that this discrepancy is due to the genetic and intrauterine influences shared by twins but not unrelated individuals; these factors may contribute heavily to global methylation differences between psychiatric disorder patients and controls (Table 5).

In synthesizing the results from global methylation studies, candidate gene studies, and EWAS in psychiatric disorders, we conclude the following:

(1) Global methylation levels mainly show increases in brain samples and decreases in peripheral tissue samples; the increased methylation probes detected in EWAS are enriched for neuron-related pathways, whereas the decreased methylation probes are enriched for immune-related pathways; the DMPs detected in EWAS with brain samples are mainly enriched for neurons, and the peripheral tissues are mainly enriched for immune-related pathways. These findings provide evidence for the role of the immune system in psychiatric disorders.

(2) Methylation levels may affect neurotransmitter systems, which may, in turn, contribute to risk of psychiatric disorders. Specifically, overactivity of dopaminergic and glutamatergic systems, as well as hypofunction of serotonergic and GABAergic systems, may be important to the etiology

Table 5 MZT Studies of Psychiatric Disorders.

Studies	Samples	Tissues	Platforms	Main Findings	Enrichments
Bönsch *et al.* (2012)[90]	20 twin pairs **discordant** for **SZ**, 8 twin pairs **concordant** for **SZ**, 42 C twin pairs (MZT and DZT)	WB	Methylation-sensitive restriction (HpaII) and PCR	Similar hypomethylation levels in the unaffected twins of discordant pairs and a mixed group of psychiatric controls	
Castellani *et al.* (2015)[237]	2 pairs of female MZT **discordant** for **SZ** and their parents	WB	MeDIP-Seq	27 genes	Cell death and survival, cellular movement, immune cell trafficking, nervous system development, and DRD4
Hannon *et al.* (2016)[238]	1714 individuals (353 **SZ**, 322 controls (phase 1); 414 **SZ**, 433 controls (phase 2); 96 MZT pairs (phase 3))	WB	HM450	76 DMPs from twin study	

Dempster *et al.* (2011)[89]	22 **discordant** MZT pairs: 11 **SZ**, 11 **BD**	WB; B for replication of TOP	HM27, EpiTYPER (Sequenom Inc., CA)	DMPs, TOP: *ST6GALNAC1*↓	**Nervous system development** and function, psychological disorders, DA and Glu receptor signaling, nervous system development
Sugawara *et al.* (2011)[135]	2 MZT pairs **discordant** for **BD** (male)	LCLs; B for replication of TOP	BS and PCR	↑*SLC6A*	Serotonin transporter
Kuratomi *et al.* (2008)[233]	1 MZT pair **discordant** for **BD** (male); replication: 16 **BD I**, 7 **BD II**, 18 C	Lymphoblastoid cells	BS and PS	*PPIEL* ↑E and ↓M in BP II	—
Byrne *et al.* (2013)[93]	12 MZT pairs **discordant** for **MDD** and 12 MZT pairs **concordant** for low neuroticism and absence of **MDD**	WB	HM450	DMPs were detected for affected vs. unaffected twins. × Discordant MZT pairs and their matched concordant MZT pairs	—

(*Continued*)

Table 5 MZT Studies of Psychiatric Disorders.—cont'd.

Studies	Samples	Tissues	Platforms	Main Findings	Enrichments
Davies *et al.* (2014)[235]	50 MZT pairs **discordant** for **MDD** (27 UK, 23 Australia); replication: 118 **MDD**, 236 C	WB	MeDIP-Seq	↑*ZBTB20*	—
Dempster *et al.* (2014)[236]	18 MZT pairs **discordant** for **MDD**; replication: 14 **MDD**, 15 C	MZT: BC, replication: CBL	HM450	Multiple DMPs, TOP: *STK32C*	Nervous system development and function, neurological disease and psychological disorders
Cordova-Palomera *et al.* (2015)[239]	17 MZT pairs (4 **concordant** for **MDD**, 6 **discordant**, 7 C)	WB	HM450	DMPs, *WDR26*	Rapid glucocorticoid signaling, DA synapse and *interleukin-2*-mediated signaling events

Oh *et al.* (2015)[240]	171 MZT pairs **MDD**	WB (100), PFC (71)	8.1 K CpG island microarrays and BS	PFC: 22 DMPs, *FOXD3*; WB: 44 DMPs	In cell proliferation on forebrain, fat cell differentiation, and Glu signaling pathway
Nguyen *et al.* (2010)[234]	3 MZT **discordant** for **ASD** (male), 2 unaffected siblings	LCLs	8.1 K CpG island microarray	73 DMPs, *RORA*	Synaptic regulation, fetal development, morphogenesis, apoptosis, inflammation, digestion, steroid biosynthesis, and mental deficiency
Wong *et al.* (2014)[109]	50 MZT pairs **discordant** and **concordant** for **ASD/ASD** severity	WB	HM27	DMPs	—

ASD, Autism spectrum disorder; *B*, brain; *BD*, bipolar disorder; *DA*, dopamine; *DZT*, dizygotic twin; *Glu*, glutamate; *HM27*, Illumina Infinium HumanMethylation 27K; *HM450*, Illumina Infinium HumanMethylation 450K; *LCLs*, leukocyte cell lines; *MDD*, major depressive disorder; *MZT*, monozygotic twin; *SZ*, schizophrenia; *WB*, white blood cells.

The bolded words pertain to immunity and neurodevelopment.

of major psychosis. Furthermore, hypomethylation of dopaminergic and glutamatergic system receptors and hypermethylation of serotonergic and GABAergic system receptors have been detected. These abnormal methylation levels may lead to receptor excitation or inhibition, giving rise to the functional changes of neurotransmission processes. Similar patterns were found for transporters and some enzymes.

(3) We listed several other candidate genes in Section 3.2.1. However, these alone may be less informative in the grand scheme of things because major psychiatric disorders are polygenic. It will be essential to detect additional related biomarkers to construct more complete models of relevant pathways and mechanisms.

(4) The biomarkers discovered in the EWAS using brain samples reveal important information regarding the importance of fetal brain development in SZ. Although a limited number of EWASs have looked at other psychiatric disorders due to differences in heterogeneity, it will be important to study the influence of fetal brain development on other disorders.

(5) Confounding factors such as smoking status, medication use, sex, age of disease onset, and varying research methods used in different studies could have an influence on any of the aforementioned results.

There are some results that cannot be explained by our current knowledge about biology and pathology. Inconsistent results could be the product of discrepancies in data quality, technical and analytical differences, and sample heterogeneity, the last of which is more likely to be a problematic factor in small sample sizes. These unresolved issues can inspire future scientific exploration.

4. FUTURE PERSPECTIVES AND CHALLENGES

In this article, we introduced DNA methylation and highlighted important findings on DNA methylation in major psychiatric disorders. Current studies still face several limitations and challenges. The major studies mentioned earlier focus on DNA methylation differences between cases and controls using bulk tissue samples. Both positive and negative results for differential methylation could be false because of confounding factors, including differences in cell composition in different tissues (e.g., brain vs. blood) used in different studies. Furthermore, more functional genomics data are still needed to help explain these epigenetic changes and their effects.

Innovative experimental and analytical approaches are expected to improve our understanding of DNA methylation and its relevance to psychiatric disorders in the coming years. Comethylation network analysis, integration of multidimensional omics data, and single-cell methylome and methylation editing techniques are a few such emerging approaches. In addition, studies on the association of DNA methylation with intermediate phenotypes other than disease diagnoses will produce more valuable findings. Lastly, DNA methylation-based therapy may serve as a future treatment for psychiatric disorders.

4.1 Network Analysis and Integration of Multidimensional Data

For a long time, DNA methylation studies have focused on individual loci or regions but failed to examine the relationships between these loci, though many methods for creating networks of loci exist. Coexpression network analysis uses gene expression correlations to construct networks; comethylation network analysis identifies CpGs with associated methylation when they are regulated by common mechanisms.[241–243] Using comethylation network analysis, Berko *et al.*[229] obtained two modules significantly associated with autism spectrum disorder case/control status after stringent Bonferroni correction.[229] Wang *et al.*[208] identified 21 AUD-associated comethylation modules in the prefrontal cortex (PFC), and found that the differentially methylated CpGs were overrepresented in two of the modules.[208] However, it should be noted that correlation does not necessarily indicate causality. Experimental validation is needed to clarify whether there is actually a regulatory relationship.

Findings on DNA methylation can be applied to various forms of methodology, as well as integrated into genetic and omics data. For example, haplotype-dependent allele-specific DNA methylation (hap-ASM) can help refine mQTL results by pinpointing DNA sequence variants that underlie genetic susceptibility to psychiatric disorders.[244] As gene expression regulation is considered to be the primary function of DNA methylation, many candidate gene methylation studies have focused on the expression–methylation relationship.[123,142,143,148] Systematic integration of transcriptome data and EWAS will be more effective and unbiased.[245] Finally, Hi-C,[246,247] ATAC-seq,[248] and DNase-seq,[249,250] which examine chromatin interaction and accessibility, could be used for investigating DNA methylation's role in regulating gene expression.[251] Building networks involving genetic variants, chromatin structure, and DNA methylation may improve the quality of gene expression regulation predictions.

4.2 Single-Cell Methylome

The current strategies for studying DNA methylation in psychiatric disorders typically use bulk tissues. As mentioned earlier, cell-specific measurement of DNA methylation is more desirable, as it is less likely to yield confounding results.

Single-cell methylome methods, such as single-cell reduced representation bisulfite sequencing (scRRBS),[252] single-cell bisulfite sequencing (scBS-seq),[253] and single-nucleus methylcytosine sequencing (snmC-seq),[254] have been developed. snmC-seq has been applied to more than 6000 single neuronal nuclei from mouse and human FC. The results identified 37 neuronal subpopulations and showed that CG and non-CG methylation exhibited cell-type-specific distributions.[254]

In addition, novel methods have been developed to measure multiomics in a single cell: single-cell NOMe-seq (scNOMe-seq)[255] and single-cell multiomics sequencing (single-cell COOL-seq)[256] allow for the simultaneous measurements of DNA methylation and chromatin accessibility from the same cell. With these single-cell techniques, we will be able to evaluate disease pathology with better resolution and confidence.

4.3 DNA Methylation Signatures of Intermediate Phenotypes

Most previous publications on DNA methylation have targeted disorder diagnoses, but endophenotypes (or intermediate phenotypes) relevant to psychiatric disorders are getting increasingly more attention. Notably, alterations in methylation have been reported to be associated with cognitive functions,[105,257] reality distortion symptoms (possibility via immune response modulation),[226] and brain volume[258–260] in psychotic disorders. Using DNA methylation signatures to study these intermediate phenotypes could help dissect the complexity of the underlying biology of psychiatric disorders.

4.4 Methylation Editing and Functional Validation

Functional validation is an important step toward translating statistical correlation into causal biological function. Gene expression change can be experimentally validated by gene editing, RNA interference silencing, and knock-in overexpression. However, validation of DNA methylation functionality through methylation editing approaches has only recently become feasible with the development of clustered regularly interspaced short palindromic repeat (CRISPR) and CRISPR-associated protein systems (CRISPR-Cas).

Several different versions of DNA methylation editing methods have been developed. DNA methylation editing can be performed using the dCas9-based system, which lacks endonuclease activity.[261] dCas9 fused to the catalytic domain of the DNA demethylase TET1 was used to induce targeted DNA demethylation.[262–264] Similarly, a fusion of the catalytic domain of DNA methyltransferase 3A (DNMT3A) to dCas9 is capable of inducing de novo methylation of the targeted region.[264–267] The modification of DNA methylation by these systems leads to changed expression of the targeted genes and cell reprogramming.[264] Using the dCas9 methylation editing toolkit, we could study the function of DNA methylation.

4.5 DNA Methylation-Based Treatments

Given that DNA methylation likely contributes to the risk of psychiatric disorders, manipulation of methylation levels is a promising method for developing novel therapeutic approaches.

Specifically, modulation of enzymes such as DNMT and TET, which are related to DNA methylation and demethylation, may produce therapeutic effects. Many chemical compounds such as 5-aza-2′-deoxycytidine (AZA), doxorubicin (DOXO), and zebularine (ZEB) have been shown to decrease the methylation and increase the expression of candidate genes such as *GAD67* and *reelin* in in vitro and in vivo experiments.[268–271] Moreover, some existing FDA-approved drugs, such as Clozapine and Sulpiride, have demonstrated the ability to activate brain DNA demethylation.[271] The CRISPR-dCas9 system holds the hope of gene therapy via DNA methylation, as well. Still, the process of therapy through methylation modification will no doubt require much time and research to achieve substantial progress.

ACKNOWLEDGMENTS

We thank Tanya Horwitz for critical reading and editing the manuscript.

REFERENCES

1. Bird AP. CpG-rich islands and the function of DNA methylation. *Nature.* 1986;321:209–213.
2. Kinney SM, et al. Tissue-specific distribution and dynamic changes of 5-hydroxymethylcytosine in mammalian genomes. *J Biol Chem.* 2011;286:24685–24693.
3. Kriaucionis S, Heintz N. The nuclear DNA base 5-hydroxymethylcytosine is present in Purkinje neurons and the brain. *Science.* 2009;324:929–930.
4. Liu C. Brain expression quantitative trait locus mapping informs genetic studies of psychiatric diseases. *Neurosci Bull.* 2011;27:123–133.
5. Tahiliani M, et al. Conversion of 5-methylcytosine to 5-hydroxymethylcytosine in mammalian DNA by MLL partner TET1. *Science.* 2009;324:930–935.
6. Wu H, Zhang Y. Charting oxidized methylcytosines at base resolution. *Nat Struct Mol Biol.* 2015;22:656–661.
7. Hahn MA, et al. Dynamics of 5-hydroxymethylcytosine and chromatin marks in Mammalian neurogenesis. *Cell Rep.* 2013;3:291–300.
8. Pinney SE. Mammalian non-CpG methylation: stem cells and beyond. *Biology (Basel).* 2014;3:739–751.
9. Guo JU, et al. Distribution, recognition and regulation of non-CpG methylation in the adult mammalian brain. *Nat Neurosci.* 2014;17:215–222.
10. Ramsahoye BH, et al. Non-CpG methylation is prevalent in embryonic stem cells and may be mediated by DNA methyltransferase 3a. *Proc Natl Acad Sci USA.* 2000;97:5237–5242.
11. Yan J, Zierath JR, Barres R. Evidence for non-CpG methylation in mammals. *Exp Cell Res.* 2011;317:2555–2561.
12. Tomizawa S, et al. Dynamic stage-specific changes in imprinted differentially methylated regions during early mammalian development and prevalence of non-CpG methylation in oocytes. *Development.* 2011;138:811–820.
13. Lister R, et al. Global epigenomic reconfiguration during mammalian brain development. *Science.* 2013;341:1237905.
14. Vertino PM, Yen RW, Gao J, Baylin SB. De novo methylation of CpG island sequences in human fibroblasts overexpressing DNA (cytosine-5-)-methyltransferase. *Mol Cell Biol.* 1996;16:4555–4565.
15. Rhee I, et al. CpG methylation is maintained in human cancer cells lacking DNMT1. *Nature.* 2000;404:1003–1007.
16. Barres R, et al. Non-CpG methylation of the PGC-1alpha promoter through DNMT3B controls mitochondrial density. *Cell Metab.* 2009;10:189–198.
17. Li J, et al. Genes with de novo mutations are shared by four neuropsychiatric disorders discovered from NPdenovo database. *Mol Psychiatry.* 2016;21:290–297.
18. Ping XL, et al. Mammalian WTAP is a regulatory subunit of the RNA N6-methyladenosine methyltransferase. *Cell Res.* 2014;24:177–189.
19. Wang X, et al. Corrigendum: structural basis of N6-adenosine methylation by the METTL3-METTL14 complex. *Nature.* 2017;542:260.
20. Wu TP, et al. DNA methylation on N(6)-adenine in mammalian embryonic stem cells. *Nature.* 2016;532:329–333.
21. Liu J, et al. A METTL3-METTL14 complex mediates mammalian nuclear RNA N6-adenosine methylation. *Nat Chem Biol.* 2014;10:93–95.
22. Luo GZ, et al. Characterization of eukaryotic DNA N(6)-methyladenine by a highly sensitive restriction enzyme-assisted sequencing. *Nat Commun.* 2016;7:11301.
23. O'Brown ZK, Greer EL. N6-Methyladenine: a conserved and dynamic DNA mark. *Adv Exp Med Biol.* 2016;945:213–246.
24. Sun Q, et al. N6-methyladenine functions as a potential epigenetic mark in eukaryotes. *Bioessays.* 2015;37:1155–1162.

25. Heyn H, Esteller M. An adenine code for DNA: a second life for N6-methyladenine. *Cell.* 2015;161:710–713.
26. Luo GZ, Blanco MA, Greer EL, He C, Shi Y. DNA N(6)-methyladenine: a new epigenetic mark in eukaryotes? *Nat Rev Mol Cell Biol.* 2015;16:705–710.
27. Wagner JR, et al. The relationship between DNA methylation, genetic and expression inter-individual variation in untransformed human fibroblasts. *Genome Biol.* 2014;15: R37.
28. Chen C, et al. Correlation between DNA methylation and gene expression in the brains of patients with bipolar disorder and schizophrenia. *Bipolar Disord.* 2014;16:790–799.
29. Maurano MT, et al. Role of DNA methylation in modulating transcription factor occupancy. *Cell Rep.* 2015;12:1184–1195.
30. Stephens DC, Poon GM. Differential sensitivity to methylated DNA by ETS-family transcription factors is intrinsically encoded in their DNA-binding domains. *Nucleic Acids Res.* 2016;44:8671–8681.
31. Hu S, et al. DNA methylation presents distinct binding sites for human transcription factors. *Elife.* 2013;2:e00726.
32. Kribelbauer JF, et al. Quantitative analysis of the DNA methylation sensitivity of transcription factor complexes. *Cell Rep.* 2017;19:2383–2395.
33. Hashimshony T, Zhang J, Keshet I, Bustin M, Cedar H. The role of DNA methylation in setting up chromatin structure during development. *Nat Genet.* 2003;34:187–192.
34. Jimenez-Useche I, et al. DNA methylation effects on tetra-nucleosome compaction and aggregation. *Biophys J.* 2014;107:1629–1636.
35. Xia J, Han L, Zhao Z. Investigating the relationship of DNA methylation with mutation rate and allele frequency in the human genome. *BMC Genomics.* 2012;13(suppl 8):S7.
36. Li E, Beard C, Jaenisch R. Role for DNA methylation in genomic imprinting. *Nature.* 1993;366:362–365.
37. Sharp AJ, et al. DNA methylation profiles of human active and inactive X chromosomes. *Genome Res.* 2011;21:1592–1600.
38. Inoue S, Oishi M. Effects of methylation of non-CpG sequence in the promoter region on the expression of human synaptotagmin XI (syt11). *Gene.* 2005;348:123–134.
39. Malone CS, et al. CmC(A/T)GG DNA methylation in mature B cell lymphoma gene silencing. *Proc Natl Acad Sci USA.* 2001;98:10404–10409.
40. Bellizzi D, et al. The control region of mitochondrial DNA shows an unusual CpG and non-CpG methylation pattern. *DNA Res.* 2013;20:537–547.
41. Yong WS, Hsu FM, Chen PY. Profiling genome-wide DNA methylation. *Epigenet Chromatin.* 2016;9:26.
42. Kurdyukov S, Bullock M. DNA methylation analysis: choosing the right method. *Biology (Basel).* 2016;5:E3.
43. Couldrey C, Cave V. Assessing DNA methylation levels in animals: choosing the right tool for the job. *Anim Genet.* 2014;45(suppl 1):15–24.
44. Song L, James SR, Kazim L, Karpf AR. Specific method for the determination of genomic DNA methylation by liquid chromatography-electrospray ionization tandem mass spectrometry. *Anal Chem.* 2005;77:504–510.
45. Armstrong KM, et al. Global DNA methylation measurement by HPLC using low amounts of DNA. *Biotechnol J.* 2011;6:113–117.
46. Yu M, et al. Base-resolution analysis of 5-hydroxymethylcytosine in the mammalian genome. *Cell.* 2012;149:1368–1380.
47. Lavoie J, Sawa A, Ishizuka K. Application of olfactory tissue and its neural progenitors to schizophrenia and psychiatric research. *Curr Opin Psychiatry.* 2017;30:176–183.
48. Walton E, et al. Correspondence of DNA methylation between blood and brain tissue and its application to schizophrenia research. *Schizophr Bull.* 2016;42:406–414.

49. Farre P, et al. Concordant and discordant DNA methylation signatures of aging in human blood and brain. *Epigenet Chromatin.* 2015;8:19.
50. Edgar RD, Jones MJ, Meaney MJ, Turecki G, Kobor MS. BECon: a tool for interpreting DNA methylation findings from blood in the context of brain. *Transl Psychiatry.* 2017;7: e1187.
51. Aberg KA, et al. Testing two models describing how methylome-wide studies in blood are informative for psychiatric conditions. *Epigenomics.* 2013;5:367–377.
52. Aberg K, et al. Methylome-wide comparison of human genomic DNA extracted from whole blood and from EBV-transformed lymphocyte cell lines. *Eur J Hum Genet.* 2012;20:953–955.
53. Zhang D, et al. Genetic control of individual differences in gene-specific methylation in human brain. *Am J Hum Genet.* 2010;86:411–419.
54. Gutierrez-Arcelus M, et al. Tissue-specific effects of genetic and epigenetic variation on gene regulation and splicing. *PLoS Genet.* 2015;11:e1004958.
55. Gibbs JR, et al. Abundant quantitative trait loci exist for DNA methylation and gene expression in human brain. *PLoS Genet.* 2010;6:e1000952.
56. Shoemaker R, Deng J, Wang W, Zhang K. Allele-specific methylation is prevalent and is contributed by CpG-SNPs in the human genome. *Genome Res.* 2010;20:883–889.
57. Meaburn EL, Schalkwyk LC, Mill J. Allele-specific methylation in the human genome: implications for genetic studies of complex disease. *Epigenetics.* 2010;5:578–582.
58. Tang A, et al. Analysis of a four generation family reveals the widespread sequence-dependent maintenance of allelic DNA methylation in somatic and germ cells. *Sci Rep.* 2016;6:19260.
59. Gamazon ER, et al. Enrichment of cis-regulatory gene expression SNPs and methylation quantitative trait loci among bipolar disorder susceptibility variants. *Mol Psychiatry.* 2013;18:340–346.
60. Hannon E, et al. Methylation QTLs in the developing brain and their enrichment in schizophrenia risk loci. *Nat Neurosci.* 2016;19:48–54.
61. van Eijk KR, et al. Identification of schizophrenia-associated loci by combining DNA methylation and gene expression data from whole blood. *Eur J Hum Genet.* 2015;23:1106–1110.
62. Jaffe AE, Eaton WW, Straub RE, Marenco S, Weinberger DR. Paternal age, de novo mutations and schizophrenia. *Mol Psychiatry.* 2014;19:274–275.
63. Malaspina D. Paternal factors and schizophrenia risk: de novo mutations and imprinting. *Schizophr Bull.* 2001;27:379–393.
64. Kong A, et al. Rate of de novo mutations and the importance of father's age to disease risk. *Nature.* 2012;488:471–475.
65. Milekic MH, et al. Age-related sperm DNA methylation changes are transmitted to offspring and associated with abnormal behavior and dysregulated gene expression. *Mol Psychiatry.* 2015;20:995–1001.
66. Kim DR, Bale TL, Epperson CN. Prenatal programming of mental illness: current understanding of relationship and mechanisms. *Curr Psychiatry Rep.* 2015;17:5.
67. Thapar A, Rutter M. Do prenatal risk factors cause psychiatric disorder? Be wary of causal claims. *Br J Psychiatry.* 2009;195:100–101.
68. Kundakovic M, Jaric I. The epigenetic link between prenatal adverse environments and neurodevelopmental disorders. *Genes (Basel).* 2017;8:E104.
69. Xu MQ, et al. Prenatal malnutrition and adult schizophrenia: further evidence from the 1959–1961 Chinese famine. *Schizophr Bull.* 2009;35:568–576.
70. Susser ES, Lin SP. Schizophrenia after prenatal exposure to the Dutch Hunger Winter of 1944–1945. *Arch Gen Psychiatry.* 1992;49:983–988.
71. Tobi EW, et al. Corrigendum: DNA methylation signatures link prenatal famine exposure to growth and metabolism. *Nat Commun.* 2015;6:7740.

72. Tobi EW, et al. DNA methylation differences after exposure to prenatal famine are common and timing- and sex-specific. *Hum Mol Genet.* 2009;18:4046–4053.
73. Pauwels S, et al. Maternal intake of methyl-group donors affects DNA methylation of metabolic genes in infants. *Clin Epigenet.* 2017;9:16.
74. Laufer BI, et al. Associative DNA methylation changes in children with prenatal alcohol exposure. *Epigenomics.* 2015;7:1259–1274.
75. Cao-Lei L, et al. DNA methylation signatures triggered by prenatal maternal stress exposure to a natural disaster: Project Ice Storm. *PLoS One.* 2014;9:e107653.
76. Serpeloni F, et al. Grandmaternal stress during pregnancy and DNA methylation of the third generation: an epigenome-wide association study. *Transl Psychiatry.* 2017;7:e1202.
77. Cottrell EC, Holmes MC, Livingstone DE, Kenyon CJ, Seckl JR. Reconciling the nutritional and glucocorticoid hypotheses of fetal programming. *FASEB J.* 2012;26:1866–1874.
78. Keunen K, van Elburg RM, van Bel F, Benders MJ. Impact of nutrition on brain development and its neuroprotective implications following preterm birth. *Pediatr Res.* 2015;77:148–155.
79. Murgatroyd C, et al. Dynamic DNA methylation programs persistent adverse effects of early-life stress. *Nat Neurosci.* 2009;12:1559–1566.
80. Numachi Y, et al. Methamphetamine alters expression of DNA methyltransferase 1 mRNA in rat brain. *Neurosci Lett.* 2007;414:213–217.
81. Numachi Y, et al. Psychostimulant alters expression of DNA methyltransferase mRNA in the rat brain. *Ann N Y Acad Sci.* 2004;1025:102–109.
82. Satta R, et al. Nicotine decreases DNA methyltransferase 1 expression and glutamic acid decarboxylase 67 promoter methylation in GABAergic interneurons. *Proc Natl Acad Sci USA.* 2008;105:16356–16361.
83. Bönsch D, et al. Lowered DNA methyltransferase (DNMT-3b) mRNA expression is associated with genomic DNA hypermethylation in patients with chronic alcoholism. *J Neural Transm (Vienna).* 2006;113:1299–1304.
84. Philibert RA, Plume JM, Gibbons FX, Brody GH, Beach SR. The impact of recent alcohol use on genome wide DNA methylation signatures. *Front Genet.* 2012;3:54.
85. Shenker NS, et al. DNA methylation as a long-term biomarker of exposure to tobacco smoke. *Epidemiology.* 2013;24:712–716.
86. Godino A, Jayanthi S, Cadet JL. Epigenetic landscape of amphetamine and methamphetamine addiction in rodents. *Epigenetics.* 2015;10:574–580.
87. Burmeister M, McInnis MG, Zollner S. Psychiatric genetics: progress amid controversy. *Nat Rev Genet.* 2008;9:527–540.
88. Craddock N, Owen MJ. The Kraepelinian dichotomy—going, going . . . but still not gone. *Br J Psychiatry.* 2010;196:92–95.
89. Dempster EL, et al. Disease-associated epigenetic changes in monozygotic twins discordant for schizophrenia and bipolar disorder. *Hum Mol Genet.* 2011;20:4786–4796.
90. Bönsch D, et al. Methylation matters? Decreased methylation status of genomic DNA in the blood of schizophrenic twins. *Psychiatry Res.* 2012;198:533–537.
91. Melas PA, et al. Epigenetic aberrations in leukocytes of patients with schizophrenia: association of global DNA methylation with antipsychotic drug treatment and disease onset. *FASEB J.* 2012;26:2712–2718.
92. Burghardt KJ, Goodrich JM, Dolinoy DC, Ellingrod VL. DNA methylation, insulin resistance and second-generation antipsychotics in bipolar disorder. *Epigenomics.* 2015;7:343–352.
93. Byrne EM, et al. Monozygotic twins affected with major depressive disorder have greater variance in methylation than their unaffected co-twin. *Transl Psychiatry.* 2013;3:e269.

94. Shimabukuro M, et al. Global hypomethylation of peripheral leukocyte DNA in male patients with schizophrenia: a potential link between epigenetics and schizophrenia. *J Psychiatr Res.* 2007;41:1042–1046.
95. Xiao Y, et al. The DNA methylome and transcriptome of different brain regions in schizophrenia and bipolar disorder. *PLoS One.* 2014;9:e95875.
96. Murphy TM, et al. Anxiety is associated with higher levels of global DNA methylation and altered expression of epigenetic and interleukin-6 genes. *Psychiatr Genet.* 2015;25:71–78.
97. Otero NK, Thomas JD, Saski CA, Xia X, Kelly SJ. Choline supplementation and DNA methylation in the hippocampus and prefrontal cortex of rats exposed to alcohol during development. *Alcohol Clin Exp Res.* 2012;36:1701–1709.
98. Nishioka M, et al. Comprehensive DNA methylation analysis of peripheral blood cells derived from patients with first-episode schizophrenia. *J Hum Genet.* 2013;58:91–97.
99. Huzayyin AA, et al. Decreased global methylation in patients with bipolar disorder who respond to lithium. *Int J Neuropsychopharmacol.* 2014;17:561–569.
100. Tseng PT, et al. Age-associated decrease in global DNA methylation in patients with major depression. *Neuropsychiatr Dis Treat.* 2014;10:2105–2114.
101. Melnyk S, et al. Metabolic imbalance associated with methylation dysregulation and oxidative damage in children with autism. *J Autism Dev Disord.* 2012;42:367–377.
102. James SJ, et al. Abnormal transmethylation/transsulfuration metabolism and DNA hypomethylation among parents of children with autism. *J Autism Dev Disord.* 2008;38:1966–1975.
103. Tremolizzo L, et al. Decreased whole-blood global DNA methylation is related to serum hormones in anorexia nervosa adolescents. *World J Biol Psychiatry.* 2014;15:327–333.
104. Frieling H, et al. Global DNA hypomethylation and DNA hypermethylation of the alpha synuclein promoter in females with anorexia nervosa. *Mol Psychiatry.* 2007;12:229–230.
105. Alelú-Paz R, et al. Epigenetics in schizophrenia: a pilot study of global DNA methylation in different brain regions associated with higher cognitive functions. *Front Psychol.* 2016;7:1496.
106. Bromberg A, Bersudsky Y, Levine J, Agam G. Global leukocyte DNA methylation is not altered in euthymic bipolar patients. *J Affect Disord.* 2009;118:234–239.
107. Rao JS, Keleshian VL, Klein S, Rapoport SI. Epigenetic modifications in frontal cortex from Alzheimer's disease and bipolar disorder patients. *Transl Psychiatry.* 2012;2:e132.
108. Soeiro-de-Souza MG, et al. Number of manic episodes is associated with elevated DNA oxidation in bipolar I disorder. *Int J Neuropsychopharmacol.* 2013;16:1505–1512.
109. Wong CC, et al. Methylomic analysis of monozygotic twins discordant for autism spectrum disorder and related behavioural traits. *Mol Psychiatry.* 2014;19:495–503.
110. Saffrey R, Novakovic B, Wade TD. Assessing global and gene specific DNA methylation in anorexia nervosa: a pilot study. *Int J Eat Disord.* 2014;47:206–210.
111. Manzardo AM, Henkhaus RS, Butler MG. Global DNA promoter methylation in frontal cortex of alcoholics and controls. *Gene.* 2012;498:5–12.
112. Rapoport JL, Giedd JN, Gogtay N. Neurodevelopmental model of schizophrenia: update 2012. *Mol Psychiatry.* 2012;17:1228–1238.
113. Khandaker GM, Zimbron J, Lewis G, Jones PB. Prenatal maternal infection, neurodevelopment and adult schizophrenia: a systematic review of population-based studies. *Psychol Med.* 2013;43:239–257.
114. Patterson PH. Immune involvement in schizophrenia and autism: etiology, pathology and animal models. *Behav Brain Res.* 2009;204:313–321.

115. Yamagata AS, et al. Selfish brain and selfish immune system interplay: a theoretical framework for metabolic comorbidities of mood disorders. *Neurosci Biobehav Rev.* 2017;72:43–49.
116. Khandaker GM, et al. Inflammation and immunity in schizophrenia: implications for pathophysiology and treatment. *Lancet Psychiatry.* 2015;2:258–270.
117. Mill J, et al. Epigenomic profiling reveals DNA-methylation changes associated with major psychosis. *Am J Hum Genet.* 2008;82:696–711.
118. Booij L, et al. DNA methylation in individuals with anorexia nervosa and in matched normal-eater controls: a genome-wide study. *Int J Eat Disord.* 2015;48:874–882.
119. Seeman MV, Seeman P. Is schizophrenia a dopamine supersensitivity psychotic reaction? *Prog Neuropsychopharmacol Biol Psychiatry.* 2014;48:155–160.
120. Seeman P. Dopamine D2 receptors as treatment targets in schizophrenia. *Clin Schizophr Relat Psychoses.* 2010;4:56–73.
121. Yoshino Y, et al. Low methylation rates of dopamine receptor D2 gene promoter sites in Japanese schizophrenia subjects. *World J Biol Psychiatry.* 2016;17:449–456.
122. Abdolmaleky HM, et al. Hypomethylation of MB-COMT promoter is a major risk factor for schizophrenia and bipolar disorder. *Hum Mol Genet.* 2006;15:3132–3145.
123. Kordi-Tamandani DM, Sahranavard R, Torkamanzehi A. Analysis of association between dopamine receptor genes' methylation and their expression profile with the risk of schizophrenia. *Psychiatr Genet.* 2013;23:183–187.
124. van Mil NH, et al. DNA methylation profiles at birth and child ADHD symptoms. *J Psychiatr Res.* 2014;49:51–59.
125. Frieling H, et al. Epigenetic dysregulation of dopaminergic genes in eating disorders. *Int J Eat Disord.* 2010;43:577–583.
126. Pjetri E, et al. Quantitative promoter DNA methylation analysis of four candidate genes in anorexia nervosa: a pilot study. *J Psychiatr Res.* 2013;47:280–282.
127. Takahashi S. Heterogeneity of schizophrenia: genetic and symptomatic factors. *Am J Med Genet B Neuropsychiatr Genet.* 2013;162B:648–652.
128. Myöhänen TT, Schendzielorz N, Männistö PT. Distribution of catechol-*O*-methyltransferase (COMT) proteins and enzymatic activities in wild-type and soluble COMT deficient mice. *J Neurochem.* 2010;113:1632–1643.
129. Nohesara S, et al. DNA hypomethylation of MB-COMT promoter in the DNA derived from saliva in schizophrenia and bipolar disorder. *J Psychiatr Res.* 2011;45:1432–1438.
130. Walton E, et al. MB-COMT promoter DNA methylation is associated with working-memory processing in schizophrenia patients and healthy controls. *Epigenetics.* 2014;9:1101–1107.
131. Murphy BC, O'Reilly RL, Singh SM. Site-specific cytosine methylation in S-COMT promoter in 31 brain regions with implications for studies involving schizophrenia. *Am J Med Genet B Neuropsychiatr Genet.* 2005;133B:37–42.
132. Chen Y, Zhang J, Zhang L, Shen Y, Xu Q. Effects of MAOA promoter methylation on susceptibility to paranoid schizophrenia. *Hum Genet.* 2012;131:1081–1087.
133. Hillemacher T, et al. Promoter specific methylation of the dopamine transporter gene is altered in alcohol dependence and associated with craving. *J Psychiatr Res.* 2009;43:388–392.
134. Nieratschker V, et al. Epigenetic alteration of the dopamine transporter gene in alcohol-dependent patients is associated with age. *Addict Biol.* 2014;19:305–311.
135. Sugawara H, et al. Hypermethylation of serotonin transporter gene in bipolar disorder detected by epigenome analysis of discordant monozygotic twins. *Transl Psychiatry.* 2011;1:e24.
136. Carrard A, Salzmann A, Malafosse A, Karege F. Increased DNA methylation status of the serotonin receptor 5HTR1A gene promoter in schizophrenia and bipolar disorder. *J Affect Disord.* 2011;132:450–453.

137. Abdolmaleky HM, et al. Epigenetic dysregulation of HTR2A in the brain of patients with schizophrenia and bipolar disorder. *Schizophr Res.* 2011;129:183–190.
138. Cheah SY, Lawford BR, Young RM, Morris CP, Voisey J. mRNA expression and DNA methylation analysis of serotonin receptor 2A (HTR2A) in the human schizophrenic brain. *Genes (Basel).* 2017;8:14.
139. Ghadirivasfi M, et al. Hypomethylation of the serotonin receptor type-2A Gene (HTR2A) at T102C polymorphic site in DNA derived from the saliva of patients with schizophrenia and bipolar disorder. *Am J Med Genet B Neuropsychiatr Genet.* 2011;156B:536–545.
140. Philibert RA, et al. The relationship of 5HTT (SLC6A4) methylation and genotype on mRNA expression and liability to major depression and alcohol dependence in subjects from the Iowa Adoption Studies. *Am J Med Genet B Neuropsychiatr Genet.* 2008;147B:543–549.
141. Kang HJ, et al. Association of SLC6A4 methylation with early adversity, characteristics and outcomes in depression. *Prog Neuropsychopharmacol Biol Psychiatry.* 2013;44:23–28.
142. Abdolmaleky HM, et al. DNA hypermethylation of serotonin transporter gene promoter in drug naive patients with schizophrenia. *Schizophr Res.* 2014;152:373–380.
143. Kordi-Tamandani DM, Dahmardeh N, Torkamanzehi A. Evaluation of hypermethylation and expression pattern of GMR2, GMR5, GMR8, and GRIA3 in patients with schizophrenia. *Gene.* 2013;515:163–166.
144. Fremeau Jr RT et al. The expression of vesicular glutamate transporters defines two classes of excitatory synapse. *Neuron.* 2001;31:247–260.
145. Jia YF, et al. Differential SLC1A2 promoter methylation in bipolar disorder with or without addiction. *Front Cell Neurosci.* 2017;11:217.
146. Zhao R, et al. Genome-wide DNA methylation patterns in discordant sib pairs with alcohol dependence. *Asia Pac Psychiatry.* 2013;5:39–50.
147. Ruzicka WB, Subburaju S, Benes FM. Circuit- and diagnosis-specific dna methylation changes at gamma-aminobutyric acid-related genes in postmortem human hippocampus in schizophrenia and bipolar disorder. *JAMA Psychiatry.* 2015;72:541–551.
148. Dong E, Ruzicka WB, Grayson DR, Guidotti A. DNA-methyltransferase1 (DNMT1) binding to CpG rich GABAergic and BDNF promoters is increased in the brain of schizophrenia and bipolar disorder patients. *Schizophr Res.* 2015;167:35–41.
149. Alcantara S, et al. Regional and cellular patterns of reelin mRNA expression in the forebrain of the developing and adult mouse. *J Neurosci.* 1998;18:7779–7799.
150. Abdolmaleky HM, et al. Hypermethylation of the reelin (RELN) promoter in the brain of schizophrenic patients: a preliminary report. *Am J Med Genet B Neuropsychiatr Genet.* 2005;134B:60–66.
151. Grayson DR, et al. The human reelin gene: transcription factors (+), repressors (-) and the methylation switch (+/-) in schizophrenia. *Pharmacol Ther.* 2006;111:272–286.
152. Tochigi M, et al. Methylation status of the reelin promoter region in the brain of schizophrenic patients. *Biol Psychiatry.* 2008;63:530–533.
153. Tamura Y, Kunugi H, Ohashi J, Hohjoh H. Epigenetic aberration of the human REELIN gene in psychiatric disorders. *Mol Psychiatry.* 2007;12:519593-600.
154. Arai KI, et al. Cytokines: coordinators of immune and inflammatory responses. *Annu Rev Biochem* 1990;59:783–836. doi:10.1146/annurev.bi.59.070190.004031.
155. Najjar S, Pearlman DM, Alper K, Najjar A, Devinsky O. Neuroinflammation and psychiatric illness. *J Neuroinflammation.* 2013;10:43.
156. Hamdani N, Doukhan R, Kurtlucan O, Tamouza R, Leboyer M. Immunity, inflammation, and bipolar disorder: diagnostic and therapeutic implications. *Curr Psychiatry Rep.* 2013;15:387.
157. Estes ML, McAllister AK. Immune mediators in the brain and peripheral tissues in autism spectrum disorder. *Nat Rev Neurosci.* 2015;16:469–486.

158. Hoge EA, et al. Broad spectrum of cytokine abnormalities in panic disorder and posttraumatic stress disorder. *Depress Anxiety.* 2009;26:447–455.
159. Michopoulos V, Powers A, Gillespie CF, Ressler KJ, Jovanovic T. Inflammation in fear- and anxiety-based disorders: PTSD, GAD, and beyond. *Neuropsychopharmacology.* 2017;42:254–270.
160. Marcos A. The immune system in eating disorders: an overview. *Nutrition.* 1997;13:853–862.
161. Smith AK, et al. Differential immune system DNA methylation and cytokine regulation in post-traumatic stress disorder. *Am J Med Genet B Neuropsychiatr Genet.* 2011;156B:700–708.
162. Rusiecki JA, et al. PTSD and DNA methylation in select immune function gene promoter regions: a repeated measures case-control study of U.S. military service members. *Front Psychiatry.* 2013;4:56.
163. Kaminsky Z, et al. A multi-tissue analysis identifies HLA complex group 9 gene methylation differences in bipolar disorder. *Mol Psychiatry.* 2012;17:728–740.
164. Pal M, et al. High precision DNA modification analysis of HCG9 in major psychosis. *Schizophr Bull.* 2016;42:170–177.
165. Provencal N, et al. Differential DNA methylation regions in cytokine and transcription factor genomic loci associate with childhood physical aggression. *PLoS One.* 2013;8: e71691.
166. Uddin M, et al. Epigenetic and inflammatory marker profiles associated with depression in a community-based epidemiologic sample. *Psychol Med.* 2011;41:997–1007.
167. Powell TR, et al. DNA methylation in interleukin-11 predicts clinical response to antidepressants in GENDEP. *Transl Psychiatry.* 2013;3:e300.
168. Cain DW, Cidlowski JA. Immune regulation by glucocorticoids. *Nat Rev Immunol.* 2017;17:233–247.
169. Oberlander TF, et al. Prenatal exposure to maternal depression, neonatal methylation of human glucocorticoid receptor gene (NR3C1) and infant cortisol stress responses. *Epigenetics.* 2008;3:97–106.
170. Radtke KM, et al. Transgenerational impact of intimate partner violence on methylation in the promoter of the glucocorticoid receptor. *Transl Psychiatry.* 2011;1:e21.
171. Perroud N, et al. Increased methylation of glucocorticoid receptor gene (NR3C1) in adults with a history of childhood maltreatment: a link with the severity and type of trauma. *Transl Psychiatry.* 2011;1:e59.
172. Steiger H, Labonte B, Groleau P, Turecki G, Israel M. Methylation of the glucocorticoid receptor gene promoter in bulimic women: associations with borderline personality disorder, suicidality, and exposure to childhood abuse. *Int J Eat Disord.* 2013;46:246–255.
173. Kim D, et al. Psychological factors and DNA methylation of genes related to immune/inflammatory system markers: the VA Normative Aging Study. *BMJ Open.* 2016;6:e009790.
174. Rogatsky I, Ivashkiv LB. Glucocorticoid modulation of cytokine signaling. *Tissue Antigens.* 2006;68:1–12.
175. Ikegame T, et al. DNA methylation of the BDNF gene and its relevance to psychiatric disorders. *J Hum Genet.* 2013;58:434–438.
176. Kordi-Tamandani DM, Sahranavard R, Torkamanzehi A. DNA methylation and expression profiles of the brain-derived neurotrophic factor (BDNF) and dopamine transporter (DAT1) genes in patients with schizophrenia. *Mol Biol Rep.* 2012;39:10889–10893.
177. Thaler L, et al. Methylation of BDNF in women with bulimic eating syndromes: associations with childhood abuse and borderline personality disorder. *Prog Neuropsychopharmacol Biol Psychiatry.* 2014;54:43–49.

178. D'Addario C, et al. Selective DNA methylation of BDNF promoter in bipolar disorder: differences among patients with BDI and BDII. *Neuropsychopharmacology.* 2012;37:1647–1655.
179. Dell'Osso B, et al. Epigenetic modulation of BDNF gene: differences in DNA methylation between unipolar and bipolar patients. *J Affect Disord.* 2014;166:330–333.
180. Keller S, et al. DNA methylation state of BDNF gene is not altered in prefrontal cortex and striatum of schizophrenia subjects. *Psychiatry Res.* 2014;220:1147–1150.
181. Li T, et al. FoxP2 is significantly associated with schizophrenia and major depression in the Chinese Han population. *World J Biol Psychiatry.* 2013;14:146–150.
182. Tolosa A, et al. FOXP2 gene and language impairment in schizophrenia: association and epigenetic studies. *BMC Med Genet.* 2010;11:114.
183. Allen NC, et al. Systematic meta-analyses and field synopsis of genetic association studies in schizophrenia: the SzGene database. *Nat Genet.* 2008;40:827–834.
184. Owen MJ, Williams NM, O'Donovan MC. Dysbindin-1 and schizophrenia: from genetics to neuropathology. *J Clin Invest.* 2004;113:1255–1257.
185. Voisey J, et al. A polymorphism in the dysbindin gene (DTNBP1) associated with multiple psychiatric disorders including schizophrenia. *Behav Brain Funct.* 2010;6:41.
186. Pae CU, et al. Effect of 5-haplotype of dysbindin gene (DTNBP1) polymorphisms for the susceptibility to bipolar I disorder. *Am J Med Genet B Neuropsychiatr Genet.* 2007;144B:701–703.
187. Raybould R, et al. Bipolar disorder and polymorphisms in the dysbindin gene (DTNBP1). *Biol Psychiatry.* 2005;57:696–701.
188. Arias B, et al. Dysbindin gene (DTNBP1) in major depression: association with clinical response to selective serotonin reuptake inhibitors. *Pharmacogenet Genomics.* 2009;19:121–128.
189. Abdolmaleky HM, et al. Antipsychotic drugs attenuate aberrant DNA methylation of DTNBP1 (dysbindin) promoter in saliva and post-mortem brain of patients with schizophrenia and psychotic bipolar disorder. *Am J Med Genet B Neuropsychiatr Genet.* 2015;168:687–696.
190. Huang XL, Zou YS, Maher TA, Newton S, Milunsky JM. A de novo balanced translocation breakpoint truncating the autism susceptibility candidate 2 (AUTS2) gene in a patient with autism. *Am J Med Genet A.* 2010;152A:2112–2114.
191. Zhang B, et al. Association study identifying a new susceptibility gene (AUTS2) for schizophrenia. *Int J Mol Sci.* 2014;15:19406–19416.
192. Schumann G, et al. Genome-wide association and genetic functional studies identify autism susceptibility candidate 2 gene (AUTS2) in the regulation of alcohol consumption. *Proc Natl Acad Sci USA.* 2011;108:7119–7124.
193. Bi W, et al. Mutations of RAI1, a PHD-containing protein, in nondeletion patients with Smith-Magenis syndrome. *Hum Genet.* 2004;115:515–524.
194. Toulouse A, Rochefort D, Roussel J, Joober R, Rouleau GA. Molecular cloning and characterization of human RAI1, a gene associated with schizophrenia. *Genomics.* 2003;82:162–171.
195. Frieling H, et al. Epigenetic downregulation of atrial natriuretic peptide but not vasopressin mRNA expression in females with eating disorders is related to impulsivity. *Neuropsychopharmacology.* 2008;33:2605–2609.
196. Bouwland-Both MI, et al. DNA methylation of IGF2DMR and H19 is associated with fetal and infant growth: the generation R study. *PLoS One.* 2013;8:e81731.
197. Groleau P, et al. Methylation of the dopamine D2 receptor (DRD2) gene promoter in women with a bulimia-spectrum disorder: associations with borderline personality disorder and exposure to childhood abuse. *J Psychiatr Res.* 2014;48:121–127.

198. Zhao J, Goldberg J, Bremner JD, Vaccarino V. Association between promoter methylation of serotonin transporter gene and depressive symptoms: a monozygotic twin study. *Psychosom Med.* 2013;75:523–529.
199. Çöpoğlu Üul., et al. DNA methylation of BDNF gene in schizophrenia. *Med Sci Monit.* 2016;22:397–402.
200. Fries GR, et al. Hypothalamic-pituitary-adrenal axis dysfunction and illness progression in bipolar disorder. *Int J Neuropsychopharmacol.* 2014;18:pyu043.
201. Nagarajan R, Hogart A, Gwye Y, Martin MR, LaSalle JM. Reduced MeCP2 expression is frequent in autism frontal cortex and correlates with aberrant MECP2 promoter methylation. *Epigenetics.* 2014;1:172–182.
202. Bönsch D, Lenz B, Kornhuber J, Bleich S. DNA hypermethylation of the alpha synuclein promoter in patients with alcoholism. *Neuroreport.* 2005;16:167–170.
203. Bleich S, et al. Epigenetic DNA hypermethylation of the HERP gene promoter induces down-regulation of its mRNA expression in patients with alcohol dependence. *Alcohol Clin Exp Res.* 2006;30:587–591.
204. Ruggeri B, et al. Association of protein phosphatase PPM1G with alcohol use disorder and brain activity during behavioral control in a genome-wide methylation analysis. *Am J Psychiatry.* 2015;172:543–552.
205. Kim YR, Kim JH, Kim MJ, Treasure J. Differential methylation of the oxytocin receptor gene in patients with anorexia nervosa: a pilot study. *PLoS One.* 2014;9:e88673.
206. Kaminsky Z, et al. DNA methylation and expression of KCNQ3 in bipolar disorder. *Bipolar Disord.* 2015;17:150–159.
207. Starnawska A, et al. Hypomethylation of FAM63B in bipolar disorder patients. *Clin Epigenet.* 2016;8:52.
208. Wang F, Xu H, Zhao H, Gelernter J, Zhang H. DNA co-methylation modules in postmortem prefrontal cortex tissues of European Australians with alcohol use disorders. *Sci Rep.* 2016;6:19430.
209. Pidsley R, et al. Methylomic profiling of human brain tissue supports a neurodevelopmental origin for schizophrenia. *Genome Biol.* 2014;15:483.
210. Jaffe AE, et al. Mapping DNA methylation across development, genotype and schizophrenia in the human frontal cortex. *Nat Neurosci.* 2016;19:40–47.
211. Nardone S, et al. DNA methylation analysis of the autistic brain reveals multiple dysregulated biological pathways. *Transl Psychiatry.* 2014;4:e433.
212. Numata S, Ye T, Herman M, Lipska BK. DNA methylation changes in the postmortem dorsolateral prefrontal cortex of patients with schizophrenia. *Front Genet.* 2014;5:280.
213. Wockner LF, et al. Genome-wide DNA methylation analysis of human brain tissue from schizophrenia patients. *Transl Psychiatry.* 2014;4:e339.
214. Zhao H, et al. Genome-wide DNA methylome reveals the dysfunction of intronic microRNAs in major psychosis. *BMC Med Genomics.* 2015;8:62.
215. Sabunciyan S, et al. Genome-wide DNA methylation scan in major depressive disorder. *PLoS One.* 2012;7:e34451.
216. Ginsberg MR, Rubin RA, Falcone T, Ting AH, Natowicz MR. Brain transcriptional and epigenetic associations with autism. *PLoS One.* 2012;7:e44736.
217. Ladd-Acosta C, et al. Common DNA methylation alterations in multiple brain regions in autism. *Mol Psychiatry.* 2014;19:862–871.
218. Kinoshita M, et al. DNA methylation signatures of peripheral leukocytes in schizophrenia. *Neuromolecular Med.* 2013;15:95–101.
219. Houtepen LC, van Bergen AH, Vinkers CH, Boks MP. DNA methylation signatures of mood stabilizers and antipsychotics in bipolar disorder. *Epigenomics.* 2016;8:197–208.
220. Li Y, et al. Genome-wide methylome analyses reveal novel epigenetic regulation patterns in schizophrenia and bipolar disorder. *Biomed Res Int.* 2015;2015:201587.

221. Numata S, et al. Blood diagnostic biomarkers for major depressive disorder using multiplex DNA methylation profiles: discovery and validation. *Epigenetics.* 2015;10:135–141.
222. Zhang R, et al. Genome-wide DNA methylation analysis in alcohol dependence. *Addict Biol.* 2013;18:392–403.
223. Wilmot B, et al. Methylomic analysis of salivary DNA in childhood ADHD identifies altered DNA methylation in VIPR2. *J Child Psychol Psychiatry.* 2016;57:152–160.
224. Sabunciyan S, Maher B, Bahn S, Dickerson F, Yolken RH. Association of DNA methylation with acute mania and inflammatory markers. *PLoS One.* 2015;10: e0132001.
225. van den Oord EJ, et al. A whole methylome CpG-SNP association study of psychosis in blood and brain tissue. *Schizophr Bull.* 2016;42:1018–1026.
226. Liu J, et al. Methylation patterns in whole blood correlate with symptoms in schizophrenia patients. *Schizophr Bull.* 2014;40:769–776.
227. Montano C, et al. Association of DNA methylation differences with schizophrenia in an epigenome-wide association study. *JAMA Psychiatry.* 2016;73:506–514.
228. Walker RM, et al. DNA methylation in a Scottish family multiply affected by bipolar disorder and major depressive disorder. *Clin Epigenet.* 2016;8:5.
229. Berko ER, et al. Mosaic epigenetic dysregulation of ectodermal cells in autism spectrum disorder. *PLoS Genet.* 2014;10:e1004402.
230. Kinoshita M, et al. Aberrant DNA methylation of blood in schizophrenia by adjusting for estimated cellular proportions. *Neuromolecular Med.* 2014;16:697–703.
231. Weder N, et al. Child abuse, depression, and methylation in genes involved with stress, neural plasticity, and brain circuitry. *J Am Acad Child Adolesc Psychiatry.* 2014;53:417–424e415.
232. Walton E, et al. Epigenetic profiling of ADHD symptoms trajectories: a prospective, methylome-wide study. *Mol Psychiatry.* 2017;22:250–256.
233. Kuratomi G, et al. Aberrant DNA methylation associated with bipolar disorder identified from discordant monozygotic twins. *Mol Psychiatry.* 2008;13:429–441.
234. Nguyen A, Rauch TA, Pfeifer GP, Hu VW. Global methylation profiling of lymphoblastoid cell lines reveals epigenetic contributions to autism spectrum disorders and a novel autism candidate gene, RORA, whose protein product is reduced in autistic brain. *FASEB J.* 2010;24:3036–3051.
235. Davies MN, et al. Hypermethylation in the ZBTB20 gene is associated with major depressive disorder. *Genome Biol.* 2014;15:R56.
236. Dempster EL, et al. Genome-wide methylomic analysis of monozygotic twins discordant for adolescent depression. *Biol Psychiatry.* 2014;76:977–983.
237. Castellani CA, et al. DNA methylation differences in monozygotic twin pairs discordant for schizophrenia identifies psychosis related genes and networks. *BMC Med Genomics.* 2015;8:17.
238. Hannon E, et al. An integrated genetic-epigenetic analysis of schizophrenia: evidence for co-localization of genetic associations and differential DNA methylation. *Genome Biol.* 2016;17:176.
239. Cordova-Palomera A, et al. Genome-wide methylation study on depression: differential methylation and variable methylation in monozygotic twins. *Transl Psychiatry.* 2015;5: e557.
240. Oh G, et al. DNA modification study of major depressive disorder: beyond locus-by-locus comparisons. *Biol Psychiatry.* 2015;77:246–255.
241. Davies MN, et al. Functional annotation of the human brain methylome identifies tissue-specific epigenetic variation across brain and blood. *Genome Biol.* 2012;13:R43.
242. Horvath S, et al. Aging effects on DNA methylation modules in human brain and blood tissue. *Genome Biol.* 2012;13:R97.

243. van Eijk KR, et al. Genetic analysis of DNA methylation and gene expression levels in whole blood of healthy human subjects. *BMC Genomics*. 2012;13:636.
244. Do C, et al. Genetic-epigenetic interactions in cis: a major focus in the post-GWAS era. *Genome Biol*. 2017;18:120.
245. Fries GR, et al. Integrated transcriptome and methylome analysis in youth at high risk for bipolar disorder: a preliminary analysis. *Transl Psychiatry*. 2017;7:e1059.
246. van Berkum NL, et al. Hi-C: a method to study the three-dimensional architecture of genomes. *J Vis Exp*. 2010;39:1869.
247. Won H, et al. Chromosome conformation elucidates regulatory relationships in developing human brain. *Nature*. 2016;538:523–527.
248. Buenrostro JD, Wu B, Chang HY, Greenleaf WJ. ATAC-seq: a method for assaying chromatin accessibility genome-wide. *Curr Protoc Mol Biol*. 2015;109:21–29.
249. Boyle AP, et al. High-resolution mapping and characterization of open chromatin across the genome. *Cell*. 2008;132:311–322.
250. Maurano MT, et al. Systematic localization of common disease-associated variation in regulatory DNA. *Science*. 2012;337:1190–1195.
251. Dixon JR, et al. Chromatin architecture reorganization during stem cell differentiation. *Nature*. 2015;518:331–336.
252. Guo H, et al. Single-cell methylome landscapes of mouse embryonic stem cells and early embryos analyzed using reduced representation bisulfite sequencing. *Genome Res*. 2013;23:2126–2135.
253. Smallwood SA, et al. Single-cell genome-wide bisulfite sequencing for assessing epigenetic heterogeneity. *Nat Methods*. 2014;11:817–820.
254. Luo C, et al. Single-cell methylomes identify neuronal subtypes and regulatory elements in mammalian cortex. *Science*. 2017;357:600–604.
255. Pott S. Simultaneous measurement of chromatin accessibility, DNA methylation, and nucleosome phasing in single cells. *Elife*. 2017;6:e23203.
256. Guo F, et al. Single-cell multi-omics sequencing of mouse early embryos and embryonic stem cells. *Cell Res*. 2017;27:967–988.
257. Grove TB, et al. Oxytocin receptor (OXTR) methylation and cognition in psychotic disorders. *Mol Neuropsychiatry*. 2016;2:151–160.
258. Liu J, et al. The association of DNA methylation and brain volume in healthy individuals and schizophrenia patients. *Schizophr Res*. 2015;169:447–452.
259. Rubin LH, et al. Sex and diagnosis specific associations between DNA methylation of the oxytocin receptor gene with emotion processing and temporal-limbic and prefrontal brain volumes in psychotic disorders. *Biol Psychiatry Cogn Neurosci Neuroimaging*. 2016;1:141–151.
260. Hass J, et al. Associations between DNA methylation and schizophrenia-related intermediate phenotypes—a gene set enrichment analysis. *Prog Neuropsychopharmacol Biol Psychiatry*. 2015;59:31–39.
261. Qi LS, et al. Repurposing CRISPR as an RNA-guided platform for sequence-specific control of gene expression. *Cell*. 2013;152:1173–1183.
262. Xu X, et al. A CRISPR-based approach for targeted DNA demethylation. *Cell Discov*. 2016;2:16009.
263. Morita S, et al. Targeted DNA demethylation in vivo using dCas9-peptide repeat and scFv-TET1 catalytic domain fusions. *Nat Biotechnol*. 2016;34:1060–1065.
264. Liu XS, et al. Editing DNA methylation in the mammalian genome. *Cell*. 2016;167:233–247e217.
265. McDonald JI, et al. Reprogrammable CRISPR/Cas9-based system for inducing site-specific DNA methylation. *Biol Open*. 2016;5:866–874.
266. Vojta A, et al. Repurposing the CRISPR-Cas9 system for targeted DNA methylation. *Nucleic Acids Res*. 2016;44:5615–5628.

267. Stepper P, et al. Efficient targeted DNA methylation with chimeric dCas9-Dnmt3a-Dnmt3L methyltransferase. *Nucleic Acids Res.* 2017;45:1703–1713.
268. Chen Y, Sharma RP, Costa RH, Costa E, Grayson DR. On the epigenetic regulation of the human reelin promoter. *Nucleic Acids Res.* 2002;30:2930–2939.
269. Kundakovic M, Chen Y, Costa E, Grayson DR. DNA methyltransferase inhibitors coordinately induce expression of the human reelin and glutamic acid decarboxylase 67 genes. *Mol Pharmacol.* 2007;71:644–653.
270. Kundakovic M, Chen Y, Guidotti A, Grayson DR. The reelin and GAD67 promoters are activated by epigenetic drugs that facilitate the disruption of local repressor complexes. *Mol Pharmacol.* 2009;75:342–354.
271. Dong E, Nelson M, Grayson DR, Costa E, Guidotti A. Clozapine and sulpiride but not haloperidol or olanzapine activate brain DNA demethylation. *Proc Natl Acad Sci USA.* 2008;105:13614–13619.

CHAPTER TEN

Understanding the Neuroepigenetic Constituents of Suicide Brain

Bhaskar Roy, Yogesh Dwivedi[1]
University of Alabama at Birmingham, Birmingham, AL, United States
[1] Corresponding author. E-mail address: ydwivedi@uab.edu

Contents

Abstract

Stressful life incidents often cause a predisposition for developing mental disorders such as major depressive disorder (MDD). Impaired neurocognitive and neuro-vegetative functions of the central nervous system are the hallmarks of this mental illness. Blunted responses from emotionally salient regions of the brain including cortex, hippocampus, and amygdala have been associated with MDD-related behavioral changes. Moreover, improper signal processing and neuronal atrophy were held responsible for the overall dysfunctionality of these vulnerable regions in the MDD brain. The prevalence of genetic susceptibility along with adverse environmental

Progress in Molecular Biology and Translational Science, volume 157
ISSN 1877-1173
http://dx.doi.org/10.1016/bs.pmbts.2018.01.007

stimuli often makes the situation worse for MDD patients, leading to an increased risk of suicidal behavior and eventually death by suicide. Despite considerable efforts to understand the complex neurobiology associated with MDD and suicidal behavior, their pathological determinants remain mostly elusive. Recent research, however, has shown that epigenetic perturbations have a formidable impact on the etiopathogenesis of MDD. Understanding the neuroepigenetic nature of this mental disorder may provide opportunities to devise more effective treatment strategies. Moreover, this can potentially lead to identifying predictive biomarkers associated with suicide risk. The present chapter critically reviews studies pertaining to epigenetic signatures of MDD and suicide brain.

1. INTRODUCTION

The brain's homeostatic ability to maintain a proper balance between resiliency and susceptibility to cope with stress is always being challenged in the central nervous system.[1] The overwhelming nature of stressful stimuli frequently results in improper functioning of the brain's stress-responsive pathways.[2] Thus, system level malfunctioning in conjunction with severe cellular and biochemical impairments in the CNS have often been attributed to serious pathophysiological consequences in depressed brain induced by stress.[2] In addition, the inability of neuronal synapses to make new connections, the atrophied nature of higher order brain structures, and reduced neurogenesis, predominantly in the hippocampus, have been considerably to be associated with the neurobiological abnormalities in MDD brain. The collective nature of these neuropathologies has been found to be overacting on the functional plasticity of neurons, thus resulting in serious neurocognitive failure associated with MDD and the suicide brain.[3,4] In this connection, the predisposing genetic make-up was earlier considered to be a critical factor, but later found limiting, to explain the impaired functionality of the depressed brain at the molecular level.[5,6] Uncovering changes associated with altered neural plasticity, coupled with defective neural signal processing, demonstrate an overall blunted response in gene expression in underlying neural networks. The dynamic nature of these changes is often associated with environmental changes and hence are called Gene × Environment (G × E) interactions.[7] Unveiling the molecular nature of G × E interactions may potentially serve as a critical tool to understand the pathophysiological basis of the depressed and suicidal brain.[8]

Over the past decade, a plethora of reports connecting the role of G × E interaction in mental disorders has shown the involvement of epigenetic interference in neuronal circuits. Epigenetic perturbations are known to cause reversible changes in genetic information processing pathways without eliciting any changes in the genetic structure. Since the changes associated with epigenetic interferences are primarily transient, a correlation of the episodic modulations observed in major depression and suicidality can be established with underlying epigenetic challenges.[8,9] An overview of the epigenetic modifications influencing the genetic information processing system at the cellular level has been documented schematically in Fig. 1. In this chapter, we critically evaluate our current understanding of the role of epigenetic regulation in the pathophysiology of major depression and

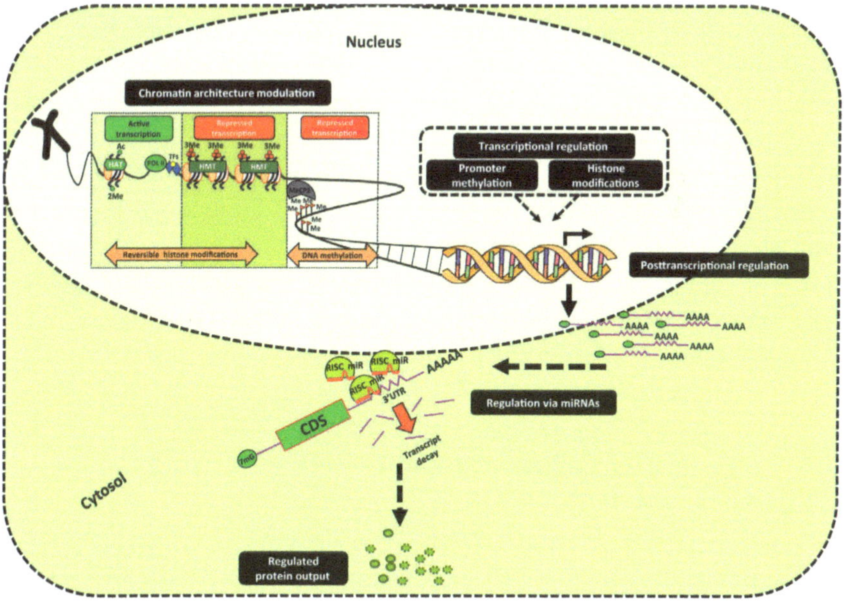

Fig. 1 ***Schematic representation of epigenetic regulatory mechanisms responsible for repatterning genetic information in MDD-suicide brain.*** **Transcriptional Regulation**: Covalent modifications involving the attachment of methyl groups to cytosine residues in DNA double strand coordinated with histone tail modifications regulate chromatin structure accessibility in a reversible fashion. **Posttranscriptional Regulation**: Functional output from transcriptional machinery could be further regulated by canonical action of miRNA at posttranscriptional level to moderate the availability of coding transcripts in cytosolic environment. Together, these three well-coordinated cellular mechanisms bring about a dynamic change in cellular phenotype.

suicidal behavior. In addition, we have the extended objective of exploring the possibility of its usefulness as a biomarker for depression and suicidality with diagnostic value and target(s) for effective treatment.

2. microRNAs AS EPIGENETIC MODIFIERS IN MAJOR DEPRESSIVE DISORDER AND SUICIDE BRAIN

In the recent past, advancements in understanding the potential involvement of regulatory RNA molecules has acquainted us with microRNAs (miRNA) as a class of small noncoding RNA molecules with a well-recognized role in gene regulation. Despite their small molecular size (~22 nt), miRNAs have shown a wide range of gene regulatory functions at the posttranscriptional level.[10,11] Moreover, they have demonstrated an overall pleiotropic nature in regulating gene functions ranging from cellular development to disease pathophysiology.[12] Over the past decade, the gene regulatory functions of miRNAs have been studied in several neuropsychiatric diseases such as schizophrenia, autism, Parkinson's disease, Huntington's disease, Tourette's syndrome, Fragile X syndrome, DiGeorge syndrome, and Alzheimer's disease.[13,14] Based on preclinical, clinical, and postmortem brain studies, compelling reports have shown miRNA's ability to interface between environmental stimuli and intracellular information processing circuits in MDD and suicidal behavior.[15–20] A comprehensive list of miRNAs involved in MDD and suicidal behavior has been presented in Table 1.

2.1 Orchestrated Cellular Dynamics Behind MicroRNA Biogenesis and Regulation

In humans, ~2500 mature miRNAs have been annotated.[28] Except for the Y chromosome, the genomic distribution of miRNA-encoding loci is generally uniform.[29] miRNAs are processed individually from independent genomic locations; however, approximately half are found to be part of clustered transcriptional units known as polycistronic miRNAs.[30] This suggests a coordinated regulatory influence on their transcription.[31] Besides this, for about 30% of miRNAs, transcription and associated regulation come from their embedded transcriptional units in the intronic region of protein-coding genes, allowing them to use their host gene promoter.[32]

Unlike other members of the small noncoding RNA family, miRNAs exhibit an evolutionary preference in using RNA polymerase II for their

Table 1 Summary of miRNA-Based Epigenetic Interference in Postmortem MDD-Suicide Brain.

Site of Epigenetic Modifications	Participating miRNA	Genes Targeted	Publication
miRNA in dorsolateral prefrontal cortex (dlPFC)	miR-142-5p, miR-33a, miR-137, miR-489, miR-148b, miR-101, miR-324-5p, miR-301a, miR-146a, miR-335, miR-494, miR-20b, miR-376a*, miR-190, miR-155, miR-660, miR-552, miR-453, miR-130a, miR-27a, miR-497, miR-10a, miR-20a, miR-142-3p	UBE2D1 and UBE2W, CAMK2G, AKAP1, SMAD5, MITF, BACH2, MYCN, GABRA4, and CACNA1C	Smalheiser et al. (2012)[17]
miRNA in frontal cortex	miR-185	TrkB.T1	Maussion et al. (2012)[21]
miRNA in prefrontal cortex	miR-139-5p, miR-320c, miR-34c-5p	SAT1, SMOX	Lopez et al. (2014a)[19]
miRNA in prefrontal cortex (2014b)[16]	miR-1202	GRM4	Lopez et al.
miRNA in basolateral amygdala (BLA)	miR-511	GFR1	Maheu et al. (2015)[22]
miRNA in anterior cingulate cortex (ACC)	miR-184 and miR-34a	NCOAR1 and NCOR2	Azevedo et al. (2016)[23]
miRNA in Edinger–Westphal nucleus (EWcp) of rostroventral midbrain area	miR-326	UCN1	Aschrafi et al. (2016)[24]
miRNA in prefrontal cortex	miR-218	Netrin-1 guidance cue receptor gene, deleted in colorectal cancer (DCC)	Torres-Berrio et al. (2017)[25]

(Continued)

Site of Epigenetic Modifications	Participating miRNA	Genes Targeted	Publication
miRNA in ventral prefrontal cortex (vPFC)	miR-146a-5p, miR-146b-5p, miR-425-3p and miR-24-3p	MAPK/WNT signaling pathway	Lopez et al. (2017)[26]
miRNA in locus coeruleus (LC)	miR-17-5p, miR-20b-5p, miR-106a-5p, miR-330-3p, miR-409-5p, miR-541-3p, miR-582-5p, miR-890, let-7g-3p, miR-99b-3p, miR-550-5p, miR-1179, miR-1197	RELN, GSK-3β, MAOA, CHRM1, PLCB1 and GRIK1	Roy et al. (2017c)[27]

transcription in eukaryotes.[33] As elaborated in Fig. 2, the sequential processing of miRNAs after their synthesis by RNA Pol II starts with endoneocleolytic cleavage in the nucleus.[34] A multiprotein complex named "Microprocessor," first crops one of the strands of primary miRNA (pri-miRNA) toward the base of stem-loop secondary structures and releases ~60–70 nts hairpin-shaped precursor miRNAs (pre-miRNAs). The site specificity and the cleavage of pre-miRNA are determined by the two major components of Microprocessor complex, RNase III-DROSHA and associated co-factor dsRNA-binding protein DiGeorge syndrome critical region 8 (DGCR8).[30] Once released, the pre-miRNA leaves the nucleus by the active participation of a nuclear transport protein, exportin 5 (XPO5). Upon arriving in the cytosolic environment, the pre-miRNA is taken up by another RNase III enzyme named DICER1 for an additional round of endonucleolytic processing. The dicing of the hairpin precursor is primarily achieved by the recruitment of transactivation-responsive RNA-binding protein (TRBP; also known as TARBP2), which helps to enhance the fidelity of DICER1 mediated cleavage. The DICER1 activity results in the cleavage of the hairpin loop of pre-miRNA and leaves a ~22 nucleotide double-stranded RNA duplex with a 2 nt protruding overhand at 3′ end. Next, TRBP1 collaborates with Argonaute (Ago) family proteins and DICER1 to build an asymmetric multiprotein assembly named RNA-Induced Silencing Complex (RISC). One of the two strands from processed

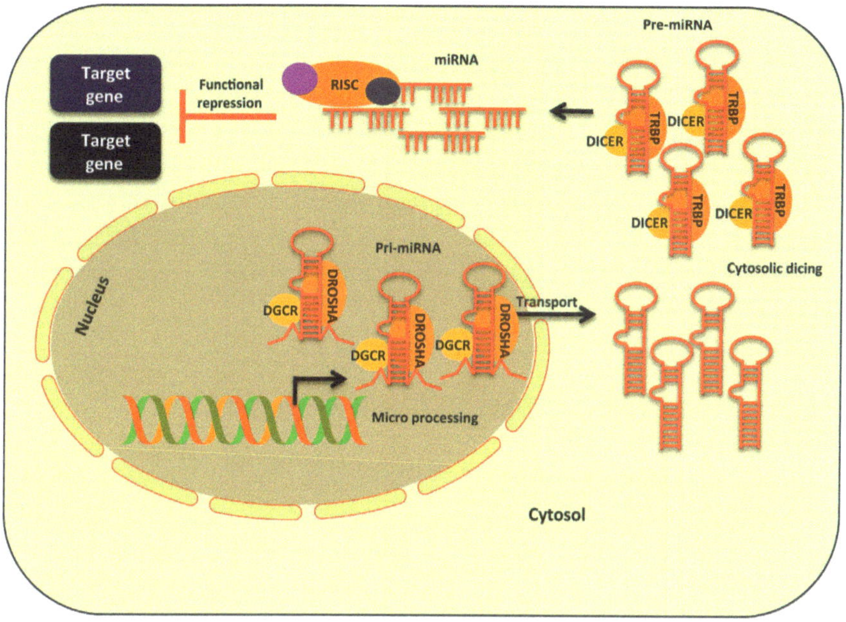

Fig. 2 ***MicroRNA biogenesis pathway.*** MiRNAs is transcribed by RNA polymerase II to synthesize pri-miRNA in the nuclear environment. Within nuclear environment the pri-miRNA is processed by nuclear RNase III enzyme Drosha to produce pre-miRNA. Pre-miRNA is transported to cytosolic environment with the help of Ran-GTP and Exportin5 transporter complex. In cytosol the pre-miRNA is processed by the RNase III enzyme Dicer to generate mature miRNAs. Mature miRNAs are incorporated into the RNA-induced silencing (RISC) complex, which regulates gene expression by pairing primarily to the 3′UTR region of protein coding mRNAs to repress target mRNA translation and/or induce target degradation.

miRNA duplex is retained in the complex as a guide strand which helps to load the silencing machinery on the 3′ region of target genes. Mature miRNAs incorporated into the RISC (miRISC), then regulates gene expression by pairing primarily with the 3′UTR region of protein-coding mRNAs to either repress their translation and/or induce target degradation.[33,35] RISC guided complementary binding between short 2–7 nts seed sequence of the miRNA and 3′UTR of target mRNA leads to translational repression. miRNA mediated translational inhibition also depends upon the 5′cap region of the target mRNA. Ago proteins can stimulate miRNA-dependent inhibition of translation by competing with elongation factor eIF4E for the 5′cap binding site, thus preventing the circularization of mRNA and lowering initiation efficiency.[36] In contrast, the binding of miRISC complemented with GW128 proteins to the 3′UTR of targets induces an exoneucleolytic cleavage from the terminal poly A region by recruiting a deadenylation factor.[37] Because of this dynamic feature, miRNAs are able to regulate genetic circuitries and thereby play a critical role in maintaining biological homeostasis.[11] Considering the fundamental role of miRNAs in mediating biological events, any perturbations in the expression of miRNAs may result in the imbalance of homeostasis, which is often reflected as an imbalance in the regulatory network that can distinguish normal versus disease states.[14,31,38]

2.2 Epigenetic Role of MicroRNAs in Major Depressive Disorder-Suicide Brain

With the increasing acceptance of miRNAs as mega controllers of gene regulatory events at cellular levels, over the past few years, several reports have shown their distinct role as critical molecular switches in the brains of MDD subjects.[39] In a study by Dwivedi and colleagues, a dorsolateral prefrontal cortex (dlPFC) specific global change in miRNA expression was found to be closely associated with depressive symptoms in suicide subjects.[17] An overall downregulation in miRNA expression was noticed in dlPFC which was linked with downstream regulation of genes known for their significant role in cellular growth, differentiation and plasticity (UBE2D1, UBE2W, CAMK2G, AKAP1, SMAD5, MITF, BACH2, MYCN, GABRA4, and CACNA1C). Half of the downregulated miRNAs from this study were parts of a miRNA co-expression network, due to the proximity of their chromosomal localization (miR-424 and 20b at Xq26.2- 3, 377 kb apart; miR-142 and 301a at 17q22, 820 kb apart; and miR-324-5p and 497 at 17p13.1, 205 kb apart) and nature of being transcribed from the same primary transcripts. Additional similarities were noted for those downregulated miRNAs, which have been shown to harbor identical seed sequences (miR-20a and 20b; miR-301a and 130a; and miR-424 and 497), to achieve an overlapping pattern of target gene regulation.[17]

Another study which utilized a candidate-specific approach, linked the overexpression of miR-185* with significant depletion of TrkB.T1 expression in the frontal cortex of suicide completers.[21] TrkB.T1 is the truncated isoform of Ntrk2 receptor gene which exhibits astrocyte-restricted expression. Acknowledging the importance of astrocyte-related pathology in mood disorder progression, a marked decline in TrkB.T1 expression provides an indication about an miRNA-mediated impaired neurotrophin-signaling pathway in depressed suicide brain.[21] In a different study, metabotropic glutamate receptor GRM4, a participating receptor protein of glutamatergic, dopaminergic, GABAergic, and serotonergic neurotransmission, was found to be affected by brain-enriched miR-1202 in depressed postmortem cortex.[16] Meanwhile, a contrasting downregulation was noticed in GRM4 expression in depressed brain with antidepressant history which was negatively correlated with miR-1202 expression. This study also indicated a negative relationship between miR-1202 expression and MDD when compared with MDD subjects who committed suicide.[16]

Dysregulated polyamine biosynthesis pathway is characteristic of MDD and suicide brain.[40,41] Earlier reports have shown aberrant changes in the expression of SAT1 and SMOX genes from this pathway in the suicide brain.[42] Later, the identification of four significantly upregulated miRNAs (miR-139-5p, miR-195, miR-320c, and miR-34c-5p) in the prefrontal cortex of depressed-suicide completers further supported a strong negative expression correlation among miR-139-5p, miR-320c, miR-34c-5p, and SAT1, SMOX genes. These reports explain the miRNA mediated possible epigenetic modulation of MDD-suicide brain by disrupting the polyamine biosynthesis pathway.[19]

In another study, epigenetic modification mediated by miR-511 has been shown to compromise the GDNF-mediated signaling cascade in glial populations of the depressed brain.[22] In the basolateral amygdala (BLA) of depressed subjects, a marked increase was observed for miR-511 with a contrasting change in GDNF family receptor alpha 1 (GFRA1) mRNA expression. This study shows the responsiveness of GFR1 to miR-511, implicating a dysfunctional glial specific GDNF-signaling axis in the depressed brain.[22]

Besides affecting genes related to either neurotransmission or neurotrophic factors, a recent report shows the involvement of miRNAs in modulating neuropeptide functions in the suicide brain. This study found a negative influence on urocortin 1 (Ucn1) gene expression mediated by miR-326 in the rostroventral midbrain area within the centrally projecting Edinger–Westphal nucleus (EWcp) of suicide subjects. This stress-sensitive brain area was also found to be deficient in Ucn1 expression in rats induced by chronic variable mild stress which was due to overexpression of miR-365. This indicates the dynamic role of miRNAs in stress-induced mental disorders.[24] In a recent study, expression variability of DCC (netrin-1 guidance cue receptor gene, deleted in colorectal cancer) gene was reported to be under the influence of miR-218 in the prefrontal cortex of MDD suicide subjects.[25] Reduced expression of DCC in PFC neurons was related to increased resiliency against stress-induced depression. As DCC receptors have the ability to control axon arborization, dendritic growth, and synapse formation, the loss of an antagonizing effect of miR-218 on DCC may possibly lead to repatterning of synaptic connections in the pyramidal neurons in the MDD brain.[25] Stress response, is also mediated indirectly by two miRNAs: miR-184 and miR-34a, whose expression is negatively correlated with Nuclear Receptor Corepressor genes (NCOR1 and NCOR2) responsible for modulating CRH gene expression.[23]

Variability in antidepressant response, especially in treatment-resistant MDD cases, is poorly understood. Thus, identifying potential markers, which correlate with treatment response, is of critical importance. In a recent report, involvement of four miRNAs (miR-146a-5p, miR-146b-5p, miR-425-3p, and miR-24-3p) in predicting the treatment responsiveness of the antidepressant (AD) duloxetine was shown. Significantly altered expressions of these four miRNAs under AD treatment and identification of downstream-signaling pathways (MAPkinase/Wnt) with known functions in depression pathophysiology, potentiate their role as predictive biomarkers for the treatment response and possible target for MDD.[26]

Recently, Dwivedi et al.[27] identified a miRNA regulatory network in the locus coeruleus (LC) of suicide subjects. This miRNA network was suicide specific and was capable of altering potential gene regulatory networks, which can yield functional clustering of predicated target genes associated with anxiety, hyperactive behavior, and stress-related disorders.[27]

3. DNA METHYLATION (5MC) BASED EPIGENETIC DYSREGULATION IN SUICIDE BRAIN

DNA methylation, the most studied epigenetic modification in the mammalian genome, involves the covalent attachment of methyl group to cytosine residues by DNA methyltransferase enzymes which primarily evokes a state of repression in mammalian gene expression process.[43] This particular type of DNA-based modification is classically found to be more stable and commonly present on adjacent CpG dinucleotides in the mammalian genome with the ability to induce a reversible change in gene expression at the transcriptional level.[44,45] A summary of reported findings on DNA methylation in postmortem brain of depressed/suicidal individuals is provided in Table 2.

Table 2 Summary of 5mC-Based DNA Modification in MDD-Suicide Brain.

Site of Epigenetic Modifications	Genes Involved	Publication
DNA methylation in frontopolar cortex	GABAA α1	Poulter et al. (2008)[46]
DNA methylation in frontal cortex (BA 8/9)	TRKB.T1	Ernst et al. (2009)[47]
DNA methylation in hippocampus	NR3C1 Exon $_{1F}$	McGowan et al. (2009)[48]
DNA methylation in prefrontal cortex (BA 8/9)	SMS Exon 1 and SMOX	Fiori and Turecki (2010)[49]
DNA methylation in Wernicke area	BDNF promoter Exon 4	Keller et al. (2010)[50]
DNA methylation in prefrontal cortex (BA 8/9)	SAT1	Fiori and Turecki (2011)[51]
DNA methylation in Wernicke area	TRKB.T1	Keller et al. (2011)[52]
DNA methylation in hippocampus	NR3C1 Exon 1_C and 1_H	Labonte et al. (2012)[53]
DNA methylation in BA 44	OAZ1, OAZ2, AMD1, and ARG2	Gross et al. (2013)[54]
DNA methylation in hippocampus (Dentate Gyrus)	NR2E1, CHRNB2, GRM7 and DBH	Lebonte et al. (2013)[55]
DNA methylation in prefrontal cortex (BA 8/9)	TRKB.T1	Maussion et al. (2014)[56]
DNA methylation in prefrontal cortex	SKA2	Guintivano et al. (2014)[57]
DNA methylation in prefrontal cortex (BA 8/9)	GRIK2 and BEGAIN	Nagy et al. (2015)[58]
DNA methylation in hippocampus	ALS2, DGKZ, HCK	Labonte et al. (2012)[53]
DNA methylation in prefrontal cortex	EYA2, MEGF11, LMNA, GLUD1, ERBB3	Haghigi et al. (2014)[59]
DNA methylation in frontal cortex (BA11/25)	DNAJC15, SNAI3, PSORS1C3, TAPBP, ATP5G2	Murphy et al. (2017)[60]
DNA methylation in anterior cingulate cortex (ACC)	POU3F1 and LINGO3	Lutz et al. (2017)[61]

3.1 GABAergic System

One of the very first reports that elaborated on the role of DNA methylation in reprogramming gene expression in suicide brains focused on the GABAergic system.[46] This report connected frontopolar cortex specific hyperactivity of the DNA methyltransferase (DNMT-3B) gene with an observed hypermethylation status of $GABA_A$ α1 subunit in the suicide brain. This relationship strongly supported the previous finding in suicide subjects where a significant $GABA_A$ α1 subunit downregulation was reported in the fronto cortical brain area.[62] The physiological relevance of this GABAergic deficit could be attributed to assessing hedonic and decision-making capabilities of prefrontal cortex which is mostly found to be underserved in suicidal behavior as part of the impaired cognitive processes.[63]

3.2 Polyamine System

The potential role of the polyamine system has gained recent acceptance in suicide neurobiology[40] for being epigenetically modulated at the DNA methylation level. The identification of downregulated enzyme "SAT1" from this polyamine system across three Brodmann Areas (BA 4, 8/9 and 11) of suicide subjects[64] of suicidal completers warranted the need to further examine the underlying regulatory mechanisms associated with this gene and other genes actively participating in the polyamine pathway. This led to the identification of SMS, SMOX and AMD1 genes which were differentially expressed in brain of suicide completers.[42,65,66] Brodmann area 8/9 based analysis of SMS and SMOX genes within 1 kb upstream of their promoter region revealed either low levels of methylation or a completely unmethylated state in suicide completers.[49] Furthermore, an overall methylation analysis found no significant difference between suicide completers and healthy controls followed by the identification of a nonsignificant correlation between gene expression downregulation and promoter DNA methylation. However, a significant relationship was observed between SMOX expression and the hypermethylation of a specific site on its promoter. In another study, selective downregulation of the SAT1 gene showed the functional impact of promoter DNA methylation in the prefrontal cortex of suicide completers.[51] The study provided further insight into the functional impact of promoter methylation on SAT1 gene expression in prefrontal cortex by mapping the presence of a polyamine responsive element (PRE) within the two heavily methylated sites. The PRE is primarily involved in increasing SAT1 expression by interacting with transcription factor Nrf2 in conjunction with polyamine-modulated factor-1 (PMF-1). Additionally, the study characterized a SNP (rs6526342) within the SAT1 promoter with high methylation incidence in suicide completers. A frequent occurrence of the C allele at this rs6526342 locus in suicide completers potentially suggests the possibility of site-specific methylation induced SAT1 repression in causing suicide pathophysiology.[51] In this regard, differential expression of few other polyamine biosynthesis pathway associated genes including ornithine decarboxylase antizymes 1 and 2 (OAZ1 and OAZ2), arginase 2 (ARG2), and AMD1 have indicated the possible involvement of potential gene regulatory mechanisms when analyzed in BA 44 area of MDD-suicide brain.[67] In fact, the identification of a hypomethylated status in the promoter area of AMD1 and ARG2 correlated well with the upregulated expression of these two genes in suicide completers; however, the report was unable to show any expression-related change in ORG1 and 2 genes due to any promoter DNA methylation. Additional information on DNA methylation-based epigenetic marks in suicidal behavior comes from a study by Gross et al.[54] As part of this study, the identification of significantly different epigenetic marks on gene promoters of 28 suicide subjects was derived from a heterogeneous population of 3 schizophrenia, 12 MDD, 7 bipolar, and 3 substance abuse cases. Despite this heterogeneity, a significant hypomethylated status of ARG2 and AMD1 gene promoters was observed with negative correlation to their respective gene expression. This study and data from other reports provide strong evidence to support DNA methylation as a potential epigenetic modulator in orchestrating the various genetic components of the polyamine system and collectively influencing the suicidal endophenotype.

3.3 Neurotrophic Factors

An overwhelming number of studies suggest compromised structural and functional plasticity in MDD and suicide brain, which could be associated with altered expression of the neuroprotective genes, brain-derived neurotrophic factor (BDNF), and its cognate receptor tropomycin receptor kinase B (TRKB).[68–72] Converging lines of evidence indicate significantly diminished expression of both BDNF and TRKB genes in neurons of the prefrontal cortex as well as the hippocampus of adult suicide subjects.[47,70,73] In another study, changes in the expression of TRKB and TRKB.T1 (truncated form of TRKB) genes were found to be associated with changes in promoter methylation in Wernicke's area of suicide brain.[52] Unlike the previous report on promoter-specific DNA methylation of the same gene (TRKB) in

Brodmann's area (BA 8/9),[47] this report was unable to show a direct correlation between methylation and suicidal behavior.[52] The unlikely outcome of this study could be due to the smaller population size (n = 18/group) and the fact that it was only limited to Caucasians. However, this finding was interesting to understand the concept of brain area specific epigenetic signatures in conferring characteristic gene expression patterns relevant to suicidal behavior. In a different study, focusing on Wernicke's area, researchers found a significant hypermethylation of the BDNF Exon IV promoter in suicide subjects.[50] Wernicke's area of the human brain grossly represents the BA 21 and 22 regions in the frontal cortex and is known for semantic thinking and decision-making capabilities associated with higher level neurocognitive functions. A progressive decline in neurocognitive functions is closely associated with suicidal behavior.[50,74–76] Therefore, identifying BDNF promoter methylation in Wernicke's area is a significant step to better understand the underlying epigenetic complexity of the suicide brain.

Recently, Maussion and colleagues provided a new perspective to understand methylation-based TRKB.T1 gene regulation in suicide brain using 3′UTR rather than promoter methylation.[56] Considering the 3′UTR coding sequence for this gene as the potential site for DNA-based methylation, the study was able to correlate astrocyte restricted TRKB.T1 isoform expression downregulation with suicide pathophysiology. The study mainly focused exploring the possibility of relative methylation contribution in regulating astrocyte-based TRKB.T1 isoform expression in the frontal cortex of suicide completers. The report showed the regulatory involvement of four CpG sites in the 3′UTR coding part of the TRKB gene. This regulatory mechanism was different from miRNA-185*-mediated posttranscriptional repression of TRKB.T1 earlier found in the frontal cortex of suicide brain.[21] This study added a new dimension by considering 3′UTR coding DNA as the target of methylation modification in frontal cortical areas of suicide completers, unlike the promoter oriented methylation regulation on TRKB. T1 isoform.[47] The mapping of a CTCF binding motif overlapped with four CpG sites further indicated the possible masking effect of DNA methylation over CTCF induced transcriptional activation of TRKB.T1 isoform. Genotypic analysis of an additional polymorphic CpG locus in close proximity to four other sites at 3′UTR of TRKB.T1, did not show a significant genetic association with suicide and ruled out the methylation-based regulatory influence on TRKB.T1 isoform expression. Overall, this interesting report is in agreement with other previous findings where methylation in the 3′UTR of a gene, other than the upstream promoter area, has been shown to be involved in transcript-specific regulation under tight spatiotemporal environment.[77,78]

3.4 Hypothalamic-Pituitary-Adrenal Axis Associated Genes

Earlier reports on blunted HPA axis response elaborated on the importance of stress-induced maladaptive neuroendocrine system in the depressed brain. Essentially, this is considered one of the major risk factors associated with suicidal behavior.[79,80] Under stressful conditions, dysregulated glucocorticoid receptor (GR) has been reported to be the primary cause behind inhibited negative feedback regulation of the HPA axis.[81,82] According to earlier reports, childhood adversity has been found to be influential in regulating HPA axis function in preclinical models. Studies have shown that deviation from normal maternal care in the form of low licking-grooming and arched back nursing in pups results in altered DNA methylation profile of GR promoter in the hippocampus.[83,84] This altered methylation at the earliest phase of life (first week after birth) works as a stable epigenetic signature for heterochromatinizing the conformation of the GR promoter and causing the inaccessibility to transcription factor NGF1.[85] A recent report found significant decrease in hippocampal NR3C1 expression in suicide victims with a history of childhood abuse.[48] Conversely, the differences in GR expression were absent in suicide victims without a history of childhood abuse. Another study indicated similar site-specific DNA methylation changes in GR promoter region of suicide brain.[53] The methylation associated changes in postmortem hippocampi were seen to be part of three noncoding transcript variants (1B, IC, and 1H) of the GR. Interestingly, these methylation associated changes were restricted to suicide subjects who had a history of childhood abuse. Moreover, the same report also indicated a strong epigenetic association between the increased methylation and diminished expression profile of the three GR transcript variants. In the contrary, examining the anterior cingulate cortex from the same suicide subjects did not reveal any significant expression associated changes in any of the GR transcript variants.[53] Together, this report represents the involvement of specific brain region in compromising the HPA axis functionality with an increased risk of committing suicide. Interestingly, another study in the hippocampi of suicide completers with a history of early life trauma showed a bidirectional pattern of DNA methylation on a genome-wide scale.[53] These bidirectional changes included hyper- and hypomethylated status of gene promoters functionally enriched for genes associated with neural plasticity. This genome-wide methylation study emphasized that traumatic incidences in early life may be perpetuated in later life with an increasing risk of suicide.[53,59]

A gene named SKA2 (spindle and kinetochore (KT)-associated protein 2), involved in the anaphasic movement of the mitotic chromosome,[86] has been identified as a molecular escort to help translocate the GR from the cytosol to nucleus where GR acts as a transcription factor.[87,88] In a recent study, Guintivano et al.[57] showed methylation-related changes on a polymorphic nucleotide (rs7208505 at cg13989295 locus) of SKA2 gene in neuronal populations of autopsy brain. Further analysis revealed the involvement of an exclusive DNA methylation signature at SKA2 gene which was functionally related to altered GR responsiveness in the prefrontal cortex of suicide subjects.[57] Moreover, in the same study, observations from peripheral blood tissue of three living cohorts were found to be aligned with the brain findings. Additionally, the role of two other methylated CpG sites located in the promoter area and 1st exon of SKA2 gene explain the diminished SKA2 expression in a cohort of PFC collected from suicide completers. Therefore, the cellular and molecular mechanisms underlying the associated risk of suicide with SKA2 genetic variation at rs7208505 site indicate a connection with defective glucocorticoid signaling, which could be increasingly driven by CpG methylation intensity in response to stress factors.

As discussed earlier, under stressful environment, a nonresponsive GR feedback mechanism is a key to understand the altered HPA functionality.[89–91] Since stress is highly comorbid with major depression and associated suicidal behavior, recent studies have unveiled the involvement of several molecular components, besides GR, in regulating HPA axis.[92] FK506-binding protein 51 (FKBP5/FKBP51) is one of the intracellular components that modulates the stress response by moderating the GR sensitivity toward cortisol binding,[93] which is generally associated with the inhibition of nuclear translocation of the GR.[94] A recent report in postmortem amygdala samples from suicide completers showed a significant expression downregulation of FKBP5 gene both at the transcript and protein levels.[95] Possibilities related to this diminished expression of FKBP5 have been recently worked out in an altered stress-gene interaction model. The results demonstrate a significant involvement of DNA methylation as an epigenetic factor in modulating FKBP5 expression.[96,97] In this context, genetic predisposition contributed by a functional polymorphism in the FKBP5 gene was shown to be fine-tuned by a DNA demethylation event to develop stress-related psychiatric disorders.[7] This study provides mechanistic evidence of early life stress exposure in individuals carrying an FKBP5 risk allele in reference to dysfunctional HPA axis. Interesting results from this study showed a differential transcriptional activation of FKBP5 gene in response to childhood trauma. The underlying cause of this altered transcriptional regulation of FKBP5 gene was associated the substantial repatterning of CpG methylation around the glucocorticoid response elements (GRE). This epigenetic insult associated with early life adversities ultimately led to the disruption of an ultra-short negative feedback loop between GR and FKBP5 genes with an increasing risk of committing suicide in adult life.

3.5 Genome-Wide Methylation Studies

Collective evidence from the above studies points out the significance of DNA methylation as the molecular interface between predisposed genetic environments and external risk factors for the assessment of suicidal behavior. These studies were primarily focused on candidate genes. Thus, an understanding of compromised neurobiological networks due to genome-wide methylomic changes in the suicide brain is not feasible. A recent study of the neuron-specific whole genome-wide DNA methylation has revealed an epigenetic influence on the neuro-cognitive functions in suicide brain. In this study, the dentate gyrus of male suicide completers indicated a methylation foot print on specific gene promoters associated with cognitive processes.[55] The functional clustering of such genes demonstrated their role in learning, memory, and other complex behavioral outcomes, which characteristically denote the possible impairment of higher order functioning in suicidal behavior. In the same study, the analysis of four genes (NR2E1, GRM7, CHRNB2, and DBH) coding for membrane receptors and membrane-associated enzyme molecules exhibited hypermethylation states. The methylation-related changes were concordant with their attenuated level of expression (except DBH). The overall deficit in the expression of these genes and their related functions in the hippocampus of suicide brains were correlated with hyper-responsive behavioral phenotypes and were considered risk factors for suicidal behavior.

Astrocyte-specific genome-wide methylation study revealed differentially methylated regions (DMRs) in BA10 area of patients who died by suicide.[58] Most interestingly, 90% of DMRs were associated with nonpromoter regions and localized in the vicinity of gene body. Two DMRs present in the intragenic regions of GRIK2 and BEGAIN genes were validated for their possible epigenetic regulatory role in suicide. Expression study for both GRIK2 and BEGAIN genes showed a significant inverse correlation with the methylation status of those two intragenic DMRs in the same BA10 area.[58]

Recently a similar study was conducted to identify genome-wide DNA methylation profile in prefrontal cortices (BA11 and BA25) of depressed-suicide completers.[60] This study identified a differentially methylated region (DMR) in close proximity to a noncoding RNA gene, PSORS1C3. The functional relevance of this DMR hypomethylation in MDD suicide brain was related to its potential to differentially regulate several inflammation-associated genes located nearby. Additionally, the study deciphered co-methylation module enriched gene networks primarily implicated in biological processes relevant to depression and suicidality.[60]

A completely different perspective of DNA methylation-based epigenomic modification has been provided in a study of suicide brains from victims of childhood abuse.[61] The study was able to associate early-life adversities with profound epigenetic and transcriptomic alterations that affected oligodendrocytes in the gray matter of the cingulate cortex. Moreover, the study identified the oligodendrocyte-specific epigenetic reprograming of POU3F1 and LINGO3 genes which were completely absent in the neuronal population from the same brain area.[61]

4. INFLUENCE OF HISTONE BASED EPIGENETIC REGULATION IN DEPRESSION AND SUICIDE PATHOPHYSIOLOGY

Histones act as the structural framework for eukaryotic chromosomes and provide a dynamic three-dimensional architecture to the genome. Histone proteins are basic in nature and are commonly synthesized as four different isoforms, that is, H2A, H2B, H3, and H4. Their basis of forming a condensed and compact chromatin assembly is fundamentally supported by their ability to form a nucleosomal core. The nucleosomal assembly is built up by spooling approximately 147 base pairs ofDNA around an octamer core made up of four different histone dimers. Adjacent histone octamers are connected with a 1.67 left-handed superhelical turn structured by an 80 bp long linker DNA thread.[98] The dynamic switching of different histone variants and posttranslation modifications of N-terminal protruding tails (PTM) of core histone molecules results in transient DNA breathing. This switching helps local nucleosomal DNA to acquire either an open (euchromatin) or a closed (heterochromatin) chromatin conformation, representative of an active or repressed state of gene transcription, respectively.[99] The N-terminal protruding tails of histones are prone to various covalent reversible modifications including acetylation, methylation, phosphorylation, and much more.[100] Distinct histone modification profiles, such as histone 3 lysine 4 (H3K4) dimethylation and trimethylation, and histone 3 lysine 27 (H3K27) acetylation at promoter regions and H3K4 monomethylation in enhancer regions of genes, are associated with active gene transcription, whereas repressed promoters are associated with H3K9 and H3K27 dimethylation and trimethylation marks. Collectively, the various histone tail modifications and associated enzymatic modifiers, such as histone methyltransferase (HMT) and histone deacetylase (HDAC), build up a potential epigenetic mechanism to transduce active changes in gene transcription.[101]

Histone-mediated reversible changes in chromatin architecture have long been studied to explain the molecular pathologies associated with mental illnesses including major depression and schizophrenia.[102–104] However, adequate knowledge is still lacking in interpreting the epigenomic complexities of suicidal behavior associated with histone modifications. An overall account of histone modifications related to epigenetic perturbations in suicide brain is summarized in Table 3.

One of the first studies related to histone-based epigenetic modifications in postmortem brain elaborated the downregulation of TRKB and TRKB.T1 in the orbital frontal cortex (Brodmann Area 10) of suicide completers. Closer examination revealed a significant increase in type 3 histone (H3) lysine (K) methylation in the core octamer of nucleosomes close to the TRKB.T1 promoter. Altogether, the experimental outcome first elucidated the role of H3K27 methylation-based repressive chromatin modification in depressed suicide brains.[47]

Suicide-related changes in polyamine genes were also found to be driven by changes in histone modifications. Brodmann area specific histone methylation turns out to be a key event in the three successive suicide-related studies.[51,107,108] However, two of the studies demonstrated lack of any significant change in the methylation of the 27th lysine residue of histone-3 tails (H3K27) in dorsolateral prefrontal cortex studies of suicide victims.[51,107] A third study indicated an association between the increased levels of H3K4 trimethylation levels with higher risk of suicide.[105] This study explained the importance of histone-based chromatin modification to

Table 3 Summary of Various Histone Modifications in MDD-Suicide Brain.

Site of Epigenetic Modifications	Type of Histone Modification	Genes Involved	Publication
Histone modification in orbitofrontal cortex (BA 10)	Promoter-based increased H3K27 trimethylation	TRKB.T1	Ernst et al. (2009)[47]
Histone modification in prefrontal cortex (BA 8/9)	Unaltered promoter-based H3K27 trimethylation	SMS and SMOX	Fiori and Turecki (2010)[49]
Histone modification in prefrontal cortex (BA 8/9)	Unaltered promoter-based H3K27 trimethylation	SAT1	Fiori and Turecki (2011)[51]
Histone modification in prefrontal cortex (BA 44)	Promoter-based increased H3K4 trimethylation in OAZ1	AMD1, ARG2, OAZ1, and OAZ2	Fiori et al. (2012)[105]
Histone modification in prefrontal cortex (BA4/17)	Promoter-based increased H3K9me3	CX30, CX40	Nagy et al. (2016)[106]

transcriptionally induce OAZ1 expression in suicide brain. Unlike the OAZ1, the other genes (AMD1, ARG2, or OAZ2) of the polyamine biosynthetic pathway did not show any responsiveness to histone modifications in the same suicide subjects. Detailed molecular analysis showed a significant correlation between the inducible chromatin modifications and upregulated OAZ1 expression in suicide brain with a significantly elevated tri-methylation level of H3K4 residues.[105]

Besides histone-based epigenetic changes in neuronal populations, recent reports suggest compromised gap junctional activity in glia due to epigenetic impairment of two connexin genes, CX30 and CX43, in suicide brain. The connexin associated changes in the astrocytic network were found to be mediated by a repressive histone modification (H3K9me3).[51,106,107] Although a negative correlation was observed between the enriched H3K9me3 modification and gene expression of both CX genes (CX30 and CX43), a significant correlation was observed only for the CX30 gene. Together, this study represented a new dimension of histone-mediated epigenetic regulation on cellular morphology and connectivity in moderating PFC function of depressed suicide brain associated with the glial lineage.

5. EPIGENETIC SIGNATURE IN PERIPHERAL BLOOD MONONUCLEAR CELLS—AN EMERGING DIAGNOSTIC APPROACH TO EXPLORE BIOMARKERS FOR SUICIDAL BEHAVIOR

The emergence of peripheral blood mononuclear cells as a reliable source to study the epigenetic signature of psychiatric disorders offers a scope to develop predictive biomarkers.[109–112] A summary of epigenetic findings in peripheral tissues of depressed and suicidal individuals is provided in Table 4. Kang et al.[113] reported an increased promoter DNA methylation of BDNF gene in MDD patients and found that this promoter methylation was significantly associated with a previous history of suicide attempt as well as suicidal ideation. This hypermethylated BDNF gene promoter paralleled to decreased BDNF gene expression in blood mononuclear cells of depressed patients with a previous history of attempted suicide.[71] The outcome of this PBMC-based study strongly suggests the possible use of observed BDNF promoter methylation as a proxy epigenetic biomarker for suicidal behavior in depressed patients. Another study was conducted in an East Asian population.[114] The objective of the study was to find a possible connection between BDNF promoter methylation and the incidence of suicidal ideation in late life. The analysis of a GC-rich region in the vicinity of the BDNF

Table 4 Summary of 5mC-Based DNA Modification in Peripheral Tissues of MDD and Suicidal Individuals.

Site of Epigenetic Modifications	Genes Involved	Publication
DNA methylation in leukocytes	BDNF	Kang et al. (2013)[113]
DNA methylation in leukocytes	SKA2	Guintivano et al. (2014)[57]
DNA methylation in leukocytes	SKA2	Sadeh et al. (2016)[88]
DNA methylation in leukocytes	BDNF	Kim et al. (2014)[114]
DNA methylation in leukocytes	CACNA1C	Kim et al. (2017)[115]
DNA methylation in PBMC	BDNF, FKBP5, NR3C1, CRHBP, CRHR1	Roy et al. (2017)[116]

transcript identified five prominent sites that were significantly hypermethylated. These hypermethylated CpG sites were strongly associated with suicidal ideation at baseline as well as over a 2-year follow-up period. This study suggests that epigenetic changes at the methylation level of BDNF gene promoter could potentially be used in suicide associated risk assessment in the elderly population.

Owing to the fact that stressful environmental stimuli and altered functionality of underlying gene networks can increase the risk of suicide in MDD patients, in a recent study, the Dwivedi group has shown significant changes in DNA methylation of susceptible genetic loci in the peripheral blood monocytes of MDD patients with strong suicidal ideation.[116] The methylation-based changes were markedly noticeable in the proximal promoter regions of stress-associated genes including BDNF, NR3C1, FKBP5, Corticotropin Releasing Hormone Binding Protein (CRHBP), and Corticotropin Releasing Hormone Receptor 1 (CRHR1). Furthermore, concomitant reductions in the expression of BDNF, FKBP5 transcript variants (1, 2, and 3), and NR3C1 genes suggest their close functional association with methylation changes in the same cohort of MDD patients.[116]

In a study population of suicide attempters blood-based analysis of 11 CpG sites in the CACNA1C gene promoter demonstrated changes in DNA methylation which were found to be influenced by exposure to different suicidal means (knives, drug, and string).[115] The respective degrees of methylation at two CpG sites (site 4 and 6) were found to be positively correlated with the presentation of knives picture to those suicide attempters. Interestingly, the knife was considered as the most violent mode of committing suicide among other two (drugs and string). Moreover, the degree of methylation of site 4 was positively correlated with strong fMRI signal intensity in the left middle and inferior frontal gyri in suicide attempters. Together, these findings suggested the possibility that differential brain activation along with methylation status of the CACNA1C gene in suicide attempters could be critical risk factors in suicidal behavior.

6. CONCLUSIONS AND FUTURE DIRECTIONS

As discussed in this review, the modulatory effects related to maladaptive changes in the MDD brain coupled with suicidal behavior are seen to be largely coordinated by CpG site-based DNA methylation, histone modifications, and miRNA mediated posttranscriptional regulation. However, certain other reversible biochemical modifications are emerging as potential regulators to cause tailor-made changes in genetic information processing.[117,118] A large repertoire of cellular RNAs with limited coding potential has shown their capacity to regulate a diverse set of biological processes. These include long non-coding RNAs (lncRNAs). Variability in length (~200 nts) made this group of a noncoding RNA class apart from miRNA (~22 nts) and seen to be attributed with discrete domains that can coordinate distinct transcription regulatory complexes.[119] Since their enriched presence in the CNS has recently been confirmed, a strong assertion could be made to connect this class of lncRNA in unraveling the complex gene regulatory circuits involved in suicide brain. A few recent reports have started showing the active participation of lncRNAs in gene regulatory functions associated with mood and anxiety disorders.[120–123] Recently, in a postmortem study, an epigenetic mechanism was reported demonstrating the influence of DNA methylation on regulating the expression of lncRNA LOC285758, which was strongly related to violent suicide traits.[124] Although further studies are warranted to explore the role of lncRNAs and to underpin their mode of action in suicidal behavior, the report cited above hints at the great potential related to this class of lncRNA to understand the complexities associated with the molecular nature of the suicide brain.

Other types of epigenetic players, such as hydroxymethylated DNA, are equally promising in repatterning genetic information.[125,126] Unlike DNA methylation, the modifications related to hydroxylation of methylated cytosine residues in gene promoters render the downstream gene to be transcriptionally active. Recent reports have linked their involvement in gene regulatory functions related to mental disorders. Variable representation of 5hmC content across different tissues in vertebrates and its relative enrichment (nearly 10-fold higher if compared with embryonic stem cells) in brain,[125,127] makes it a pertinent choice to explore its relevance in regulating higher order chromatin structures in brain regions such as the frontal cortex,[125] which is most susceptible to suicide.[128] In a recent report, the role of 5hmC methylation was well appreciated due to its ability to bring a change in BDNF IX and FGF1B gene expression. This 5hmC-mediated epigenomic change was associated with the overall dysfunctionality of neurocognitive system and neural plasticity of the suicide brain.[129] Thus, understanding the intricate cellular and molecular changes confronted with various adverse environmental insults might be useful in further understanding the neurobiology of MDD and suicidal behavior.

ACKNOWLEDGMENTS

This research was supported by grants from the National Institute of Mental Health (R01MH082802; R21MH081099; 1R01MH101890; R01MH100616; 1R01MH107183) to Dr. Y. Dwivedi.

REFERENCES

1. McEwen BS. Physiology and neurobiology of stress and adaptation: central role of the brain. *Physiol Rev.* 2007;87(3):873–904.
2. Gold PW, Machado-Vieira R, Pavlatou MG. Clinical and biochemical manifestations of depression: relation to the neurobiology of stress. *Neural Plast.* 2015;2015:581976.
3. Liu W, Ge T, Leng Y, et al. The role of neural plasticity in depression: from hippocampus to prefrontal cortex. *Neural Plast.* 2017;2017:6871089.
4. Otte C, Gold SM, Penninx BW, et al. Major depressive disorder. *Nat Rev Dis Primers.* 2016;2:16065.
5. Cohen-Woods S, Craig IW, McGuffin P. The current state of play on the molecular genetics of depression. *Psychol Med.* 2013;43(4):673–687.
6. Lohoff FW. Overview of the genetics of major depressive disorder. *Curr Psychiatry Rep.* 2010;12(6):539–546.
7. Klengel T, Binder EB. Gene-environment interactions in major depressive disorder. *Can J Psychiatry.* 2013;58(2):76–83.
8. Turecki G. Epigenetics and suicidal behavior research pathways. *Am J Prev Med.* 2014;47 (3 Suppl. 2):S144–S151.
9. Autry AE, Monteggia LM. Epigenetics in suicide and depression. *Biol Psychiatry.* 2009;66(9):812–813.
10. Bartel DP. MicroRNAs: genomics, biogenesis, mechanism, and function. *Cell.* 2004;116(2):281–297.
11. Jonas S, Izaurralde E. Towards a molecular understanding of microRNA-mediated gene silencing. *Nat Rev Genet.* 2015;16(7):421–433.
12. Sayed D, Abdellatif M. MicroRNAs in development and disease. *Physiol Rev.* 2011;91 (3):827–887.
13. Xu B, Hsu PK, Karayiorgou M, Gogos JA. MicroRNA dysregulation in neuropsychiatric disorders and cognitive dysfunction. *Neurobiol Dis.* 2012;46(2):291–301.
14. Miller BH, Wahlestedt C. MicroRNA dysregulation in psychiatric disease. *Brain Res.* 2010;1338:89–99.
15. Ma K, Guo L, Xu A, Cui S, Wang JH. Molecular mechanism for stress-induced depression assessed by sequencing miRNA and mRNA in medial prefrontal cortex. *PLoS One.* 2016;11(7):e0159093.
16. Lopez JP, Lim R, Cruceanu C, et al. miR-1202 is a primate-specific and brain-enriched microRNA involved in major depression and antidepressant treatment. *Nat Med.* 2014;20(7):764–768.
17. Smalheiser NR, Lugli G, Rizavi HS, Torvik VI, Turecki G, Dwivedi Y. MicroRNA expression is down-regulated and reorganized in prefrontal cortex of depressed suicide subjects. *PLoS One.* 2012;7(3):e33201.
18. Gururajan A, Naughton ME, Scott KA, et al. MicroRNAs as biomarkers for major depression: a role for let-7b and let-7c. *Transl Psychiatry.* 2016;6(8):e862.
19. Lopez JP, Fiori LM, Gross JA, et al. Regulatory role of miRNAs in polyamine gene expression in the prefrontal cortex of depressed suicide completers. *Int J Neuropsychopharmacol.* 2014;17(1):23–32.
20. Roy B, Dunbar M, Shelton RC, Dwivedi Y. Identification of MicroRNA-124-3p as a putative epigenetic signature of major depressive disorder. *Neuropsychopharmacology.* 2017;42(4):864–875.

21. Maussion G, Yang J, Yerko V, et al. Regulation of a truncated form of tropomyosin-related kinase B (TrkB) by Hsa-miR-185* in frontal cortex of suicide completers. *PLoS One*. 2012;7(6):e39301.
22. Maheu M, Lopez JP, Crapper L, Davoli MA, Turecki G, Mechawar N. MicroRNA regulation of central glial cell line-derived neurotrophic factor (GDNF) signalling in depression. *Transl Psychiatry*. 2015;5:e511.
23. Azevedo JA, Carter BS, Meng F, et al. The microRNA network is altered in anterior cingulate cortex of patients with unipolar and bipolar depression. *J Psychiatric Res*. 2016;82:58–67.
24. Aschrafi A, Verheijen JM, Gordebeke PM, et al. MicroRNA-326 acts as a molecular switch in the regulation of midbrain urocortin 1 expression. *J Psychiatry Neurosci*. 2016;41(5):342–353.
25. Torres-Berrio A, Lopez JP, Bagot RC, et al. DCC confers susceptibility to depression-like behaviors in humans and mice and is regulated by miR-218. *Biol Psychiatry*. 2017;81 (4):306–315.
26. Lopez JP, Fiori LM, Cruceanu C, et al. MicroRNAs 146a/b-5 and 425-3p and 24-3p are markers of antidepressant response and regulate MAPK/Wnt-system genes. *Nat Commun*. 2017;8:15497.
27. Roy B, Wang Q, Palkovits M, Faludi G, Dwivedi Y. Altered miRNA expression network in locus coeruleus of depressed suicide subjects. *Sci Rep*. 2017;7(1):4387.
28. Huntley RP, Sitnikov D, Orlic-Milacic M, et al. Guidelines for the functional annotation of microRNAs using the gene ontology. *RNA (New York NY)*. 2016;22 (5):667–676.
29. Sotiropoulou G, Pampalakis G, Lianidou E, Mourelatos Z. Emerging roles of microRNAs as molecular switches in the integrated circuit of the cancer cell. *RNA (New York NY)*. 2009;15(8):1443–1461.
30. Kim VN. MicroRNA biogenesis: coordinated cropping and dicing. *Nat Rev Mol Cell Biol*. 2005;6(5):376–385.
31. Bracken CP, Scott HS, Goodall GJ. A network-biology perspective of microRNA function and dysfunction in cancer. *Nat Rev Genet*. 2016;17(12):719–732.
32. Lin S, Gregory RI. MicroRNA biogenesis pathways in cancer. *Nat Rev Cancer*. 2015;15 (6):321–333.
33. Winter J, Jung S, Keller S, Gregory RI, Diederichs S. Many roads to maturity: microRNA biogenesis pathways and their regulation. *Nat Cell Biol*. 2009;11 (3):228–234.
34. Davis BN, Hata A. Regulation of MicroRNA biogenesis: a miRiad of mechanisms. *Cell Commun Signaling*. 2009;7:18.
35. Dwivedi Y. Emerging role of microRNAs in major depressive disorder: diagnosis and therapeutic implications. *Dialogues Clin Neurosci*. 2014;16(1):43–61.
36. Mathonnet G, Fabian MR, Svitkin YV, et al. MicroRNA inhibition of translation initiation in vitro by targeting the cap-binding complex eIF4F. *Science*. 2007;317 (5845):1764–1767.
37. Pasquinelli AE. MicroRNAs and their targets: recognition, regulation and an emerging reciprocal relationship. *Nat Rev Genet*. 2012;13(4):271–282.
38. Luoni A, Riva MA. MicroRNAs and psychiatric disorders: from aetiology to treatment. *Pharmacol Ther*. 2016;167:13–27.
39. Dwivedi Y. Pathogenetic and therapeutic applications of microRNAs in major depressive disorder. *Prog Neuropsychopharmacol Biol Psychiatry*. 2016;64:341–348.
40. Fiori LM, Turecki G. Implication of the polyamine system in mental disorders. *J Psychiatry Neurosci*. 2008;33(2):102–110.
41. Gross JA, Turecki G. Suicide and the polyamine system. *CNS Neurol Disord Drug Targets*. 2013;12(7):980–988.

42. Sequeira A, Gwadry FG, Ffrench-Mullen JM, et al. Implication of SSAT by gene expression and genetic variation in suicide and major depression. *Arch Gen Psychiatry.* 2006;63(1):35–48.
43. Moore LD, Le T, Fan G. DNA methylation and its basic function. *Neuropsychopharmacology.* 2013;38(1):23–38.
44. Jones PA. Functions of DNA methylation: islands, start sites, gene bodies and beyond. *Nat Rev Genet.* 2012;13(7):484–492.
45. Jin B, Li Y, Robertson KD. DNA methylation: superior or subordinate in the epigenetic hierarchy? *Genes Cancer.* 2011;2(6):607–617.
46. Poulter MO, Du L, Weaver IC, et al. GABAA receptor promoter hypermethylation in suicide brain: implications for the involvement of epigenetic processes. *Biol Psychiatry.* 2008;64(8):645–652.
47. Ernst C, Deleva V, Deng X, et al. Alternative splicing, methylation state, and expression profile of tropomyosin-related kinase B in the frontal cortex of suicide completers. *Arch Gen Psychiatry.* 2009;66(1):22–32.
48. McGowan PO, Sasaki A, D'Alessio AC, et al. Epigenetic regulation of the glucocorticoid receptor in human brain associates with childhood abuse. *Nat Neurosci.* 2009;12 (3):342–348.
49. Fiori LM, Turecki G. Genetic and epigenetic influences on expression of spermine synthase and spermine oxidase in suicide completers. *Int J Neuropsychopharmacol.* 2010;13(6):725–736.
50. Keller S, Sarchiapone M, Zarrilli F, et al. Increased BDNF promoter methylation in the Wernicke area of suicide subjects. *Arch Gen Psychiatry.* 2010;67(3):258–267.
51. Fiori LM, Turecki G. Epigenetic regulation of spermidine/spermine N1-acetyltransferase (SAT1) in suicide. *J Psychiatric Res.* 2011;45(9):1229–1235.
52. Keller S, Sarchiapone M, Zarrilli F, et al. TrkB gene expression and DNA methylation state in Wernicke area does not associate with suicidal behavior. *J Affect Disord.* 2011;135 (1–3):400–404.
53. Labonte B, Suderman M, Maussion G, et al. Genome-wide epigenetic regulation by early-life trauma. *Arch Gen Psychiatry.* 2012;69(7):722–731.
54. Gross JA, Fiori LM, Labonte B, Lopez JP, Turecki G. Effects of promoter methylation on increased expression of polyamine biosynthetic genes in suicide. *J Psychiatric Res.* 2013;47(4):513–519.
55. Labonte B, Suderman M, Maussion G, et al. Genome-wide methylation changes in the brains of suicide completers. *Am J Psychiatry.* 2013;170(5):511–520.
56. Maussion G, Yang J, Suderman M, et al. Functional DNA. methylation in a transcript specific 3'UTR region of TrkB associates with suicide. *Epigenetics.* 2014;9 (8):1061–1070.
57. Guintivano J, Brown T, Newcomer A, et al. Identification and replication of a combined epigenetic and genetic biomarker predicting suicide and suicidal behaviors. *Am J Psychiatry.* 2014;171(12):1287–1296.
58. Nagy C, Suderman M, Yang J, et al. Astrocytic abnormalities and global DNA methylation patterns in depression and suicide. *Mol Psychiatry.* 2015;20(3):320–328.
59. Haghighi F, Xin Y, Chanrion B, et al. Increased DNA methylation in the suicide brain. *Dialogues Clin Neurosci.* 2014;16(3):430–438.
60. Murphy TM, Crawford B, Dempster EL, et al. Methylomic profiling of cortex samples from completed suicide cases implicates a role for PSORS1C3 in major depression and suicide. *Transl Psychiatry.* 2017;7(1):e989.
61. Lutz PE, Tanti A, Gasecka A, et al. Association of a history of child abuse with impaired myelination in the anterior cingulate cortex: convergent epigenetic, transcriptional, and morphological evidence. *Am J Psychiatry.* 2017. appiajp201716111286.

62. Merali Z, Du L, Hrdina P, et al. Dysregulation in the suicide brain: mRNA expression of corticotropin-releasing hormone receptors and GABA(A) receptor subunits in frontal cortical brain region. *J Neurosci.* 2004;24(6):1478–1485.
63. Pabba M, Sibille E, GABA, depression and suicide.Kaschka WP, Rujescu D, eds. *Biological Aspects of Suicidal Behavior.* vol. 30. Basel: Karger; 2016: 37–50.
64. Fiori LM, Bureau A, Labbe A, et al. Global gene expression profiling of the polyamine system in suicide completers. *Int J Neuropsychopharmacol.* 2011;14(5):595–605.
65. Guipponi M, Deutsch S, Kohler K, et al. Genetic and epigenetic analysis of SSAT gene dysregulation in suicidal behavior. *Am J Med Genet Part B Neuropsychiatr Genet.* 2009;150 (b6):799–807.
66. Sequeira A, Klempan T, Canetti L, et al. Patterns of gene expression in the limbic system of suicides with and without major depression. *Mol Psychiatry.* 2007;12(7):640–655.
67. Gross JA, Pacis A, Chen GG, et al. Gene-body 5-hydroxymethylation is associated with gene expression changes in the prefrontal cortex of depressed individuals. *Transl Psychiatry.* 2017;7(5):e1119.
68. Dwivedi Y. Brain-derived neurotrophic factor: role in depression and suicide. *Neuropsychiatr Dis Treat.* 2009;5:433–449.
69. Dwivedi Y. Brain-derived neurotrophic factor and suicide pathogenesis. *Ann Med.* 2010;42(2):87–96.
70. Dwivedi Y, Rizavi HS, Conley RR, Roberts RC, Tamminga CA, Pandey GN. Altered gene expression of brain-derived neurotrophic factor and receptor tyrosine kinase B in postmortem brain of suicide subjects. *Arch Gen Psychiatry.* 2003;60(8):804–815.
71. Lee BH, Kim YK. BDNF mRNA expression of peripheral blood mononuclear cells was decreased in depressive patients who had or had not recently attempted suicide. *J Affect Disord.* 2010;125(1–3):369–373.
72. Pandey GN, Ren X, Rizavi HS, Conley RR, Roberts RC, Dwivedi Y. Brain-derived neurotrophic factor and tyrosine kinase B receptor signalling in post-mortem brain of teenage suicide victims. *Int J Neuropsychopharmacol.* 2008;11(8):1047–1061.
73. Karege F, Vaudan G, Schwald M, Perroud N, La Harpe R. Neurotrophin levels in postmortem brains of suicide victims and the effects of antemortem diagnosis and psychotropic drugs. *Brain Res Mol Brain Res.* 2005;136(1–2):29–37.
74. Jollant F, Bellivier F, Leboyer M, et al. Impaired decision making in suicide attempters. *Am J Psychiatry.* 2005;162(2):304–310.
75. Jollant F, Guillaume S, Jaussent I, Castelnau D, Malafosse A, Courtet P. Impaired decision-making in suicide attempters may increase the risk of problems in affective relationships. *J Affect Disord.* 2007;99(1–3):59–62.
76. Jollant F, Lawrence NS, Olie E, et al. Decreased activation of lateral orbitofrontal cortex during risky choices under uncertainty is associated with disadvantageous decision-making and suicidal behavior. *Neuroimage.* 2010;51(3):1275–1281.
77. Choi JK, Bae JB, Lyu J, Kim TY, Kim YJ. Nucleosome deposition and DNA methylation at coding region boundaries. *Genome Biol.* 2009;10(9):R89.
78. Zheleznyakova GY, Nilsson EK, Kiselev AV, et al. Methylation levels of SLC23A2 and NCOR2 genes correlate with spinal muscular atrophy severity. *PLoS One.* 2015;10(3): e0121964.
79. Carballo JJ, Akamnonu CP, Oquendo MA. Neurobiology of suicidal behavior. An integration of biological and clinical findings. *Arch Suicide Res.* 2008;12(2):93–110.
80. Melhem NM, Keilp JG, Porta G, et al. Blunted HPA axis activity in suicide attempters compared to those at high risk for suicidal behavior. *Neuropsychopharmacology.* 2016;41 (6):1447–1456.
81. Arnett MG, Muglia LM, Laryea G, Muglia LJ. Genetic approaches to hypothalamic-pituitary-adrenal axis regulation. *Neuropsychopharmacology.* 2016;41(1):245–260.

82. Pandey GN. Biological basis of suicide and suicidal behavior. *Bipolar Disord.* 2013;15 (5):524–541.
83. Weaver IC, Cervoni N, Champagne FA, et al. Epigenetic programming by maternal behavior. *Nat Neurosci.* 2004;7(8):847–854.
84. Weaver IC, D'Alessio AC, Brown SE, et al. The transcription factor nerve growth factor-inducible protein a mediates epigenetic programming: altering epigenetic marks by immediate-early genes. *J Neurosci.* 2007;27(7):1756–1768.
85. Hellstrom IC, Dhir SK, Diorio JC, Meaney MJ. Maternal licking regulates hippocampal glucocorticoid receptor transcription through a thyroid hormone-serotonin-NGFI-A signalling cascade. *Philos Trans R Soc London Ser B Biol Sci.* 2012;367(1601):2495–2510.
86. Hanisch A, Sillje HH, Nigg EA. Timely anaphase onset requires a novel spindle and kinetochore complex comprising Ska1 and Ska2. *EMBO J.* 2006;25(23):5504–5515.
87. Rice L, Waters CE, Eccles J, et al. Identification and functional analysis of SKA2 interaction with the glucocorticoid receptor. *J Endocrinol.* 2008;198(3):499–509.
88. Sadeh N, Wolf EJ, Logue MW, et al. Epigenetic variation at SKA2 predicts suicide phenotypes and internalizing psychopathology. *Depression Anxiety.* 2016;33 (4):308–315.
89. Herman JP, Ostrander MM, Mueller NK, Figueiredo H. Limbic system mechanisms of stress regulation: hypothalamo-pituitary-adrenocortical axis. *Prog Neuro-Psychopharmacol Biol Psychiatry.* 2005;29(8):1201–1213.
90. Jankord R, Herman JP. Limbic regulation of hypothalamo-pituitary-adrenocortical function during acute and chronic stress. *Ann N Y Acad Sci.* 2008;1148:64–73.
91. Smith SM, Vale WW. The role of the hypothalamic-pituitary-adrenal axis in neuroendocrine responses to stress. *Dialogues Clin Neurosci.* 2006;8(4):383–395.
92. Currier D, Mann JJ. Stress, genes and the biology of suicidal behavior. *Psychiatr Clin North Am.* 2008;31(2):247–269.
93. Binder EB. The role of FKBP5, a co-chaperone of the glucocorticoid receptor in the pathogenesis and therapy of affective and anxiety disorders. *Psychoneuroendocrinology.* 2009;34(Suppl. 1):S186–S195.
94. Tatro ET, Everall IP, Kaul M, Achim CL. Modulation of glucocorticoid receptor nuclear translocation in neurons by immunophilins FKBP51 and FKBP52: implications for major depressive disorder. *Brain Res.* 2009;1286:1–12.
95. Perez-Ortiz JM, Garcia-Gutierrez MS, Navarrete F, Giner S, Manzanares J. Gene and protein alterations of FKBP5 and glucocorticoid receptor in the amygdala of suicide victims. *Psychoneuroendocrinology.* 2013;38(8):1251–1258.
96. Ewald ER, Wand GS, Seifuddin F, et al. Alterations in DNA methylation of Fkbp5 as a determinant of blood-brain correlation of glucocorticoid exposure. *Psychoneuroendocrinology.* 2014;44:112–122.
97. Hohne N, Poidinger M, Merz F, et al. FKBP5 genotype-dependent DNA methylation and mRNA regulation after psychosocial stress in remitted depression and healthy controls. *Int J Neuropsychopharmacol.* 2015;18(4).
98. Venkatesh S, Workman JL. Histone exchange, chromatin structure and the regulation of transcription. *Nat Rev Mol Cell Biol.* 2015;16(3):178–189.
99. Luger K, Dechassa ML, Tremethick DJ. New insights into nucleosome and chromatin structure: an ordered state or a disordered affair? *Nat Rev Mol Cell Biol.* 2012;13 (7):436–447.
100. Li B, Carey M, Workman JL. The role of chromatin during transcription. *Cell.* 2007;128(4):707–719.
101. Maze I, Noh KM, Allis CD. Histone regulation in the CNS: basic principles of epigenetic plasticity. *Neuropsychopharmacology.* 2013;38(1):3–22.
102. Gavin DP, Sharma RP. Histone modifications, DNA methylation, and schizophrenia. *Neurosci Biobehav Rev.* 2010;34(6):882–888.

103. Sun H, Kennedy PJ, Nestler EJ. Epigenetics of the depressed brain: role of histone acetylation and methylation. *Neuropsychopharmacology.* 2013;38(1):124–137.
104. Tsankova N, Renthal W, Kumar A, Nestler EJ. Epigenetic regulation in psychiatric disorders. *Nat Rev Neurosci.* 2007;8(5):355–367.
105. Fiori LM, Gross JA, Turecki G. Effects of histone modifications on increased expression of polyamine biosynthetic genes in suicide. *Int J Neuropsychopharmacol.* 2012;15 (8):1161–1166.
106. Nagy C, Torres-Platas SG, Mechawar N, Turecki G. Repression of astrocytic connexins in cortical and subcortical brain regions and prefrontal enrichment of H3K9me3 in depression and suicide. *Int J Neuropsychopharmacol.* 2016.
107. Fiori LM, Turecki G. Association of the SAT1 in/del polymorphism with suicide completion. *Am J Med Genet Part B Neuropsychiatr Genet.* 2010;153b(3):825–829.
108. Fiori LM, Turecki G. Broadening our horizons: gene expression profiling to help better understand the neurobiology of suicide and depression. *Neurobiol Dis.* 2012;45 (1):14–22.
109. Arosio B, D'Addario C, Gussago C, et al. Peripheral blood mononuclear cells as a laboratory to study dementia in the elderly. *Biomed Res Int.* 2014;2014:169203.
110. Fan HM, Sun XY, Guo W, et al. Differential expression of microRNA in peripheral blood mononuclear cells as specific biomarker for major depressive disorder patients. *J Psychiatr Res.* 2014;59:45–52.
111. Munkholm K, Peijs L, Kessing LV, Vinberg M. Reduced mRNA expression of PTGDS in peripheral blood mononuclear cells of rapid-cycling bipolar disorder patients compared with healthy control subjects. *Int J Neuropsychopharmacol.* 2015;18(5).
112. Sharma RP. Blood chromatin as a biosensor of the epigenetic milieu: a tool for studies in living psychiatric patients. *Epigenomics.* 2012;4(5):551–559.
113. Kang HJ, Kim JM, Lee JY, et al. BDNF promoter methylation and suicidal behavior in depressive patients. *J Affect Disord.* 2013;151(2):679–685.
114. Kim JM, Kang HJ, Bae KY, et al. Association of BDNF promoter methylation and genotype with suicidal ideation in elderly Koreans. *Am j Geriatr Psychiatry.* 2014;22 (10):989–996.
115. Kim YJ, Park HJ, Jahng GH, et al. A pilot study of differential brain activation to suicidal means and DNA methylation of CACNA1C gene in suicidal attempt patients. *Psychiatry Res.* 2017;255:42–48.
116. Roy B, Shelton RC, Dwivedi Y. DNA methylation and expression of stress related genes in PBMC of MDD patients with and without serious suicidal ideation. *J Psychiatr Res.* 2017;89:115–124.
117. Qureshi IA, Mattick JS, Mehler MF. Long non-coding RNAs in nervous system function and disease. *Brain Res.* 2010;1338:20–35.
118. Breiling A, Lyko F. Epigenetic regulatory functions of DNA modifications: 5-methylcytosine and beyond. *Epigenet Chromatin.* 2015;8:24.
119. Engreitz JM, Ollikainen N, Guttman M. Long non-coding RNAs: spatial amplifiers that control nuclear structure and gene expression. *Nat Rev Mol Cell Biol.* 2016;17 (12):756–770.
120. Liu Z, Li X, Sun N, et al. Microarray profiling and co-expression network analysis of circulating lncRNAs and mRNAs associated with major depressive disorder. *PLoS One.* 2014;9(3):e93388.
121. Spadaro PA, Flavell CR, Widagdo J, et al. Long noncoding RNA-directed epigenetic regulation of gene expression is associated with anxiety-like behavior in mice. *Biol Psychiatry.* 2015;78(12):848–859.
122. Goff LA, Groff AF, Sauvageau M, et al. Spatiotemporal expression and transcriptional perturbations by long noncoding RNAs in the mouse brain. *Proc Natl Acad Sci USA.* 2015;112(22):6855–6862.

123. Cui X, Niu W, Kong L, et al. Long noncoding RNA expression in peripheral blood mononuclear cells and suicide risk in Chinese patients with major depressive disorder. *Brain Behav.* 2017;7(6):e00711.
124. Punzi G, Ursini G, Shin JH, Kleinman JE, Hyde TM, Weinberger DR. Increased expression of MARCKS in post-mortem brain of violent suicide completers is related to transcription of a long, noncoding, antisense RNA. *Mol Psychiatry.* 2014;19 (10):1057–1059.
125. Kinde B, Gabel HW, Gilbert CS, Griffith EC, Greenberg ME. Reading the unique DNA methylation landscape of the brain: non-CpG methylation, hydroxymethylation, and MeCP2. *Proc Natl Acad Sci USA.* 2015;112(22):6800–6806.
126. Lunnon K, Hannon E, Smith RG, et al. Variation in 5-hydroxymethylcytosine across human cortex and cerebellum. *Genome Biol.* 2016;17:27.
127. Kato T, Iwamoto K. Comprehensive DNA methylation and hydroxymethylation analysis in the human brain and its implication in mental disorders. *Neuropharmacology.* 2014;80:133–139.
128. Furczyk K, Schutova B, Michel TM, Thome J, Buttner A. The neurobiology of suicide – a review of post-mortem studies. *J Mol Psychiatry.* 2013;1(1):2.
129. Sun W, Zang L, Shu Q, Li X. From development to diseases: the role of 5hmC in brain. *Genomics.* 2014;104(5):347–351.

CHAPTER ELEVEN

Epigenetic Mechanisms of Traumatic Brain Injuries

Nagalakshmi B., Sneha Sagarkar, Amul J. Sakharkar[1]
Savitribai Phule Pune University, Pune, Maharashtra, India
[1] Corresponding author. E-mail address: amul.sakharkar@unipune.ac.in

Contents

Abstract

Of all types of injuries, traumatic brain injuries (TBIs) are most likely to result in death or permanent physical or mental disabilities. With the increased frequency of military operations, terror attacks, sports activities, and road mishaps, TBIs are increasingly becoming a serious public health concern. Patients who meet with moderate-to-severe TBI suffer from a constellation of cognitive deficits and phenotypes due to diffuse axonal injury. On the contrary, minimal TBI precipitates in long-term behavioral complications commonly referred to as post-traumatic stress disorders, which consist of depression, anxiety, hallucinations, and so on. During the last decade, epigenetic

Progress in Molecular Biology and Translational Science, volume 157
ISSN 1877-1173
http://dx.doi.org/10.1016/bs.pmbts.2017.12.013

mechanisms, such as histone posttranslational modifications, DNA methylation, and noncoding RNAs, have received reasonable attention by scientists studying the neurobiology of TBI. Chromatin remodeling sculpted by various epigenetic factors in different parts of the brain controls many intricate processes involved in important neural functions. Although stable and dogmatic, the reversible nature of epigenetic modifications has attracted special attention for their emerging potential in therapeutic development. Herein, we offer an overview of recent observations on epigenetic pathways that are likely involved in shaping neural function and maintain the state of neuropathology in response to traumatic events.

1. INTRODUCTION

Traumatic brain injury (TBI) is one of the leading devastating types of injuries that affect around 10 million people worldwide.[1] The outcome of TBI in terms of clinical manifestation is generally complex and thus cannot be regarded as a single clinical entity. TBI involves a spectrum of brain injuries from mild concussion to severe injury, from acute to chronic, such as chronic traumatic encephalopathy, a neurodegenerative disorder linked to repetitive head trauma. A recent epidemiological survey on TBI across 29 different countries reports 85% of TBI as mild to moderate, whereas 15% is categorized as severe TBI based on the Glasgow Coma Scale.[2] In yet another study, mild TBI is found to be the most profound type of injury that comprises about 80% of total TBI.[3] Although men are placed at a higher risk to experience TBI than women, the age of patients at the time of TBI ranged from 27 to 60 years.[2]

Growing evidences suggest that the spectrum of TBI and repeated mild TBI are highly associated with complex neurological and neurodegenerative disorders such as Alzheimer's disease,[4] chronic traumatic encephalopathy,[5] and Parkinson's disease.[6] Recent reviews narrate the emerging role of epigenetics in the pathophysiology of CNS injury, stroke, and spinal cord injury.[7–10] The significance of epigenetic mechanisms in neural plasticity and in learning and memory is also explained.[11] These studies fortify our current understanding about structural and functional changes induced by TBI and could provide innovative approaches for TBI recovery and therapeutic intervention. Herein, we highlight the available evidence supporting the role for epigenetic mechanisms in the pathophysiologies of TBI.

1.1 Traumatic Brain Injury

TBIs are complex with a broad spectrum of symptoms and psychosomatic disabilities. This occurs when a bump, blow, or other head injury causes damage to the brain. Symptoms of mild-to-moderate TBI include headaches, dizziness, nausea, and amnesia. Moreover, these injuries usually resolve within a few days, but when repeated, occasionally may result in long-term cognitive and behavioral deficits. As TBI is an incredibly complex disorder of the brain, the classification is precarious. Based on clinical trials, different systems of classification of TBI are devised and several indices are constructed to quantitate the severity of TBI. Mainly, TBIs are classified in terms of severity and types of injuries.

1.1.1 Severity of Injuries

Glasgow Coma Scale: The TBI is graded into mild, moderate, severe, and vegetative according to the Glasgow Coma Scale (GCS). The GCS is the most common and accepted 15-point scale used to measure coma and impaired consciousness after TBI. The GCS measures three different components: eye opening (E), verbal responses (V), and motor responses (M). The summation of the individual score (i.e., E + V + M) classifies the person into mild (score = 13–15), moderate (score = 9–12), severe (score = 3–8), and vegetative state (score <3).[12]

Post-Concussion Symptom Scale: Sports-related TBI is mainly characterized based on the Post-Concussion Symptom Scale (PCSS). The PCSS is a 6-point scale, which accesses concussion based on various physical (headache, nausea, vomiting, balance problems, dizziness, fatigue, visual problems, light sensitivity, noise sensitivity, numbness), sleep-related (trouble falling to sleep, excessive sleep, loss of sleep, drowsiness), emotional (irritability, sadness, nervousness, more emotional), and attentive (feeling slowdown, feeling mentally foggy, difficulty concentrating, difficulty remembering) symptoms. It classifies concussions into no symptom (score = 0), mild (score = 1–2), moderate (score = 3–4), and severe (score = 5–6) categories.[13–15]

Loss of consciousness and post-traumatic amnesia: The LOC (loss of consciousness) and PTA (post-traumatic amnesia) are other indices commonly used to assess the severity of TBI.[16] The LOC classifies mild (LOC <30 min), moderate (LOC >30 min to 6 h), and severe (LOC >6 h) TBI. The PTA measures the time elapsed from injury until the patient is oriented and starts recollecting memories. The PTA classifies TBI as mild (PTA <24 h), moderate (PTA <1 week), and severe (PTA >1 week).[17]

TBIs, therefore, can be diagnosed based on the above indices and the results of neuroimaging. For example, based on the indices, the diagnostic criterion for mild TBI is the GCS score of 13–15, LOC <30 min, and the PTA <24 h after the injury.

1.1.2 Types of Injuries

Focal and diffuse injuries: TBI can also be classified based on the presence of focal lesions in focal and diffuse injuries. The focal injury is usually caused by contact, whereas diffuse injury is triggered by acceleration and deceleration of the brain. The focal injury includes contusion, intraparenchymal hemorrhage, and subdural and epidural hematoma. Diffuse injury includes axonal, hypoxic-ischemic, and microvascular injuries affecting various anatomical regions during single event.[18–20] The mortality rate with focal and diffuse injuries is reported to be 40 and 25%, respectively.[21] Current studies suggest that mild TBI mainly induces diffuse axonal-type injuries, whereas moderate-to-severe injuries produce focal lesions as well as induce diffuse-type injuries.

Primary and secondary injuries: The resultant injury after the traumatic insult may appear visibly direct or the damage can be indirect. The primary injuries are generally caused as a direct result of the external mechanical forces such as acceleration and deceleration linear forces, rotational forces, forces generated by blast winds associated with blast injury and damage the neurons, axons, dendrites, glia, and blood vessels. On the contrary, the secondary neurologic damage is produced by a cascade of pathological events that include hypoxia/ischemia, cerebral edema, raised intracranial pressure, excitotoxicity through glutamate release, calcium dysregulation, cytoskeletal proteolysis, free radical oxidative damage, altered synaptic physiology, cytokine release causing inflammation, and ultimately neuronal cell death.[21–29] The secondary injuries are largely considered to be therapeutically reversible and may offer targets for preventing the consequential neuropathologies that result after traumatic insult.

In view of studying the neurobiology of TBI, several preclinical models are developed which attempt to simulate the clinical situations of brain trauma. For example, blast induced, fluid percussion, controlled cortical impact (CCI), and closed-head weight drop injury models are commonly used in rodents to study the pathology of TBI. These models are modulated to generate injuries with characteristic grades of TBI, that is, from mild to severe.[20]

1.2 Epigenetics

Epigenetics refers to the functional heritable changes in the genome that alter gene expression without changing the DNA sequence.[30] DNA is generally packaged into repeated units known as nucleosomes, which consist of a stretch of 146 base pairs of DNA wrapped around an octamer of histone proteins (two units of each H2A, H2B, H3, and H4). Chemical modifications in both the components of nucleosomes, that is, DNA and histone proteins, contribute to changes in chromatin architecture. DNA methylation and posttranslational modifications (PTMs) of histone tails, such as acetylation, methylation, and phosphorylation, are major epigenetic changes. DNA methylation and histone modifications either relax or condense the chromatin architecture, thereby providing or curtailing the access of promoter regions to transcription factors. Another layer of complexity in epigenetic regulation of gene expression is added by noncoding RNAs such as microRNAs (miRNAs) which modulate functional levels of gene expressions posttranscriptionally. These epigenetic changes are specific to different cell types and thus maintain tissue-specific gene expression, despite identical DNA in all cells of an organism.

2. DNA METHYLATION

DNA methylation in mammalian cells refers to the addition of a methyl group from S-adenosyl methionine (SAM) to the 5th carbon atom of cytosine forming 5-methylcytosine (5-mC). The reaction is catalyzed by the family of DNA methyltransferases (DNMTs). The opposing mechanism of demethylation is carried out by a set of enzymes which maintain the homeostasis of DNA methylation in concert with DNMTs. Cytosine methylation is carried out in two ways: maintenance and de novo methylation. DNMT1 is the key enzyme involved in maintenance methylation which replicates methylation patterns from parental strand onto newly synthesized daughter strands during DNA replication. On the contrary, DNMT3a and DNMT3b are involved in de novo introduction of methyl group to cytosine. Methylation primarily occurs at cytosine residues with adjoining guanosine, commonly referred to as CpG sites. Long stretches of high density of CpG sites are found in eukaryotic genomes and are commonly referred to as CpG islands. CpG islands are found to be more frequently present in promoter sequences of genes, thereby suggesting a role in gene expression. DNA methylation in CpG islands prevents binding of essential regulatory factors and thus modulates the gene expression. However, many reports also suggest that methylation of non-CpG cytosines in mouse and human embryonic stem cells is absent in mature tissues.[31,32] Moreover, recently non-CpG methylation is observed in the murine frontal cortex.[33] Despite the evidence showing methylation at non-CpG sites, the functional consequences are still controversial. The DNA methylation pattern in postmitotic cells is supposed to be stable. However, postmitotic neurons within the mature brain expresses substantial levels of DNMTs, raising the likelihood that DNA methylation likely plays a role in regulating gene expression in brain.[34,35] The activities of enzymes involved in DNA methylation and demethylation are altered in several experimental models of neuronal injury, suggesting a role for this mechanism in the pathophysiology of brain trauma.

2.1 Writers (DNMTs)

DNMTs function as writers of DNA methylation catalyzing the addition of a methyl group onto cytosine residue. DNMT3b is highly expressed in embryonic and neural precursor cells during early development, whereas DNMT3a is expressed later in development and is required for normal cellular differentiation, maturation of neurons, oligodendrocytes, and astrocytes.[36,37] DNMT3L is another enzyme in the DNMT family, which lacks a catalytic domain and is expressed early during development, and is restricted to germ cells and thymus in adulthood.[38,39] Despite the lack of a catalytic unit, DNMT3L assists DNMT3a and 3b in its activity.[40–42] In contrast, DNMT1 is expressed in postmitotic neurons and maintains the methylation status after DNA replication.[43]

2.2 Erasers

Active demethylation in dividing and nondividing cells is a process where the 5mC is converted to naked cytosine by a series of enzymatic reactions. The conversion takes place either by oxidation or deamination of 5mC into a product that gets easily recognized by enzymes of base excision repair (BER) mechanisms. The ten-eleven translocation methylcytosine dioxygenase (TET—TET1, TET2, TET3) catalyzes a series of conversion of 5mC to 5hmC (hydroxymethyl cytosine); 5fc (formyl cytosine); 5caC (carboxyl cytosine), which is then cleaved and converted to naked cytosine by BER mechanisms.[44–46] Another enzyme AID/APOBEC (activation-induced cytidine deaminase/apolipoprotein B mRNA-editing enzyme complex) is also involved in active DNA demethylation, which either deaminates 5mC to thymine or oxidizes 5hmC to 5hmU (hydroxymethyl uracil) which is then recognized by BER enzymes (TDG/SMUG) and is converted into naked cytosine.[47,48] Apart from the maintenance of genomic stability, DNA repair, and suppression of cell growth, Gadd45 (growth arrest and DNA-damage-inducible protein 45) proteins are involved in active DNA demethylation.[49,50] Li *et al.* for the first time implicated the actual mechanism of Gadd45 family proteins in active DNA demethylation.[51] Gadd45 demethylates DNA in a thymine DNA glycosylase (TDG)-dependent manner, where it interacts physically and functionally with TDG and contributes to DNA demethylation. Likewise, Gadd45 family of proteins is also reported to be involved in activity-dependent neuronal demethylation in the mouse brain.[52]

2.3 DNA Methylation and Demethylation in TBI

The implications of DNA methylation in gene expression are well studied in nonpathological conditions, whereas its role in pathological conditions is only recently emerging. The mechanism is poorly understood in brain pathologies due to TBI. Here, we catalogue some recent reports of altered DNA methylation after TBI and a few preclinical studies showing the role of DNA methylation in regulating gene expression (Table 1).

2.3.1 DNA Methylation in Blast-Induced Neurotrauma

Bailey and his team have shown the involvement of epigenetic mechanisms in blast-induced neurotrauma (BINT) in two brain regions (prefrontal cortex and hippocampus) and reported that the hippocampus is the most vulnerable region.[63] The expression of DNMT enzymes is dependent on the magnitude of the blast (10 psi, 17 psi, 23 psi) and the brain regions affected. In the hippocampus, DNMTs (1, 3b), TETs (2, 3), and TDG are dysregulated after BINT, whereas only DNMT3b and TET2 are altered in the prefrontal cortex. The study further reports a positive correlation between global DNA methylation and magnitude of the blast overpressure exposure. This differential expression of the enzymes in the two different regions is suggested to be due to the varying glial to neuron ratio in the hippocampus versus cortex.[70] Studies have also suggested alterations in cell-specific global DNA methylation in neurons and glial cells after BINT and TBI.[64] These perturbations in methylation remain sustained for 8 months after the blast injury, suggesting the implication of methylation in long-term TBI-associated phenotypic changes such as anxiety and symptoms of post-traumatic stress disorder. Furthermore, the study further screened methylation in genes relevant to several neuropsychiatric disorders resulting from TBI. Most of the genes methylated were involved in the regulation of the sleep cycle such as aralkylamine *N*-acetyltransferase (Aanat), nitric oxide synthase 1 (Nos1),

Table 1 Summary of Preclinical Models of Traumatic Brain Injuries Used to Study the Implications of Epigenetic Modifications in Different Brain Areas in Trauma-Induced Behavioral Outcomes.

Animal Model	Injury Model	Brain Region/Other Tissues	Methods of Molecular Analysis	Behavior Tests	References
Histone modifications					
Mouse	Weight drop	Frontal cortex	Western blotting		53
	Controlled cortical impact	Hippocampus	Western blotting	Morris water maze test, fear conditioning task	54
			Histofluorescence (Fluoro Jade C), Western blotting	Beam walking test	55
		Hippocampus, cortex	Western blotting	Foot faults test, wire hanging test, cylinder test, morris water maze test	56
Rat	Controlled cortical impact	Hippocampus	Real time PCR, chromatin immunoprecipitation, bisulfite sequencing	—	57
			Immunohistochemistry, Western blotting	Morris water maze test, beam walking test	58
		Hippocampus (CA1 and CA3)	Immunohistochemistry, Western blotting		59
		Frontal cortex	Western blotting, immunohistochemistry	Skilled forelimb reaching test	60
	Fluid percussion	Hippocampus	Histofluorescence (Fluoro Jade C and Cresyl violet histochemistry)	Morris water maze test	61

DNA methylation					
Mouse	Controlled cortical impact	Hippocampus (dentate gyrus)	Immunohistochemistry, Western blot	Cat-walk assisted test, Barnes maze test	62
Rat	Blast induced	Prefrontal cortex, hippocampus	Global DNA methylation using ELISA, real-time PCR	—	63
		Frontal cortex	Expanded reduced representation bisulfite sequencing (ERRBS), nanostring nConuter elements technology	—	64
	Weight drop	Brain	Immunohistochemistry and confocal microscopy	—	65
		Somatosensory cortex	Immunohistochemistry	—	66
		Amygdala	Real-time PCR, methylated DNA immunoprecipitation (MeDIP)	Light dark box exploration test	67
	Fluid percussion	Hippocampus (CA3 and dentate gyrus)	MeDIP-sequencing, transcriptomics	—	68
		Hippocampus, leucocytes	Reduced representation bisulfite sequencing (RRBS), transcriptomics	Barnes maze test	69

interleukin-1 receptor type 1 (Il1r1), Homer scaffolding protein 1 (Homer1), cholinergic receptor nicotinic alpha 3 subunit (Chrna3), and period circadian clock 3 (Per3), and showed downregulation in the frontal cortex of the blast-induced animals.[64] The Aanat gene encodes an enzyme which catalyzes the conversion of serotonin to melatonin. Low levels of melatonin are related to sleep deprivation and it is also an important protein in maintaining circadian rhythms linked to depression susceptibility. The study for the first time suggests that sleep disturbances in the war veterans may be the result of TBI-specific epigenetic changes in the Aanat gene.[71–73]

2.3.2 Spatiotemporal Relationship of DNMT With TBI

In a study of open head injury using the weight drop model, immunohistochemical staining shows DNMT1 reactivity in the nucleus of NeuN-positive cells in the normal rat brain. DNMT1 was localized in the nucleus, as well as in cytoplasm, of the cells stained with Nestin and GFAP in the perilesional and neocortical regions of the brain subjected to TBI.[65] The cytoplasmic DNMT1 staining disappeared eventually at 14th day after TBI, suggesting a temporal relationship with TBI. The DNMT1 positive reaction, along with nestin, was also observed in both the normal and TBI brains in the periventricular region at all the time points. This suggests that the cytoplasmic localization of DNMT1 might possess a role in cell differentiation, and is induced in response to TBI. Expression of nestin generally indicates the transition of cells to a less differentiated state. The JAK-STAT pathway is involved in astrocyte differentiation, and DNA methylation controls the timing of astrocyte differentiation by inhibiting STAT activity.[74] Thus, relocalization of DNMT1 activates the JAK-STAT pathway genes which are important in astrocyte differentiation when induced by TBI.

2.3.3 DNA Methylation and TBI-Induced Anxiety-Like Behaviors

A recent study from our laboratory observed the induction of anxiety-like behaviors in rats after repeated mild trauma induced by the weight drop model.[67] DNA methylation at the promoter regions of brain-derived neurotrophic factor (BDNF) was concurrently altered at early and late time points in amygdala after trauma. BDNF maintains the neuronal architecture by its vital role in neural development, synaptic plasticity, survival, and growth.[75] The trauma induced DNMT activity with concomitant increase in DNMT3a and 3b levels and reduced GADD45 expression at 48 h after minimal traumatic brain injury (MTBI). Most of the changes related to DNA methylation persisted for 30 days. The increase in DNMT activity resulted in increased 5mC levels at the BDNF IV and IX promoters with decreases in corresponding mRNA levels (IV and IX), which might be linked to anxiety-like behaviors. These molecular and behavioral changes were reversed by treatment with 5-azacytidine, a pan DNMT inhibitor.[67] These results strongly implicate DNA methylation-induced BDNF gene regulation in the amygdala in trauma-induced anxiety-like behaviors. This notion is further supported by previous observations wherein stress-related DNA methylation of BDNF promoters is linked to behavioral anomalies.[76–78]

2.3.4 DNA Methylation and the Inflammatory Response Following TBI

The inflammatory response after TBI originates in the form of microglial activation and monocyte accumulation in brain lesions.[79] The accumulation of microglia and macrophages upregulates selective membrane proteins and they express different cytokines and transform their morphology after TBI.[80] However, the prolonged activation of these immune cells can cause deleterious effects in CNS tissue.[81] Therefore, gene expression in such cells might be regulated periodically according to cellular demands. In this regard, epigenetic changes are essential to coordinate inflammatory and angiogenic gene expression to repair and recover from CNS lesions.[82] Immediate effects of TBI due to weight drop reveal global hypomethylation within the first day after injury in regions of widespread necrosis (pannecrosis) in somatosensory cortex and slightly delayed (day 2) in peripheral regions of the lesions (peripannecrosis). Further analysis with double immunostaining demonstrates hypomethylation specifically in microglia/macrophages. This suggests that hypomethylation at early stages after injury defies a subpopulation in response to inflammation. Dexamethasone, a widely used glucocorticoid, exerts an early and transient anti-inflammatory activity and attenuates the hypomethylation at day 2 in both pannecrosis and peripannecrosis regions.[66] These studies emphasize the role of corticosterioids in the management of DNA methylation caused by TBI.

2.3.5 Fluoxetine and Epigenetic Factors

CCI injury in mice leads to a reduction in doublecortin positive cells in the dentate gyrus of the hippocampus. TBI reduces methyl-binding domain protein (MBD1), a key transcriptional regulator involved in DNA methylation, and also causes global histone hypoacetylation. Treatment with fluoxetine, a serotonin reuptake inhibitor, not only improves hippocampal neurogenesis by increasing the numbers of doublecortin cells, but also increases MBD1 gene expression and acetylated histones. However, the impaired gait function and spatial memory behaviors due to CCI injury are unaffected after chronic fluoxetine treatment. Thus, fluoxetine increases neuronal plasticity in the hippocampus, but does not restore locomotor function and neither improves memory deficits.[62]

2.3.6 DNA Methylation and TBI-Induced Functional Changes in the Enteric Nervous System

The primary and secondary inflammatory response after TBI results in several pathological conditions, including intestinal dysfunction.[83,84] One of the intestinal dysfunctions includes impairment of permeability and motility of the small intestine. Repeated MTBI in rats provokes anxiety-like behaviors as analyzed using the light dark box exploration test.[85] However, the anxiety-like behaviors are also linked to deficits in gut motility, nutrient absorption, and also to inflammatory bowel syndrome.[86,87] Neuropeptide Y (NPY), a major gut neuropeptide released by neuronal cells of the enteric nervous system, modulates several gut functions including intestinal gut motility. Previous studies reveal the role of NPY in gastrointestinal motility.[88,89] A recent study from our laboratory implicates DNA methylation in MTBI-induced NPY gene regulation in the jejunum. NPY in the small intestine is decreased after 48 h of repeated MTBI, which is subsequently increased at 30 days after injury. In yet another study, NPY is heightened at 24 h after brain injury.[83] The models used in both studies are identical. Therefore, the discrepancies in the results in these studies suggest different traumatic injury regimes and time points. The increase in NPY in the small intestine correlates with the reductions in gut motility at the protracted time point of 30 days. These results suggest that NPY might play a crucial role in maintaining long-term changes in gut motility after repeated MTBI. The expression of DNA demethylating factors including GADD45b in intestine correlates with the expression of NPY after MTBI. On the contrary, DNMT1 is increased at both time points after MTBI. Furthermore, 5-azacytidine treatment restores NPY levels and gut motility in MTBI-induced animals.[85] These results suggest altered DNA methylation in brain trauma-induced disabilities in gut function which warrants further investigation as causative mechanisms.

3. HISTONE MODIFICATIONS

The principal component of chromatin other than DNA is the histone proteins, which forms chromatin architecture along with DNA. Histones possess highly basic amino acids that protrude from the histones commonly referred to as histone tails. Histone tails undergo various PTMs, manifesting in altered sculpting of nucleosomes affecting internucleosomal interactions. Some of the major histone modifications include acetylation, methylation, phosphorylation, sumoylation, and so on. The different states of chromatin architecture achieved via these modifications on selected amino acids in histone tails can result in altered gene expression.

3.1 Histone Acetylation

One of the most prominent and dynamic modifications of histone tails is acetylation. Histone acetylation is governed by two families of enzymes with opposing functions, that is, histone acetyltransferases (HATs) and histone deacetylases (HDACs). Generally, HATs utilize acetyl CoA as a cofactor and transfer the acetyl group to lysine residues of histone tails, thereby altering the ionic charge of the affected residue. This change in charge loosens the chromatin architecture providing for open chromatin conformations which facilitates the binding of transcription factors on promoter regions of the genes.[90] Reciprocally, HDACs oppose the effects of HATs by deacetylating acetylated lysine residues, thus imparting positive charges. These changes result in more condensed chromatin which restricts access of transcription factors to promoter regions. In general, acetylation is believed to activate gene expression, whereas deacetylation by HDACs is commonly known to repress gene expression. HDACs are classified into four classes: class I (HDACs 1, 2, 3, and 8), class II (HDACs 4, 5, 6, 7, 9, and 10), class III (Sirtuins-SIRT 1–7), and class IV (HDAC11). However, class I, II, and IV are zinc-dependent classical HDACs, and class III HDACs require NAD^+ for their enzyme activity.[91,92]

3.2 Histone Methylation

Histone methylation occurs at lysine and arginine residues in histone tails and unlike histone acetylation, it does not impart a charge to the histone proteins. However, the degree of methylation at a single lysine residue can occur in three different levels by the addition of successive methyl groups, commonly referred to as mono-, di-, and trimethylation, respectively. Therefore, methylation offers more complex chromatin architectural regulation than acetylation.[93] The state of chromatin offered by different levels of methylation is complex. For example, trimethylation of lysine 4 (K4) on histone 3 (H3-K4Me3) is known to be an active mark for transcription, whereas dimethylation of lysine 9 on histone 3 (H3-K9Me2) facilitates transcriptional silencing.[94]

Enzymes involved in histone methylation are identified as protein lysine-specific histone methyltransferases and arginine methyltransferases. Suppressor of variegation, enhancer of zeste, and trithorax (SET) domain containing proteins and disruptor of telomeric silencing 1 (DOT1) like proteins methylate lysine. Seven families of SET domain proteins (EZ, SET1, SET2, SMYD, SUV39, SUV4-20, and RIZ) are reported in several studies, whereas SET8/PR-SET7 and SET7/9 families were reported less. Similarly, the protein arginine *N*-methyltransferase (PRMT) family of proteins has been shown to methylate arginines. These enzymes transfer the methyl group from SAM to lysine and arginine amino acid residues, respectively. Histone methylation via these methyltransferases is involved in the maintenance of chromatin structure by recruiting heterochromatin protein (HP1) to initiate the process of heterochromatinisation.[95] On the contrary, lysine demethylases are classified into two types based on their catalytic activity. Lysine-specific demethylases mediate oxidative demethylation by reducing FAD. For example, LSD1 acts on H3-K4Me1/2 or H3-K9Me2. Jumonji histone demethylases catalyze demethylation through various steps involving Fe (IV)-oxo species on arginine H3-R2, H4-R3[95] and lysine H3-K9me3, H3-K36me3.[96]

3.3 Histone Modifications in TBI

3.3.1 Global Histone Methylation and Acetylation in the Hippocampus

Compact cortical impact injury in rat pups of postnatal day (PND) 17 causes alterations in histone H3 acetylation and methylation levels at multiple time points after injury.[59] The H3 acetylation levels in the CA3 hippocampal region are reduced during the initial 6 h and 24 h after injury. These are subsequently normalized by 72 h. However, dimethylated histone levels in the CA3 region are lower at all the time points, that is, 6, 24, and 72 h. These studies clearly suggest that histone modification alterations due to traumatic events may play a major role in gene expression resulting into neuropathologies.[59] This is supported by the observation that several HDAC inhibitors have shown neuroprotective effects in various neurodegenerative disorders.[97] With the demonstrated potential of HDAC inhibitors in correcting neural mechanisms in various neuropsychiatric disorders, studies into evaluating the possibility of histone acetylation as a therapeutic target are warranted.

3.3.2 Cross Talk Between Histone Modifications and DNA Methylation in TBI-Induced Alternative Splicing

DNA methylation has been believed to play a physical role in determining the boundaries of genomic sequences to delineate exons or introns.[98] Evidence suggests that dynamic interactions between DNA methylation and histone modifications account for alternative splicing.[99,100] For example, alternative splicing of insulin-like growth factor-1 (IGF-1) primary transcript can produce two variants, IGF-1A and IGF-1B, which is dependent on interactions between DNA methylation and histone modifications. The IGF-1B includes an extra exon 5, whereas the IGF-1A transcript lacks exon 5. The IGF gene is regulated by two promoters, P1 and P2, and by an exon splicing enhancer element, which is upstream of exon 5. Schober *et al.* reported that altered IGF-1 gene expression in traumatic pathology is possibly regulated epigenetically.[57] IGF-1B mRNA levels peak in the hippocampus at 2 days after injury and normalize after 2 weeks of time. On the contrary, IGF-1A mRNA levels heighten on the third day after injury, which persists after 2 weeks. DNA methylation at the P1 promoter is increased on the third day and is eventually normalized by 2 weeks, whereas methylation is not altered at P2 in either time point. Moreover, the DNA methylation is increased on the third day upstream and is reduced downstream of exon 5, which correlates with increased levels of IGF-1B mRNA levels on the third day. On the contrary, methylation is increased on both regions of exon 5, with a decrease in IGF-1B mRNA levels at 2 weeks of time. Intriguingly, the gene activation histone modifications, such as H3-K9ac, H3-K14ac, H3-K4Me3, and H3-K36Me3, are increased at both time points at promoter P1, whereas these marks are increased at promoter P2 at the third day, with the exception of a persistent increase in H3-K4Me3 after 2 weeks. In exon 5, the activation marks (H3-K9ac, H3-K36Me3, and H3-K4Me3) are increased at the third day, whereas none of these lasted for 2 weeks. The relaxed chromatin provides access to the spliceosome complex to bind at the exon splicing enhancer element and alternative splicing is initiated resulting in reductions in IGF-1B mRNA levels at protracted time points.[57]

3.3.3 Role of HDAC Inhibitors in TBI Pathology

HDAC inhibitors such as sodium butyrate (SB), valproic acid, trichostatin A, suberoylanilide hydroxamic acid, and so on have long been studied for their role in normalizing the deleterious consequences of environmental factors including addictive drugs and stress.[101] Similarly, recent studies also show the neuroprotective effects of HDAC inhibitors via restoration of histone acetylation in the frontal cortex and hippocampus of TBI-induced animals.

Sodium butyrate: Sodium butyrate, the sodium salt of butyric acid, is a well-known HDAC inhibitor, and competitively binds to the zinc sites in class I and II HDACs. The administration of SB on the 14th day after CCI injury increases global histone acetylation in the hippocampus. In addition, SB also improves CCI-induced deficits in spatial learning and memory, as well as fear conditioning due to TBI.[54]

Valproic acid (2-propylpentanoic acid): Valproic acid is a widely used antiepileptic drug that acts on several targets in the central nervous system, including GABA transaminase, glycogen synthase kinase-3, voltage-gated sodium channels, and HDACs. As TBI initiates several pathological events such as excessive glutamate release, increase in blood–brain barrier (BBB) permeability, and ultimately neuronal dysfunction, it is believed that valproic acid may be an effective strategy to treat TBI via its multiple targets. Systemic administration of valproate (400 mg/kg) increases H3 and H4 histone acetylation in the hippocampus after CCI injury in mice.[94] These changes coincide with reduction in TBI-induced BBB permeability and contusion volume, further normalizing deficits in motor function, spatial learning, and memory behaviors. These behavioral impairments are not improved when CCI-exposed animals are treated with suberoylanilide hydroxamic acid.[58] Considering 400 mg/kg of valproate as a high dose, Tai *et al.*[60] studied the effect of 30 mg/kg on CCI-induced behavioral consequences. The administration of 30 mg/kg of valproate reduces contusion volume, number of inflammatory cells and astrocytes, and BBB permeability. However, the histone (H3) acetylation levels increase only after administering 200 and 400 mg/kg doses, as previously reported by Dash *et al.*[58] In addition, phosphorylated cAMP response element-binding factor and phosphorylated extracellular signal regulated kinase are increased at 200 or 400 mg/kg and 100 mg/kg doses, respectively.[60] In yet another study, the cotreatment of lithium (1 mg/kg) and valproate (200 mg/kg) in TBI-exposed animals reduces the contusion, BBB permeability, neurodegeneration, and improves motor coordination. Concomitantly, the cotreatment also causes a robust increment in histone H3 acetylation levels.[55] These results reasonably support the possibility of valproate as an effective drug in alleviating TBI-induced long-term neuropsychiatric pathologies.

ITF2357: Shein *et al.* demonstrated that a single dose of ITF2357, a pan HDAC inhibitor, at 24 h after injury improves the neurobehavioral recovery from the sixth day onward in a mouse subjected to a closed head injury. The neurobehavioral outcome is also accompanied by reduced neurodegeneration, decreased lesion volume, and is preceded with increased histone (H3) acetylation in the frontal cortex.[53]

Scriptaid: The neuroprotective effects of Scriptaid in terms of decreased lesion volume and a concomitant attenuation of motor and cognition deficits are observed when administered within 30 min after CCI injury. The normalization of behavioral deficits after injury is also associated with an increase in H3 and H4 acetylation levels in cortical tissues.[56]

MS-275: The CCI injury-induced neurodegeneration in the CA2 and CA3 regions of the hippocampus after 24 h of impact is restored after the administration of MS-275 in rats. The reduction in CCI-induced neurodegeneration due to MS-275 also results in neuronal survival correcting cognitive deficits as tested using the Morris water maze test on the 14th day of after injury.[61]

4. microRNAs

miRNAs are small noncoding RNAs of about 20–22 nucleotides in length, known for their role in regulating gene expression by degrading mRNAs or inhibiting translation via binding complementarily to 3′-UTR regions.[102] miRNAs are important as mRNAs in maintaining the homeostasis of functional gene expression in the CNS and therefore play pivotal roles in CNS pathologies.[103,104] Single miRNA can target several mRNAs, whereas single mRNA can be targeted by several miRNAs bringing complexity to the gene regulation by miRNA.[105] miRNAs are therefore meant to be potent regulators of neural function in pathophysiological conditions.[106] In this context, we discuss recent data showing circulating miRNAs as causative agents for manifesting TBI in neuropsychiatric conditions (Table 2). Clinical studies recently report circulating miRNAs as candidates for biomarkers in combination with imaging, neurocognitive approaches, and motor functions, which may help improve TBI patient's gradation and management.

4.1 Preclinical Studies

4.1.1 miRNAs in the Blast Injury Model

miRNA-132 (miR-132) and miR-183 are found to be upregulated in the cerebellum 6 h after repeated brain injury, target proteins in the cholinergic

Table 2 Summary of miRNAs in Various Brain Areas of Rodents Affected by Different Paradigms of Traumatic Brain Injury.

Animal Model	Injury Model	Brain Region/Other Tissues	miRNA	References
Mouse	Blast	Cerebellum	miR-132 and miR-183	107
	Weight drop	Parietal cortex, hippocampus, thalamus, substantia nigra, amygdala	miR-155, miR-150, miR-222, miR-30d, miR-449a, miR-18a-5p, miR-125b, miR-449, miR-296-5p	108
		Serum	miR-376a, has-miR-214, -miR-214, miR-337-5p, miR-574-3p, miR-434-3p, miR-671-3p, miR-218, miR-199a-3p, has miR-106b, miR-106b, miR-31, miR-196c	109
	Controlled cortical impact	Cortex	miR-21	110
			miR-23a and miR-27a	111
		Extracellular vesicles from cerebral region	miR-212, miR-7a, miR-7b, miR-146, miR-21	112
		Hippocampus	miR-21	113
	Fluid percussion	Cerebral cortex	miR-21, miR-92a, miR-874, miR-138, let-7c, miR-124	114

Rat	Blast	Serum, cerebrospinal fluid	miRNA let-7i	115
	Weight drop	Hippocampus	miR-27a, miR-23b	116
	Controlled cortical impact	Hippocampus	miR-142-3p and miR-221	117
			miR-107, miR-130a, miR-223, miR-292-5p, miR-433-3p, miR-451, miR-541 and miR-711	118
			miR-21	119
			miR-144, miR-136, miR-148b-5p, miR-342-5p, miR-23a*, miR-135a. miR-135b, miR-190, miR-130b, miR-363	120
			miR-21, miR-34a	121
	Fluid percussion	Hippocampus	miR-21	122
		Microvessels, cerebral cortex	miR-21	123
		Cortex	miR-21	124
		Dentate gyrus	miR-139-5p, miR-127, miR-136, miR-341, miR-376a, miR-369	125

inflammatory signaling pathway.[107] miRNA let-7i is detected in serum and cerebrospinal fluid in animals subjected to TBI, hence serves as ideal biomarker in neurotrauma.[115]

4.1.2 miRNAs in the Weight Drop Model

Several brain regions (parietal cortex, hippocampus, thalamus, substantia nigra, and amygdala) are studied for differential miRNA expression after MTBI. The analysis reveals that the dysregulated miRNAs target proteins involved in the axon guidance pathway, calcium signaling, and synaptic signaling.[108] Several circulating serum miRNAs dysregulate with different grades of MTBI. Regardless of the spectrum of brain injury, 13 miRNAs (miR-376a, has-miR-214, miR-214, miR-337-5p, miR-574-3p, miR-434-3p, miR-671-3p, miR-218, miR-199a-3p, has miR 106b, miR-106b, miR-31, and miR-196c) are commonly reported as potential targets of neurotrauma.[109] Sun *et al.* demonstrate a putative role of downregulated miR-27a and miR-23b in cognitive impairment and also suppression of TBI-induced neuronal autophagy via FoxO3a and ATG gene expression in the hippocampus.[116]

4.1.3 miRNAs in the Controlled Cortical Impact Injury Model

miRNA profiling in the rat hippocampus of TBI-induced animals suggested differential expression of miRNAs with respect to different grades of trauma.[111,112,118,120] Seventeen miRNAs were found to be consistent among samples from animals, which received different degrees of TBI. Especially, miR-142-3p and miR-221 were emerged as promising biomarkers for TBI.[117]

miR-21: miR-21 was observed to regulate various TBI-induced cellular mechanisms in the brain.[119] Elevated levels of miRNA-21 was observed in rat hippocampus after 3 days of injury which persisted for15 days. TBI-induced upregulation in miR-21 was also reported in the cerebral cortex of adult mice after 24 h and was normalized by 7 days. On the contrary, it remained unaltered in the cerebral cortex of aged mice subjected to TBI. The alleviated miRNA-21-induced genes involved in the process of apoptosis such as PDCD4, TIMP3, RECK, and PTEN.[110] Recently, alterations in miR-21 are linked to neuronal plasticity after TBI. miR-21 upregulates in the hippocampus of animals that showed TBI-induced cognitive impairments. Inhibition of miR-21 increases axonal extension and spine density in hippocampal neurons with improved cognitive abilities in the Morris water maze test.[113] It shall be noted that the miR-21 and miR-34 are suggested to be involved in the recovery process such as cognitive impairments.[121]

4.1.4 miRNAs in the Fluid Percussion Injury Model

The levels of some miRNAs such as miR-21, miR-92a, miR-874, miR-138, let-7c, and miR-124 are upregulated in TBI-induced animals which receive voluntary exercise using running wheels, as compared with that in nonrunners. These miRNAs can be involved in the prevention and protection afforded by voluntary exercise in a TBI model.[114] Studies also emphasize the role of miRNAs in long-term implications of TBI. For example, downregulation of 12 miRNAs and upregulation of 39 miRNAs in the dentate gyrus are observed 3 months after injury. Among these miRNAs, downregulation of miR-139-5p is responsible for increasing NOTCH1, a transcription factor and positive enrichment of NOTCH1 target genes. These findings implicate the role of miRNA-targeted regulation of transcription as a chronic manifestation of TBI.[125]

miR-21: Microarray analysis reveals differential expression of several miRNAs at different time points, that is, 6, 24, 48, and 72 h after injury caused by using the fluid percussion technique. miR-21 is found to be highly expressed in cerebral cortex at all time points after TBI.[124] In contrast to previous reports on miR-21, a few studies have also implicated its neuroprotective role during TBI. Ge *etal.* show that the upregulation of miR-21 gives a better neurological outcome after the TBI by improving long-term neurobehavioral functions such as reduction in the lesion volume and prevention of damage to BBB. miR-21 inhibits apoptosis and also promotes angiogenesis by inhibiting PTEN and activating Akt signaling pathway in the hippocampus. In addition, miR-21 can promote the expression of Ang-1 and Tie-2 microvascular endothelial cells (BMVECs) in cortex after TBI, which is involved in the expression of tight junction proteins and amplifies BBB stabilization.[122,123]

4.2 Clinical Studies

Circulating biomarkers in the plasma, serum, or cerebrospinal fluid may be helpful in characterization of the TBI. However, clinical evidence is scanty accounting miRNA as a biomarker to screen the various grades of TBI. As the diagnosis of mild TBI using imaging and other clinical practices is very poor, it is pertinent to deduce a screening parameter to diagnose and provide appropriate treatment for mild TBI. miRNAs may hold the potential to screen the spectrum of TBI. Here, we discuss a few studies suggesting the potential of circulating miRNAs in diagnosing the spectrum of TBI, which may prove to be a value addition regarding the management and treatment of the pathological outcomes of TBI.

miR-93, miR-191, and miR-499 levels are significantly increased in the serum of TBI patients. These miRNA levels in serum are observed to be higher in severe TBI as compared with that in mild and moderate type of brain injuries.[126] The miRNAs circulating in the cerebrospinal fluid may be more important than serum miRNAs. In this regard, a study prominently shows upregulation of 10 common miRNAs (miR-505, miR-195, miR-328, miR-362-3p miR-486, miR-451, miR-151-5p, miR-30d, and miR-20a) in mild, moderate, and severe TBI of which only four (miR-328, miR-362-3p miR-486, and miR-451) are increased in CSF.[127] In addition, Di Pietro *et al.* recently report the downregulation of miR-425-5p, miR-502, miR-21, and miR-335 as possible biomarkers to screen mild-to-severe TBI.[128] It is found that plasma miRNAs have a different profile in TBI patients compared with healthy ones. Of the 108 miRNAs studied, 52 are dysregulated and an additional 8 miRNAs are only in plasma of TBI patients 24 h after injury. As miR-765 is unaltered in mild TBI, the analysis of miR-16, miR-92a, and miR-765 represents biomarkers to diagnose severe TBI.[129] Further studies in evaluating the importance of these miRNAs are clinically relevant to diagnose and characterize the severity of TBI.

5. MITOEPIGENETICS

5.1 Mitochondrial Dysfunction in TBI

Multiple studies implicate mitochondrial dysfunction as one of the noticeable secondary injuries after TBI.[130–133] Likewise, mitochondrial dysfunction is also a prominent feature of many neurodegenerative disorders, such as Alzheimer's disease,[134–137] Parkinson's disease,[138–140] and amyotropic lateral sclerosis.[141–144] As neurons are known to have lower endogenous antioxidant activity compared with other cell types, the cells within the brain seem to have a higher vulnerability to oxidative damage.[145] It should be noted that the brain is known to utilize approximately 20% of the body's total oxygen supply leaving cells within this tissue more prone to oxidative stress.[146] When combined with the fact that neurons are postmitotic, this gives the cells within the brain a lifetime to accumulate oxidative damage compared with other cell types. Of several underlying mechanisms governing pathology due to MTBI, one of the least explored mechanisms is mitochondrial epigenetics.

5.2 Mitochondrial Epigenetics

The discovery of a mitochondrial localized DNA methyltransferase 1 (mtDNMT1) and methylation marks in mtDNA demonstrates a role for epigenetic mechanisms in regulating mitochondrial gene regulation.[147] Except mtDNMT1, no other specific isoforms of DNMTs or enzymes regulating hydroxymethylation are identified for functional activity in this organelle. Nevertheless, DNMT3b, TET1, and TET2 at very low levels are detected in the mitochondrial protein fraction of mouse fibroblasts and HeLa cells.[148] Likewise, TET1 and TET2 are also detected in mitochondrial fraction of primary neuronal cultures of 5 days old mice.[149] The displacement loop (D loop) in mtDNA plays an important role in replication and transcription. Therefore, it is likely that methylation in D loop may influence mitochondrial gene expression, either directly or indirectly, via modulation of the mtDNA copy number. Despite the absence of CpG islands, methylation within the D loop and other loci in the mtDNA correlates with the expression of mitochondrial genes.[150,151] Several studies have reported differential methylation in the D loop region; however, a precise function for these molecular marks is not yet ascribed.[148,152–154] Besides CpG methylation, a peculiar non-CpG methylation is also observed in mtDNA (CpC, CpA, and CpT), which is a characteristic of prokaryotic genomes such as those of bacteria.[148,155] A recent study describes D loop methylation in replicative and senescent endothelial cells and identified an uneven distribution of 5mC (both CpG and non-CpG) between the L and H strands.[152] In addition to 5mC, 5hmC also occurs in mtDNA.[147,148,156,157] However, it is important to note that despite harboring their own genetic material, mitochondrial function is heavily dependent on the expression of nuclear DNA (ncDNA), which encodes the bulk of mitochondrial proteins. Not much is known about the complex coordination that exists between nuclear epigenetics and mitochondrial function. Although nuclear gene expression is meticulously regulated via different epigenetic mechanisms, it still remains unclear whether mtDNA adheres to the same principles. Mitochondria are devoid of histones, and therefore mtDNA is packaged into protein–DNA complexes forming nucleoid-like structures. These nucleoid structures are dynamic and actively redistributed when the mitochondria undergo fission and fusion, or when mtDNA undergoes transcription and replication.[158] Therefore, the protein–mtDNA complex tends to undergo varying aberrations under different pathophysiological conditions.[159] A recent review addressed the complex role of various mitochondrial proteins and the importance of mitochondrial nucleoid structures in the process of mtDNA transcription.[160] Briefly, the core machinery for the regulation of mitochondrial transcription includes regulators of initiation, composed of three components: POLRMT (mitochondrial RNA polymerase), TFB2 M (mitochondrial transcription factor B), and TFAM (mitochondrial transcription factor A). Apart from this, several additional components including the MTERF (mitochondrial terminator factor) family also play an important role in mtDNA transcription.[161] Depending on the protein concentration ratio of TFAM and POLRMT/TFB2 M, transcription preferentially occurs from LSP (intermediate ratio), HSP1 (high ratio), or HSP2 (low ratio).[162] Several nuclear transcription factors also are capable of binding to mtDNA. It is believed that these transcription factors may play tissue-specific roles in the regulation of mtDNA expression and apoptosis.[163] TFAM is a major component of the nucleoid structure,[164,165] which fulfills an architectural[166] or scaffolding[167] role via its strong affinity for DNA sequences. TFAM contains two DNA-binding, high-mobility group (HMG) box domains, which are essential for its maximal DNA bending property.[168–172] The binding of TFAM at sites across the genome exhibits positive cooperativity spreading over mtDNA.[173,174] These proteins, which are required for mitochondrial biogenesis, are encoded by nuclear genome and are therefore regulated by network of ncDNA-binding transcription factors and coregulators. The peroxisome proliferator-activated receptor (PPAR) gamma coactivator (PGC) family of transcriptional activators and PGC-related coactivator (PRC) are the primary regulators for mitochondrial biogenesis. PGC-1α activates nuclear respiratory factor 2 (NRF2), which subsequently activates the related gene NRF1. NRF1 is known to induce TFAM, TFB2 M, and TFB1 M, which are essential transcription factors in mitochondria. In addition, NRF1 also activates several nuclear-encoded genes responsible for mitochondria replication, transcription, translation, and repair mechanisms.[175–177] Mitochondrial superoxide dismutase, SOD2, is encoded by ncDNA, which plays a prominent role in relieving the oxidative stress in mitochondria. Brain trauma is known to increase oxidative stress and SOD2 helps scavenge the trauma-induced oxidative stress in brain. SOD2 increases at 24 h after injury induced by mild-blast overpressure, which is normalized after 30 days of injury. These observations signify the role of SOD2 in correcting the oxidative stress during the initial stages of trauma.[178]

Interestingly, there is evidence of the presence of miRNAs in the mitochondrial subcellular fraction of the hippocampus of rats, which are altered after CCI injury. In addition, proteins involved in miRNA processing such as

Argonaute (AGO) and Dicer are also localized in the mitochondria. Immunoprecipitation of AGO with miRNAs from the mitochondrial fraction suggests the presence of RNA-induced silencing complex (RISC) in mitochondria. At 12 h after TBI, several mitochondrial miRNAs are significantly altered in the hippocampus. Furthermore, the analysis reveals a subset of miRNAs (miR-155 and miR-223) that plays important roles in inflammatory responses, and is profoundly elevated in the cytoplasm as well in the mitochondria after TBI.[179] These reports suggest that the miRNA-mediated gene regulation occurs in the mitochondria and might play a crucial role in mitochondrial function during TBI. In view of the scanty information available, more studies are warranted to explore the role of miRNA in mitochondrial function in the neuropathologies caused due to different levels of TBI.

6. CONCLUDING REMARKS

TBIs are classified from mild to severe based on degree of injuries and its consequences. However, their complexity in diagnosis creates difficulties for clinicians to offer proper treatments. Therefore, future studies are warranted to understand more specific neural and molecular mechanisms that can pinpoint the severity of TBI. We summarize here recent findings about the potential of myriad epigenetic mechanisms including DNA methylation and histone PTMs, mitochondrial epigenetics, and miRNAs that have been implicated in diverse neuropsychiatric consequences induced by TBI. Although various preclinical models have been used to simulate the clinical situations of TBI, results from these studies often complicate the understanding about manifestations of trauma. Nonetheless, some of the common outcomes of TBI include long-term deficits in locomotion, motor coordination, attention, learning and memory, and emotions. Histone modifications and DNA methylation are found to be altered in brain areas relevant to these neural functions (e.g., amygdala, hippocampus, and cortex). However, the effects of trauma in many regions of the brain such as hypothalamus, ventral tegmental area, and nucleus accumbens, relevant to other brain functions, for example, reward and feeding, are yet to be explored.

The studies on epigenetic aspects of TBI are variable in terms of modifications (acetylation, methylation), location of modifications (H3-K9, H3-14), targets (BDNF, Aanat), brain areas (cortex, hippocampus), and neuropsychiatric outcomes (anxiety, depression, learning, memory). In these multitudes of complexities, the use of different models for TBI adds to the caveats in drawing precise conclusions. Although some mechanistic studies provide relevant clues to the epigenetic marks and their downstream targets, lacunas are still existed in using currently available information in furthering the studies for the development of targets from therapeutic and diagnostic standpoint. In this disarray, some encouragement can be garnered from the studies showing the potential of HDAC inhibitors in normalizing the deleterious effects of TBI on concurrent molecular and behavioral deficits. In terms of DNA methylation, 5-azacytidine reverses BDNF expression in the amygdala concomitantly normalizing anxiety in TBI-induced animals. Another intriguing candidate is miR-21, which shows overlapping observations in two different TBI models. HDAC and DNMT inhibitors utilized in these studies are nonspecific. In view of various isoforms of HDACs and DNMTs with cell- and region-specific expression in the brain, the use of isoform-targeted drugs or molecular strategies could provide better insights in establishing the underlying causes of TBI-induced pathologies. In addition, the use of peripheral routes of administration of these drugs do not provide correct information on amount of the drug and the location of actions in the brain. More specifically, drugs and manipulation strategies may help in circumventing these shortcomings and provide concrete clues toward the mechanisms and targets. Nonetheless, available data provide important clues toward development of possible drugs and biomarkers for the treatment of TBI-induced pathological outcomes.

ACKNOWLEDGMENTS

The preparation of the manuscript is supported by University Grant Commission (UGC), Government of India, New Delhi, India through UGC-FRP program to AS. NS is grateful to UGC (India) for the award of junior research fellowship. The authors also acknowledge the Department Research and Development Program (DRDP), Department of Biotechnology, Savitribai Phule Pune University, Pune, India.

REFERENCES

1. Hyder AA, Wunderlich CA, Puvanachandra P, Gururaj G, Kobusingye OC. The impact of traumatic brain injuries: a global perspective. *NeuroRehabilitation*. 2007;22:341–353.
2. Li M, Zhao Z, Yu G, Zhang J. Epidemiology of traumatic brain injury over the world: a systematic review. *Gen Med*. 2016;4:e275.
3. Fischer H. *US Military Casualty Statistics: Operation New Dawn, Operation Iraqi Freedom, and Operation Enduring Freedom*. Washington, DC: Congressional Research Service, Library of Congress; 2013.
4. Lye TC, Shores EA. Traumatic brain injury as a risk factor for Alzheimer's disease: a review. *Neuropsychol Rev*. 2000;10:115–129.

5. McKee AC, Cantu RC, Nowinski CJ, Hedley-Whyte ET, Gavett BE, Budson AE, Santini VE, Lee HS, Kubilus CA, Stern RA. Chronic traumatic encephalopathy in athletes: progressive tauopathy after repetitive head injury. *J Neuropathol Exp Neurol.* 2009;68:709–735.
6. Hutson CB, Lazo CR, Mortazavi F, Giza CC, Hovda D, Chesselet MF. Traumatic brain injury in adult rats causes progressive nigrostriatal dopaminergic cell loss and enhanced vulnerability to the pesticide paraquat. *J Neurotrauma.* 2011;28:1783–1801.
7. Elder J, Cortes M, Rykman A, Hill J, Karuppagounder S, Edwards D, Ratan RR. The epigenetics of stroke recovery and rehabilitation: from polycomb to histone deacetylases. *Neurotherapeutics.* 2013;10:808–816.
8. Felling RJ, Song H. Epigenetic mechanisms of neuroplasticity and the implications for stroke recovery. *Exp Neurol.* 2015;268:37–45.
9. Qureshi IA, Mehler MF. Emerging role of epigenetics in stroke: part 1: DNA methylation and chromatin modifications. *Arch Neurol.* 2010;67:1316–1322.
10. York EM, Petit A, Roskams AJ. Epigenetics of neural repair following spinal cord injury. *Neurotherapeutics.* 2013;10:757–770.
11. Woldemichael BT, Bohacek J, Gapp K, Mansuy IM. Epigenetics of memory and plasticity. *Prog Mol Biol Transl Sci.* 2014;122:305–340.
12. Teasdale G, Jennett B. Assessment of coma and impaired consciousness: a practical scale. *Lancet.* 1974;304:81–84.
13. Lau B, Lovell MR, Collins MW, Pardini J. Neurocognitive and symptom predictors of recovery in high school athletes. *Clin J Sport Med.* 2009;19:216–221.
14. Lovell MR, Collins MW, Iverson GL, Field M, Maroon JC, Cantu R, Podell K, Powell JW, Belza M, Fu FH. Recovery from mild concussion in high school athletes. *J Neurosurg.* 2003;98:296–301.
15. Lovell MR, Collins MW, Iverson GL, Johnston KM, Bradley JP. Grade 1 or "ding" concussions in high school athletes. *Am J Sports Med.* 2004;32:47–54.
16. Cantu RC. Posttraumatic retrograde and anterograde amnesia: pathophysiology and implications in grading and safe return to play. *J Athl Train.* 2001;36:244.
17. Forde CT, Karri SK, Young AM, Ogilvy CS. Predictive markers in traumatic brain injury: opportunities for a serum biosignature. *Br J Neurosurg.* 2014;28:8–15.
18. Coronado VG, Xu L, Basavaraju SV, McGuire LC, Wald MM, Faul MD, Guzman BR, Hemphill JD. Surveillance for traumatic brain injury-related deaths: United States, 1997-2007. *MMWR Surveill Summ.* 2011;60:1–32.
19. Eucker SA, Smith C, Ralston J, Friess SH, Margulies SS. Physiological and histopathological responses following closed rotational head injury depend on direction of head motion. *Exp Neurol.* 2011;227:79–88.
20. Namjoshi DR, Good C, Cheng WH, Panenka W, Richards D, Cripton PA, Wellington CL. Towards clinical management of traumatic brain injury: a review of models and mechanisms from a biomechanical perspective. *Dis Model Mech.* 2013;6:1325–1338.
21. Marshall LF, Gautille T, Klauber MR, Eisenberg HM, Jane JA, Luerssen TG, Marmarou A, Foulkes MA. The outcome of severe closed head injury. *J Neurosurg.* 1991;75:S28–S36.
22. Barkhoudarian G, Hovda DA, Giza CC. The molecular pathophysiology of concussive brain injury. *Clin Sports Med.* 2011;30:33–48.
23. Bullock R, Zauner A, Myseros JS, Marmarou A, Woodward JJ, Young HF. Evidence for prolonged release of excitatory amino acids in severe human head trauma. *Ann N Y Acad Sci.* 1995;765:290–297.
24. Bullock R, Zauner A, Woodward JJ, Myseros J, Choi SC, Ward JD, Marmarou A, Young HF. Factors affecting excitatory amino acid release following severe human head injury. *J Neurosurg.* 1998;89:507–518.

25. Greig NH, Tweedie D, Rachmany L, Li Y, Rubovitch V, Schreiber S, Chiang YH, Hoffer BJ, Miller J, Lahiri DK, Sambamurti K. Incretin mimetics as pharmacologic tools to elucidate and as a new drug strategy to treat traumatic brain injury. *Alzheimers Dement.* 2014;10:S62–S75.
26. Greve MW, Zink BJ. Pathophysiology of traumatic brain injury. *Mt Sinai J Med.* 2009;76:97–104.
27. Maas AI, Stocchetti N, Bullock R. Moderate and severe traumatic brain injury in adults. *Lancet Neurol.* 2008;7:728–741.
28. Merlo L, Cimino F, Angileri FF, La Torre D, Conti A, Cardali SM, Saija A, Germano A. Alteration in synaptic junction proteins following traumatic brain injury. *J Neurotrauma.* 2014;31:1375–1385.
29. Raghupathi R. Cell death mechanisms following traumatic brain injury. *Brain Pathol.* 2004;14:215–222.
30. Hughes V. The sins of the father. *Nature.* 2014;507:522.
31. Lister R, Pelizzola M, Dowen RH, Hawkins RD, Hon G, Tonti-Filippini J, Nery JR, Lee L, Ye Z, Ngo QM, Edsall L. Human DNA methylomes at base resolution show widespread epigenomic differences. *Nature.* 2009;462:315.
32. Ramsahoye BH, Biniszkiewicz D, Lyko F, Clark V, Bird AP, Jaenisch R. Non-CpG methylation is prevalent in embryonic stem cells and may be mediated by DNA methyltransferase 3a. *Proc Natl Acad Sci.* 2000;97:5237–5242.
33. Xie W, Barr CL, Kim A, Yue F, Lee AY, Eubanks J, Dempster EL, Ren B. Base-resolution analyses of sequence and parent-of-origin dependent DNA methylation in the mouse genome. *Cell.* 2012;148:816–831.
34. Feng J, Chang H, Li E, Fan G. Dynamic expression of de novo DNA methyltransferases Dnmt3a and Dnmt3b in the central nervous system. *J Neurosci Res.* 2005;79:734–746.
35. Goto K, Numata M, Komura JI, Ono T, Bestor TH, Kondo H. Expression of DNA methyltransferase gene in mature and immature neurons as well as proliferating cells in mice. *Differentiation.* 1994;56:39–44.
36. Okano M, Bell DW, Haber DA, Li E. DNA methyltransferases Dnmt3a and Dnmt3b are essential for de novo methylation and mammalian development. *Cell.* 1999;99: 247–257.
37. Watanabe D, Uchiyama K, Hanaoka K. Transition of mouse de novo methyltransferases expression from Dnmt3b to Dnmt3a during neural progenitor cell development. *Neuroscience.* 2006;142:727–737.
38. Aapola U, Shibuya K, Scott HS, Ollila J, Vihinen M, Heino M, Shintani A, Kawasaki K, Minoshima S, Krohn K, Antonarakis SE. Isolation and initial characterization of a novel zinc finger gene, DNMT3L, on 21q22. 3, related to the cytosine-5-methyltransferase 3 gene family. *Genomics.* 2000;65:293–298.
39. Aapola U, Lyle R, Krohn K, Antonarakis SE, Peterson P. Isolation and initial characterization of the mouse Dnmt3l gene. *Cytogenet Cell Genet.* 2001;92:122–126.
40. Hata K, Okano M, Lei H, Li E. Dnmt3L cooperates with the Dnmt3 family of de novo DNA methyltransferases to establish maternal imprints in mice. *Development.* 2002;129:1983–1993.
41. Jia D, Jurkowska RZ, Zhang X, Jeltsch A, Cheng X. Structure of Dnmt3a bound to Dnmt3L suggests a model for de novo DNA methylation. *Nature.* 2007;449:248.
42. Suetake I, Shinozaki F, Miyagawa J, Takeshima H, Tajima S. DNMT3L stimulates the DNA methylation activity of Dnmt3a and Dnmt3b through a direct interaction. *J Biol Chem.* 2004;279:27816–27823.
43. Chestnut BA, Chang Q, Price A, Lesuisse C, Wong M, Martin LJ. Epigenetic regulation of motor neuron cell death through DNA methylation. *J Neurosci.* 2011;31: 16619–16636.

44. Ito S, D'Alessio AC, Taranova OV, Hong K, Sowers LC, Zhang Y. Role of Tet proteins in 5mC to 5hmC conversion, ES cell self-renewal, and ICM specification. *Nature.* 2010;466:1129.
45. Ito S, Shen L, Dai Q, Wu SC, Collins LB, Swenberg JA, He C, Zhang Y. Tet proteins can convert 5-methylcytosine to 5-formylcytosine and 5-carboxylcytosine. *Science.* 2011;333:1300–1303.
46. Tahiliani M, Koh KP, Shen Y, Pastor WA, Bandukwala H, Brudno Y, Agarwal S, Iyer LM, Liu DR, Aravind L, Rao A. Conversion of 5-methylcytosine to 5-hydroxymethylcytosine in mammalian DNA by MLL partner TET1. *Science.* 2009;324:930–935.
47. Cortellino S, Xu J, Sannai M, Moore R, Caretti E, Cigliano A, Le Coz M, Devarajan K, Wessels A, Soprano D, Abramowitz LK. Thymine DNA glycosylase is essential for active DNA demethylation by linked deamination-base excision repair. *Cell.* 2011;146:67–79.
48. Guo JU, Su Y, Zhong C, Ming GL, Song H. Hydroxylation of 5-methylcytosine by TET1 promotes active DNA demethylation in the adult brain. *Cell.* 2011;145:423–434.
49. Fornace AJ, Alamo I, Hollander MC. DNA damage-inducible transcripts in mammalian cells. *Proc Natl Acad Sci.* 1988;85:8800–8804.
50. Niehrs C, Schäfer A. Active DNA demethylation by Gadd45 and DNA repair. *Trends Cell Biol.* 2012;22:220–227.
51. Li Z, Gu TP, Weber AR, Shen JZ, Li BZ, Xie ZG, Yin R, Guo F, Liu X, Tang F, Wang H. Gadd45a promotes DNA demethylation through TDG. *Nucleic Acids Res.* 2015;43:3986–3997.
52. Ma DK, Jang MH, Guo JU, Kitabatake Y, Chang ML, Pow-Anpongkul N, Flavell RA, Lu B, Ming GL, Song H. Neuronal activity-induced Gadd45b promotes epigenetic DNA demethylation and adult neurogenesis. *Science.* 2009;323:1074–1077.
53. Shein NA, Grigoriadis N, Alexandrovich AG, Simeonidou C, Lourbopoulos A, Polyzoidou E, Trembovler V, Mascagni P, Dinarello CA, Shohami E. Histone deacetylase inhibitor ITF2357 is neuroprotective, improves functional recovery, and induces glial apoptosis following experimental traumatic brain injury. *FASEB J.* 2009;23:4266–4275.
54. Dash PK, Orsi SA, Moore AN. Histone deactylase inhibition combined with behavioral therapy enhances learning and memory following traumatic brain injury. *Neuroscience.* 2009;163:1–8.
55. Yu F, Wang Z, Tanaka M, Chiu CT, Leeds P, Zhang Y, Chuang DM. Posttrauma cotreatment with lithium and valproate: reduction of lesion volume, attenuation of blood-brain barrier disruption, and improvement in motor coordination in mice with traumatic brain injury. *J Neurosurg.* 2013;119:766–773.
56. Wang G, Jiang X, Pu H, Zhang W, An C, Hu X, Liou AK, Leak RK, Gao Y, Chen J. Scriptaid, a novel histone deacetylase inhibitor, protects against traumatic brain injury via modulation of PTEN and AKT pathway. *Neurotherapeutics.* 2013;10:124–142.
57. Schober ME, Ke X, Xing B, Block BP, Requena DF, McKnight R, Lane RH. Traumatic brain injury increased IGF-1B mRNA and altered IGF-1 exon 5 and promoter region epigenetic characteristics in the rat pup hippocampus. *J Neurotrauma.* 2012;29:2075–2085.
58. Dash PK, Orsi SA, Zhang M, Grill RJ, Pati S, Zhao J, Moore AN. Valproate administered after traumatic brain injury provides neuroprotection and improves cognitive function in rats. *PLoS One.* 2010;5:e11383.
59. Gao WM, Chadha MS, Kline AE, Clark RS, Kochanek PM, Dixon CE, Jenkins LW. Immunohistochemical analysis of histone H3 acetylation and methylation-evidence for altered epigenetic signaling following traumatic brain injury in immature rats. *Brain Res.* 2006;1070:31–34.

60. Tai YT, Lee WY, Lee FP, Lin TJ, Shih CL, Wang JY, Chiu WT, Hung KS. Low dose of valproate improves motor function after traumatic brain injury. *BioMed Res Int.* 2014;2014:980657.
61. Cao P, Liang Y, Gao X, Zhao MG, Liang GB. Retracted: administration of MS-275 improves cognitive performance and reduces cell death following traumatic brain injury in rats. *CNS Neurosci Ther.* 2013;19:337–345.
62. Wang Y, Neumann M, Hansen K, Hong SM, Kim S, Noble-Haeusslein LJ, Liu J. Fluoxetine increases hippocampal neurogenesis and induces epigenetic factors but does not improve functional recovery after traumatic brain injury. *J Neurotrauma.* 2011;28:259–268.
63. Bailey ZS, Grinter MB, Campos DD, VandeVord PJ. Blast induced neurotrauma causes overpressure dependent changes to the DNA methylation equilibrium. *Neurosci Lett.* 2015;604:119–123.
64. Haghighi F, Ge Y, Chen S, Xin Y, Umali MU, De Gasperi R, Gama Sosa MA, Ahlers ST, Elder GA. Neuronal DNA methylation profiling of blast-related traumatic brain injury. *J Neurotrauma.* 2015;32:1200–1209.
65. Lundberg J, Karimi M, von Gertten C, Holmin S, Ekström TJ, Sandberg-Nordqvist AC. Traumatic brain injury induces relocalization of DNA-methyltransferase 1. *Neurosci Lett.* 2009;457:8–11.
66. Zhang ZY, Zhang Z, Fauser U, Schluesener HJ. Global hypomethylation defines a subpopulation of reactive microglia/macrophages in experimental traumatic brain injury. *Neurosci Lett.* 2007;429:1–6.
67. Sagarkar S, Bhamburkar T, Shelkar G, Choudhary A, Kokare DM, Sakharkar AJ. Minimal traumatic brain injury causes persistent changes in DNA methylation at BDNF gene promoters in rat amygdala: a possible role in anxiety-like behaviors. *Neurobiol Dis.* 2017;106:101–109.
68. Dębski KJ, Pitkanen A, Puhakka N, Bot AM, Khurana I, Harikrishnan KN, Ziemann M, Kaspi A, El-Osta A, Lukasiuk K, Kobow K. Etiology matters—genomic DNA methylation patterns in three rat models of acquired epilepsy. *Sci Rep.* 2016;9(6):25668.
69. Meng Q, Zhuang Y, Ying Z, Agrawal R, Yang X, Gomez-Pinilla F. Traumatic brain injury induces genome-wide transcriptomic, methylomic, and network perturbations in brain and blood predicting neurological disorders. *EBioMedicine.* 2017;16:184–194.
70. Bandeira F, Lent R, Herculano-Houzel S. Changing numbers of neuronal and non-neuronal cells underlie postnatal brain growth in the rat. *Proc Natl Acad Sci.* 2009;106:14108–14113.
71. Shekleton JA, Parcell DL, Redman JR, Phipps-Nelson J, Ponsford JL, Rajaratnam SM. Sleep disturbance and melatonin levels following traumatic brain injury. *Neurology.* 2010;74:1732–1738.
72. Soria V, Martínez-Amorós Ègae, Escaramís G, Valero J, Crespo JM, Gutiérrez-Zotes A, Bayés M, Martorell L, Vilella E, Estivill X, Menchón JM. Resequencing and association analysis of arylalkylamine N-acetyltransferase (AANAT) gene and its contribution to major depression susceptibility. *J Pineal Res.* 2010;49:35–44.
73. Zheng W, Cole PA. Serotonin N-acetyltransferase: mechanism and inhibition. *Curr Med Chem.* 2002;9:1187–1199.
74. Fan G, Martinowich K, Chin MH, He F, Fouse SD, Hutnick L, Hattori D, Ge W, Shen Y, Wu H, ten Hoeve J. DNA methylation controls the timing of astrogliogenesis through regulation of JAK-STAT signaling. *Development.* 2005;132:3345–3356.
75. Karpova NN. Role of BDNF epigenetics in activity-dependent neuronal plasticity. *Neuropharmacology.* 2014;76:709–718.
76. Mitchelmore C, Gede L. Brain derived neurotrophic factor: epigenetic regulation in psychiatric disorders. *Brain Res.* 2014;1586:162–172.

77. Roth TL, Sweatt JD. Epigenetic marking of the BDNF gene by early-life adverse experiences. *Horm Behav.* 2011;59:315–320.
78. Roth TL, Zoladz PR, Sweatt JD, Diamond DM. Epigenetic modification of hippocampal BDNF DNA in adult rats in an animal model of post-traumatic stress disorder. *J Psychiatric Res.* 2011;45:919–926.
79. Aihara N, Hall JJ, Pitts LH, Fukuda K, Noble LJ. Altered immunoexpression of microglia and macrophages after mild head injury. *J Neurotrauma.* 1995;12:53–63.
80. Kreutzberg GW. Microglia: a sensor for pathological events in the CNS. *Trends Neurosci.* 1996;19:312–318.
81. Kempermann G, Neumann H. Microglia: the enemy within? *Science.* 2003;302: 1689–1690.
82. Carmichael ST. Gene expression changes after focal stroke, traumatic brain and spinal cord injuries. *Curr Opin Neurol.* 2003;16:699–704.
83. Duan H, Hao C, Fan Y, Wang H, Liu Y, Hao J, Xu C, Liu X, Zhang H. The role of neuropeptide Y and aquaporin 4 in the pathogenesis of intestinal dysfunction caused by traumatic brain injury. *J Surg Res.* 2013;184:1006–1012.
84 Olsen AB, Hetz RA, Xue H, Aroom KR, Bhattarai D, Johnson E, Bedi S, Cox CS, Uray K. Effects of traumatic brain injury on intestinal contractility. *Neurogastroenterol Motil.* 2013;25:593.
85. Sagarkar S, Mahajan S, Choudhary AG, Borkar CD, Kokare DM, Sakharkar AJ. Traumatic stress-induced persistent changes in DNA methylation regulate neuropeptide Y expression in rat jejunum. *Neurogastroenterol Motil.* 2017;29:e13074.
86. Lin HC. Small intestinal bacterial overgrowth: a framework for understanding irritable bowel syndrome. *J Am Med Assoc.* 2004;292:852–858.
87. Zhang H, Yan Y, Shi R, Lin Z, Wang M, Lin L. Correlation of gut hormones with irritable bowel syndrome. *Digestion.* 2008;78:72–76.
88. Hubel KA, Renquist KS. Effect of neuropeptide Y on ion transport by the rabbit ileum. *J Pharmacol Exp Ther.* 1986;238:167–169.
89. Saria A, Beubler E. Neuropeptide Y (NPY) and peptide YY (PYY) inhibit prostaglandin E2-induced intestinal fluid and electrolyte secretion in the rat jejunum in vivo. *Eur J Pharmacol.* 1985;119:47–52.
90. Masumoto H, Hawke D, Kobayashi R, Verreault A. A role for cell-cycle-regulated histone H3 lysine 56 acetylation in the DNA damage response. *Nature.* 2005;436:294–298.
91. Blander G, Guarente L. The Sir2 family of protein deacetylases. *Annu Rev Biochem.* 2004;73:417–435.
92. Haberland M, Montgomery RL, Olson EN. The many roles of histone deacetylases in development and physiology: implications for disease and therapy. *Nat Rev Genet.* 2009;10:32–42.
93. Kouzarides T. Chromatin modifications and their function. *Cell.* 2007;128:693–705.
94. Gupta S, Kim SY, Artis S, Molfese DL, Schumacher A, Sweatt JD, Paylor RE, Lubin FD. Histone methylation regulates memory formation. *J Neurosci.* 2010;30: 3589–3599.
95. Chang B, Chen Y, Zhao Y, Bruick RK. JMJD6 is a histone arginine demethylase. *Science.* 2007;318:444–447.
96. Whetstine JR, Nottke A, Lan F, Huarte M, Smolikov S, Chen Z, Spooner E, Li E, Zhang G, Colaiacovo M, Shi Y. Reversal of histone lysine trimethylation by the JMJD2 family of histone demethylases. *Cell.* 2006;125:467–481.
97. Hahnen E, Hauke J, Tränkle C, Eyüpoglu IY, Wirth B, Blümcke I. Histone deacetylase inhibitors: possible implications for neurodegenerative disorders. *Expert Opin Investig Drugs.* 2008;17:169–184.

98. Chodavarapu RK, Feng S, Bernatavichute YV, Chen PY, Stroud H, Yu Y, Hetzel J, Kuo F, Kim J, Cokus SJ, Casero D. Relationship between nucleosome positioning and DNA methylation. *Nature.* 2010;466:388.
99. Lyko F, Foret S, Kucharski R, Wolf S, Falckenhayn C, Maleszka R. The honey bee epigenomes: differential methylation of brain DNA in queens and workers. *PLoS Biol.* 2010;8:e1000506.
100. Jin B, Li Y, Robertson KD. DNA methylation: superior or subordinate in the epigenetic hierarchy? *Genes Cancer.* 2011;2:607–617.
101. Landgrave-Gómez J, Mercado-Gómez O, Guevara-Guzmán R. Epigenetic mechanisms in neurological and neurodegenerative diseases. *Front Cell Neurosci.* 2015;9.
102. Carthew RW, Sontheimer EJ. Origins and mechanisms of miRNAs and siRNAs. *Cell.* 2009;136:642–655.
103. Qureshi IA, Mehler MF. Emerging roles of non-coding RNAs in brain evolution, development, plasticity and disease. *Nat Rev Neurosci.* 2012;13:528.
104. Salta E, De Strooper B. Non-coding RNAs with essential roles in neurodegenerative disorders. *Lancet Neurol.* 2012;11:189–200.
105. Gurtan AM, Sharp PA. The role of miRNAs in regulating gene expression networks. *J Mol Biol.* 2013;425:3582–3600.
106. Meza-Sosa KF, Valle-García D, Pedraza-Alva G, Pérez-Martínez L. Role of microRNAs in central nervous system development and pathology. *J Neurosci Res.* 2012;90:1–2.
107. Valiyaveettil M, Alamneh YA, Miller SA, Hammamieh R, Arun P, Wang Y, Wei Y, Oguntayo S, Long JB, Nambiar MP. Modulation of cholinergic pathways and inflammatory mediators in blast-induced traumatic brain injury. *Chem Biol Interact.* 2013;203:371–375.
108. Chandran R, Sharma A, Bhomia M, Balakathiresan NS, Knollmann-Ritschel BE, Maheshwari RK. Differential expression of microRNAs in the brains of mice subjected to increasing grade of mild traumatic brain injury. *Brain Inj.* 2017;31:106–119.
109. Sharma A, Chandran R, Barry ES, Bhomia M, Hutchison MA, Balakathiresan NS, Grunberg NE, Maheshwari RK. Identification of serum microRNA signatures for diagnosis of mild traumatic brain injury in a closed head injury model. *PLoS One.* 2014;9:e112019.
110. Sandhir R, Gregory E, Berman NE. Differential response of miRNA-21 and its targets after traumatic brain injury in aging mice. *Neurochem Int.* 2014;78:117–121.
111. Sabirzhanov B, Zhao Z, Stoica BA, Loane DJ, Wu J, Borroto C, Dorsey SG, Faden AI. Downregulation of miR-23a and miR-27a following experimental traumatic brain injury induces neuronal cell death through activation of proapoptotic Bcl-2 proteins. *J Neurosci.* 2014;34:10055–10071.
112. Harrison EB, Hochfelder CG, Lamberty BG, Meays BM, Morsey BM, Kelso ML, Fox HS, Yelamanchili SV. Traumatic brain injury increases levels of miR-21 in extracellular vesicles: implications for neuroinflammation. *FEBS J.* 2016;6:835–846.
113. Hu T, Zhou FJ, Chang YF, Li YS, Liu GC, Hong Y, Chen HL, Xiyang YB, Bao TH. miR21 is associated with the cognitive improvement following voluntary running wheel exercise in TBI mice. *J Mol Neurosci.* 2015;57:114–122.
114. Miao W, Bao TH, Han JH, Yin M, Yan Y, Wang WW, Zhu YH. Voluntary exercise prior to traumatic brain injury alters miRNA expression in the injured mouse cerebral cortex. *Braz J Med Biol Res.* 2015;48:433–439.
115. Balakathiresan N, Bhomia M, Chandran R, Chavko M, McCarron RM, Maheshwari RK. MicroRNA let-7i is a promising serum biomarker for blast-induced traumatic brain injury. *J Neurotrauma.* 2012;29:1379–1387.
116. Sun J, Liu Y, Tran J, O'Neal P, Baudry M, Bi X. mTORC1-S6K1 inhibition or mTORC2 activation improves hippocampal synaptic plasticity and learning in Angelman syndrome mice. *Cell Mol Life Sci.* 2016;73:4303–4314.

117. Sun TY, Chen XR, Liu ZL, Zhao LL, Jiang YX, Qu GQ, Wang RS, Huang SZ, Liu L. Expression profiling of microRNAs in hippocampus of rats following traumatic brain injury. *J Huazhong Univ Sci Technol Med Sci.* 2014;34:548–553.
118. Redell JB, Liu Y, Dash PK. Traumatic brain injury alters expression of hippocampal microRNAs: potential regulators of multiple pathophysiological processes. *J Neurosci Res.* 2009;87:1435–1448.
119. Redell JB, Zhao J, Dash PK. Altered expression of miRNA-21 and its targets in the hippocampus after traumatic brain injury. *J Neurosci Res.* 2011;89:212–221.
120. Hu Z, Yu D, Almeida-Suhett C, Tu K, Marini AM, Eiden L, Braga MF, Zhu J, Li Z. Expression of miRNAs and their cooperative regulation of the pathophysiology in traumatic brain injury. *PLoS One.* 2012;7:e39357.
121. Bao TH, Miao W, Han JH, Yin M, Yan Y, Wang WW, Zhu YH. Spontaneous running wheel improves cognitive functions of mouse associated with miRNA expressional alteration in hippocampus following traumatic brain injury. *J Mol Neurosci.* 2014;54:622–629.
122. Ge XT, Lei P, Wang HC, Zhang AL, Han ZL, Chen X, Li SH, Jiang RC, Kang CS, Zhang JN. miR-21 improves the neurological outcome after traumatic brain injury in rats. *Sci Rep.* 2014;4.
123. Ge X, Han Z, Chen F, Wang H, Zhang B, Jiang R, Lei P, Zhang J. miR-21 alleviates secondary blood–brain barrier damage after traumatic brain injury in rats. *Brain Res.* 2015;1603:150–157.
124. Lei P, Li Y, Chen X, Yang S, Zhang J. Microarray based analysis of microRNA expression in rat cerebral cortex after traumatic brain injury. *Brain Res.* 2009;1284: 191–201.
125. Puhakka N, Bot AM, Vuokila N, Debski KJ, Lukasiuk K, Pitkänen A. Chronically dysregulated NOTCH1 interactome in the dentate gyrus after traumatic brain injury. *PLoS One.* 2017;12:e0172521.
126. Yang T, Song J, Bu X, Wang C, Wu J, Cai J, Wan S, Fan C, Zhang C, Wang J. Elevated serum miR-93, miR-191, and miR-499 are noninvasive biomarkers for the presence and progression of traumatic brain injury. *J Neurochem.* 2016;137:122–129.
127. Bhomia M, Balakathiresan NS, Wang KK, Papa L, Maheshwari RK. A panel of serum MiRNA biomarkers for the diagnosis of severe to mild traumatic brain injury in humans. *Sci Rep.* 2016;6:28148.
128. Di Pietro V, Ragusa M, Davies D, Su Z, Hazeldine J, Lazzarino G, Hill LJ, Crombie N, Foster M, Purrello M, Logan A. MicroRNAs as novel biomarkers for the diagnosis and prognosis of mild and severe traumatic brain injury. *J Neurotrauma.* 2017;34:1948–1956.
129. Redell JB, Moore AN, Ward III NH, Hergenroeder GW, Dash PK. Human traumatic brain injury alters plasma microRNA levels. *J Neurotrauma.* 2010;27:2147–2156.
130. Abdul-Muneer PM, Chandra N, Haorah J. Interactions of oxidative stress and neurovascular inflammation in the pathogenesis of traumatic brain injury. *Mol Neurobiol.* 2015;51:966–979.
131. Blennow K, Hardy J, Zetterberg H. The neuropathology and neurobiology of traumatic brain injury. *Neuron.* 2012;76:886–899.
132. Lifshitz J, Sullivan PG, Hovda DA, Wieloch T, McIntosh TK. Mitochondrial damage and dysfunction in traumatic brain injury. *Mitochondrion.* 2004;4:705–713.
133. Sullivan PG, Keller JN, Mattson MP, Scheff SW. Traumatic brain injury alters synaptic homeostasis: implications for impaired mitochondrial and transport function. *J Neurotrauma.* 1998;15:789–798.
134. Cadonic C, Sabbir MG, Albensi BC. Mechanisms of mitochondrial dysfunction in Alzheimer's disease. *Mol Neurobiol.* 2016;53:6078–6090.
135. Moreira PI, Carvalho C, Zhu X, Smith MA, Perry G. Mitochondrial dysfunction is a trigger of Alzheimer's disease pathophysiology. *Biochim Biophys Acta.* 2010;1802:2–10.

136. Santos RX, Correia SC, Wang X, Perry G, Smith MA, Moreira PI, Zhu X. Alzheimer's disease: diverse aspects of mitochondrial malfunctioning. *IntJ ClinExpPathol.* 2010;3:570.
137. Wang X, Wang W, Li L, Perry G, Lee HG, Zhu X. Oxidative stress and mitochondrial dysfunction in Alzheimer's disease. *Biochim Biophys Acta.* 2014;1842:1240–1747.
138. Mandemakers W, Morais VA, De Strooper B. A cell biological perspective on mitochondrial dysfunction in Parkinson disease and other neurodegenerative diseases. *J Cell Sci.* 2007;120:1707–1716.
139. Moon HE, Paek SH. Mitochondrial dysfunction in Parkinson's disease. *Exp Neurobiol.* 2015;24:103–116.
140. Winklhofer KF, Haass C. Mitochondrial dysfunction in Parkinson's disease. *Biochim Biophys Acta.* 2010;1802:29–44.
141. Cozzolino M, Carrì MT. Mitochondrial dysfunction in ALS. *Prog Neurobiol.* 2012;97:54–66.
142. Jiang Z, Wang W, Perry G, Zhu X, Wang X. Mitochondrial dynamic abnormalities in amyotrophic lateral sclerosis. *Transl Neurodegener.* 2015;4:14.
143. Muyderman H, Chen T. Mitochondrial dysfunction in amyotrophic lateral sclerosis—a valid pharmacological target? *BrJ Pharmacol.* 2014;171:2191–2205.
144. Shi P, Gal J, Kwinter DM, Liu X, Zhu H. Mitochondrial dysfunction in amyotrophic lateral sclerosis. *Biochim Biophys Acta.* 2010;1802:45–51.
145. Shulman RG, Rothman DL, Behar KL, Hyder F. Energetic basis of brain activity: implications for neuroimaging. *Trends Neurosci.* 2004;27:489–495.
146. Buono MJ, Kolkhorst FW. Estimating ATP resynthesis during a marathon run: a method to introduce metabolism. *Adv Physiol Educ.* 2001;25:70–71.
147. Shock LS, Thakkar PV, Peterson EJ, Moran RG, Taylor SM. DNA methyltransferase 1, cytosine methylation, and cytosine hydroxymethylation in mammalian mitochondria. *Proc Natl Acad Sci.* 2011;108:3630–3635.
148. Bellizzi D, D'Aquila P, Scafone T, Giordano M, Riso V, Riccio A, Passarino G. The control region of mitochondrial DNA shows an unusual CpG and non-CpG methylation pattern. *DNA Res.* 2013;20:537–547.
149. Dzitoyeva S, Chen H, Manev H. Effect of aging on 5-hydroxymethylcytosine in brain mitochondria. *Neurobiol Aging.* 2012;33:2881–2891.
150. Mishra M, Kowluru RA. Epigenetic modification of mitochondrial DNA in the development of diabetic retinopathy. *Invest Ophthalmol Vis Sci.* 2015;56:5133–5142.
151. Pirola CJ, Gianotti TF, Burgueño AL, Rey-Funes M, Loidl CF, Mallardi P, San Martino J, Castaño GO, Sookoian S. Epigenetic modification of liver mitochondrial DNA is associated with histological severity of nonalcoholic fatty liver disease. *Gut.* 2013;62:1356–1363.
152. Bianchessi V, Vinci MC, Nigro P, Rizzi V, Farina F, Capogrossi MC, Pompilio G, Gualdi V, Lauri A. Methylation profiling by bisulfite sequencing analysis of the mtDNA Non-Coding Region in replicative and senescent Endothelial Cells. *Mitochondrion.* 2016;27:40–47.
153. Feng S, Xiong L, Ji Z, Cheng W, Yang H. Correlation between increased ND2 expression and demethylated displacement loop of mtDNA in colorectal cancer. *Mol Med Rep.* 2012;6:125–130.
154. Zheng LD, Linarelli LE, Liu L, Wall SS, Greenawald MH, Seidel RW, Estabrooks PA, Almeida FA, Cheng Z. Insulin resistance is associated with epigenetic and genetic regulation of mitochondrial DNA in obese humans. *Clin Epigenet.* 2015;7:60.
155. Campbell A, Mrazek J, Karlin S. Genome signature comparisons among prokaryote, plasmid, and mitochondrial DNA. *Proc Natl Acad Sci.* 1999;96:9184–9189.
156. Ghosh S, Sengupta S, Scaria V. Comparative analysis of human mitochondrial methylomes shows distinct patterns of epigenetic regulation in mitochondria. *Mitochondrion.* 2014;18:58–62.

157. Ghosh S, Sengupta S, Scaria V. Hydroxymethyl cytosine marks in the human mitochondrial genome are dynamic in nature. *Mitochondrion.* 2016;27:25–31.
158. Garrido N, Griparic L, Jokitalo E, Wartiovaara J, van der Bliek AM, Spelbrink JN. Composition and dynamics of human mitochondrial nucleoids. *Mol Biol Cell.* 2003;14:1583–1596.
159. Wang Y, Bogenhagen DF. Human mitochondrial DNA nucleoids are linked to protein folding machinery and metabolic enzymes at the mitochondrial inner membrane. *J Biol Chem.* 2006;281:25791–25802.
160. Rebelo AP, Dillon LM, Moraes CT. Mitochondrial DNA transcription regulation and nucleoid organization. *J Inherit Metab Dis.* 2011;34:941–951.
161. Shutt TE, Shadel GS. A compendium of human mitochondrial gene expression machinery with links to disease. *Environ Mol Mutagen.* 2010;51:360–379.
162. Lodeiro MF, Uchida A, Bestwick M, Moustafa IM, Arnold JJ, Shadel GS, Cameron CE. Transcription from the second heavy-strand promoter of human mtDNA is repressed by transcription factor A in vitro. *Proceedings of the National Academy of Sciences.* 2012;109:6513–6518.
163. Psarra AM, Sekeris CE. Nuclear receptors and other nuclear transcription factors in mitochondria: regulatory molecules in a new environment. *Biochim Biophys Acta.* 2008;1783:1–11.
164. Bogenhagen DF, Wang Y, Shen EL, Kobayashi R. Protein components of mitochondrial DNA nucleoids in higher eukaryotes. *Mol Cell Proteomics.* 2003;2:1205–1216.
165. Kukat C, Wurm CA, Spåhr H, Falkenberg M, Larsson NG, Jakobs S. Super-resolution microscopy reveals that mammalian mitochondrial nucleoids have a uniform size and frequently contain a single copy of mtDNA. *Proc Natl Acad Sci.* 2011;108: 13534–13539.
166. Kanki T, Ohgaki K, Gaspari M, Gustafsson CM, Fukuoh A, Sasaki N, Hamasaki N, Kang D. Architectural role of mitochondrial transcription factor A in maintenance of human mitochondrial DNA. *Mol Cell Biol.* 2004;24:9823–9834.
167. Ekstrand MI, Falkenberg M, Rantanen A, Park CB, Gaspari M, Hultenby K, Rustin P, Gustafsson CM, Larsson NG. Mitochondrial transcription factor A regulates mtDNA copy number in mammals. *Hum Mol Genet.* 2004;13:935–944.
168. Dairaghi DJ, Shadel GS, Clayton DA. Addition of a 29 residue carboxyl-terminal tail converts a simple HMG box-containing protein into a transcriptional activator. *J Mol Biol.* 1995;249:11–28.
169. Gangelhoff TA, Mungalachetty PS, Nix JC, Churchill ME. Structural analysis and DNA binding of the HMG domains of the human mitochondrial transcription factor A. *Nucleic Acids Res.* 2009;37:3153–3164.
170. Ngo HB, Kaiser JT, Chan DC. The mitochondrial transcription and packaging factor Tfam imposes a U-turn on mitochondrial DNA. *Nat Struct Mol Biol.* 2011;18:1290–1296.
171. Rubio-Cosials A, Sydow JF, Jiménez-Menéndez N, Fernández-Millán P, Montoya J, Jacobs HT, Coll M, Bernadó P, Solà M. Human mitochondrial transcription factor A induces a U-turn structure in the light strand promoter. *Nat Struct Mol Biol.* 2011;18:1281–1289.
172. Wong TS, Rajagopalan S, Freund SM, Rutherford TJ, Andreeva A, Townsley FM, Petrovich M, Fersht AR. Biophysical characterizations of human mitochondrial transcription factor A and its binding to tumor suppressor p53. *Nucleic Acids Res.* 2009;37:6765–6783.
173. Kaufman BA, Durisic N, Mativetsky JM, Costantino S, Hancock MA, Grutter P, Shoubridge EA. The mitochondrial transcription factor TFAM coordinates the assembly of multiple DNA molecules into nucleoid-like structures. *Mol Biol Cell.* 2007;18:3225–3236.

174. Malarkey CS, Bestwick M, Kuhlwilm JE, Shadel GS, Churchill ME. Transcriptional activation by mitochondrial transcription factor A involves preferential distortion of promoter DNA. *Nucleic Acids Res.* 2011;40:614–624.
175. Kelly DP, Scarpulla RC. Transcriptional regulatory circuits controlling mitochondrial biogenesis and function. *Genes Dev.* 2004;18:357–368.
176. Lin J, Handschin C, Spiegelman BM. Metabolic control through the PGC-1 family of transcription coactivators. *Cell Metab.* 2005;1:361–370.
177. Scarpulla RC. Transcriptional paradigms in mammalian mitochondrial biogenesis and function. *Physiol Rev.* 2008;88:611–638.
178. Huber BR, Meabon JS, Martin TJ, Mourad PD, Bennett R, Kraemer BC, Cernak I, Petrie EC, Emery MJ, Swenson ER, Mayer C. Blast exposure causes early and persistent aberrant phospho-and cleaved-tau expression in a murine model of mild blast-induced traumatic brain injury. *J Alzheimers Dis.* 2013;37:309–323.
179. Wang WX, Visavadiya NP, Pandya JD, Nelson PT, Sullivan PG, Springer JE. Mitochondria-associated microRNAs in rat hippocampus following traumatic brain injury. *Exp Neurol.* 2015;265:84–93.

CHAPTER TWELVE

Nonprotein-coding RNAs in Fetal Alcohol Spectrum Disorders

Amanda H. Mahnke*,, Nihal A. Salem*, Alexander M. Tseng*, Dae D. Chung*, Rajesh C. Miranda*,**,[1]**

* Texas A&M University Health Science Center, Bryan, TX, United States

** Women's Health in Neuroscience Program, Texas A&M University Health Science Center, Bryan, TX, United States

[1] Corresponding author. E-mail address: miranda@medicine.tamhsc.edu

Contents

Progress in Molecular Biology and Translational Science, volume 157
ISSN 1877-1173
http://dx.doi.org/10.1016/bs.pmbts.2017.11.024

Abstract

Early developmental exposure to ethanol, a known teratogen, can result in a range of neurodevelopmental disorders, collectively referred to as Fetal Alcohol Spectrum Disorders (FASDs). Changes in the environment, including exposure to teratogens, can result in long term alterations to the epigenetic landscape of a cell, thereby altering gene expression. Non-coding RNAs (ncRNAs) can affect transcription and translation of networks of genes. ncRNAs are dynamically expressed during development and have been identified as a target of alcohol. ncRNAs therefore make for attractive targets for novel therapeutics to address the developmental deficits associated with FASDs.

1. INTRODUCTION

In 1973, Kenneth Jones and colleagues reported that children who were prenatally exposed to high levels of alcohol display developmental and growth delays, as well as cranial and midface structural abnormalities.[1] These effects of prenatal exposure to alcohol were termed "Fetal Alcohol Syndrome" (FAS).[2] Since then, researchers and clinicians have shown that prenatal alcohol exposure (PAE) can result in a range of structural, neurodevelopmental, and cognitive delays and disorders, collectively termed Fetal Alcohol Spectrum Disorders (FASDs).[3] A recent metaanalysis estimated the worldwide prevalence of FASD at 22.7 per 1000 births,[4] and active case ascertainment studies indicate that between 2% and 5% of US school-aged children are likely to have a FASD.[5] However, PAE, particularly in vulnerable populations, is likely to be underreported.[6,7] Estimates of PAE, based on the presence of a unique alcohol metabolite, phosphatidylethanol (PEth), in infant dried blood spots, suggest that the prevalence of FASD in the US may be higher than estimated. Recent studies in New Mexico[8] and Texas[9] reported local and state-wide incidence of PAE at 6.5% and 8.5%, respectively. Estimates for PAE and FASD are likely to meet or even exceed the prevalence estimates for other common sources of developmental disability like Autism Spectrum Disorders (ASDs).[10] Converging sociocultural factors, including the high prevalence of unplanned pregnancies[11] and binge patterns of alcohol consumption[12] mean that PAE remains difficult to prevent by messaging alone. Consequently, there is a significant need to find ways to intervene, to disconnect PAE from its effects on early development.

The effects of PAE depend on a number of factors, including the timing, amount, and duration of alcohol exposure. Studies have shown that the number of FASD-associated deficits, as well as the severity of these deficits, can be correlated with increased levels of alcohol consumption.[13–15] Some neurobehavioral deficits can occur with very low alcohol consumption, while other neurobehavioral deficits require a threshold of alcohol exposure for impairment, after which the degree of deficit can increase linearly with exposure.[16] Binge-type alcohol consumption, that is, the consumption of four or more standard drinks at one time for women and five or more for men,[17] may be more harmful than other consumption patterns. Binge consumption patterns have been shown to increase the rate of subnormal verbal IQ and delinquent behavior.[17,18] Animal studies have shown that binge consumption can increase cell death across brain regions[19,20] and that binge consumption at an overall lower dose of ethanol can be more damaging than higher doses of ethanol administered over a longer period of time.[21] Additionally, timing is an important component in alcohol exposure.[22,23] Both primate and mouse studies have shown that binge exposure during gastrulation (mouse gestational day (GD) 7, primate GD 19 and 20) most closely recapitulates the facial features of FAS.[24,25] In the mouse, binge ethanol exposure just a single day later no longer recapitulates the smooth philtrum of the FAS face and shows less severe midface hypoplasia.[26,27] These data suggest a critical period of ethanol exposure during which facial development is particularly vulnerable to the teratogenic effects of ethanol. PAE outside of this window may not cause the canonical FAS facial phenotypes, but may result in other structural defects and developmental disabilities.

Interestingly, not all children exposed to prenatal alcohol have similar outcomes and monozygotic twins are more likely to have a concordant FASD diagnosis than dizygotic twins.[28] These types of data suggest that genetic background modifies the effects of PAE. Studies in a South African population found that allelic variations in the alcohol dehydrogenase gene, *ADH1B*, contribute to the risk for FAS,[29] suggesting, not surprisingly, that alcohol bioavailability contributes to the risk for adverse birth outcomes. Evidence from animal models suggests a more complex and nuanced contribution of the genetic landscape to susceptibility to PAE.[30] In these models, loss of function gene mutations that do not result in an identifiable phenotype in heterozygous, unexposed animals, nevertheless, result in severe dysmorphologies following developmental exposure to low levels of alcohol. For example, heterozygous zebrafish mutants for *Vangl2*, a gene locus that controls early embryo patterning, do not exhibit overt midface defects. However, ethanol, at a dose which by itself does not lead to a statistically significant increase in embryo defects, results in severe midface defects in *Vangl2* heterozygous mutant fish.[31] While genetic variation does confer susceptibility to PAE, widespread alterations in gene expression are more likely to be facilitated by epigenetic mechanisms that control the gene transcription and translation landscape. Consequently, the epigenome may constitute a significant linkage between PAE and FASD.

2. EPIGENETICS

The term 'Epigenetics' refers to the modifying effects of life experiences on genome structure and its transcriptional and translational output. Covalent modification of DNA and histones can lead to the condensation of chromatin, and therefore the repression of gene expression, or conversely, to the relaxation of chromatin to allow for, or enhance, gene expression. A newly discovered network of regulatory RNA molecules serves as an additional layer of epigenetic control. Such regulatory RNAs are now known to control gene transcription, by interacting with DNA and chromatin modification machinery, and by regulating gene translation as well. These regulatory, nonprotein-coding RNAs have increasingly been found to mediate alcohol effects, and will be the focus of this chapter.

2.1 DNA Modifications

DNA methylation in eukaryotic organisms occurs principally on cytosine bases, particularly where cytosine and guanine are present in tandem (CpG sites).[32] DNA methylation is facilitated by the methyl donor intermediary, S-adenosyl methionine, which is generated from essential nutrients such as folic acid and choline.[33] CpG methylation is facilitated by a class of enzymes termed DNA methyltransferases (DNMTs). DNMT1, which has high affinity for hemimethylated regions (where one DNA strand is methylated),[34] is responsible for maintaining existing methylation marks, while DNMT3a and DNMT3b place new methylation marks on the DNA.[35] After methylation, through the binding of methyl-binding domain proteins (MBDs), promoter regions of genes are inactivated, resulting in nucleosome remodeling, chromatin condensation, and consequently, loss of gene transcription.[36] As DNMT1 is able to re-establish hemimethylated regions, DNA methylation can be a stable mark that persists through cell division, but can also be dynamically modulated to affect gene expression.[37] Overall, methylation of DNA has been associated with repression of gene expression while hypomethylation has been associated with enhanced gene expression.[38]

2.2 Histone Modifications

Unlike DNA methylation, which is associated with chromatin condensation and repressed gene expression, histones can be posttranslationally modified on their N-terminal positively charged tails to either open or close the chromatin. These marks result in a complex histone modification language, referred to as the "histone code".[39] Acetylation of the histone tail, by histone acetyltransferases (HATs), relaxes the spacing of nucleosomes to allow for more open chromatin.[40] Acetylation can also be actively removed by histone deacetylases (HDACs), to condense the chromatin.[41] Methylation, on the other hand, can both encourage chromatin relaxation and condensation depending on the location and amount of methylation. Methylation of histone H3 on lysine 4 (H3K4) and single methylation on H3K9 and H3K27 promote open chromatin while trimethylation at H3K9 and H3K27 promote closed chromatin conformations.[42] While acetylation and methylation are commonly studied posttranslational modifications, histones can also be phosphorylated, ubiquitinylated, sumoylated, among others, adding to the overall complexity of the histone code.[43]

2.3 Transgenerational Effects

Epimutations are heritable, nongenomic changes in gene expression and are thought to be due mainly to pervasive alterations to DNA methylation and ncRNAs.[44] How are these parental experiences transmitted to offspring? Research in the fields of pesticides and plasticizers have shown transgenerational transmission of traits through DNA methylation.[45] Additionally, research in animal models has shown that the paternal germ cell (sperm) miRNA (discussed in detail in the forthcoming paragraph) content can be altered by a variety of environmental factors, and such male germ cell-associated modifications may be particularly impactful because sperm miRNA have been shown to influence, and in some cases be required for, zygote development.[46] Early life paternal stress, during postnatal and pubertal periods, as well as adult onset stress, has been shown to change offspring behavior. For example, early life (postnatal) paternal stress results in increased depressive-like behaviors and reduced avoidance behavior in the next generation of males.[47] Similarly, pubertal and adult paternal stress cause blunted corticosterone responses to an acute restraint stress in the next generation, in both male and female offspring.[48] Both of these studies also showed altered miRNA content in the sperm of the males which were subjected to stress. For example, Mansuy and colleagues showed that stress results in significant elevation of a number of miRNAs and piRNAs, including miR-375-3p and -5p, miR-200b-3p, miR-672-5p, and miR-466-5p.[47] Others, Bale and colleagues for example, showed that both pubertal and adult stress increases the expression of miRNAs, including miR-29c, miR-30a, miR-30c, miR-32, miR-193-5p, miR-204, miR-532–3p, miR-698, as well as miR-375-3p.[48] Moreover, both groups went on to show a causal link between the miRNA content in sperm and the following generation's outcomes. The Mansuy group isolated the whole RNA content from sperm and injected it into fertilized oocytes and found that the offspring have the same behavioral phenotype of depressive-like symptoms and reduced avoidance. The Bale group injected their pool of nine miRNAs into a fertilized oocyte and found that the offspring recapitulated the blunted corticosterone response and that this pooled set of miRNAs decreases a number of maternal mRNAs 24 h after injection.[49] These studies provide evidence of parental transmission of traits through germ cell ncRNAs.

2.4 Noncoding RNAs

2.4.1 Small Nucleolar and Small Nuclear RNAs

In 1958, Francis Crick hypothesized that there was a "template" class of RNA, which we now know as mRNA, and a "soluble" RNA, which we now know as transfer RNA (tRNA),[50] suggesting an early recognition that not all RNA molecules were translated into proteins. Over the years, the catalog of this class of nonprotein-coding RNAs (ncRNAs) has expanded, and research has shown that a number of these ncRNA have regulatory functions (Fig. 1).[51] Isolated in nucleoli, small nucleolar RNAs (snoRNAs) recruit protein complexes to process preribosomal RNA (rRNA) and to covalently modify other RNAs, including methylating and pseudouridylating tRNAs, rRNAs, and small nuclear RNAs (snRNA).[52–54] snRNAs, in turn, localize to the nucleus in ribonucleoprotein complexes, to form the spliceosome, which removes introns to create mature mRNA.[55,56]

2.4.2 PIWI-associated RNAs

Some ncRNAs interact directly with chromatin remodeling epigenetic mechanisms. For example, the Piwi family of proteins, which are required

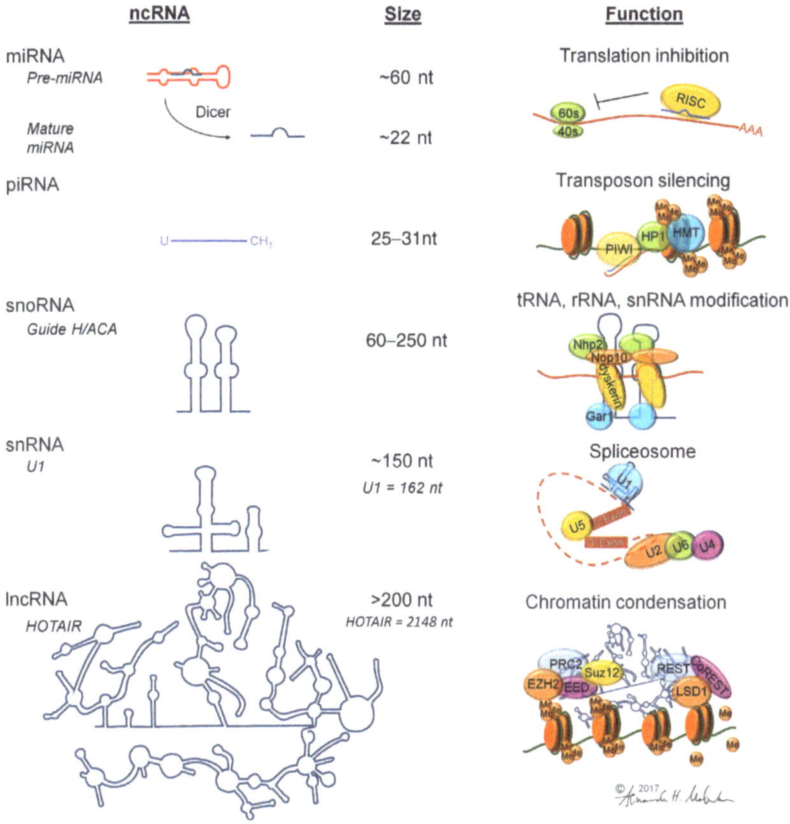

Fig. 1 ***Families of noncoding RNA (ncRNA) and their major intracellular functions.*** microRNAs (miRNAs) are small ncRNA transcripts which associate with the RNA-induced silencing complex (RISC) to inhibit the translation of mRNA. piwi-interacting RNAs (piRNAs) can be modified including a 5′ monophosphate uridine and 3′ methylation. piRNAs bind with the PIWI protein and to nascent transposon transcripts. Subsequently, heterochromatin protein 1 (HP1) and histone methyltransferase (HMT) are recruited, leading to H3K27me3 and chromatin condensation. Small nucleolar RNAs (snoRNAs) process other ncRNA transcripts. Members of the guide H/ACA class of snoRNA associate with Nhp2 and Nop10 to trap the RNA in a pocket of the snoRNA to be pseudouridinylated by dyskerin with Gar1 binding to catalyze the reaction. snRNAs associate with pre-mRNAs and act to cleave introns and join exons. The mature mRNA is formed in the spliceosome where five snRNAs, and associated proteins, interact to catalyze the creation of mature mRNA. The lncRNA *HOTAIR* has an elaborate two-dimensional structure.[57] Within the nucleus, *HOTAIR* assembles with the polycomb repressive complex containing PRC2, EED, Suz12, and EZH2 which catalyzes H3K27me3 and condenses the chromatin. Additionally, the 3′ end of *HOTAIR* can associate with REST/CoREST/LSD1 to demethylate H3K4 which also contributes to chromatin condensation.

for the maintenance of germ cells, utilize a class of ncRNAs, piwi-interacting RNAs (piRNAs), to identify retrotransposons and prevent transposition by condensing the chromatin around the encoded retrotransposon. Epigenetic suppression of retrotransposons is protective, as active transposition has been shown to be damaging for germ cell development.[58,59] PIWI/piRNAs are thought to epigenetically silence transposons through DNA methylation[60,61] and H3K9 trimethylation through recruitment of heterochromatin protein 1 (HP1) and histone methyltransferases (HMTs).[62] More recently, evidence has emerged in support of additional roles for PIWI/piRNA. For example, PIWI/piRNA have been found in mouse mesenchymal stem cells,[63] suggesting a role for PIWI/piRNA in somatic stem cell biology.[64] Functional piRNAs have also been recently identified in adult neurons.[65,66] piRNAs have been implicated in synaptogenesis in mouse hippocampal neurons[67] and in memory consolidation in the central nervous system of the invertebrate sea snail (*Aplysia californica*).[68]

2.4.3 microRNAs

The most well studied epigenetic interactors of the noncoding RNAs are the microRNAs (miRNAs) and the long noncoding RNAs (lncRNAs) (Fig. 2). Such ncRNAs play a special role in the epigenetic landscape. They can act both within the nucleus, to influence covalent DNA and histone epigenetic modifications, and within the cytosol, to affect translation of mRNA. miRNAs are small sequences of RNA, ~22 nucleotides (nt) in length, that regulate the expression of mRNA either by inhibiting translation or triggering mRNA degradation.[69,70] These short regulatory RNAs affect the expression of networks of up to several hundred protein-coding genes.[71,72] miRNAs are constructed from long primary miRNA transcripts (pri-miRNAs), which are cleaved down to shorter stem loop structures known as pre-miRNAs, by the endonuclease, Drosha.[73–75] This pre-miRNA is then exported from the nucleus by Exportin-5[76,77] where it is further processed to mature miRNA by Dicer.[78] Dicer creates a double stranded RNA sequence, one of which is the guide strand, or the mature miRNA, the other is the passenger strand, or miRNA*, which can also act as a mature miRNA or be degraded.[69] In the cytoplasm, the mature miRNAs interact with the argonaute family of proteins to form a RNA-induced silencing complex (RISC), which binds to regions within the 3′-untranslated region (3′-UTR) of target mRNAs[79] to inhibit mRNA translation or cause mRNA destabilization by 5′-decapping or poly-A tail degradation.[80,81] Even though the total level of repression of an individual mRNA transcript may be

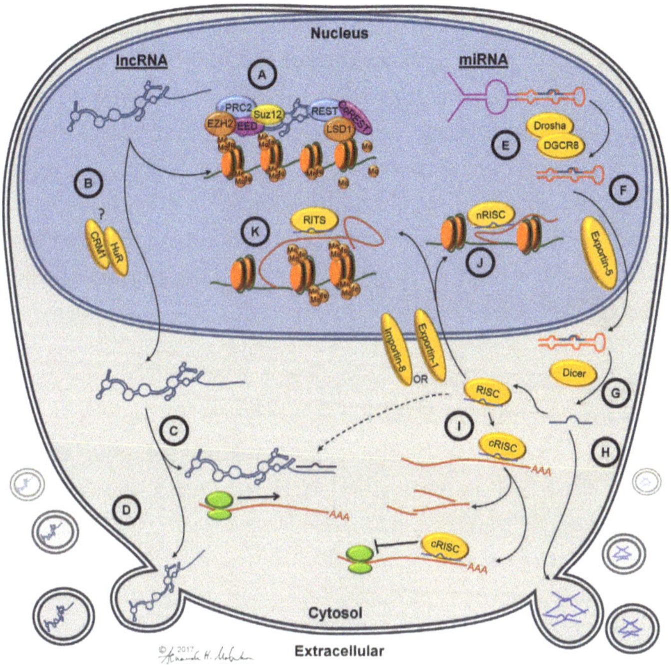

Fig. 2 ***Nuclear, cytosolic, and extracellular function of miRNA and lncRNA.*** After lncRNA is transcribed it can act locally in the nucleus, to trigger chromatin condensation through mechanisms discussed in Fig. 1 (**A**), or it can be exported from the nucleus, perhaps through the action of RNA binding protein human antigen R (HuR/ELAVL1) and the export factor chromosome region maintenance-1 (CRM1) (**B**).[88] In the cytosol, lncRNAs can modulate translation through a number of mechanisms, including acting as a sponge for miRNAs, releasing translational repression of mRNAs (**C**). lncRNAs can also be packaged into extracellular vesicles and transmitted to other cells (**D**). miRNAs are derived from pri-miRNA transcripts which are cleaved by Drosha and DGCR8 into pre-miRNA transcripts (**E**). Pre-miRNAs are then exported to the cytosol, mainly through an exportin-5 mechanism (**F**). In the cytosol, pre-miRNAs are cleaved by Dicer to produce mature miRNAs (**G**). These mature miRNAs then can be packaged into vesicles for extracellular release (**H**). miRNAs can also be loaded into the RISC complex where they can either stay in the cytosol to inhibit the transcription of mRNA (**I**) or be shuttled into the nucleus through exportin-1 or importin-8. In the nucleus, miRNAs can target nascent RNA transcripts, including lncRNAs, to trigger degradation (**J**) or can interact with the RNA-induced transcriptional silencing complex (RITS) to change chromatin conformation (**K**).

modest, it can be sufficient to fine tune the mRNA transcript level to a threshold at which translation is effectively halted.[82] Even when translation is not halted, the ability of miRNAs to target gene networks can result in alteration to overall cellular function.[72] Mature miRNAs have also been found in the nucleus, with some present at a higher concentration in the nucleus than the cytoplasm, and are thought to play a role in regulating gene transcription.[83] Research suggests that miRNAs associated with the RISC complex can also be trafficked back into the nucleus and have been found to associate with the nuclear epigenetic machinery,[84] and to associate with nascent RNA transcripts, including lncRNAs.[85] It is thought that these actions result in silencing of gene transcription, although data from analysis of miR-373[86] and miR-589[87] function suggest that nuclear miRNA/RISC complexes can also activate gene transcription.

2.4.4 Long Noncoding RNAs

The interactions of miRNA with lncRNA are thought to constitute a significant and novel layer of nuclear regulation of gene transcription. lncRNAs are classified as transcribed ncRNA molecules larger than 200 nucleotides in length. lncRNAs can arise from a variety of genomic regions, including within introns, from the antisense strand of protein-coding genes, and from regions between protein-coding genes (sometimes referred to as long intergenic/intervening RNA, lincRNA).[89] Many lncRNAs are tissue-specific, or enriched in specific types of tissues, at specific stages of development, and their dysregulation has been associated with a number of diseases.[90] lncRNAs play a number of roles within cells, including chromatin remodeling, transcriptional regulation, posttranscriptional modification, and translational regulation.[91,92] The lincRNA, *HOTAIR* (*Homo sapiens HOX transcript antisense RNA*), has been shown to lead to transcriptional repression and serves as a model for the determination of function of other lncRNAs.[93] *HOTAIR* is encoded in the antisense orientation within the genomic locus that also encodes a cluster of Homeobox (Hox)-C transcription factors. The temporal expression of the *Hox* gene family, grouped into 4 separate chromosomal clusters, *HOXA* to *HOXD*, is important for the appropriate spatial organization of the embryo body plan.[94] *HOTAIR* is transcribed from an intergenic area of the *HOXC* locus and acts to repress expression at the *HOXD* locus. This repression occurs by the recruitment of the polycomb remodeling complex 2 (PRC2), containing a H3K27 methyltransferase (Ezh2), leading to trimethylation and chromatin condensation at the *HOXD* locus.[95] A number of other lncRNA have been shown to have similar associations with PRC2 to trigger chromatin condensation.[93] More recent evidence has shown that *HOTAIR* can also recruit H3K4 demethylation factors, through interactions with the LSD1/CoREST/REST complex, thereby further contributing to chromatin condensation.[96]

lncRNA can also influence translation by stabilizing or destabilizing mRNA transcripts,[97] through antisense inhibition of mRNA, acting as decoys for miRNA, by modulating alternative splicing of mRNA, or by serving as a source of endogenous small interfering RNAs (siRNAs).[98] The ability of lncRNAs to act as sponges for miRNAs,[99] thereby protecting target mRNA from miRNA-mediated translation repression, is a well-recognized means for posttranscriptional regulation. An added complexity is that sponge lncRNAs may also serve as precursor transcripts for miRNAs. For example, the *H19* lncRNA (which will be discussed in the forthcoming paragraphs in relation to FASD) has been shown to act as a miRNA sponge for let-7.[100] However, *H19* also serves as a precursor for miR-675-3p and -5p,[101] miRNAs which have been shown to mediate some of *H19*'s effects on growth.[102,103] Therefore on one hand, sponge lncRNAs may inhibit RISC function, but when alternatively processed into miRNAs may also facilitate RISC function. Here we will discuss lncRNA as a target of PAE, in the context of lessons that can be learned from research on lncRNA in other neurodevelopmental disorders.

3. A CASE FOR EPIGENETICS AND FASD

Given that PAE can result in a heterogeneity of disabilities and defects across multiple body tissues and organs, it is not surprising that epigenetics has been implicated as having a causal role in FASD.[104–107] Histone modifications and DNA methylation have been found to be dysregulated in response to ethanol exposure.[108] The common alterations to histone modifications after PAE include decreased trimethylation of H3K27 and increased acetylation of H3K9, which relax the chromatin, and increased demethylation of H3K9 and H3K27, which condense the chromatin.[109] For DNA modification there is a similar, contradictory pattern. Overall, PAE in the mouse has been shown to result in global hypomethylation of DNA,[110] and specific hypomethylation of promoters of neural development genes,[111] yet selected gene promoters, such as the promotor for brain-derived neurotrophic factor (BDNF), may show hypermethylation.[112] Data suggest that some of the alterations to methylation are specific[113] and last throughout development[114] while other evidence suggests that ethanol exposure more generally dysregulates the relationship between chromatin remodeling and gene transcription.[115]

The case for epigenetics is particularly strong when considering the role of paternal preconception drinking behavior in contributing to FASDs.[116]

Epidemiological evidence suggests that paternal drinking behavior can contribute to low birth weight,[117] heart defects,[118] as well as cognitive deficits.[119] Animal models have also shown that paternal alcohol exposure contributes to decreased birth weight.[120,121] Paternal alcohol consumption has been shown to result in increased sensitivity to alcohol in offspring during adulthood, including increased anxiolytic properties, decreased alcohol preference, and increased BDNF expression in the ventral tegmental area of the brain in male offspring only.[121,122] Additionally, paternal drinking appears to blunt the corticosterone response to acute restraint stress.[121] In response to moderate or heavy drinking, research has shown reduced DNMT expression and hypomethylation in the sperm of rats and humans, respectively.[120,123] Similarly, research has shown that paternal drinking can cause hypomethylation at the imprinted *H19* locus (discussed further in the forthcoming paragraph),[123,124] although this effect of alcohol can occur in the PAE embryo as well.[125] Additionally, these changes are not just restricted to the F1 generation. PAE-induced hypomethylation of the *H19* locus, has been shown to persist in sperm in the F2 generation,[126] and increased methylation at another locus, encoding the pro-opiomelanocortin (*POMC*) gene have been found into the F3 generation.[127] Alterations to imprinting can have direct results on miRNA and other ncRNA expression. In one study, ~20% of miRNAs that are affected in PAE can be localized to three commonly imprinted regions, *Sfmbt2*, *Snrpn-Ube3a*, and *Dlk1-Dio3*, which also encode a number of other ncRNAs.[128] Therefore paternal preconception drinking may also directly affect ncRNA expression.

Ethanol clearly produces persistent epigenetic changes that can last multiple generations. In this chapter, we further examine how noncoding RNAs may act to contribute to the FASD phenotype, particularly as they are positioned to both interact with the genome and interact with translation. We will also highlight how new data is showing that ncRNAs may prove to be a valuable site of therapeutic intervention.

4. miRNA AND FASD

4.1 Alterations to miRNA Expression in FASD

Ethanol-induced patterns of disruption to miRNAs have been found in the developing brain (Table 1). Because of the widespread pattern of miRNA dysregulation, miRNAs have been suggested to be "master regulators" of

Table 1 Studies Showing Alteration Specifically to the miRNAs Highlighted in This Chapter (Shown in Bold).

Publication	Cells/Tissue	Model	miRNA Regulation
Sathyan *et al.*, 2007	Mouse neural progenitors	PAE	↓**miR-9**, **miR-21**, **miR-153**, miR-335
Pappalardo-Carter *et al.*, 2013	Zebrafish	PAE	↓**miR-9**
Wang *et al.*, 2009	Mouse fetal brain	PAE	↑**miR-9**, **miR-10a**, **miR-10b**, miR-30a-3p, miR145, miR-152 ↓miR-29c, miR-30e-5p, miR-154, miR-200a, miR-296, miR-362, miR-496
Balaraman *et al.*, 2012	Mouse neural progenitors	PAE/prenatal nicotine	↓**miR-9**, **miR-21**, miR-140-3p, **miR-153**, miR-335,
Soares *et al.*, 2012	Zebrafish	PAE	↑let-7k, miR-30d, miR-100, **miR-153a**, miR-725, miR732, miR-738
Gangisetty *et al.*, 2017	Rat adolescent pituitary	PAE	↑**miR-9**, miR-326
Tal *et al.*, 2012	Zebrafish	PAE	↓**miR-9**, **miR-153c**, miR-204
Guo *et al.*, 2012	Mouse cortical neuron culture	PAE	↓**miR-10a**, miR-142-3p, miR-146a, miR-197, miR-200a, miR-202b, miR-223, miR-509-3p, miR-669, miR-680, miR-682, miR-687, miR-692, miR-704, miR-804 ↑miR-122, miR-193, miR-206, miR-485, miR-501-5p
Beech *et al.*, 2014	Human peripheral blood	AUD/stress	↑**miR-10a**, **miR-21**
Nunez *et al.*, 2013[a]		AUD	↑let-7, miR-7, miR-15, miR-17, miR-34,

(Continued)

Publication	Cells/Tissue	Model	miRNA Regulation
	Human PFC and mouse FCtx		miR-101, miR-135, miR-140, miR-144, miR-152 miR-146, miR-301, miR-339, miR-368
Nunez *et al.*, 2013[a]	Mouse FCtx and rat mPFC	AUD	↑**miR-10**, miR-17, miR-29, miR-339, miR-368, ↓(Rat)↑(Mouse) miR-7, **miR-9**, miR-15, miR-26, miR-30, miR-101, miR-130, miR-204, miR-340, miR-434 ↑(Mouse), altered rat, miR-181, miR-467
Tapocik *et al.*, 2013	Rat mPFC	AUD	↑miR-18a, miR-30c-2*, miR-106b*, miR-125a-3p, miR-151*, miR-204*, miR-206, miR-210, miR-218*, miR-324-39, miR-351, miR-378*, miR-503 ↓miR-7b, **miR-9**, miR-22*, miR-26b, miR-29c, miR-101a, miR-101b, miR-137, miR-186, miR-204, miR-218, miR-301b, miR-340-5p, miR-350, miR-374, miR-376b-3p, miR-376-5p, miR-376c, miR-382*, miR-466b
van Steenwyk *et al.*, 2013	Human cell lines (HEK293T, SH SY5Y, 1321 N1)	AUD	↑**miR-153** (chronic intermittent ethanol), miR-7, **miR-153**, miR-152, miR-15B (chronic ethanol withdrawal)
Lewohl *et al.*, 2011[b]	Human PFC	AUD	↑let-7g, let-7f, miR-1, miR-7, miR-203, miR-454, miR-586, miR-301a, miR-144, miR-152, miR-380, miR-374b, miR-101,

(*Continued*)

Publication	Cells/Tissue	Model	miRNA Regulation
			miR-519b-3p, miR-18a, miR-196a, miR-15b, **miR-153**, miR-580, miR-665, miR-376c, miR-194, miR-135b, miR-92a, miR-369-3p

AUD, Alcohol use disorder; PAE, prenatal alcohol exposure.

[a]These data are from a study presenting the miRNA profile from the frontal cortex (FCtx) of adult ethanol-exposed mice. Altered miRNAs depicted in table are those which were shown in both the mouse FCtx and preexisting data from the human prefrontal cortex (PFC)[134] or from the mouse FCtx and rat medial prefrontal cortex (mPFC).[135]

[b]These data are the miRNA that also showed decreased expression of putative mRNA targets and do not contain the full number of significantly changed miRNA.

ethanol toxicity.[129] Early evidence from Sulik's research group supports this hypothesis of miRNA perturbation as "master regulators" of FASD phenotype. Her group reported that early embryonic exposure to high levels of ethanol can result in facial phenotypes in mice that mimic DiGeorge Syndrome.[27] This was an important observation, because one of the implicated gene loci, DiGeorge Syndrome Critical Region 8 (*DGCR8*), was subsequently found to control the biogenesis of miRNA, cleaving pri-miRNA into pre-miRNA transcripts.[130] Perhaps not surprisingly, the developmental timing of ethanol exposure determines which miRNAs are altered. For example, Singh and colleagues showed that PAE can result in differing profiles of miRNA expression in the adult mouse brain based on the trimester-equivalent period of ethanol exposure, with 80% or more of PAE-sensitive miRNAs showing sensitivity within a specific developmental exposure window.[128] PAE specifically during the third trimester equivalent period was reported to result in increased miR-26b expression in the adult. Importantly, the increase in miRNA expression was associated with decreased expression in mRNA for a presumptive target, cannabinoid receptor 1 (*Cnr1*),[131] implying a causal link between miRNA and target gene expression in FASD adults. Additional evidence indicates that the effects of ethanol on miRNA expression are dependent on dose.[132] Moreover, expression varies based on the cell and tissue type examined.[133] These data collectively show that PAE results in persistent effects on miRNAs and their target genes in a dose, timing, and tissue dependent manner.

Comprehensive reviews have been published specifically examining alterations to miRNAs in alcohol use disorders (AUDs),[136] prenatal alcohol exposure,[137] and in ethanol toxicity.[129] In this section, we focus on a few key miRNAs, miR-9, miR-153, miR-21, and miR-10, which exhibit perturbed expression in response to alcohol exposure across development and into adulthood, in models of AUDs (Table 1). These miRNAs serve as examples of how ethanol-induced perturbations of miRNA expression can have lasting consequences for development. Additionally, we examine current work on usefulness of miRNAs as biomarkers of prenatal alcohol exposure and therapeutic efficacy.

4.1.1 miR-9 as a Contributor to Craniofacial Deficits

The miRNA, miR-9, is evolutionarily conserved and an important regulator of neurodevelopmental processes including stem cell proliferation and maturation.[138] Research has shown that miR-9 can play a dual role in neurogenesis, at times promoting progenitor proliferation, and at other times promoting differentiation.[139] These opposing actions occur because miR-9 targets the 3′-UTR of neural progenitor proliferation factors, such as *Hes1* and members of the zinc finger transcription family like *Zic5*, as well as neural differentiation promoting factors, such as *elavl3/HuD*.[139,140] miR-9, along with its guide strand miR-9* (miR-9-3p), and miR-124 can reprogram human fibroblasts into neurons[141] and encourage neuronal differentiation of embryonic stem cells.[142] Yet miR-9 has also been shown to encourage proliferation of embryonic stem cell derived neural progenitor cells and repress neural cell migration,[143] further supporting the dual role of miR-9 in neurodevelopment, in both progenitor cell renewal and neuronal formation.

In the developing brain, ethanol exposure has been shown to affect miR-9 expression, which is thought to directly contribute to morphological deficits seen in FASD. In a murine model, ethanol was shown to decrease miR-9 expression in neural progenitor cells[144] and in zebrafish, early embryonic ethanol exposure also decreased miR-9 expression.[145,146] Furthermore, in zebrafish, miR-9 knockdown recapitulates some of the cranial deficits, like microcephaly, due to developmental alcohol exposure,[145] and in a mouse double knockout model, deletion of two of the miR-9 genomic loci, *miR-9-2* and *miR-9-3*, also results in thinning of the cortical plate,[147] a feature also observed in mouse models of PAE.[148] These data suggesting that the biology of miR-9 and its mediating role in ethanol teratology is conserved through vertebrate evolution.

The function of miR-9 changes with developmental stage. During development, a major role of miR-9 is regulation of neurogenesis, while in emerging neurons, miR-9 modulates neuronal migration and functional activity. In the neurons of the developing cortex, miR-9 targets *Foxp2*. Ectopic expression of Foxp2 results in aberrant radial migration, stalling of newly formed neurons in lower cortical layers, and perturbation of neurite outgrowth. These data suggest that miR-9 plays an important role in the formation and maturation of the brain.[149] Interestingly, while ethanol suppresses miR-9 in neural progenitor cells, it appears to have opposite and inductive effects on miR-9 expression during later periods of brain maturation. In a mouse model, in which ethanol exposure occurred from implantation into the period for the development of deep cortical layers (GD 6–GD 15), PAE results in increased miR-9 expression in the fetal mouse brain at GD 17.5.[150] At GD 17.5 the deep cortical layers V and VI have formed,[151] therefore this contrasting increase in miR-9 expression may be a specific result of increased expression in maturing neurons that outweighs the decreased expression in neural progenitors. Likewise, in a rat model, in which ethanol exposure occurred from implantation through in utero development (GD 7 through GD 21), PAE increases miR-9 in young adulthood (~PD 90). In this study, miR-9 was found to increase in the pituitary, where it targets the dopamine D2 receptor (*D2R*) 3′-UTR.[152] miR-9 action on *D2R* is thought to increase prolactin release and contribute to hyperprolactemia after PAE.[152,153] Upregulation of miR-9 has also been found in the adult striatum and supraoptic nucleus in a rodent model of AUD. This upregulation of miR-9 changes the composition of calcium- and voltage-dependent potassium channels, BK channels. The BK isoforms which are more sensitive to alcohol are suppressed leading to preferential expression of BK channels which have reduced alcohol sensitivity, thereby implicating miR-9 in tolerance to alcohol in adulthood.[154]

While the opposing effects of ethanol on miR-9 expression may be linked to cellular developmental stage, even in early development, intracellular adaptations may prevent the complete loss of miR-9 expression. In neuroprogenitor cells, even though miR-9 levels are decreased after ethanol exposure, there is enhanced binding of the BAF/BRM nucleosome remodeling complex to the *miR-9-2* locus.[155] This mechanism may be part of a feedback loop to recover miR-9 expression after ethanol insult.

4.1.2 miR-153, Neuronal Development and Hyperactivity

miR-153 is downregulated in neural progenitors after exposure to ethanol.[144,156] miR-153 has been shown to target the nuclear factor 1 family of transcription factors, namely, *NFIA* and *NFIB*[148,157] by binding to specific sites on the 3′-UTRs of targeted transcripts. These transcription factors play a role in neurogenesis[158,159] and gliogenesis.[160,161] miR-153 overexpression in neural progenitors decreases neurite outgrowth during differentiation into early migratory neurons in vitro and doublecortin (DCX, a marker for differentiating neurons) in vivo.[148] Likewise, overexpression of miR-153 has also been shown to inhibit glial fibrillary acidic protein expression,[157] an indicator of decreased gliogenesis. These miR-153 overexpression results agree with data from NFIA and NFIB knockout mice in which the thickness of the ventricular zone is increased along with the number of neural progenitor cells.[162,163] Suppression of miR-153 in the developing cortex increases NFIA and NFIB expression,[148,157] sensitizes cells to gliogenic cytokines, and decreases progenitor cell numbers.[157] These data support a hypothesis that ethanol affects the developing cortex by increasing early maturation of neural progenitor cells leading to inappropriate cortical development.[164–166]

Within the developing zebrafish, decreased miR-153 levels from ethanol exposure have been shown to contribute to changes in behavior. Tanguay and colleagues found that ethanol exposure results in hyperactivity in the dark at the larval stage, and at the juvenile stage, hypoactivity in the dark and hyperactivity in the light. Reduction of the expression of a particular zebrafish miR-153 isoform, miR-153c, was able to recapitulate the behavioral phenotype of ethanol exposure.[146] Interestingly, miR-153 shows varying evolutionary expansion, with zebrafish having 3 genomic copies (miR-153a, b, c), mice, a single genomic copy, and humans, 2 genomic copies (miR-153-1 and miR-153-2). Basic local alignment search tool analysis (BLAST, https://blast.ncbi.nlm.nih.gov/Blast.cgi) of the pre-miRNA-153 sequence from the mouse shows that this isoform is 97% homologous to miR-153a in the zebrafish, 99% homologous to miR-153-2 in the human, and 88% homologous to miR-153-1 in the human. Zebrafish miR-153b and miR-153c have much longer pre-miRNA stem loop structures (mouse miR-153: 69 bp; zebrafish miR-153a: 81 bp; miR-153b: 148 bp; and miR-153c: 151 bp) and sequence homology of the pre-miRNA was undetectable using BLAST. In the study from the Tanguay group, ethanol exposure from gastrulation through somitogenesis downregulates expression of miR-153c[146] while the Santos research group reported increased expression of miR-153a in response to ethanol exposure in zebrafish during the same developmental window.[132] While the expression data for miR-153a is contradictory to the evidence from neural progenitor cells, given that zebrafish miR-153a is the closest homologue to mouse miR-153, these data suggest that the isoforms of miR-153 in the zebrafish are controlled in a nonredundant manner and that the expansion of expression may serve an important compensatory function in zebrafish.

As with miR-9 (as mentioned in the previous paragraphs), miR-153 can also affect the function of mature neurons. miR-153 has been shown to target mediators of synaptic development, including the synaptic vesicle complex protein, SNAP-25.[167] Likewise, miR-153 is also increased postmortem, in the forebrain of adults with chronic AUDs.[134] miR-153 is also upregulated in response to chronic intermittent alcohol exposure in a human neuroblastoma cell line (SH SY5Y) with dopaminergic, neural crest characteristics, and returns to control levels after five days of withdrawal.[133] Interestingly, the same study showed that in a human brain-derived astrocytoma cell line (1321 N1), chronic ethanol exposure downregulates miR-153 expression while withdrawal significantly upregulates expression. These data indicate that the effects of ethanol on miR-153 expression may depend on the cell-type, differentiation stage, and/or tissue of origin. Additionally, miR-153 has been shown to target α-synuclein,[168] a presynaptic protein which plays a role in vesicle release.[169] α-Synuclein is increased in the plasma of alcoholic patients and associated with alcohol dependence and craving[170–172] and is increased in alcohol preferring rats compared to nonpreferring rats.[173] While evidence from human tissues show that in the adult, AUDs increase miR-153 expression, and should thereby reduce α-synuclein, a single nucleotide polymorphism in the 3′-UTR coding region of *Snca*, present in a strain of alcohol preferring rats, actually increases the half-life of the α-synuclein mRNA.[174] This polymorphism occurs at site +679 (from translational start site) in the 3′-UTR, which does not overlap the putative miR-153 binding site (+890)[175] but this mutation, along with other mutations which are closer to the miR-153 binding site,[173] may affect the binding dynamics of miR-153, thereby indicating a possible interaction between genetics, miR-153, α-synuclein, and alcohol.

4.1.3 miRNAs and Other PAE Phenotypes

While the neurodevelopmental and craniofacial deficits from PAE are some of the most commonly studied teratogenic effects of ethanol, the brain and face are not the only regions affected by PAE. For instance, congenital heart defects occur in an estimated 28.5% of FASD children, a majority of which are atrial and septal valve defects.[176] miR-21 has been shown to be downregulated with PAE[144,156] but upregulated in peripheral blood in stress-induced drinking in heavy drinkers.[177] miR-21 has also been implicated in heart disease including cardiac hypertrophy[178,179] and fibrogenic epithelial to mesenchymal transition (EMT, see more on developmental EMT in *miRNA and placental phenotype*)[180,181] by inhibiting expression of sprouty homolog 1 (Spry1). Likewise, miR-10 also exhibits altered expression after PAE (increased in the fetal brain)[150] and may also affect cardiac development pathways. Both mouse isoforms, miR-10a and miR-10b, are genomically located within the *Hox* gene domains and, like the lncRNA *HOTAIR* (discussed earlier), also appear to regulate Hox expression. miR-10a targets *Hoxa1*, which has been shown to be transiently expressed in cardiac progenitors[182] and truncation mutations of *Hoxa1* are known to cause heart deficits, including conotruncal heart deficits, in humans.[183] Therefore it is likely that peripheral alterations to miR-21 and miR-10 expression after PAE could have consequences for heart development.

The aforementioned miRNAs have also been implicated alcohol-induced organ injury, including pancreatitis and alcoholic liver disease.[136] Human and animal studies show that PAE contributes to altered liver function.[184,185] Therefore miR-21 and miR-10 may also contribute to these FASD liver phenotypes. Additionally, stress is a confounding variable in drinking behavior, increasing craving and anxiety in heavy drinkers and reducing efficacy of treatment.[186,187] As mentioned earlier, in a study of heavy and moderate drinkers, increased expression of miR-10 and miR-21 was associated with increased stressed-induced drinking in heavy, but not moderate, drinkers,[177] suggesting that miR-10 and miR-21 may be indicators of alcohol by stress interactions. Prenatal stress is known to have long-term consequences for neurodevelopment in the offspring as well as disrupt the maternal-fetal placental interface.[188,189] In the following section, we will see how prenatal alcohol is also able to disrupt placental function, some of which may be contributed to, or exacerbated by, maternal stress.

4.2 miRNA and Placental Function in FASD

Prenatal and postnatal growth restriction is a diagnostic criterion for FAS.[3] Indeed, one study profiling birth outcomes in South Africa revealed that prenatal (or intrauterine) growth restriction (IUGR), defined as a gestational weight $<$10th percentile, occurs in 41% of infants with a history of heavy alcohol exposure during in utero development.[190] A major cause of IUGR is placental dysfunction.[191] A metaanalysis of human clinical studies from 1996 to 2006, conducted by Burd and colleagues, reported that heavy alcohol consumption during pregnancy correlated with smaller placental sizes, increased rates of placental abruption, and impaired placental vascular function.[192] Preeclampsia is also associated with PAE and is thought to be caused by poor placental invasion and inadequate remodeling of uterine spiral arteries.[193] Because of the potential for adverse infant and child health consequences, there has been significant recent emphasis on identifying early biomarkers for FASD-associated placental dysfunction. One emerging class of biomarkers are circulating miRNAs.[194,195] A recent study by Balaraman *et al.* (also further discussed in the forthcoming paragraphs) identified 11 miRNAs elevated in maternal plasma of infants with FAS.[196] Prediction algorithms suggested these elevated miRNAs may interfere with key placental developmental pathways, including the epithelial-mesenchymal transition pathway crucial for placental migration and invasion. Interestingly, two of these elevated miRNAs, miR-519a-3p and miR-518f-3p, belong to the placental-restricted and primate-specific C19MC cluster of miRNAs, whose members have previously been shown to be altered in the plasma and placenta of women whose pregnancies were complicated by preeclampsia and IUGR.[197–199] In turn, several C19MC members are known regulate placental trophoblast invasion, suggesting that alteration in these miRNA levels are not merely biomarkers of placental pathology but also contribute to the development of pathology.[200]

Follow-up studies in human placentas and rodent models of IUGR indicate that the placenta of growth-restricted infants have reduced trophoblast proliferation and impaired nutrient transport capacity.[201] This growth dysfunction and altered amino acid transport has been recapitulated in in vivo and in vitro studies on the effect of alcohol on the placenta.[202–204] Profiles of placental miRNAs in growth restricted infants reveal a general increase in miRNA expression. For example, miR-10 and miR-363 were found to be increased in placentas of IUGR infants.[205] In vitro studies show that these miRNAs in turn inhibit Snat1 and 2 sodium dependent amino acid transporters, potentially explaining the impaired amino acid transport in IUGR placentas and alcohol exposed placental trophoblasts.[205] In contrast, miR-16 and miR-21 were found to be decreased in placentas of IUGR infants.[206] Nicotine exposure, which commonly co-occurs with alcohol use during pregnancy, also correlates with lower miR-16 and miR-21 expression.[207] Interestingly, placental miR-16 levels have been shown to negatively correlate with attention scores during early postnatal development.[208] It is tempting to speculate that this and other placental miRNAs may mediate some of the adverse neurobehavioral outcomes of PAE.

4.3 miRNAs as Biomarkers of Prenatal Alcohol Exposure

Circulating miRNAs can be transferred between cells as a potentially novel form of intercellular communication[209,210] and can exist in several different forms: free-floating, protein-bound, or encapsulated in membranous vesicles.[209,211,212] Human and animal model studies have shown that alcohol intake during pregnancy can alter maternal serum miRNA levels. Early evidence of the association between circulating miRNA and PAE in maternal and fetal compartments came from a study in pregnant sheep, exposed to a binge model of ethanol consumption over a human three-trimester equivalent period. This study reported that plasma miRNA profiles in the pregnant mother and newborn lamb reflected ethanol exposure status.[213] Human studies have also shown distinct circulating miRNA profiles in pregnant mothers who have a history of alcohol use. For instance, in a study from New Mexico on pregnant women with a self-reported history of drinking, 55 miRNAs were found to be significantly altered in the serum of women who had consumed alcohol compared to women who had abstained from alcohol during pregnancy.[214] Interestingly, pathway analysis revealed that these altered miRNAs may control a number of cardiovascular development pathways, and thereby may contribute to the heart development deficits in FASD (discussed earlier). However, plasma miRNA in pregnant women may also be useful in predicting variations in birth outcomes following PAE. For example, in a study of pregnant women from Ukraine, who self-reported to be either abstinent or heavy drinkers, 11 miRNAs were found to be significantly elevated in plasma at mid-pregnancy in mothers who were heavily exposed to alcohol and subsequently gave birth to infants with craniofacial and other deficits.[196] Interestingly, equally exposed mothers who subsequently gave birth to an apparently unaffected infant did not exhibit a similar elevation in those plasma miRNAs. Using a random forest machine learning-based classification model, a set of miRNAs were shown to be highly predictive of infant outcome after PAE and was correctly able to classify children into high exposed with physical characteristics of FASD or unexposed groups with 86.7% accuracy. This classification model was also shown to potentially subclassify infants who were heavily exposed but apparently unaffected into subpopulations with varying degrees of similarity to FASD infants. These studies suggest that miRNAs in the pregnant mother may be useful biomarkers not only for exposure, but also for predicting subsequent birth outcomes.

4.4 miRNA as Biomarkers of Therapeutic Efficacy and Potential Therapeutic Agents

Studies in animal models have shown that miRNA expression may also be a marker of therapeutic efficacy. Research in rat models of PAE has shown that choline administration after prenatal alcohol exposure can be beneficial to ameliorate learning deficits.[215,216] Clinical studies have also shown that choline supplementation in FASD patients at 2.5–5 yrs of age can improve performance in some memory tasks,[217] but this is not recapitulated in children 5–10 yrs old.[218] Interestingly, in a rat model of PAE, third-trimester equivalent exposure resulted in increased hippocampal miR-200c expression, but choline supplementation was able to decrease miR-200c expression levels.[219] These data from clinical trials and preclinical studies show that more needs to be done to determine the best window for choline treatment. Moreover, animal studies have shown that suboptimal dosing of choline can be detrimental in PAE models, such as by delaying motor development.[220] Examination of alteration to expression of miRNAs, such as miR-200c, could assist in determining the most efficacious choline supplementation regimen.

Other therapeutics can also restore miRNA expression after PAE. Wang *et al.* showed elevation of miR-10a and miR-10b in the fetal brain after PAE (discussed earlier).[150] They further examined the role for folic acid in ameliorating this PAE phenotype. They found in whole embryo culture models, that concurrent exposure to folic acid during ethanol administration increased growth measurements suppressed miR-10 expression in a dose-dependent manner while increasing expression of *Hoxa1*, miR-10's target transcript. Tsai *et al.* also examined varenicline, a nicotine receptor agonist, as a possible therapeutic agent.[148] They found that varenicline after PAE not only increased miR-153, which is decreased after ethanol exposure, but also restored expression of a number of miR-153's target mRNA transcripts which had reduced expression after ethanol exposure. These data indicate that a panel of miRNAs, perhaps including miR-10 and miR-200 isoforms as well as miR-153, could serve as a readout of therapeutic efficacy in FASD.

miRNAs also have the potential to serve as therapeutics in FASD. Tsai *et al.* also examined how manipulating miR-153 expression may affect outcome after PAE. They found that transfection of miR-153 constructs into neural progenitor cells prior to ethanol exposure was preventative, blocking the ethanol-induced increase in NFI family transcripts. miR-153 expression after ethanol exposure was also able to restore normal expression of some of the altered mRNA transcripts but also led to an unexpected additional increase in other transcripts, including the NFI family. These data suggest that miRNAs have potential to be preventative in PAE and that further study is needed to assess their therapeutic potential. Moreover, as mentioned earlier, miRNAs can be used as endocrine signaling factors, particularly when packaged in exosomes. Studies show that the miRNA population in exosomes can be distinct from the miRNA profile in the cell of origin.[221] Sequence motifs present in miRNAs can control specific loading of RNA species into exosomes by the chaperone protein heterogeneous nuclear ribonucleoprotein A2B1 (hnRNPA2B1).[222] Exosomes can be targeted to recipient cells through antigen presentation and receptor binding. Antigen-driven exosome uptake has been shown to result in the unidirectional transfer of miRNAs from T cells to antigen-presenting cells.[223] Therefore by using exosomal biology, there is potential for therapeutic miRNAs to be directly targeted to vulnerable cellular populations across the body.

5. lncRNA AND FASD

Researchers are just beginning to implicate lncRNAs in FASD outcomes. In particular, the lncRNA, *H19*, plays an important role in determining fetal size and has been implicated in prenatal growth restriction in FASD. *H19* and insulin-like growth factor 2 (IGF2) are encoded in a contiguous genomic locus, but show different maternal and paternal imprinting. On the maternal chromosome, a lack of methylation of *H19* locus leads to expression of *H19* and loss of IGF2 expression, resulting in decreased fetal size. On the paternal chromosome, in contrast, methylation of a region upstream of *H19* blocks transcription of *H19* and facilitates transcription of *IGF2*, increasing fetal size.[224,225] Ethanol exposure resulted in hypomethylation at the paternal *H19*/*IGF2* locus in the placenta of mid-gestation embryos which also displayed placental and embryonic growth restriction.[226] Preconception paternal drinking was found to decrease postnatal growth and methylation at the *H19* region.[123,124] *H19* serves as an epigenetic factor, in that it binds methyl-CpG-binding domain protein 1 (MBD-1) and recruits histone methyltransferases to a number of growth and development genes, including *IGF2*, thereby decreasing transcription.[227] *H19* can also serve as the primary transcript for miR-675, which targets the insulin-like growth factor 1 receptor (*IGF1R*) mRNA.[103] Both of these mechanisms serve to decrease prenatal growth, suggesting that factors in the fetal environment like ethanol which result in hypomethylation and increased transcription from the *H19* locus, can contribute to decreased prenatal growth. *H19* is likely not the only lncRNA to contribute to FASD. Research on additional developmental disorders, including AUDs, implicates additional mechanisms of ethanol-induced lncRNA disruption, as well as possible lncRNA mediators of the FASD phenotype.

5.1 lncRNA and Lessons from Alcohol Use Disorders and Development

5.1.1 Alcohol Use Disorder

AUDs in the adult can alter lncRNA expression. In the postmortem brain of chronic alcoholics, the lncRNA *MALAT1* was shown to be upregulated in the cerebellum, hippocampus, and brain stem, but was not changed in the cortex. While the cerebellar phenotype was not recapitulated in adult rats exposed to chronic vapor ethanol exposure, after 24 h of withdrawal, *Malat1* was increased in the cortex of the ethanol-exposed rats.[228] Interestingly, *MALAT1* has been found to be upregulated both in human-induced pluripotent stem cells and in neurons differentiated from these stem cells[229] and can be targeted for degradation within the nucleus by miR-9.[230] Within the postmortem prefrontal cortex of alcoholics, some lncRNAs were upregulated while others were downregulated, indicating heterogeneity in the lncRNA response to AUDs.[231] In the sperm of adult drinking males, there is increased demethylation at normally hypermethylated imprinted differentially methylated regions. The degree of demethylation correlates with alcohol consumption levels.[123] One locus (*IG-DMR*) harbors the lncRNA *MEG3/GTL2* (referred to as *MEG3*), a maternally expressed noncoding RNA, with high levels of expression in the developing central nervous system.[232] *MEG3* overexpression has been shown to downregulate miR-21 and act as a sponge for miR-9. *MEG3* overexpression, and the subsequent decrease in miR-21, results in suppressed endothelial cell proliferation and migration,[233] and decreased proliferation and increased apoptosis in two cervical cancer cell lines (HeLa and CaSki).[234] Through interactions as a miR-9 sponge, *MEG3* overexpression decreases vascular endothelial cell proliferation and angiogenesis.[235] These data indicate that lncRNA expression can be altered by alcohol exposure in a region-specific manner and that the effect of alcohol is not a global up or down regulation of lncRNAs, but a fine tuning of relative lncRNA expression. Additionally, the lncRNAs,

MALAT1 and *MEG3*, are implicated in neurodevelopment, and interact with ethanol-sensitive miR-9 and miR-21 (as discussed earlier, Table 1), further implicating a role for these lncRNAs in FASD.

5.1.2 Developmental Disorders

The genetic locus implicated in the HELLP syndrome, a pregnancy-associated disorder which induces maternal Hemolysis, Elevated Liver enzymes, and Low Platelets, has been shown to encode a lncRNA, important for placental function. A 205 kb noncoding RNA transcript was identified from an intergenic region between pro-melanocortin concentrating hormone (*PMCH*) and insulin-like growth factor 1 (*IGF1*) genes. Polymorphisms in this region are associated with HELLP syndrome.[236,237] This lncRNA (*HELLP* lncRNA) was shown to be expressed in the extravillous trophoblast of the placenta. In HELLP, extravillous trophoblasts show a deficient differentiation into an invasive phenotype, with HELLP placenta showing reduced invasion into maternal decidua. Using morpholinos to mask sites in the *HELLP* lncRNA that are commonly mutated in the disease, there was a decrease in lncRNA degradation and decreased extravillous trophoblast invasion, implicating these mutations in the trophoblast phenotype in HELLP.[237] This *HELLP* lncRNA, like *H19*, is important for placental development and may also play an as yet unascertained role in other placental growth restriction diseases including in PAE.

5.1.3 Neurodevelopmental Disorders

RNA sequencing (RNAseq) data from the postmortem cortex shows that ASD patients differentially express 263 lncRNAs compared to non-ASD samples.[238] Of the differentially expressed lncRNA, 41% are antisense to protein-coding genes, five of which are antisense to known ASD risk genes, including Dlx6. Antisense lncRNAs can interact with DNA, H3K9, and H3K27 methylation machinery, to condense chromatin structure at the sense protein-coding strand.[239] The DLX6 antisense transcript (*DLXas1/EVF2*) is downregulated in the cortex of autism patients.[238] DLX6 is required for craniofacial development[240] while *Evf2* has been shown in the developing ventral forebrain to interact with important DNA regulatory elements in the Dlx5/6 intergenic region,[241] indicating a feedback regulation for this antisense lncRNA. *Evf2* has also been shown to localize with BRG1/BAF170 (part of the BAF/BRM nucleosome remodeling complex, discussed earlier) and inhibit BRG1 ATPase and chromatin remodeling activities.[242] These data suggest that *EVF2* may play a role in regulating chromatin remodeling and DLX5/6 expression during development, and may be thereby poised to affect craniofacial development.

6. SMALL NUCLEOLAR RNA AND FASD

Small nucleolar RNAs (snoRNAs) have been implicated in FASD. Brains of adult PAE mice show alterations in the expression of both miRNAs and snoRNAs.[128] These changes in ncRNA expression levels are treatment-paradigm specific, that is, dependent on the timing of exposure. For example, snoRNA *HBII-52* (*SNORD115*) is altered irrespective of whether ethanol exposure occurred in the first, second, or third trimester-equivalent periods or throughout the entire three trimester-equivalent period. In contrast, *SNORD113* is downregulated in the third trimester exposure model and upregulated in the continuous exposure paradigm, and members of the HBII-85 family (*SNORD116*) are downregulated in the first and second trimester-equivalent models. Therefore the *Dlk1-Dio3* and *Snrpn-Ube3a* imprinted loci, which contains these snoRNAs, may hold a long-term foot print for PAE.

Significant alterations to methylation at the *Snrpn-Ube3a* locus were further shown in buccal swabs from FASD patients.[243] Moreover, changes in expression of *SNORD115* and *SNORD116* (encoded in *Snrpn-Ube3a*) have been shown to recapitulate FASD phenotypes. Deletion of *SNORD116* in mice causes early onset postnatal growth deficiency, motor learning deficiency, increased anxiety, and hyperphagia,[244] which are all symptoms overlapping with FASDs.[245–247] *SNORD115* was shown to regulate alternative splicing of the serotonin 2C receptor (*5-HT2C*). Loss of *SNORD115* in Prader-Willi Syndrome patients causes the mis-splicing of *5-HT2C* RNA leading to unique isoforms of this serotonin receptor not present in normal individuals.[248] Serotonergic systems are also affected in FASD children. For example, magnetic resonance imaging and single-photon emission computed tomography show that children prenatally exposed to alcohol have reduced serotonin in the medial frontal cortex.[249] Altered imprinting at the *Snrpn-Ube3a* locus, particularly through changes to *SNORD115* and *SNORD116* expression, may contribute to behavioral and neurochemical alterations in FASD.

7. FUTURE DIRECTIONS

7.1 Speciation, Differentiation, and Teratogenicity

ncRNAs are thought to be drivers of speciation.[250] They may arise at distinct branches of the phylogenetic tree, but once they appear, may be evolutionarily conserved. For example, both miR-9 and miR-153 first appear with Nephrozoa taxon,[251] which are triploblastic, when early embryonic development began to have three lineages, adding in the mesoderm to the endoderm and ectoderm, and are also bilaterian, having a bilateral, symmetrical body plan.[252] The mesoderm is important for the subcompartmentalization of body functions through the creation of organ structures.[253] Therefore it is not surprising that miR-9 and miR-153, are both important for differentiation of function within a developing organ, that is, they are both needed for neurogenesis within the brain.[138,167] miR-21 was originally identified as an "onco-miR" due to upregulation of expression in cancers, but given that it first appeared with the taxon Vertebrata, who have vertebral columns which has allowed for enclosure of a central nervous system, it is not surprising that miR-21 is also implicated in degeneration and neuroprotection,[144,254,255] and is highly expressed in the developing brain as well as in tumors of neural origin.[256,257] These data suggest that we can infer likely targets and functions of miRNAs from when they arose during evolution. The phenotypic outcome of teratogen exposure can be the loss of an evolutionarily acquired phenotype, such as the loss of appropriate cortical and facial development that can occur in FASD. Therefore examination of the role of miRNAs, and indeed, other ncRNAs across evolution and in directing speciation, could direct us to likely targets for teratogenesis.

7.2 Role of Sex Differences in FASD and FASD-associate ncRNAs

Sex differences in gene expression can vary in tissues throughout the body (Here we are referring to genetic sex and how the XY chromosomes contribute to differences in males and females. Gender, or how societal norms affect the expression of genetic sex, while also an important variable, adds an additional complexity to potential male and female differences but is difficult to further investigate in animal models, and therefore not discussed.).[258] These differences can arise before the onset of any sex hormones,[259] or in response to sex hormones both perinatally, where they organize the sexually dimorphic regions of the body, and at puberty, where these underlying sex differences are activated.[260] Importantly for FASD, these differences can contribute to differential vulnerability to environmental stressors. For example, we know that males are more vulnerable to developing a number of neurodevelopmental disorders, and that the contribution of maternal stress to these disorders is sexually dimorphic for the offspring due to differences in male and female placental dynamics.[261] As discussed earlier, prenatal exposure to ethanol can cause a placental phenotype, but the role of sex in the effects of ethanol on the placenta are currently unknown.

In FASD, sex differences can be seen in various aspects of the disease phenotype. These include differences in hyperresponsivity to stressors, decreased organizational sex hormone content in males (testosterone) at birth, delay of puberty in females, and differences in social recognition.[262,263] While the underlying mechanisms of some of these differences are beginning to be elucidated, the contribution of cellular sex differences to the FASD phenotype is not well known. Moreover, sex differences in expression are not just in protein-coding genes, but also in miRNAs[264] and likely lncRNAs, though the lncRNA for X-chromosome inactivation (XIST) is the only widely examined sexually dimorphic lncRNA so far.[265] With these ncRNAs serving as drivers of many of the FASD phenotypes, there is likely a role of ncRNAs in the sexual dimorphic phenotypes.

7.3 Role for Other ncRNA Families

As discussed earlier, data has implicated other ncRNAs outside of miRNAs and lncRNAs as perturbed by ethanol exposure. More investigation is needed to determine the underlying role of these ncRNAs in FASD. For example, piRNAs, snoRNAs, and lncRNAs have been found to be capable of novel functions and may also mediate many structural and functional anomalies associated with PAE. piRNAs have recently been shown to have functions in the brain, including potential roles in hippocampal synaptogenesis[67] and synaptic plasticity.[68] snoRNAs have been shown to break down into miRNA sized species and some pre-miRs have been shown to have snoRNA-like processing domains.[266,267] While the purpose of this functional redundancy has not been elucidated, there is evidence that snoRNA-derived small RNAs may be upregulated in the response stressful environmental conditions.[266] lncRNA can take on a circular shape from the back-splicing and circularization of pre-mRNAs.[268] Circular RNAs were shown to be highly abundant in mouse and human central nervous system and have been indicated to be specific for neuronal, not glial, identity.[269] Additionally, loss of a circular RNA (*Cdr1as*) has been shown to cause deficits in sensorimotor gating and synaptic transmission, resembling deficits seen in several human neuropsychiatric disorders, through circular RNA-miRNA interactions.[270] As we understand more about these alternative mechanisms of ncRNAs controlling cellular function, we may find novel pathways that are disrupted in response to ethanol exposure and novel sites for intervention.

7.4 Targeting lncRNA as Potential Therapeutics

As discussed earlier, lncRNAs are important players in brain development, disruption of their expression levels can cause neurodevelopmental disorders, or they can be targeted by teratogens, such as ethanol. One study showed elevated *H19* expression at GD 14.5 in mice after treatment with 2,3,7,8-tetrachlorodibenzo-p-dioxin mediates the formation of cleft palate.[271] lncRNAs which are disrupted early in prenatal life and have later developmental consequences can be targeted therapeutically to rescue development. For example, some of the effects of Angelman syndrome have been ameliorated by using antisense oligonucleotides to *Ube3a-ATS* lncRNA.[272] In Angelman syndrome there is maternal disruption in *UBE3A*, which results in elevation of *Ube3a-ATS* and, consequently, silencing of the paternal *UBE3A* gene expression, leading to severe brain developmental deficits.[273] Antisense oligonucleotide targeting of *Ube3a-ATS* decreases this lncRNA and relieves the paternal suppression of the *UBE3A* gene, rescuing some of the cognitive deficits in Angelman syndrome.[272] Similar targeting of lncRNAs perturbed by ethanol exposure may ameliorate deficits found in FASD.

8. CONCLUSIONS

Genetics have been shown to be one determinant of outcomes in FASD. However, DNA sequence is not the only determinant of the effects of PAE, as epigenetic factors add an extra layer of complexity to shape how the genome is expressed. Here, we show that epigenetics can be shaped by ncRNAs which can act both in the nucleus and the cytoplasm to regulate gene expression and are themselves targeted by PAE. Because of their ability to regulate the expression of gene networks, ncRNAs are attractive therapeutic targets for ameliorating the developmental effects of PAE and preventing FASD.

REFERENCES

1. Jones KL, Smith DW, Ulleland CN, Streissguth P. Pattern of malformation in offspring of chronic alcoholic mothers. *Lancet*. 1973;1(7815):1267–1271.
2. Jones KL. The fetal alcohol syndrome. *Addict Dis*. 1975;2(1–2):79–88.
3. Hoyme HE, May PA, Kalberg WO, et al. A practical clinical approach to diagnosis of fetal alcohol spectrum disorders: clarification of the 1996 institute of medicine criteria. *Pediatrics*. 2005;115:39–47.
4. Roozen S, Peters GJ, Kok G, Townend D, Nijhuis J, Curfs L. Worldwide prevalence of fetal alcohol spectrum disorders: a systematic literature review including meta-analysis. *Alcohol Clin Exp Res*. 2016;40(1):18–32.

5. May PA, Baete A, Russo J, et al. Prevalence and characteristics of fetal alcohol spectrum disorders. *Pediatrics.* 2014;134(5):855–866.
6. Chasnoff IJ, Wells AM, King L. Misdiagnosis and missed diagnoses in foster and adopted children with prenatal alcohol exposure. *Pediatrics.* 2015;135(2):264–270.
7. Ernhart CB, Morrow-Tlucak M, Sokol RJ, Martier S. Underreporting of alcohol use in pregnancy. *Alcohol Clin Exp Res.* 1988;12(4):506–511.
8. Bakhireva LN, Savich RD, Raisch DW, et al. The feasibility and cost of neonatal screening for prenatal alcohol exposure by measuring phosphatidylethanol in dried blood spots. *Alcohol Clin Exp Res.* 2013;37(6):1008–1015.
9. Bakhireva LN, Sharkis J, Shrestha S, Miranda-Sohrabji TJ, Williams S, Miranda RC. Prevalence of prenatal alcohol exposure in the state of Texas as assessed by phosphatidylethanol in newborn dried blood spot specimens. *Alcohol Clin Exp Res.* 2017;41 (5):1004–1011.
10. Zablotsky B, Black LI, Maenner MJ, Schieve LA, Blumberg SJ. Estimated prevalence of autism and other developmental disabilities following questionnaire changes in the 2014 National Health Interview Survey. *Natl Health Stat Rep.* 2015;87:1–20.
11. Finer LB, Zolna MR. Declines in unintended pregnancy in the United States, 2008–2011. *N Engl J Med.* 2016;374(9):843–852.
12. Tan CH, Denny CH, Cheal NE, Sniezek JE, Kanny D. Alcohol use and binge drinking among women of childbearing age – United States, 2011–2013. *MMWR Morb Mortal Wkly Rep.* 2015;64(37):1042–1046.
13. Streissguth AP, Sampson PD, Barr HM. Neurobehavioral dose–response effects of prenatal alcohol exposure in humans from infancy to adulthood. *Ann N Y Acad Sci.* 1989;562:145–158.
14. Sood B, Delaney-Black V, Covington C, et al. Prenatal alcohol exposure and childhood behavior at age 6 to 7 years: I. Dose–response effect. *Pediatrics.* 2001;108(2):E34.
15. Patra J, Bakker R, Irving H, Jaddoe VW, Malini S, Rehm J. Dose–response relationship between alcohol consumption before and during pregnancy and the risks of low birthweight, preterm birth and small for gestational age (SGA) – a systematic review and meta-analyses. *BJOG.* 2011;118(12):1411–1421.
16. Jacobson JL, Jacobson SW. Prenatal alcohol exposure and neurobehavioral development. *Alcohol Health Res World.* 1994;18(1):30–36.
17. National Institute of Alcohol Abuse and Alcoholism. *NIAAA Council Approves Definition of Binge Drinking.* vol. 3. 2004.
18. Bailey BN, Delaney-Black V, Covington CY, et al. Prenatal exposure to binge drinking and cognitive and behavioral outcomes at age 7 years. *Am J Obstet Gynecol.* 2004;191 (3):1037–1043.
19. Ikonomidou C, Bittigau P, Ishimaru MJ, et al. Ethanol-induced apoptotic neurodegeneration and fetal alcohol syndrome. *Science.* 2000;287(5455):1056–1060.
20. Bonthius DJ, Goodlett CR, West JR. Blood alcohol concentration and severity of microencephaly in neonatal rats depend on the pattern of alcohol administration. *Alcohol.* 1988;5(3):209–214.
21. Bonthius DJ, West JR. Alcohol-induced neuronal loss in developing rats: increased brain damage with binge exposure. *Alcohol Clin Exp Res.* 1990;14:107–118.
22. Sulik KK. Genesis of alcohol-induced craniofacial dysmorphism. *Exp Biol Med.* 2005;230(6):366–375.
23. Guerri C, Bazinet A, Riley EP. Foetal Alcohol Spectrum Disorders and alterations in brain and behaviour. *Alcohol Alcohol.* 2009;44(2):108–114.
24. Astley SJ, Magnuson SI, Omnell LM, Clarren SK. Fetal alcohol syndrome: changes in craniofacial form with age, cognition, and timing of ethanol exposure in the macaque. *Teratology.* 1999;59:163–172.

25. Sulik KK, Johnston MC, Webb MA. Fetal alcohol syndrome: embryogenesis in a mouse model. *Science.* 1981;214:936–938.
26. Lipinski RJ, Hammond P, O'Leary-Moore SK, et al. Ethanol-induced face-brain dysmorphology patterns are correlative and exposure-stage dependent. *PLoS ONE.* 2012;7:e43067.
27. Sulik KK, Johnston MC, Daft PA, et al. Fetal alcohol syndrome and DiGeorge anomaly. Critical ethanol exposure periods for craniofacial malformations as illustrated in an animal model. *Am J Med Genet.* 1986;25(S2):97–112.
28. Streissguth AP, Dehaene P. Fetal alcohol syndrome in twins of alcoholic mothers: concordance of diagnosis and IQ. *Am J Med Genet.* 1993;47(6):857–861.
29. Warren KR, Li TK. Genetic polymorphisms: impact on the risk of fetal alcohol spectrum disorders. *Birth Defects Res A Clin Mol Teratol.* 2005;73(4):195–203.
30. Eberhart JK, Parnell SE. The genetics of fetal alcohol spectrum disorders. *Alcohol Clin Exp Res.* 2016;40(6):1154–1165.
31. Swartz ME, Wells MB, Griffin M, et al. A screen of zebrafish mutants identifies ethanol-sensitive genetic loci. *Alcohol Clin Exp Res.* 2014;38(3):694–703.
32. Yoder JA, Soman NS, Verdine GL, Bestor TH. DNA (cytosine-5)-methyltransferases in mouse cells and tissues. Studies with a mechanism-based probe. *J Mol Biol.* 1997;270 (3):385–395.
33. Mentch SJ, Locasale JW. One-carbon metabolism and epigenetics: understanding the specificity. *Ann N Y Acad Sci.* 2016;1363:91–98.
34. Pradhan S, Bacolla A, Wells RD, Roberts RJ. Recombinant human DNA (cytosine-5) methyltransferase. I. Expression, purification, and comparison of de novo and maintenance methylation. *J Biol Chem.* 1999;274(46):33002–33010.
35. Okano M, Bell DW, Haber DA, Li E. DNA methyltransferases Dnmt3a and Dnmt3b are essential for de novo methylation and mammalian development. *Cell.* 1999;99 (3):247–257.
36. Bird AP, Wolffe AP. Methylation-induced repression – belts, braces, and chromatin. *Cell.* 1999;99(5):451–454.
37. Métivier R, Gallais R, Tiffoche C, et al. Cyclical DNA methylation of a transcriptionally active promoter. *Nature.* 2008;452(7183):45–50.
38. Klose RJ, Bird AP. Genomic DNA methylation: the mark and its mediators. *Trends Biochem Sci.* 2006;31(2):89–97.
39. Strahl BD, Allis CD. The language of covalent histone modifications. *Nature.* 2000;403 (6765):41–45.
40. Wang Z, Zang C, Rosenfeld JA, et al. Combinatorial patterns of histone acetylations and methylations in the human genome. *Nat Genet.* 2008;40(7):897–903.
41. Shahbazian MD, Grunstein M. Functions of site-specific histone acetylation and deacetylation. *Annu Rev Biochem.* 2007;76:75–100.
42. Barski A, Cuddapah S, Cui K, et al. High-resolution profiling of histone methylations in the human genome. *Cell.* 2007;129(4):823–837.
43. Kouzarides T. Chromatin modifications and their function. *Cell.* 2007;128(4):693–705.
44. Bohacek J, Mansuy IM. Molecular insights into transgenerational non-genetic inheritance of acquired behaviours. *Nat Rev Genet.* 2015;16(11):641–652.
45. Manikkam M, Guerrero-Bosagna C, Tracey R, Haque MM, Skinner MK. Transgenerational actions of environmental compounds on reproductive disease and identification of epigenetic biomarkers of ancestral exposures. *PLoS ONE.* 2012;7(2): e31901.
46. Jodar M, Selvaraju S, Sendler E, Diamond MP, Krawetz SA, Reproductive Medicine Network. The presence, role and clinical use of spermatozoal RNAs. *Hum Reprod Update.* 2013;19(6):604–624.

47. Gapp K, Jawaid A, Sarkies P, et al. Implication of sperm RNAs in transgenerational inheritance of the effects of early trauma in mice. *Nat Neurosci.* 2014;17(5):667–669.
48. Rodgers AB, Morgan CP, Bronson SL, Revello S, Bale TL. Paternal stress exposure alters sperm microRNA content and reprograms offspring HPA stress axis regulation. *J Neurosci.* 2013;33(21):9003–9012.
49. Rodgers AB, Morgan CP, Leu NA, Bale TL. Transgenerational epigenetic programming via sperm microRNA recapitulates effects of paternal stress. *Proc Natl Acad Sci U S A.* 2015;112(44):13699–13704.
50. Crick FH. On protein synthesis. *Symp Soc Exp Biol.* 1958;12:138–163.
51. Morris KV, Mattick JS. The rise of regulatory RNA. *Nat Rev Genet.* 2014;15 (6):423–437.
52. Filipowicz W, Pogacic V. Biogenesis of small nucleolar ribonucleoproteins. *Curr Opin Cell Biol.* 2002;14(3):319–327.
53. Meier UT. The many facets of H/ACA ribonucleoproteins. *Chromosoma.* 2005;114 (1):1–14.
54. Reichow SL, Hamma T, Ferre-D'Amare AR, Varani G. The structure and function of small nucleolar ribonucleoproteins. *Nucleic Acids Res.* 2007;35(5):1452–1464.
55. Wang Z, Burge CB. Splicing regulation: from a parts list of regulatory elements to an integrated splicing code. *RNA.* 2008;14(5):802–813.
56. Will CL, Luhrmann R. Spliceosome structure and function. *Cold Spring Harb Perspect Biol.* 2011;3(7):a003707.
57. Somarowthu S, Legiewicz M, Chillon I, Marcia M, Liu F, Pyle AM. HOTAIR forms an intricate and modular secondary structure. *Mol Cell.* 2015;58(2):353–361.
58. Aravin AA, Hannon GJ, Brennecke J. The Piwi-piRNA pathway provides an adaptive defense in the transposon arms race. *Science.* 2007;318(5851):761–764.
59. Bourc'his D, Bestor TH. Meiotic catastrophe and retrotransposon reactivation in male germ cells lacking Dnmt3L. *Nature.* 2004;431(7004):96–99.
60. Aravin AA, Sachidanandam R, Bourc'his D, et al. A piRNA pathway primed by individual transposons is linked to de novo DNA methylation in mice. *Mol Cell.* 2008;31(6):785–799.
61. Manakov SA, Pezic D, Marinov GK, Pastor WA, Sachidanandam R, Aravin AA. MIWI2 and MILI have differential effects on piRNA biogenesis and DNA methylation. *Cell Rep.* 2015;12(8):1234–1243.
62. Luteijn MJ, Ketting RF. PIWI-interacting RNAs: from generation to transgenerational epigenetics. *Nat Rev Genet.* 2013;14(8):523–534.
63. Wu Q, Ma Q, Shehadeh LA, et al. Expression of the Argonaute protein PiwiL2 and piRNAs in adult mouse mesenchymal stem cells. *Biochem Biophys Res Commun.* 2010;396(4):915–920.
64. Bamezai S, Rawat VP, Buske C. Concise review: the Piwi-piRNA axis: pivotal beyond transposon silencing. *Stem Cells.* 2012;30(12):2603–2611.
65. Peng JC, Lin H. Beyond transposons: the epigenetic and somatic functions of the Piwi-piRNA mechanism. *Curr Opin Cell Biol.* 2013;25(2):190–194.
66. Iwasaki YW, Siomi MC, Siomi H. PIWI-interacting RNA: its biogenesis and functions. *Annu Rev Biochem.* 2015;84:405–433.
67. Lee EJ, Banerjee S, Zhou H, et al. Identification of piRNAs in the central nervous system. *RNA.* 2011;17(6):1090–1099.
68. Rajasethupathy P, Antonov I, Sheridan R, et al. A role for neuronal piRNAs in the epigenetic control of memory-related synaptic plasticity. *Cell.* 2012;149(3):693–707.
69. Bartel DP. MicroRNAs: genomics, biogenesis, mechanism, and function. *Cell.* 2004;116(2):281–297.
70. Ambros V. The functions of animal microRNAs. *Nature.* 2004;431(7006):350–355.

71. Selbach M, Schwanhausser B, Thierfelder N, Fang Z, Khanin R, Rajewsky N. Widespread changes in protein synthesis induced by microRNAs. *Nature.* 2008;455 (7209):58–63.
72. Baek D, Villen J, Shin C, Camargo FD, Gygi SP, Bartel DP. The impact of microRNAs on protein output. *Nature.* 2008;455(7209):64–71.
73. Han J, Lee Y, Yeom KH, Kim YK, Jin H, Kim VN. The Drosha-DGCR8 complex in primary microRNA processing. *Genes Dev.* 2004;18(24):3016–3027.
74. Han J, Lee Y, Yeom KH, et al. Molecular basis for the recognition of primary microRNAs by the Drosha-DGCR8 complex. *Cell.* 2006;125(5):887–901.
75. Lee Y, Ahn C, Han J, et al. The nuclear RNase III Drosha initiates microRNA processing. *Nature.* 2003;425(6956):415–419.
76. Bohnsack MT, Czaplinski K, Gorlich D. Exportin 5 is a RanGTP-dependent dsRNA-binding protein that mediates nuclear export of pre-miRNAs. *RNA.* 2004;10 (2):185–191.
77. Yi R, Qin Y, Macara IG, Cullen BR. Exportin-5 mediates the nuclear export of pre-microRNAs and short hairpin RNAs. *Genes Dev.* 2003;17(24):3011–3016.
78. Bernstein E, Caudy AA, Hammond SM, Hannon GJ. Role for a bidentate ribonuclease in the initiation step of RNA interference. *Nature.* 2001;409(6818):363–366.
79. Bartel DP. MicroRNAs: target recognition and regulatory functions. *Cell.* 2009;136 (2):215–233.
80. Jonas S, Izaurralde E. Towards a molecular understanding of microRNA-mediated gene silencing. *Nat Rev Genet.* 2015;16(7):421–433.
81. Ameres SL, Zamore PD. Diversifying microRNA sequence and function. *Nat Rev Mol Cell Biol.* 2013;14(8):475–488.
82. Mukherji S, Ebert MS, Zheng GX, Tsang JS, Sharp PA, van Oudenaarden A. MicroRNAs can generate thresholds in target gene expression. *Nat Genet.* 2011;43 (9):854–859.
83. Roberts TC. The microRNA biology of the mammalian nucleus. *Mol Ther Nucleic Acids.* 2014;3:e188.
84. Ameyar-Zazoua M, Rachez C, Souidi M, et al. Argonaute proteins couple chromatin silencing to alternative splicing. *Nat Struct Mol Biol.* 2012;19(10):998–1004.
85. Chi SW, Zang JB, Mele A, Darnell RB. Argonaute HITS-CLIP decodes microRNA-mRNA interaction maps. *Nature.* 2009;460(7254):479–486.
86. Place RF, Li LC, Pookot D, Noonan EJ, Dahiya R. MicroRNA-373 induces expression of genes with complementary promoter sequences. *Proc Natl Acad Sci U S A.* 2008;105 (5):1608–1613.
87. Matsui M, Chu Y, Zhang H, et al. Promoter RNA links transcriptional regulation of inflammatory pathway genes. *Nucleic Acids Res.* 2013;41(22):10086–10109.
88. Noh JH, Kim KM, Abdelmohsen K, et al. HuR and GRSF1 modulate the nuclear export and mitochondrial localization of the lncRNA RMRP. *Genes Dev.* 2016;30 (10):1224–1239.
89. Guttman M, Amit I, Garber M, et al. Chromatin signature reveals over a thousand highly conserved large non-coding RNAs in mammals. *Nature.* 2009;458 (7235):223–227.
90. D'haene E, Jacobs EZ, Volders PJ, De Meyer T, Menten B, Vergult S. Identification of long non-coding RNAs involved in neuronal development and intellectual disability. *Sci Rep.* 2016;6:28396.
91. Mercer TR, Dinger ME, Mattick JS. Long non-coding RNAs: insights into functions. *Nat Rev Genet.* 2009;10(3):155–159.
92. Kung JT, Colognori D, Lee JT. Long noncoding RNAs: past, present, and future. *Genetics.* 2013;193(3):651–669.

93. Khalil AM, Guttman M, Huarte M, et al. Many human large intergenic noncoding RNAs associate with chromatin-modifying complexes and affect gene expression. *Proc Natl Acad Sci U S A.* 2009;106(28):11667–11672.
94. Kmita M, Duboule D. Organizing axes in time and space; 25 years of colinear tinkering. *Science.* 2003;301(5631):331–333.
95. Rinn JL, Kertesz M, Wang JK, et al. Functional demarcation of active and silent chromatin domains in human HOX loci by noncoding RNAs. *Cell.* 2007;129(7):1311–1323.
96. Tsai MC, Manor O, Wan Y, et al. Long noncoding RNA as modular scaffold of histone modification complexes. *Science.* 2010;329(5992):689–693.
97. Rashid F, Shah A, Shan G. Long non-coding RNAs in the cytoplasm. *Genomics Proteomics Bioinform.* 2016;14(2):73–80.
98. Wu P, Zuo X, Deng H, Liu X, Liu L, Ji A. Roles of long noncoding RNAs in brain development, functional diversification and neurodegenerative diseases. *Brain Res Bull.* 2013;97:69–80.
99. Ebert MS, Sharp PA. Emerging roles for natural microRNA sponges. *Curr Biol CB.* 2010;20(19):R858–R861.
100. Kallen AN, Zhou XB, Xu J, et al. The imprinted H19 lncRNA antagonizes let-7 microRNAs. *Mol Cell.* 2013;52(1):101–112.
101. Cai X, Cullen BR. The imprinted H19 noncoding RNA is a primary microRNA precursor. *RNA.* 2007;13(3):313–316.
102. Dey BK, Pfeifer K, Dutta A. The H19 long noncoding RNA gives rise to microRNAs miR-675-3p and miR-675-5p to promote skeletal muscle differentiation and regeneration. *Genes Dev.* 2014;28(5):491–501.
103. Keniry A, Oxley D, Monnier P, et al. The H19 lincRNA is a developmental reservoir of miR-675 that suppresses growth and Igf1r. *Nat Cell Biol.* 2012;14(7):659–665.
104. Haycock PC. Fetal alcohol spectrum disorders: the epigenetic perspective. *Biol Reprod.* 2009;81(4):607–617.
105. Kleiber ML, Diehl EJ, Laufer BI, et al. Long-term genomic and epigenomic dysregulation as a consequence of prenatal alcohol exposure: a model for fetal alcohol spectrum disorders. *Front Genet.* 2014;5:161.
106. Basavarajappa BS, Subbanna S. Epigenetic mechanisms in developmental alcohol-induced neurobehavioral deficits. *Brain Sci.* 2016;6(2):12.
107. Kobor MS, Weinberg J. Focus on: epigenetics and fetal alcohol spectrum disorders. *Alcohol Res Health.* 2011;34(1):29–37.
108. Laufer BI, Chater-Diehl EJ, Kapalanga J, Singh SM. Long-term alterations to DNA methylation as a biomarker of prenatal alcohol exposure: from mouse models to human children with fetal alcohol spectrum disorders. *Alcohol.* 2017;60:67–75.
109. Chater-Diehl EJ, Laufer BI, Singh SM. Changes to histone modifications following prenatal alcohol exposure: an emerging picture. *Alcohol.* 2017;60:41–52.
110. Garro AJ, McBeth DL, Lima V, Lieber CS. Ethanol consumption inhibits fetal DNA methylation in mice: implications for the fetal alcohol syndrome. *Alcohol Clin Exp Res.* 1991;15(3):395–398.
111. Zhou FC, Balaraman Y, Teng M, Liu Y, Singh RP, Nephew KP. Alcohol alters DNA methylation patterns and inhibits neural stem cell differentiation. *Alcohol Clin Exp Res.* 2011;35(4):735–746.
112. Maier SE, Cramer JA, West JR, Sohrabji F. Alcohol exposure during the first two trimesters equivalent alters granule cell number and neurotrophin expression in the developing rat olfactory bulb. *J Neurobiol.* 1999;41(3):414–423.
113. Liu Y, Balaraman Y, Wang G, Nephew KP, Zhou FC. Alcohol exposure alters DNA methylation profiles in mouse embryos at early neurulation. *Epigenetics.* 2009;4(7):500–511.

114. Kaminen-Ahola N, Ahola A, Maga M, et al. Maternal ethanol consumption alters the epigenotype and the phenotype of offspring in a mouse model. *PLoS Genet.* 2010;6(1): e1000811.
115. Veazey KJ, Wang H, Bedi YS, Skiles WM, Chang RC, Golding MC. Disconnect between alcohol-induced alterations in chromatin structure and gene transcription in a mouse embryonic stem cell model of exposure. *Alcohol.* 2017;60:121–133.
116. Abel E. Paternal contribution to fetal alcohol syndrome. *Addict Biol.* 2004;9 (2):127–133. [discussion 135-126].
117. Little RE, Sing CF. Father's drinking and infant birth weight: report of an association. *Teratology.* 1987;36(1):59–65.
118. Savitz DA, Schwingl PJ, Keels MA. Influence of paternal age, smoking, and alcohol consumption on congenital anomalies. *Teratology.* 1991;44(4):429–440.
119. Hegedus AM, Alterman AI, Tarter RE. Learning achievement in sons of alcoholics. *Alcohol Clin Exp Res.* 1984;8(3):330–333.
120. Bielawski DM, Zaher FM, Svinarich DM, Abel EL. Paternal alcohol exposure affects sperm cytosine methyltransferase messenger RNA levels. *Alcohol Clin Exp Res.* 2002;26 (3):347–351.
121. Rompala GR, Finegersh A, Slater M, Homanics GE. Paternal preconception alcohol exposure imparts intergenerational alcohol-related behaviors to male offspring on a pure C57BL/6J background. *Alcohol.* 2017;60:169–177.
122. Finegersh A, Homanics GE. Paternal alcohol exposure reduces alcohol drinking and increases behavioral sensitivity to alcohol selectively in male offspring. *PLOS ONE.* 2014;9:e99078.
123. Ouko LA, Shantikumar K, Knezovich J, Haycock P, Schnugh DJ, Ramsay M. Effect of alcohol consumption on CpG methylation in the differentially methylated regions of H19 and IG-DMR in male gametes: implications for fetal alcohol spectrum disorders. *Alcohol Clin Exp Res.* 2009;33(9):1615–1627.
124. Knezovich JG, Ramsay M. The effect of preconception paternal alcohol exposure on epigenetic remodeling of the H19 and Rasgrf1 imprinting control regions in mouse offspring. *Front Genet.* 2012;3:10.
125. Resendiz M, Chen Y, Ozturk NC, Zhou FC. Epigenetic medicine and fetal alcohol spectrum disorders. *Epigenomics.* 2013;5(1):73–86.
126. Stouder C, Somm E, Paoloni-Giacobino A. Prenatal exposure to ethanol: a specific effect on the H19 gene in sperm. *Reprod Toxicol.* 2011;31(4):507–512.
127. Govorko D, Bekdash RA, Zhang C, Sarkar DK. Male germline transmits fetal alcohol adverse effect on hypothalamic proopiomelanocortin gene across generations. *Biol Psychiatry.* 2012;72:378–388.
128. Laufer BI, Mantha K, Kleiber ML, Diehl EJ, Addison SM, Singh SM. Long-lasting alterations to DNA methylation and ncRNAs could underlie the effects of fetal alcohol exposure in mice. *Dis Model Mech.* 2013;6(4):977–992.
129. Miranda RC, Pietrzykowski AZ, Tang Y, et al. MicroRNAs: master regulators of ethanol abuse and toxicity? *Alcohol Clin Exp Res.* 2010;34(4):575–587.
130. Landthaler M, Yalcin A, Tuschl T. The human DiGeorge syndrome critical region gene 8 and its *D. melanogaster* homolog are required for miRNA biogenesis. *Curr Biol CB.* 2004;14(23):2162–2167.
131. Stringer RL, Laufer BI, Kleiber ML, Singh SM. Reduced expression of brain cannabinoid receptor 1 (Cnr1) is coupled with an increased complementary micro-RNA (miR-26b) in a mouse model of fetal alcohol spectrum disorders. *Clin Epigenet.* 2013;5 (1):14.
132. Soares AR, Pereira PM, Ferreira V, et al. Ethanol exposure induces upregulation of specific microRNAs in zebrafish embryos. *Toxicol Sci.* 2012;127(1):18–28.

133. van Steenwyk G, Janeczek P, Lewohl JM. Differential effects of chronic and chronic-intermittent ethanol treatment and its withdrawal on the expression of miRNAs. *Brain Sci.* 2013;3(2):744–756.
134. Lewohl JM, Nunez YO, Dodd PR, Tiwari GR, Harris RA, Mayfield RD. Up-regulation of microRNAs in brain of human alcoholics. *Alcohol Clin Exp Res.* 2011;35 (11):1928–1937.
135. Tapocik JD, Solomon M, Flanigan M, et al. Coordinated dysregulation of mRNAs and microRNAs in the rat medial prefrontal cortex following a history of alcohol dependence. *Pharmacogenomics J.* 2013;13(3):286–296.
136. Natarajan SK, Pachunka JM, Mott JL. Role of microRNAs in alcohol-induced multi-organ injury. *Biomolecules.* 2015;5(4):3309–3338.
137. Mandal C, Halder D, Jung KH, Chai YG. Maternal alcohol consumption and altered miRNAs in the developing fetus: context and future perspectives. *J Appl Toxicol.* 2018;38 (1):100–107.
138. Coolen M, Katz S, Bally-Cuif L. miR-9: a versatile regulator of neurogenesis. *Front Cell Neurosci.* 2013;7:220.
139. Coolen M, Thieffry D, Drivenes O, Becker TS, Bally-Cuif L. miR-9 controls the timing of neurogenesis through the direct inhibition of antagonistic factors. *Dev Cell.* 2012;22(5):1052–1064.
140. Leucht C, Stigloher C, Wizenmann A, Klafke R, Folchert A, Bally-Cuif L. MicroRNA-9 directs late organizer activity of the midbrain-hindbrain boundary. *Nat Neurosci.* 2008;11(6):641–648.
141. Yoo AS, Sun AX, Li L, et al. MicroRNA-mediated conversion of human fibroblasts to neurons. *Nature.* 2011;476(7359):228–231.
142. Krichevsky AM, Sonntag KC, Isacson O, Kosik KS. Specific microRNAs modulate embryonic stem cell-derived neurogenesis. *Stem Cells.* 2006;24(4):857–864.
143. Delaloy C, Liu L, Lee JA, et al. MicroRNA-9 coordinates proliferation and migration of human embryonic stem cell-derived neural progenitors. *Cell Stem Cell.* 2010;6 (4):323–335.
144. Sathyan P, Golden HB, Miranda RC. Competing interactions between micro-RNAs determine neural progenitor survival and proliferation after ethanol exposure: evidence from an ex vivo model of the fetal cerebral cortical neuroepithelium. *J Neurosci.* 2007;27:8546–8557.
145. Pappalardo-Carter DL, Balaraman S, Sathyan P, Carter ES, Chen W-JA. Miranda RC. Suppression and epigenetic regulation of miR-9 contributes to ethanol teratology: evidence from zebrafish and murine fetal neural stem cell models. *Alcohol Clin Exp Res.* 2013;37:1657–1667.
146. Tal TL, Franzosa JA, Tilton SC, et al. MicroRNAs control neurobehavioral development and function in zebrafish. *FASEB J.* 2012;26(4):1452–1461.
147. Shibata M, Nakao H, Kiyonari H, Abe T, Aizawa S. MicroRNA-9 regulates neurogenesis in mouse telencephalon by targeting multiple transcription factors. *J Neurosci.* 2011;31(9):3407–3422.
148. Tsai PC, Bake S, Balaraman S, et al. miR-153 targets the nuclear factor-1 family and protects against teratogenic effects of ethanol exposure in fetal neural stem cells. *Biol Open.* 2014;3(8):741–758.
149. Clovis YM, Enard W, Marinaro F, Huttner WB, De Pietri Tonelli D. Convergent repression of Foxp2 3′UTR by miR-9 and miR-132 in embryonic mouse neocortex: implications for radial migration of neurons. *Development.* 2012;139(18):3332–3342.
150. Wang L-L, Zhang Z, Li Q, et al. Ethanol exposure induces differential microRNA and target gene expression and teratogenic effects which can be suppressed by folic acid supplementation. *Hum Reprod.* 2009;24:562–579.

151. Molyneaux BJ, Arlotta P, Menezes JR, Macklis JD. Neuronal subtype specification in the cerebral cortex. *Nat Rev Neurosci.* 2007;8(6):427–437.
152. Gangisetty O, Jabbar S, Wynne O, Sarkar DK. microRNA-9 regulates fetal alcohol-induced changes in D2 receptor to promote prolactin production. *J Endocrinol.* 2017;235(1):1–14.
153. Gangisetty O, Wynne O, Jabbar S, Nasello C, Sarkar DK. Fetal alcohol exposure reduces dopamine receptor D2 and increases pituitary weight and prolactin production via epigenetic mechanisms. *PLOS ONE.* 2015;10(10):e0140699.
154. Pietrzykowski AZ, Friesen RM, Martin GE, et al. Posttranscriptional regulation of BK channel splice variant stability by miR-9 underlies neuroadaptation to alcohol. *Neuron.* 2008;59:274–287.
155. Burrowes SG, Salem NA, Tseng AM, et al. The BAF (BRG1/BRM-Associated Factor) chromatin-remodeling complex exhibits ethanol sensitivity in fetal neural progenitor cells and regulates transcription at the miR-9-2 encoding gene locus. *Alcohol.* 2017;60:149–158.
156. Balaraman S, Winzer-Serhan UH, Miranda RC. Opposing actions of ethanol and nicotine on microRNAs are mediated by nicotinic acetylcholine receptors in fetal cerebral cortical-derived neural progenitor cells. *Alcohol Clin Exp Res.* 2012;36 (10):1669–1677.
157. Tsuyama J, Bunt J, Richards LJ, et al. microRNA-153 regulates the acquisition of gliogenic competence by neural stem cells. *Stem Cell Rep.* 2015;5(3):365–377.
158. Shin J, Berg DA, Zhu Y, et al. Single-cell RNA-Seq with waterfall reveals molecular cascades underlying adult neurogenesis. *Cell Stem Cell.* 2015;17(3):360–372.
159. Wang W, Crandall JE, Litwack ED, Gronostajski RM, Kilpatrick DL. Targets of the nuclear factor I regulon involved in early and late development of postmitotic cerebellar granule neurons. *J Neurosci Res.* 2010;88(2):258–265.
160. Namihira M, Kohyama J, Semi K, et al. Committed neuronal precursors confer astrocytic potential on residual neural precursor cells. *Dev Cell.* 2009;16(2):245–255.
161. Deneen B, Ho R, Lukaszewicz A, Hochstim CJ, Gronostajski RM, Anderson DJ. The transcription factor NFIA controls the onset of gliogenesis in the developing spinal cord. *Neuron.* 2006;52(6):953–968.
162. Piper M, Barry G, Hawkins J, et al. NFIA controls telencephalic progenitor cell differentiation through repression of the Notch effector Hes1. *J Neurosci.* 2010;30 (27):9127–9139.
163. Barry G, Piper M, Lindwall C, et al. Specific glial populations regulate hippocampal morphogenesis. *J Neurosci.* 2008;28(47):12328–12340.
164. Santillano DR, Kumar LS, Prock TL, Camarillo C, Tingling JD, Miranda RC. Ethanol induces cell-cycle activity and reduces stem cell diversity to alter both regenerative capacity and differentiation potential of cerebral cortical neuroepithelial precursors. *BMC Neurosci.* 2005;6:59.
165. Camarillo C, Miranda RC. Ethanol exposure during neurogenesis induces persistent effects on neural maturation: evidence from an ex vivo model of fetal cerebral cortical neuroepithelial progenitor maturation. *Gene Express.* 2008;14(3):159–171.
166. Miranda RC. microRNAs and fetal brain development: implications for ethanol teratology during the second trimester period of neurogenesis. *Front Genet.* 2012;3:77.
167. Wei C, Thatcher EJ, Olena AF, et al. miR-153 regulates SNAP-25, synaptic transmission, and neuronal development. *PLOS ONE.* 2013;8(2):e57080.
168. Doxakis E. Post-transcriptional regulation of alpha-synuclein expression by mir-7 and mir-153. *J Biol Chem.* 2010;285(17):12726–12734.
169. Lashuel HA, Overk CR, Oueslati A, Masliah E. The many faces of alpha-synuclein: from structure and toxicity to therapeutic target. *Nat Rev Neurosci.* 2013;14(1):38–48.

170. Bonsch D, Lederer T, Reulbach U, Hothorn T, Kornhuber J, Bleich S. Joint analysis of the NACP-REP1 marker within the alpha synuclein gene concludes association with alcohol dependence. *Hum Mol Genet.* 2005;14(7):967–971.
171. Bonsch D, Reulbach U, Bayerlein K, Hillemacher T, Kornhuber J, Bleich S. Elevated alpha synuclein mRNA levels are associated with craving in patients with alcoholism. *Biol Psychiatry.* 2004;56(12):984–986.
172. Bonsch D, Greifenberg V, Bayerlein K, et al. Alpha-synuclein protein levels are increased in alcoholic patients and are linked to craving. *Alcohol Clin Exp Res.* 2005;29 (5):763–765.
173. Liang T, Spence J, Liu L, et al. alpha-Synuclein maps to a quantitative trait locus for alcohol preference and is differentially expressed in alcohol-preferring and -nonpreferring rats. *Proc Natl Acad Sci U S A.* 2003;100(8):4690–4695.
174. Liang T, Carr LG. Regulation of alpha-synuclein expression in alcohol-preferring and -non preferring rats. *J Neurochem.* 2006;99(2):470–482.
175. Agarwal V, Bell GW, Nam JW, Bartel DP. Predicting effective microRNA target sites in mammalian mRNAs. *Elife.* 2015;4:e05005.
176. Burd L, Deal E, Rios R, Adickes E, Wynne J, Klug MG. Congenital heart defects and fetal alcohol spectrum disorders. *Congenit Heart Dis.* 2007;2:250–255.
177. Beech RD, Leffert JJ, Lin A, et al. Stress-related alcohol consumption in heavy drinkers correlates with expression of miR-10a, miR-21, and components of the TAR-RNA-binding protein-associated complex. *Alcohol Clin Exp Res.* 2014;38 (11):2743–2753.
178. Espinoza-Lewis RA, Wang DZ. microRNAs in heart development. *Curr Top Dev Biol.* 2012;100:279–317.
179. Thum T, Gross C, Fiedler J, et al. microRNA-21 contributes to myocardial disease by stimulating MAP kinase signalling in fibroblasts. *Nature.* 2008;456(7224):980–984.
180. Yan S, Jiao K. Functions of miRNAs during mammalian heart development. *Int J Mol Sci.* 2016;17(5):789.
181. Bronnum H, Andersen DC, Schneider M, et al. miR-21 promotes fibrogenic epithelial-to-mesenchymal transition of epicardial mesothelial cells involving Programmed Cell Death 4 and Sprouty-1. *PLOS ONE.* 2013;8(2):e56280.
182. Makki N, Capecchi MR. Hoxa1 lineage tracing indicates a direct role for Hoxa1 in the development of the inner ear, the heart, and the third rhombomere. *Dev Biol.* 2010;341 (2):499–509.
183. Tischfield MA, Bosley TM, Salih MA, et al. Homozygous HOXA1 mutations disrupt human brainstem, inner ear, cardiovascular and cognitive development. *Nat Genet.* 2005;37(10):1035–1037.
184. Liu Q, Gao F, Liu X, et al. Prenatal alcohol exposure and offspring liver dysfunction: a systematic review and meta-analysis. *Arch Gynecol Obstet.* 2016;294(2):225–231.
185. Hofer R, Burd L. Review of published studies of kidney, liver, and gastrointestinal birth defects in fetal alcohol spectrum disorders. *Birth Defects Res A Clin Mol Teratol.* 2009;85:179–183.
186. Sinha R, Fox HC, Hong KA, Bergquist K, Bhagwagar Z, Siedlarz KM. Enhanced negative emotion and alcohol craving, and altered physiological responses following stress and cue exposure in alcohol dependent individuals. *Neuropsychopharmacology.* 2009;34(5):1198–1208.
187. Sinha R, Fox HC, Hong KI, Hansen J, Tuit K, Kreek MJ. Effects of adrenal sensitivity, stress- and cue-induced craving, and anxiety on subsequent alcohol relapse and treatment outcomes. *Arch Gen Psychiatry.* 2011;68(9):942–952.
188. Bale TL. Sex differences in prenatal epigenetic programming of stress pathways. *Stress.* 2011;14(4):348–356.

189. Bronson SL, Bale TL. Prenatal stress-induced increases in placental inflammation and offspring hyperactivity are male-specific and ameliorated by maternal antiinflammatory treatment. *Endocrinology.* 2014;155(7):2635–2646.
190. Carter RC, Jacobson JL, Molteno CD, Dodge NC, Meintjes EM, Jacobson SW. Fetal alcohol growth restriction and cognitive impairment. *Pediatrics.* 2016;138(2):e20160775.
191. Krishna U, Bhalerao S. Placental insufficiency and fetal growth restriction. *J Obstet Gynaecol India.* 2011;61(5):505–511.
192. Burd L, Roberts D, Olson M, Odendaal H. Ethanol and the placenta: a review. *J Matern Fetal Neonatal Med.* 2007;20(5):361–375.
193. Chabenne A, Moon C, Ojo C, Khogali A, Nepal B, Sharma S. Biomarkers in fetal alcohol syndrome. *Biomark Genomic Med.* 2014;6(1):12–22.
194. Munaut C, Tebache L, Blacher S, Noël A, Nisolle M, Chantraine F. Dysregulated circulating miRNAs in preeclampsia. *Biomed Rep.* 2016;5(6):686–692.
195. Jairajpuri DS, Malalla ZH, Mahmood N, Almawi WY. Circulating microRNA expression as predictor of preeclampsia and its severity. *Gene.* 2017;627:543–548.
196. Balaraman S, Schafer JJ, Tseng AM, et al. Plasma miRNA profiles in pregnant women predict infant outcomes following prenatal alcohol exposure. *PLOS ONE.* 2016;11(11): e0165081.
197. Wang D, Na Q, Song W-W, Song G-Y. Altered expression of miR-518b and miR-519a in the placenta is associated with low fetal birth weight. *Am J Perinatol.* 2014;31 (09):729–734.
198. Zhu XM, Han T, Sargent IL, Yin GW, Yao YQ. Differential expression profile of microRNAs in human placentas from preeclamptic pregnancies vs normal pregnancies. *Am J Obstet Gynecol.* 2009;200(6). 661.e1–7.
199. Hromadnikova I, Kotlabova K, Ondrackova M, et al. Expression profile of C19MC microRNAs in placental tissue in pregnancy-related complications. *DNA Cell Biol.* 2015;34(6):437–457.
200. Xie L, Sadovsky Y. The function of miR-519d in cell migration, invasion, and proliferation suggests a role in early placentation. *Placenta.* 2016;48:34–37.
201. Lean SC, Heazell AEP, Dilworth MR, Mills TA, Jones RL. Placental dysfunction underlies increased risk of fetal growth restriction and stillbirth in advanced maternal age women. *Sci Rep.* 2017;7(1):9677.
202. Gordon BH, Streeter ML, Rosso P, Winick M. Prenatal alcohol exposure: abnormalities in placental growth and fetal amino acid uptake in the rat. *Biol Neonate.* 1985;47(2):113–119.
203. Clave S, Joya X, Salat-Batlle J, Garcia-Algar O, Vall O. Ethanol cytotoxic effect on trophoblast cells. *Toxicol Lett.* 2014;225(2):216–221.
204. Lui S, Jones RL, Robinson NJ, Greenwood SL, Aplin JD, Tower CL. Detrimental effects of ethanol and its metabolite acetaldehyde, on first trimester human placental cell turnover and function. *PLOS ONE.* 2014;9(2):e87328.
205. Thamotharan S, Chu A, Kempf K, et al. Differential microRNA expression in human placentas of term intra-uterine growth restriction that regulates target genes mediating angiogenesis and amino acid transport. *PLOS ONE.* 2017;12(5):e0176493.
206. Maccani MA, Padbury JF, Marsit CJ. miR-16 and miR-21 expression in the placenta is associated with fetal growth. *PLOS ONE.* 2011;6(6):e21210.
207. Wang J, Cui Q. Specific roles of microRNAs in their interactions with environmental factors. *J Nucleic Acids.* 2012;2012:10.
208. Maccani MA, Padbury JF, Lester BM, Knopik VS, Marsit CJ. Placental miRNA expression profiles associated with measures of infant neurobehavioral outcomes. *Pediatr Res.* 2013;74(3):272–278.
209. Etheridge A, Lee I, Hood L, Galas D, Wang K. Extracellular microRNA: a new source of biomarkers. *Mutat Res.* 2011;717(1–2):85–90.

210. Valadi H, Ekstrom K, Bossios A, Sjostrand M, Lee JJ, Lotvall JO. Exosome-mediated transfer of mRNAs and microRNAs is a novel mechanism of genetic exchange between cells. *Nat Cell Biol.* 2007;9(6):654–659.
211. Gibbings DJ, Ciaudo C, Erhardt M, Voinnet O. Multivesicular bodies associate with components of miRNA effector complexes and modulate miRNA activity. *Nat Cell Biol.* 2009;11(9):1143–1149.
212. Thery C, Ostrowski M, Segura E. Membrane vesicles as conveyors of immune responses. *Nat Rev Immunol.* 2009;9(8):581–593.
213. Balaraman S, Lunde ER, Sawant O, Cudd TA, Washburn SE, Miranda RC. Maternal and neonatal plasma microRNA biomarkers for fetal alcohol exposure in an ovine model. *Alcohol Clin Exp Res.* 2014;38(5):1390–1400.
214. Gardiner AS, Gutierrez HL, Luo L, et al. Alcohol use during pregnancy is associated with specific alterations in microRNA levels in maternal serum. *Alcohol Clin Exp Res.* 2016;40(4):826–837.
215. Thomas JD, Garrison M, O'Neill TM. Perinatal choline supplementation attenuates behavioral alterations associated with neonatal alcohol exposure in rats. *Neurotoxicol Teratol.* 2004;26(1):35–45.
216. Ryan SH, Williams JK, Thomas JD. Choline supplementation attenuates learning deficits associated with neonatal alcohol exposure in the rat: effects of varying the timing of choline administration. *Brain Res.* 2008;1237:91–100.
217. Wozniak JR, Fuglestad AJ, Eckerle JK, et al. Choline supplementation in children with fetal alcohol spectrum disorders: a randomized, double-blind, placebo-controlled trial. *Am J Clin Nutr.* 2015;102(5):1113–1125.
218. Nguyen TT, Risbud RD, Mattson SN, Chambers CD, Thomas JD. Randomized, double-blind, placebo-controlled clinical trial of choline supplementation in school-aged children with fetal alcohol spectrum disorders. *Am J Clin Nutr.* 2016;104(6):1683–1692.
219. Balaraman S, Idrus NM, Miranda RC, Thomas JD. Postnatal choline supplementation selectively attenuates hippocampal microRNA alterations associated with developmental alcohol exposure. *Alcohol.* 2017;60:159–167.
220. Idrus NM, Breit KR, Thomas JD. Dietary choline levels modify the effects of prenatal alcohol exposure in rats. *Neurotoxicol Teratol.* 2017;59:43–52.
221. Guduric-Fuchs J, O'Connor A, Camp B, O'Neill CL, Medina RJ, Simpson DA. Selective extracellular vesicle-mediated export of an overlapping set of microRNAs from multiple cell types. *BMC Genomics.* 2012;13:357.
222. Villarroya-Beltri C, Gutiérrez-Vázquez C, Sánchez-Cabo F, et al. Sumoylated hnRNPA2B1 controls the sorting of miRNAs into exosomes through binding to specific motifs. *Nat Commun.* 2013;4:2980.
223. Mittelbrunn M, Gutiérrez-Vázquez C, Villarroya-Beltri C, et al. Unidirectional transfer of microRNA-loaded exosomes from T cells to antigen-presenting cells. *Nat Commun.* 2011;2:282.
224. Bell AC, Felsenfeld G. Methylation of a CTCF-dependent boundary controls imprinted expression of the Igf2 gene. *Nature.* 2000;405(6785):482–485.
225. Chao W, D'Amore PA. IGF2: epigenetic regulation and role in development and disease. *Cytokine Growth Factor Rev.* 2008;19(2):111–120.
226. Haycock PC, Ramsay M. Exposure of mouse embryos to ethanol during preimplantation development: effect on DNA methylation in the h19 imprinting control region. *Biol Reprod.* 2009;81(4):618–627.
227. Monnier P, Martinet C, Pontis J, Stancheva I, Ait-Si-Ali S, Dandolo L. H19 lncRNA controls gene expression of the Imprinted Gene Network by recruiting MBD1. *Proc Natl Acad Sci U S A.* 2013;110(51):20693–20698.

228. Kryger R, Fan L, Wilce PA, Jaquet V. MALAT-1, a non protein-coding RNA is upregulated in the cerebellum, hippocampus and brain stem of human alcoholics. *Alcohol.* 2012;46(7):629–634.
229. Lin M, Pedrosa E, Shah A, et al. RNA-Seq of human neurons derived from iPS cells reveals candidate long non-coding RNAs involved in neurogenesis and neuropsychiatric disorders. *PLoS ONE.* 2011;6(9):e23356.
230. Leucci E, Patella F, Waage J, et al. microRNA-9 targets the long non-coding RNA MALAT1 for degradation in the nucleus. *Sci Rep.* 2013;3:2535.
231. Farris SP, Mayfield RD. RNA-Seq reveals novel transcriptional reorganization in human alcoholic brain. *Int Rev Neurobiol.* 2014;116:275–300.
232. Schuster-Gossler K, Bilinski P, Sado T, Ferguson-Smith A, Gossler A. The mouse Gtl2 gene is differentially expressed during embryonic development, encodes multiple alternatively spliced transcripts, and may act as an RNA. *Dev Dyn.* 1998;212 (2):214–228.
233. Wu Z, He Y, Li D, et al. Long noncoding RNA MEG3 suppressed endothelial cell proliferation and migration through regulating miR-21. *Am J Transl Res.* 2017;9 (7):3326–3335.
234. Zhang J, Yao T, Wang Y, Yu J, Liu Y, Lin Z. Long noncoding RNA MEG3 is downregulated in cervical cancer and affects cell proliferation and apoptosis by regulating miR-21. *Cancer Biol Ther.* 2016;17:104–113.
235. He C, Yang W, Yang J, et al. Long noncoding RNA MEG3 negatively regulates proliferation and angiogenesis in vascular endothelial cells. *DNA Cell Biol.* 2017;36 (6):475–481.
236. Lachmeijer AM, Arngrimsson R, Bastiaans EJ, et al. A genome-wide scan for preeclampsia in the Netherlands. *Eur J Hum Genet EJHG.* 2001;9(10):758–764.
237. van Dijk M, Thulluru HK, Mulders J, et al. HELLP babies link a novel lincRNA to the trophoblast cell cycle. *J Clin Invest.* 2012;122:4003–4011.
238. Gudenas BL, Srivastava AK, Wang L. Integrative genomic analyses for identification and prioritization of long non-coding RNAs associated with autism. *PLOS ONE.* 2017;12 (5):e0178532.
239. Villegas VE, Zaphiropoulos PG. Neighboring gene regulation by antisense long non-coding RNAs. *Int J Mol Sci.* 2015;16(2):3251–3266.
240. Robledo RF, Rajan L, Li X, Lufkin T. The Dlx5 and Dlx6 homeobox genes are essential for craniofacial, axial, and appendicular skeletal development. *Genes Dev.* 2002;16 (9):1089–1101.
241. Bond AM, VanGompel MJ, Sametsky EA, et al. Balanced gene regulation by an embryonic brain non-coding RNA is critical for GABA circuitry in adult hippocampus. *Nat Neurosci.* 2009;12(8):1020–1027.
242. Cajigas I, Leib DE, Cochrane J, et al. Evf2 lncRNA/BRG1/DLX1 interactions reveal RNA-dependent inhibition of chromatin remodeling. *Development.* 2015;142 (15):2641–2652.
243. Laufer BI, Kapalanga J, Castellani CA, Diehl EJ, Yan L, Singh SM. Associative DNA methylation changes in children with prenatal alcohol exposure. *Epigenomics.* 2015;7 (8):1259–1274.
244. Ding F, Li HH, Zhang S, et al. snoRNA Snord116 (Pwcr1/MBII-85) deletion causes growth deficiency and hyperphagia in mice. *PLoS ONE.* 2008;3(3):e1709.
245. Danis RP, Newton N, Keith L. Pregnancy and alcohol. *Curr Probl Obstet Gynecol.* 1981;4 (6):2–48.
246. Clarke ME, Gibbard WB. Overview of fetal alcohol spectrum disorders for mental health professionals. *Can Child Adolesc Psychiatr Rev.* 2003;12:57–63.
247. Amos-Kroohs RM, Fink BA, Smith CJ, et al. Abnormal eating behaviors are common in children with fetal alcohol spectrum disorder. *J Pediatr.* 2016;169:194–200.e191.

248. Kishore S, Stamm S. The snoRNA HBII-52 regulates alternative splicing of the serotonin receptor 2C. *Science (New York NY).* 2006;311(5758):230–232.
249. Riikonen RS, Nokelainen P, Valkonen K, et al. Deep serotonergic and dopaminergic structures in fetal alcoholic syndrome: a study with nor-beta-CIT-single-photon emission computed tomography and magnetic resonance imaging volumetry. *Biol Psychiatry.* 2005;57(12):1565–1572.
250. Pang KC, Frith MC, Mattick JS. Rapid evolution of noncoding RNAs: lack of conservation does not mean lack of function. *Trends Genet.* 2006;22(1):1–5.
251. Heimberg AM, Sempere LF, Moy VN, Donoghue PC, Peterson KJ. microRNAs and the advent of vertebrate morphological complexity. *Proc Natl Acad Sci U S A.* 2008;105 (8):2946–2950.
252. Baguna J, Martinez P, Paps J, Riutort M. Back in time: a new systematic proposal for the Bilateria. *Philos Trans R Soc Lond B Biol Sci.* 2008;363(1496):1481–1491.
253. Kimelman D. Mesoderm induction: from caps to chips. *Nat Rev Genet.* 2006;7:360–372.
254. Strickland IT, Richards L, Holmes FE, Wynick D, Uney JB, Wong LF. Axotomy-induced miR-21 promotes axon growth in adult dorsal root ganglion neurons. *PLoS ONE.* 2011;6(8):e23423.
255. Ge XT, Lei P, Wang HC, et al. miR-21 improves the neurological outcome after traumatic brain injury in rats. *Sci Rep.* 2014;4:6718.
256. Chan JA, Krichevsky AM, Kosik KS. microRNA-21 is an antiapoptotic factor in human glioblastoma cells. *Cancer Res.* 2005;65(14):6029–6033.
257. Polajeva J, Swartling FJ, Jiang Y, et al. miRNA-21 is developmentally regulated in mouse brain and is co-expressed with SOX2 in glioma. *BMC Cancer.* 2012;12:378.
258. Yang X, Schadt EE, Wang S, et al. Tissue-specific expression and regulation of sexually dimorphic genes in mice. *Genome Res.* 2006;16(8):995–1004.
259. Dewing P, Shi T, Horvath S, Vilain E. Sexually dimorphic gene expression in mouse brain precedes gonadal differentiation. *Brain Res Mol Brain Res.* 2003;118(1–2):82–90.
260. Kelly SJ, Ostrowski NL, Wilson MA. Gender differences in brain and behavior: hormonal and neural bases. *Pharmacol Biochem Behav.* 1999;64(4):655–664.
261. Bale TL. The placenta and neurodevelopment: sex differences in prenatal vulnerability. *Dialogues Clin Neurosci.* 2016;18(4):459–464.
262. Weinberg J, Sliwowska JH, Lan N, Hellemans KG. Prenatal alcohol exposure: foetal programming, the hypothalamic–pituitary–adrenal axis and sex differences in outcome. *J Neuroendocrinol.* 2008;20(4):470–488.
263. Terasaki LS, Gomez J, Schwarz JM. An examination of sex differences in the effects of early-life opiate and alcohol exposure. *Philos Trans R Soc Lond B Biol Sci.* 2016;371 (1688):20150123.
264. Morgan CP, Bale TL. Sex differences in microRNA regulation of gene expression: no smoke, just miRs. *Biol Sex Differ.* 2012;3(1):22.
265. Qureshi IA, Mehler MF. Genetic and epigenetic underpinnings of sex differences in the brain and in neurological and psychiatric disease susceptibility. *Prog Brain Res.* 2010;186:77–95.
266. Wang Y, Li H, Sun Q, Yao Y. Characterization of small RNAs derived from tRNAs, rRNAs and snoRNAs and their response to heat stress in wheat seedlings. *PLOS ONE.* 2016;11(3):e0150933.
267. Scott MS, Ono M. From snoRNA to miRNA: dual function regulatory non-coding RNAs. *Biochimie.* 2011;93(11):1987–1992.
268. Huang S, Yang B, Chen BJ, et al. The emerging role of circular RNAs in transcriptome regulation. *Genomics.* 2017;109:401–407.
269. Rybak-Wolf A, Stottmeister C, Glažar P, et al. Circular RNAs in the mammalian brain are highly abundant. *Conserv Dyn Express Mol Cell.* 2015;58(5):870–885.

270. Piwecka M, Glazar P, Hernandez-Miranda LR, et al. Loss of a mammalian circular RNA locus causes miRNA deregulation and affects brain function. *Science (New York NY)*. 2017;357(6357):eaam8526.
271. Wu Q, Ohsako S, Ishimura R, Suzuki JS, Tohyama C. Exposure of mouse preimplantation embryos to 2,3,7,8-tetrachlorodibenzo-p-dioxin (TCDD) alters the methylation status of imprinted genes H19 and Igf2. *Biol Reprod*. 2004;70(6):1790–1797.
272. Meng L, Ward AJ, Chun S, Bennett CF, Beaudet AL, Rigo F. Towards a therapy for Angelman syndrome by reduction of a long non-coding RNA. *Nature*. 2015;518 (7539):409–412.
273. Landers M, Calciano MA, Colosi D, Glatt-Deeley H, Wagstaff J, Lalande M. Maternal disruption of Ube3a leads to increased expression of Ube3a-ATS in trans. *Nucleic Acids Res*. 2005;33(13):3976–3984.

FURTHER READINGS

Guo, Y.Guo., Chen, Y.Chen., Carreon, S.Carreon., & Qiang, M.Qiang. (). Chronic intermittent ethanol exposure and its removal induce a different miRNA expression pattern in primary cortical neuronal cultures. *Alcohol Clin Exp Res*, *36*(6), 1058–1066.

Nunez, Y.O.Nunez., Truitt, J.M.Truitt., & Gorini, G.Gorini., et al. (). Positively correlated miRNA-mRNA regulatory networks in mouse frontal cortex during early stages of alcohol dependence. *BMC Genomics*, *14*, 725.

CHAPTER THIRTEEN

The Epigenetics of Early Life Adversity: Current Limitations and Possible Solutions

Sarah Barnett Burns*, Daniel Almeida*, Gustavo Turecki*,†,1
* Douglas Mental Health University Institute, McGill University, Montreal, QC, Canada
† McGill University, Montreal, QC, Canada
1 Corresponding author. E-mail address: gustavo.turecki@mcgill.ca

Contents

Progress in Molecular Biology and Translational Science, volume 157
ISSN 1877-1173
http://dx.doi.org/10.1016/bs.pmbts.2018.01.008

Abstract

Adverse experiences during sensitive postnatal developmental periods can disrupt the calibration of fundamental systems and increase the risk of a wide range of adult disease states, including psychiatric illnesses. Accumulating evidence suggests that epigenetic modifications involving DNA methylation, posttranslational histone modifications, and noncoding RNAs may be a key mediating factor in this disruption. Accumulating evidence from both animal models and human studies suggests that early life adversity alters the epigenome at multiple loci across the genome, but that the specific alterations, and the associated transcriptomic and psychiatric outcomes may be dependent on multiple individual factors. Here we will review the current findings on early life adversity associated alterations of the epigenome in animal models and humans, discuss current limitations, and highlight possible solutions and future directions.

NOMENCLATURE

Human genes	*OXTR*
Human proteins	OXTR
Rodent genes	*Oxtr*
Rodent proteins	OXTR
Nonhuman primate	Human conventions

1. INTRODUCTION

The postnatal environment can have a lasting influence, considerably impacting the risk for adulthood diseases, including psychiatric illnesses. Studies in animals that have been exposed to variations in maternal care, and studies in humans who have been exposed to early life adversity (ELA), such as child abuse, suggest that these early experiences can stably modify DNA methylation at multiple sites across the genome. In addition, there is growing evidence that other epigenetic processes are altered as well, such as histone posttranslational modifications (PTMs) and noncoding RNA (ncRNA) expression. However, depending on the gene locus, the exact modification patterns tend to differ based on a number of individual factors such as genetic background, sex, age, and type of stressor. Although more research is needed to tease out these individual factors and to determine the functional and psychiatric consequences of these modifications, overall the literature points to the possibility that epigenetic alterations can lead to lasting changes in gene expression and neural function, as well as an increased risk for psychopathology in adulthood for many individuals.

Here we will briefly outline ELA in humans and animal models, the stress response, and epigenetic mechanisms. We then review in detail the current evidence on the epigenetic impact of ELA and the individual variations observed. At the end, we discuss the current technical limitations of the field and offer future research avenues to possibly address these limitations.

2. STRESS: DEFINITIONS AND SYSTEMS

Although the term "stress" is used colloquially, arriving at an exact definition is not simple. Stress can refer to threatening or aversive environments, in other words, "stressors,"[1] or it can refer to the physiological response to a stressor.[2] In addition, as the physiological response to stress involves varying increases in nervous system arousal, the differences between neutral experiences, such as alertness, positive experiences, such as excitement, and negative experiences, such as agitation, are not clearly defined. The inverted-U dose–response curve of nervous system arousal describes this continuum, where in the middle of the curve optimal levels of physiological arousal result in high performance, and either ends of the curve (either too much, or too little) is less than ideal.[3] Congruently, Hans Selye, who originally coined the term stress in 1936,[4] used the term "eustress" to indicate positive stressors, and/or stress responses, and "distress" to describe the negative.[5] However, whether one uses the inverted-U description of optimal/suboptimal stress or Selye's eustress/distress, the line between a positive and negative physiological response is difficult to define. Nervous system acclimations to environmental stimuli and the homeostatic processes that maintain set points are often not wholly positive or negative and are highly variable between individuals and contexts.[6] Today, most researchers accept that the term stress can describe both the environmental stimuli and the physiological response to those stimuli, and that this can encompass both negative and positive outcomes.[7] Thus, in stress research it has become essential to define the scope of study.

2.1 Early Life Adversity

In this article, we will be reviewing the literature on ELA, which refers to specific threatening experiences during postnatal development, such as child abuse, as well as experiences of deprivation, such as neglect. During development, optimal levels of stimulation are needed for neural development and learning. Although ELA is often conflated with the overstimulation that is characteristic of toxic stress, it can also encompass the other extreme of inadequate stimulation if the environment or caregiver does not provide any learning challenges at all. This lack of exposure to stimuli can take multiple forms, including caregiver neglect, institutionalized orphanages, and low socioeconomic status. Although such deprivation may constitute a type of threat, it may not be acutely stressful or operate though the activation of the stress response systems. However, it still can have similarly profound and long-lasting impacts on brain development as exposure to overstimulation that illicits an acute stress response. For a review on the specific neurodevelopmental impacts of deprivation, see McLaughlin *et al.*[8]

It is also important to note that ELA exposure is not deterministic of adult outcomes. In most ELA studies there exists a percentage of the exposed population that display resilience. Although this is well established, and it is clear that responses can be highly individualistic, it is not yet known what factors provide protection against ELA in cases of resilience. Thus, this is an important topic of ongoing study and something we will discuss later in this review.

2.1.1 Animal Models

To complement the study of ELA in human populations, multiple animal models have been developed. Although for ease of discussion we will refer to them as animal models of ELA, they more accurately model variations in the developmental environment. They are often based on a decrease in the quality or quantity of maternal care, such as the naturally occurring low versus high licking and grooming and arched back nursing model (LG-ABN), and the experimentally induced maternal separation (MS) or deprivation (MD) models. See a detailed list of ELA animal models in Table 1. These well-studied

Table 1 Overview of Animal Models of Early Life Adversity.

Paradigm	Description
Low vs. high licking, grooming, arched back nursing (LG-ABN)	In rodents: from P1 to P6–10 naturally occurring maternal care is observed daily for 4–8 periods of 60–75 min in 3–4 min intervals. Frequency of licking and grooming (LG) and different types of arched back nursing (ABN) are recorded. Offspring from mothers who are 1–2 SDs above or below the mean for these behaviors are considered high and low LG-ABN pups, respectively.[9–23] Variation 1: short-term protocol: on P4, P7, and P10, mothers are moved to a new cage and observed with a single pup every 30 s for 10 min. Maternal care is scored the same as above.[15]
Maternal observations (MO)	In primates: from birth to 12 weeks old, mother–infant interactions are observed 3 times a week. Protective/affiliative interactions were scored such as touching, grooming, pulling the infant close, accommodating infant needs, and so on. Aggressive interactions were also scored including biting, chasing, scratching, pressing to the ground, dragging, pushing away, and grabbing infants. Maternal scores were calculated by subtracting aggressive scores from protective scores.[24]
Maternal separation (MS)	In rodents: from P1/P2 to P10/15, rodent mothers are separated from all or some of their litter and placed in new cages and separate rooms for 3–4 h/day.[25–33] *Variation 1: In most cases the litter is placed in a new cage with fresh bedding to remove the maternal scent; however, in some cases the litter remains in the home cage with the same bedding.[14] Variation 2: single-event: on P9 mothers are removed from the litter and home cage and placed in a new cage for 24 h, pups remain in home cage.[34]

(*Continued*)

Paradigm	Description
Maternal deprivation (MD)	In rodents: from P1 to P13, rodent offspring are separated from the mother and each other and placed in individual cages for 6 h/day.[35] *Variation 1: shorter separation: from P1 to P14, mice offspring are separated from their mother and each other in individual cups or cages for 3 h/day[36] In primates: from P1 to 7 months of age, primate offspring are reared in a nursery without their mother, usually with an inanimate surrogate, with either intermittent or continuous socialization with peers after the first month. In all paradigms they are housed in social groups at 7 months.[37–39]
Handling (H)	In rats: from P0 to P21 rodent pups are removed from the nest for 3–15 min/day and then reunited with the mother.[11,40]
Maternal unpredictable chronic stress (MUCS)	In mice: from P1 to P14 mothers are separated from their litter for 3 h/day, and subjected to chronic unpredictable stressors (constant light, predator odor, novel object, restraint, novel noise, frequent cage changes, flooded bedding, etc.).[41,42] *Variation 1: Maternal and litter UCS in mice: from P12 to P18 both rodent mothers and litters are subjected to the above chronic unpredictable stressors together.[43]
Limited bedding and nesting (LBN)	In rats: from P2 to P9 most of the cage bedding and nesting materials are removed and only a small amount of nesting cotton is supplied. Mom and pups are returned to standard nesting conditions at P9. Maternal behavior is scored every 3 min over an hour 3 times a day as described in the low vs. high LG-ABN model with the addition of scores of aggressive behavior and nest departures.[44] In mice: from P2 to P9 mother and pups are placed on an elevated grate with the

(*Continued*)

Paradigm	Description
	nesting out of reach. They are given a small piece of nesting cotton from P2 to P9, then returned to standard cages and bedding. Maternal behavior is scored as above.[45]
Variable foraging demand (VFD)	In primates: from 3 to 6 months of age infants are reared with their mothers in small social groups with 3–6 other mother–infant pairs. All mothers are given a foraging device within which they have to forage for food. Mothers in the VFD condition are given a device that requires the expenditure of more time and effort in foraging, whereas control mothers can readily retrieve food.[46]
Maternal maltreatment (limited bedding and nonbiological fostering) (MM)	In rats: from P1 to P7 pups are exposed for 30 min a day to nonbiological mothers that have been stressed with a limited bedding condition. Maternal behavior is scored as in LG-ABN with addition scoring for aggressive behaviors.[47–49]
Communal nest (CN)	In mice: from P0 to P25 three litters and their mothers born on the same day share a large cage.[50]

Day of birth is considered postnatal day 0 (P0).

models have been established in both rodents and nonhuman primates and are thought to somewhat represent early life neglect in humans.

Newer models, such as maternal maltreatment (MM) and limited bedding and nesting (LBN), are perhaps more representative of child abuse, as these paradigms increase maternal aggression toward pups[51] and induce a chronically stressful environment for offspring.[52,53] Many of these models involve resource manipulations and thus also have parallels with low socioeconomic status (SES) deprivation, which has been similarly associated with adverse impacts on the brain and stress response systems, and negative health outcomes in adulthood, including increased risk for psychopathology.[54] Some ELA models have started to incorporate adult paradigms of depression to reflect maternal psychopathology in humans such as using maternal unpredictable chronic stress (MUCS). Future studies may also consider using powerful adult paradigms such as adapted social defeat paradigms that work in female rodents[55] and apply them to a maternal context, or combining paradigms to more accurately model the mixture of stressors that are often present in ELA, such as the maternal separation and limited bedding combination that was recently used by Peña *et al.*[56] to show that age-specific ELA exposure can alter transcriptional processes in the ventral tegmental area and increase adult stress vulnerability in mice.

Another recent direction involves the use of postnatal enrichment paradigms. In adults, enriched environments have been associated with increased neurogenesis[57] and synaptogenesis,[58] possibly via epigenetic regulation of brain-derived neurotropic factor (BDNF).[59] Thus, enrichment may alter postnatal development as well. In addition, early life laboratory conditions are deficient of many of the elements that are common in natural environments, thus enrichment conditions may provide a way to compare developmental epigenetic outcomes from a more stimulating environment to those of ELA paradigms to observe a more naturalistic effect size.

3. THE STRESS RESPONSE PATHWAYS

The most studied stress response pathway in the context of ELA is the slow-acting hypothalamic–pituitary–adrenal (HPA) axis and the fast-acting autonomic nervous system (ANS). Here we give a brief overview of these two most studied pathways (for a detailed review, see De Kloet *et al.*).[60] However, there are many other systems involved in the physiological response to stressful stimuli that are starting to receive attention in the context of ELA, such as the enteric nervous system,[61] the polyamine system,[62] and the immune system.[63,64] Although these systems are often described separately for ease of illustration, it is important to note that there are so many points of interaction that they should be thought of as one coordinated response.

3.1 Hypothalamic–Pituitary–Adrenal Axis

Perceived threats from the environment activate the amygdala, which combined with modulatory input from the prefrontal cortex (PFC), the bed nucleus of the stria terminalis, and the hippocampus sends signals to the paraventricular nucleus (PVN) of the hypothalamus. The PVN incorporates these inputs with information coming from the physical body via brainstem nuclei, such as the locus coeruleus (LC) and the solitary nucleus (SN), and if necessary, triggers the release of corticotrophin-releasing hormone (CRH) and arginine vasopressin (AVP). CRH and AVP bind to receptors in the anterior pituitary and stimulate the release of adrenocorticotrophic hormone (ACTH). ACTH stimulates the adrenal cortex to release glucocorticoids (GCs), such as cortisol in humans, into the circulation, where they have a wide range of impacts on all the major systems of the body. Glucocorticoid receptors (GRs) and mineralocorticoid receptors (MRs) in the hippocampus, PVN, and pituitary detect the levels of circulating GCs and repress the release of CRH, AVP, and ACTH in a negative feedback loop. In addition, on a cellular level, the chaperone protein FKBP5 regulates the translocation of GR to the nucleus where it acts as a transcription factor for the expression of multiple genes, including FKBP5 itself,[65] and thus further regulates HPA reactivity.

3.2 Autonomic Nervous System

Although it has not received as much attention in the context of ELA, activation of the HPA axis does not occur without a response from the sympathetic arm of the ANS. A positive feedback loop exists between the PVN of the hypothalamus and the LC in the brain stem, such that amygdala activation of the PVN leads to LC activation and visa versa. The LC releases norepinephrine, which is released into the circulation and also signals back to multiple other sites in the brain, including the amygdala. The positive feedback loop between the LC and the amygdala also impacts the hippocampus and striatum, such that increased norepinephrine signaling enhances the long-term memory of stressful experiences.[66] In addition, stress-induced amygdala activation of CRH from the PVN is also involved in the activation of the sympathetic nerves, which results in the release of norepinephrine from their terminals and, in turn, the release of epinephrine from the adrenal medulla.[2] The activation of sympathetic nerves leads to the "fight or flight" response, such as increased heart rate, pupil dilation, and constriction of blood away from digestive system and toward the muscles.[66] The other half of the ANS, the parasympathetic nervous system, counteracts this fight or flight response and is often called the "rest and digest" reflex and is primarily regulated by the release of acetylcholine (ACh) via the cranial nerves. The balance between the two arms of the ANS maintains multiple homeostatic functions in the body and is what determines the optimum arousal set point that is described in the inverted-U dose response curve. It is thought that this ANS set point can be shifted up or down based on a number of individual factors and environmental exposures[6] and may contribute to the wide variety in individual response to stress.

4. EARLY LIFE ADVERSITY AND PSYCHIATRIC DISEASE

ELA exposure has been associated with chronic health issues, such as cardiovascular and respiratory diseases, diabetes, obesity, and cancer[67,68]; neurodevelopmental disorders, such as ADHD and autism; and psychiatric disorders, such as post-traumatic stress disorder (PTSD), addiction, anxiety disorders, depression, borderline personality disorder, dissociative identity disorder, phobias, schizophrenia, and suicide.[69–73] In addition, it is well established that there is a dose–response relationship between increased exposure and severity and the likelihood of developing physical and psychiatric diseases later in life.[68,74] The wide variety of outcomes may be partly due to the developmental timing of exposure, as distressful experiences during sensitive developmental windows can become biologically embedded in fundamental systems.[75–77] Although this review is focused on post-natal development, it is important to note that prenatal stress exposure is also closely associated with dysregulation of fundamental systems and negative adult health outcomes (see Bale and Weinstock[78,79] for in-depth reviews). Taken together, this large body of literature strongly suggests that exposure to stress during development can shape a wide variety of biological processes in lasting ways.

ELA is generally associated with alterations in both gray and white matter volume across the brain, as well as changes in connectivity (for in-depth reviews on the structural impact of ELA, see Bick and Nelson and Teicher *et al.*[80,81]). Children exposed to ELA display increased cortical thinning as well as alterations in the volume and morphology of the amygdala,[82,83] hippocampus (although this may be sex specific), striatum, anterior cingulate cortex (ACC), orbitalfrontal cortex, prefrontal cortex, and many other areas.[73] In addition, the reactivity of the amygdala is generally increased,[84] with a lowered set point for threat perception[85–87] that can persist into adulthood.[88]

However, this increased threat perception does not always translate into increased HPA activity. Interestingly, both hyperactive and hypoactive HPA axis activity at baseline and in response to stress challenges are associated with ELA exposure as well as psychopathology. Depending on the severity, type, or timing of ELA, exposed children can either display a blunted diurnal cortisol pattern with low levels in the morning and higher levels in the evening, or conversely can display a more pronounced slope with higher levels in the morning (see Fisher *et al.*[89] for a review of studies). This same variability of cortisol dysregulation is evident in adults with psychopathologies, such as major depression. For example, some individuals diagnosed with MDD have chronic increased levels of cortisol that are even resistant to a dexamethasone challenge. As dexamethasone normally lowers cortisol, sustained cortisol levels after administration are associated with a dysregulated HPA negative feedback mechanism and is a reliable biomarker for treatment resistance.[90] At the same time, other individuals have decreased levels of basal cortisol and/or a blunted cortisol response to a stress challenge.[91] Interestingly, it has been shown that GC signaling is important in maintaining balance between the sympathetic and parasympathetic arms of the ANS,[92] suggesting one way in which both low and high levels of basal cortisol might result in stress hyperreactivity.

As this literature has expanded, there have been multiple schools of thought outlining how early stress exposure might lead to psychopathology, such as the diathesis-stress model,[93] the two (and three) hit model,[94] the stress inoculation model,[95] HPA and SNS asymmetry,[96] differential susceptibility,[97,98] the adaptive calibration model,[99] and more recently the neuroimmune network hypothesis[67] and the gut–brain axis.[100–102] Although these approaches have much in common, including alterations of HPA activity, they have been increasingly expanding the frame of reference to include multiple other systems, for example, the ANS (HPA/ANS asymmetry), genome by environment interactions (two hit model, differential susceptibility, adaptive calibration), and most recently the immune system, the enteric nervous system, and even the gut microbiome (neuroimmune network and the gut–brain axis).

Overall, this collective literature represents strong correlative evidence of an association between ELA and systemic dysregulation of neural, hormonal, humoral, and immune signaling that contributes to the development of psychopathologies. However, the underlying mechanisms have yet to be fully described. Growing evidence suggests that epigenetic processes, which serve as a molecular link between environmental cues and gene expression, may be the mechanism by which ELA alters fundamental systems and increases the risk for psychiatric illness.

5. THE ROLE OF EPIGENETICS DURING DEVELOPMENT

"Epigenetics" refers to a collection of gene regulatory processes that do not change the DNA sequence, but together determine chromatin conformation and transcriptional accessibility. Although the various mechanisms included in this list are continuously expanding, the epigenome currently includes modifications at multiple levels of the genome, from the residues themselves (DNA methylation) to nucleosomes (posttranslational histone modifications), regulatory transcripts (ncRNAs), and chromatin interactions (long range chromatin loops).[103] It is well established that these epigenetic mechanisms are critical for cell differentiation during development[104]; however, they are also one of the key mechanisms by which the environment dynamically modulates gene expression, even in postmitotic neurons.[105,106] Multiple experience-dependent events occur during early postnatal development where the genome is especially sensitive to environmental cues.[81] Normal mammalian development requires these environmental inputs to coordinate the quantitative and temporal regulation of gene expression. These epigenetic processes allow organisms to integrate this incoming information and calibrate systems in alignment with specific environments.[107,108] However, early life epigenetic programming can also become a liability over the lifespan if the early environment is deficient of a required cue, or is otherwise extreme, and thus modifies fundamental systems in a way that becomes detrimental in the context of the later environment. Epigenetic processes such as DNA methylation, histone PTMs, and ncRNA expression have all been shown to be altered by ELA and thus may contribute to variations in psychopathology risk. In addition, many epigenetic modifications, especially those occurring during development, can be stable over the lifespan, and although still controversial, may even be passed to the next generation in some cases (see Nagy and Turecki[109] for a review). However, at the same time, epigenetic modifications in the brain are, in theory, reversible as they dynamically respond to ongoing environmental signals, such as learning.[110] It is this unique capacity for both stability and reversibility that makes epigenetic processes a promising target for future psychiatric interventions.

5.1 DNA Methylation

One of the most intensively studied epigenetic marks is cytosine methylation, which refers to the covalent addition of a methyl group to the 5′ position of a cytosine (5mC) residue via the activity of de novo and maintenance DNA methyltransferases (DNMTs).[111] DNMT1, DNMT3A, and DNMT3B all play a role in maintaining methylation patterns during DNA replication, whereas DNMT3A and DNMT3B are responsible for de novo marks.[112] On the contrary, DNA demethylation is a process that can be either active, via the ten-eleven translocation (TET) family of enzymes, or passively lost through genetic replication (for an in depth review, see Iurlaro *et al.*[113]).

DNA methylation most often, but not always, exists in a CpG context. Many studies have focused on CpG dense regions of the genome, known as CpG islands that are frequently associated with gene promoters. Methylation in these regions is thought to recruit methyl-binding proteins that can result in transcriptional repression through chromatin condensation of regulatory regions[114] and/or by interference with the binding of transcription factors.[115–117] However, transcriptional repression is not always the final result of cytosine modifications, as studies identifying DNA methylation within gene bodies have found that in this context it leads to transcriptional activation and alternative splicing.[118] Results entailing a functional role of DNA methylation on gene expression have therefore been mixed and are likely explained by other cytosine modifications such as hydroxymethylation (5hmC) and non-CpG (mCH) methylation.

For instance, many studies investigating cytosine modifications have used methods that are incapable of differentiating between 5mC and 5hmC. 5hmC is formed during the demethylation process, converting 5mC to 5hmC, via TET methylcytosine dioxygenase enzymes. Whether 5hmC is a possible stepping-stone to demethylation or may serve as a separately important functional mark remains to be elucidated. However, 5hmC is often localized to gene bodies and is positively associated with gene transcription, thus variability in previous literature may be explained by a lack of 5hmC detection.[119,120] 5hmC is especially important when working with mammalian brain because it is particularly abundant in this tissue type.[121,122]

DNA methylation in a non-CpG context (mCH) has also been described (i.e., methylation at a cytosine followed by A, T, or C).[123] Interestingly, brain tissue is characterized by higher concentrations of mCH compared with other tissues.[124,125] In fact, mCH plays an important role in neurodevelopment as it accumulates in neurons during the first few years of life.[126]

5.2 Histone Posttranslational Modifications

Chromatin is not an inert structure, but is responsive to the environment, and has an important functional impact on gene expression. One way in which chromatin structure is itself regulated is via histone modifications. Histones form ocatameric protein complexes referred to as nucleosomes, around which DNA is coiled.[127] The configuration of nucleosomes determines whether chromatin is open (euchromatin) and actively transcribed, or closed (heterochromatin) and repressed. PTMs such as covalent additions of acetyl, methyl, phosphoryl, and other chemical groups to the N-terminal tails of histones regulate nucleosome position and tightness of coiling, and thus chromatin configuration.[128] Although various types of tags exist, methylation and acetylation have received the most characterization. The enzymes responsible for the addition or removal of these groups include histone methyltransferases (HMTs)/histone acetyltransferases (HATs) and histone demethylases (HDMTs)/deacetylases (HDACs), respectively (as reviewed in Koch *et al.*[129]). Generally, increased acetylation is associated with euchromatin states, whereas increased histone methylation produces variable effects depending on the residue modified.

5.3 Noncoding RNAs

Although it is estimated that only about 1% of the mammalian genome is protein coding,[130] the vast majority of the whole genome is nevertheless transcribed, far beyond the boundaries of known genes. The RNA species that show no evidence of coding for a protein are collectively known as ncRNAs and play important roles in chromatin remodeling, transcriptional and posttranscriptional regulation, as well as other functions such as chromatin looping.[131] ncRNAs can be divided into constitutively expressed infrastructural RNAs that include transfer, ribosomal, small nuclear, and small nucleolar RNAs; the regulatory class of ncRNAs is made up of microRNAs (miRNAs), long noncoding RNAs (lncRNAs), piwi-interacting RNAs (piRNAs), and small interfering RNAs (siRNAs).[131] Each of these RNAs works through a variety of mechanisms and has various functions. For instance, lncRNAs have an important function as *cis*- and *trans*-acting modulators of protein-coding genes through their recruitment of chromatin remodeling complexes.[132,133] ncRNAs also play important roles in posttranslational processing, such as posttranslational gene silencing (PTGS) and transcript degradation mediated by siRNA and miRNAs.[134] Interestingly, preliminary evidence suggests that ncRNAs are central to neurodevelopment (see reviews Chen *et al.* and Obiols-Guardia and Guil[135,136]), and thus may be impacted by ELA.

6. ANIMAL STUDIES ON THE EPIGENETIC IMPACT OF ELA

HPA axis reactivity was the focus for the first investigations into the epigenetic impact of ELA using animal models. However, since then the field has rapidly expanded to explore other candidate genes in other signaling pathways, and recently, genome-wide studies. Here, we will review in detail the current evidence from animal models of ELA on epigenetic patterns at multiple sites across the genome. See Table 2 for a list of the animal studies included in this review.

6.1 HPA Axis

The first study in the investigation of the epigenetic impact of ELA interrogated the *glucocorticoid receptor* (*Nr3c1*) gene. Weaver *et al.*[9] used the natural variations in licking and grooming (LG-ABN) model (Table 1) to demonstrate that reduced maternal care in Long Evans rats is associated with increased levels of DNA methylation at exon 17 and a transcription factor (NGFI-A)-binding site in the hippocampal *Nr3c1*, along with a decrease of nearby H3K9Ac.[9] They also correlated this with a decrease in NGFI-A binding and hippocampal NR3C1 expression. In subsequent studies, this group also showed that low levels of LG-ABN were associated with increased HPA axis reactivity to adult stress challenges[10,149] as well as increases in anxiety-like behaviors.[11]

Since then, multiple other groups have investigated ELA-induced epigenetic alterations to the GR gene in a wide variety of rodent strains and ELA models (Table 2). Overall, the effects of ELA on DNA methylation patterns at the GR locus appear to be species, strain, sex, and age dependent. The original findings have been replicated in multiple studies using Long Evans male rats and the LG-ABN model.[12,147] In addition, the use of the LG-ABN model or maternal separation (MS) (Table 1) in Balb/c and DBA/2J mice also found similar methylation profiles.[13,34] However, studies using Sprague Dawley or Wistar rats and either LG-ABN or MS did not see an effect of ELA on DNA methylation in the hippocampus.[14,15,25,140] C57BL/6J (B6) mice also appear to be resilient to MS at least at the *Nr3c1* locus in the hippocampus.[34,45,140] However, the picture is more complicated. One study did find an effect at the *Nr3c1* locus in the cerebellum,[15] and another study using the maternal unpredictable chronic stress (MUCS) ELA model (Table 1) found increased methylation of the *Nr3c1* promoter in the hippocampus of B6 mice

Table 2 Animal Studies on ELA-Associated Alterations in Epigenetic Processes.

References	Species	ELA Model	Gene	Tissue	Sex	Age	Epigenetic Alteration	Epigenetic Method	Expression of Target Gene
29	Rats (Wistar)	MS	*Dnmt1* *Dnmt3a* *Dnmt3b* and GW	NAc	M	P15 P120	Decreased methylation on a global level Increased methylation of *Pp1c* and *A2ar* promoters	MeDIP (methylated DNA immunoprecipitation)	Increased DNMTs Decreased PP1C and A2AR
35	Rats (Sprague Dawley)	MD	*Bdnf*	HPC	M	>P91	Increased miR-16	qPCR for miR-16	Decreased BDNF VI
137	Rats (Long Evans)	LG-ABN	*mGluR1*	HPC	M	Adult NR	Increased methylation Increased H3K9ac	Bisulfite sequencing ChIP (H3K9Ac, H3K4me)	Decreased
138	Rats (Long Evans)	MM	*Bdnf* *Reln*	mPFC	M	P8 P30 P90	*Bdnf:* Increased methylation of exon I in males at P30 Decreased methylation of IV in females at P30 Decreased methylation of I in both males and females at P90 Increased methylation of IV in females at P90	Bisulfite qPCR	NC in bdnf Decreased reeling in females at P90

							Reln: Decreased methylation in females at P8 NC at P30 No effect of maltreatment at P90 (effects of nonbiological care (both MM and fostercare compared with control) at P90		
139	Rats (Long Evans)	MM	*Bdnf*	mPFC		P90	Increased methylation in females at CpG 4 and 6 in exon IV in neurons only	FACS Bisulfite sequencing	NR
21	Rats (Long Evans)	LG-ABN	*Oxtr*	HPC STRY HYP PBMC	M	NR	Decreased methylation in PBMC only	Bisulfite Sequenom (mass spec) Bisulfite sequencing	NC
50	Mice (CD-1)	CN	*Bdnf*	HPC	M	NR	Increased H3ac	ChIP (H3K27me3, H3K9Ac, H3K14Ac)	Increased expression after stress challenge
23	Rats (Long Evans)	LG-ABN	GW *Dmnts*	HPC	M	P90	No differences in global methylation between groups Increased global methylation in CG	Immunohistochemistry (anti-5mC, NeuN)	DNMT expression by group NR Increased

(*Continued*)

Table 2 Animal Studies on ELA-Associated Alterations in Epigenetic Processes.—cont'd.

References	Species	ELA Model	Gene	Tissue	Sex	Age	Epigenetic Alteration	Epigenetic Method	Expression of Target Gene
							and CA1 compared to CA2 and CA3		DNMT1 in DG and CA1 Increased DNMT3b in CA1
19	Rats (Long Evans)	LG-ABN	ERα	MPOA	F	P6 P90	Increased methylation	Bisulfite sequencing	Increased at P6 and P90
26	Rats (Sprague Dawley)	MS	*Crh*	PVN AMYG	B	P61	Decreased methylation at CpG 1 and 2 in PVN of male and females NC in CeA	Bisulfite sequencing	Increased basal primary CRH transcripts in PVN in females and altered stimulated responses NC in CeA
14	Rats (Sprague Dawley)	MS variation 1	*Nr3c1*	HPC	M	P21	NC	Bisulfite sequencing	NR
140	Mice (C57BL/6J)	MS	*Nr3c1*	HPC	M	P25	NC	Bisulfite sequencing	Decreased

49	Rats (Long Evans)	MM	*Bdnf* GW	AMYG HPC	B	P30	Decreased global 5hmC in AMYG of males Increased global methylation in the dorsal HPC of males Increased global methylation in the ventral HPC of males Increased methylation in *Bdnf* exon IV in AMYG of females Increased methylation in *Bdnf* exon IV in ventral HPC of females (validated by BS)	MethylFlash (antibody + absorbance) (Global methylation and hydroxymethylation percentages) Bisulfite qPCR for BDNF Bisulfite sequencing	NR
42	Mice (C57Bl/6J)	MUCS	*Mecp2* *Maoa* *Cb1* *Crh-r2*	GC (male) Brain (female)	B	NR F1 F2 F3	F1 male germ cells: Increased methylation of *Mecp2* and *Cb1* promoters Decreased methylation of *Crh-r2* NC in *Maoa* F2 female brain: Increased methylation of *Mecp2* and *Cb1* Decreased methylation of *Crh-r2*	Bisulfite sequencing	F2 female brain: decreased MeCP2, CB1, CRH-R2

(*Continued*)

Table 2 Animal Studies on ELA-Associated Alterations in Epigenetic Processes.—cont'd.

References	Species	ELA Model	Gene	Tissue	Sex	Age	Epigenetic Alteration	Epigenetic Method	Expression of Target Gene
25	Rats (Wister)	LG-ABN	*Nr3c1*	DG (HPC)	M	P84	NC	Bisulfite sequencing	NR
34	Mice (C57BL/6 and DBA/2J)	MS variation 2	*Nr3c1* *Avp* *Nr4a1*	HPC	M	P98–105	*Nr3c1*: Increased methylation at CpG 13, 14, 17 in DBA/2J only NC in B6 *Avp*: Increased methylation at CpG 1 in both strains *Nr4a1*: Decreased methylation at CpG 2 in B6 only	Bisulfite sequencing	NR
38	Rhesus macaques	MD	*Slc6a4*	Blood	B	P90–120 infants	Increased methylation in carriers of the rh5-HTTLPR risk allele	Bisulfite Sequencing	Decreased
46	Bonnet macaques	VFD	*Slc6a4* GW	Blood	B	M = 8.4 years	No change in global or 5HTT methylation by VFD Increased global and 5HTT methylation was associated with increased stress reactivity	Bisulfite Sequencing (SLC6A4) Methylation ELISA (GW)	NR

24	Rhesus macaques	MO	*Slc6a4*	Blood	B	P90–120 93 months	Increased methylation	Bisulfite Sequencing	NR
15	Rats (Sprague-Dawley)	LG-ABN (short-term)	*Nr3c1*	HPC CRBL	B	P35	Increased methylation in CRBL only	Bisulfite sequencing	NC
41	Mice (Balb/c and C57BL/6J)	MUCS	*Nr3c1* *Bdnf*	HPC PFC	B	P40	B6: Increased methylation of *Nr3c1* in HPC of males only NC in *Bdnf* Balb/c: NC in *Nr3c1* Increased methylation of *Bdnf* promoter region IV in HPC of both males and females	Bisulfite sequencing	B6: No change in NR3C1 Decreased BDNF in HPC of females Balb/c: No change in NR3C1 Increased BDNF in PFC in males and females
31	Mice (BALB/c and C57BL/6J)	MS	GW	HPC FB STRY	M	P21 P60	Balb/: H4K12Ac decreased in FB at P21 H4K12Ac increased in FB at P60 H4K5Ac increased at in FB at P60	Western blot (H5K5Ac, H4K8Ac, H4K12Ac, H4K16Ac, H3, H4, H4K9me2, H3K4me3)	Increased HDAC (1,3,7,8,10) in FB at P21 Decreased in FB at P35

(Continued)

Table 2 Animal Studies on ELA-Associated Alterations in Epigenetic Processes.—cont'd.

References	Species	ELA Model	Gene	Tissue	Sex	Age	Epigenetic Alteration	Epigenetic Method	Expression of Target Gene
									and P60 in Balb/c
13	Mice (Balb/c)	LG-ABN	*Nr3c1*	HPC Feces	B	P35	Increased methylation at CpG 1 and 2, and NGFI-A site in HPC only No change in feces	Bisulfite sequencing	NR
39	Rhesus macaques	MD	*Dr-d3* *Slc6a4* *Grm8* *α1-ar* *Itpr2* *Gaba-aα2* *Rest* *Mapk1* GW	PFC Blood	M		Decreased 5hmC at *Dr-d3*, *Slc6a4*, *Grm8*, and *α1-ar* Increased 5hmC at *Itpr2*, *Gaba-aα2*, *Rest* *Mapk1* MAPK1	MeDIP + 5hmC microarray Hydroxymethyl collector kit + qPCR	5hmC correlated with expression at sites where promoter is at least 80%.
32	Rats (Wistar-Kyoto)	MS	GW	HPC PVN mPFC AMYG BNST SEP	M	P343–350	Increased global methylation in HPC (MethylFlash) Hypomethylation of Insr, Map3k5 Hypermethylation of Igf1r and Grik4 in HPC (MethylCap-Seq)	MethylFlash MethylCap-Seq (on HPC) Methylated DNA is captured with protein bound beads, enriched and purified, then sequenced on HiSeq.	Increase in Insr in HPC

							NC in PVN, mPFC, AMYG, BNST, SEP (MethylFlash)		
12	Rats (Long Evans)	LG-ABN	★*Nr3c1* (3.5 MB region)	HPC	M	P90	Increased methylation in *Nr3c1* Differential methylation and H3K9Ac in multiple genes, especially *Pcdh* genes	Custom 44K microarray qChIP Bisulfite sequencing	Decreased
141	Mice (C57BL/6J)	MS	*Avp*	HPC PVN	M	P42 P84 P364	Decreased methylation of CGI3 enhancer at all ages	Bisulfite sequencing	Increased expression of AVP
45	Mice (C57BL/6J)	LBN	*Nr3c1*	HPC	M	P9 P120	Increased methylation at P9 NC at P120	Bisulfite sequencing	NC
43	Mice (C57BL/6J)	MUCS variation 1	GW	HYP	F	P84	Differential methylation at many sites including Nr3c2, Nrxn1, Nfia, Clip1	ChIP-seq (5hmC)	Correlated differential expression and isoforms
20	Rats (Long Evans)	LG-ABN	ERα	MPOA	F	P0 P6 P21 P66	P21: Increased methylation, H3K9me3 Decreased H3K4me3	Bisulfite sequencing ChIP (H3K4me3, H3K9me3)	Decreased ERa and ERB by P21
142	Rhesus macaques	MD	GW	PFC	M		PFC: Differential methylation at 1373 promoters Genes associated with	MeDIP + microarray	NR

(*Continued*)

Table 2 Animal Studies on ELA-Associated Alterations in Epigenetic Processes.—cont'd.

References	Species	ELA Model	Gene	Tissue	Sex	Age	Epigenetic Alteration	Epigenetic Method	Expression of Target Gene
							nervous system development, immune function, RNA metabolism, stimulus response, axonogenesis, leukocyte differentiation, transcription regulation, nucleosome assembly, and stress response Blood: Differential methylation at 138 promoters. Increased methylation at 46. Decreased methylation at 92. Genes involved in transcription, fatty acid oxidation, and immune response, proteolysis, chromatin modification, and stimulus response		

47	Rats (Long Evans)	MM	*Bdnf*	HPC PFC	B	P8 P30 P90	Increased methylation at IV in PFC	Bisulfite sequencing 5-methylcytosine immunoprecipitation	Decreased in PFC
143	Rats (Long Evans	MM	*Bdnf*	HPC AMYG	B	P90	P90: Increased methylation of exon I in ventral HPC of females Increased methylation of exon IV in ventral HPC of males Decreased methylation of exon I and IV in amygdala of males Infants: NC	Bisulfite sequencing	NR
36	Mice (C57BL/6N)	MS	*Drd1a*	VTA NAc	B	P77	Increased methylation in females in NAc	Bisulfite Sequencing	Decreased Drd1a in females in NAc
144	Mice (Balb/c and C57BL/6J)	MS	*Gαq*	FB	B	NR F1 F2	Balb/c: Increased H4K5Ac, H4K8Ac, and H4K12Ac in F1 Increased H4K12Ac in F2 from MS exposed F1 females B6 raised by balb/c: NC	qChIP (H4K12Ac)	

(*Continued*)

Table 2 Animal Studies on ELA-Associated Alterations in Epigenetic Processes.—cont'd.

References	Species	ELA Model	Gene	Tissue	Sex	Age	Epigenetic Alteration	Epigenetic Method	Expression of Target Gene
28	Rats (Sprague Dawley)	MS	*Bdnf*	HPC	M	P21 P56 P420	Decreased in H3K9me2 at P21 and P56 Increased H3K9me2 at P420	ChIP (H3K9ac, H3K4me2, H3K9me2, H3K9me3, H3K27me3)	Increased IV and IX at P21 and P56 Decreased IV at P420
30	Rats (Sprague Dawley)	MS	*Dmd* *Prkacb* *Kitlg*, *Akap7* *Ryr3*, *Akap9*, *Btg2*, *Rcan3*	HPC	M	P60 P450	At P60: Increased H3ac at *Dmd*, *Prkacb*, and *Akap7* promoters Increased H3K9ac at *Kitlg* and *Akap7* promoters Decreased H3K9me2 at *Dmd* and *Kitlg* Decreased H3K9me2, H3K9me3, H3K9me2 at *Prkacb* At P450: Increased H3K9me2 at *Ryr3*, *Btg2*, and *Rcan3* Decreased H3K9ac at *Ryr3* and *Akap9* promoters	ChIP (H3ac, H3K9ac, H3K9me2, H3K9me3)	Decreased Hdac2, Hdac8, Hdac7, Hdac9 and Suv39h1at P60 Increased Hdac3 at P60 Increased Hdac2, Hdac6, Hdac8 and Suv39h1at P450 NC in G9a and Ehmt2

145	Rats (Sprague Dawley)	MS	*Nts* *Ntsr1* *Ntsr2*	AMYG			Increased methylation in *Ntsr1* promoter	MIRA (Methylated CpG Island Recovery Assay) + qPCR	Decreased NTSR1 only
146	Rats (Sprague Dawley)	MS	*Crh*	HPC	NR	NR	Decreased methylation Increased H3Ac	MeDIP	Increased
9	Rats (Long Evans)	LG-ABN	*Nr3c1*	HPC	M	P6 P21 P90	Increased methylation at exon 17 and NGF1-A-binding site Increased H3K9Ac	Bisulfite sequencing ChIP qPCR	Decreased
147	Rats (Long Evans)	LG-ABN	*Nr3c1*	HPC	M	P90	Increased methylation at exon 17 and NGF1-A-binding site	Bisulfite sequencing	Decreased
27	Mice (C57BL/6J)	MS	*Pomc*	PIT	M	P42 P84 P364	Decreased methylation of promoter at P42 and P84	Bisulfite sequencing	Increased expression (in vitro)
148	Mice (C57Bl/6J)	MS	*Egr1* *Arc*	HPC	M	P16	Increased H3Ac H4Ac (Western Blot) Increased H4Ac only at *Egr1* and *Arc* promoter (ChIP)	qWestern Blot nChIP-qPCR	Increased ARC and EGR1
17		LG-ABN	*Gad1* *Dmnt1*	HPC	NR	Adult NR	Increased methylation of *Gad1*		Decreased GAD1 at P4

(*Continued*)

Table 2 Animal Studies on ELA-Associated Alterations in Epigenetic Processes.—cont'd.

References	Species	ELA Model	Gene	Tissue	Sex	Age	Epigenetic Alteration	Epigenetic Method	Expression of Target Gene
	Rats (Long Evans)						Decrease in H3K9Ac in *Gad1*	Bisulfite sequencing ChIP (H3K9Ac)	and adult increased DNMT1 expression

Age: P, postnatal day #. NR, not reported. NT, not tested. NC, no changes. ELA model: LG-ABN, licking, grooming, and arched back nursing model; MS, maternal separation; MUCS, maternal unpredictable chronic stress; MM, maternal maltreatment; LBN, limited bedding and nesting; MD, maternal deprivation; CN, communal nest. Genes: NR3C1, glucocorticoid receptor; AVP, arginine vasopressin; BDNF, brain-derived neurotropic factor; CRH, corticotropin-releasing hormone; CRH-R2, CRH receptor-2; POMC, pro-opiomelanocortin (precursor for ACTH); mGluR1, metabotropic glutamate receptor 1; GAD1, glutamate decarboxylase 1; DNMT, DNA methytransferase; ERα, estrogen receptor alpha; OXTR, oxytocin receptor; HDAC, histone deacetylases; PP1C, protein phosphatase 1 catalytic subunit; A2AR, adenosine A2A receptor; Insr, insulin receptor preprotein; Map3k5, mitogen-activated protein kinase kinase kinase 5; Grik4, glutamate receptor ionotropic kainite 4 precursor; Igf1r, insulin-like growth factor 1 receptor; MeCP2, methyl-CpG-binding protein 2; MAOA, monoamine oxidase A; CB1, cannabinoid receptor type; Gαq, G-protein alpha subunit; Egr1, early growth response protein 1; Arc, activity-regulated cytoskeleton-associated protein; Reln, Reelin; Dmd, dystrophin; Prkacb, protein kinase A subunit b; Kitlg, kit ligand; Akap7, kinase A anchoring protein 7; Ryr3, ryanodine receptor 3; Btg2, B-cell translocation gene2; Rcan3, repressor of calcineurin; DRD3, dopaminergic receptor 3; α1-AR, adrenoreceptor alpha 1; Grm8, glutamatergic receptor Grm8; ITPR2, inositol 1,4,5-trisphosphate receptor; type 2; GABA-A; GABA A receptor alpha 2, REST; RE1-Silencing Transcription factor; MAPK1; mitogen-activated protein kinase 1. Tissues: HPC; hippocampus; DG, dentate gyrus; CRBL, cerebellum; HYP, hypothalamus; PVN, paraventricular nucleus of the hypothalamus; MPOA, medial preoptic area of the hypothalamus; FB, forebrain; PFC, prefrontal cortex; mPFC, medial prefrontal cortex; AMYG, amygdala; NAc, nucleus accumbens; STRY, striatum; SEP, septum; BNST, bed nucleus of the stria terminalis; PBMC, peripheral blood mononuclear cell; GW, genome wide. Sex: M, male; F, female; B, both.

that persisted into adulthood.[41] Together this suggests that genetic background interacts with specific types of ELA to determine the DNA methylation profile. In addition, studies using both male and female mice suggest that ELA-induced differential methylation at the *Nr3c1* locus is more common in males.[34,41] These results suggest that although specific combinations of ELA exposure, rodent strain, and sex interact to produce stable increases in DNA methylation at the *Nr3c1* locus, this is not necessarily universal across all individuals. Thus, more work needs to be done to determine how individual differences interact with ELA to alter DNA methylation processes at the *Nr3c1* locus.

In addition, epigenetic alterations at other points along the HPA axis have been investigated in animal models, such as genes encoding for CRH, AVP, and ACTH. As previously described, CRH signaling is crucial for activation of the HPA axis, the sympathetic nervous system, and the reactivity of the amygdala. In the context of HPA axis hyperactivity, that is largely suggested by studies on the *Nr3c1* locus, one would expect CRH signaling to be increased. Indeed, DNA methylation of the *Crh* promoter was shown to be decreased in the paraventricular nucleus (PVN) of both male and female rats exposed to maternal deprivation (MD) (Table 1). In addition, there was a corresponding increase in CRH expression in response to a stressor.[26] Maternal separation (MS) in rats also increases H3ac and decreases DNA methylation at the *Crh* promoter in the CA1 region of the hippocampus.[146] This was shown to be due to a decrease in HDAC2 and DNMT1 binding. Impacts of ELA also appear to impact CRH signaling across generations. After B6 mice were exposed to MUCS (Table 1), the germ cells of F1 and F2 males and the PFC of F2 females display a decrease in DNA methylation in the CRH receptor-2 (*Crh*-r2) gene.[42] Altogether this indicates that ELA increases CRH signaling in multiple areas of the brain by decreasing DNA methylation, and that, strikingly, this may impact subsequent generations in complex ways.

The first study of epigenetic regulation at the *Avp* locus found decreased methylation of the hypothalamic *Avp* promoter in male mice exposed to MD, and sustained increased in AVP expression (Table 1). They also found increased despair-like behaviors and memory deficits.[141] A subsequent in vitro study showed that polycomb repressive complex 2 (PRC2) binds to this site to decrease AVP expression in advance of any increase in DNA methylation, possibly priming cells for experience-dependent DNA methylation.[150] Another study also found ELA-associated alterations in methylation at the *Avp* locus, but also demonstrated multiple sex- and strain-dependent effects. A variation in the MS model, this time a single event (Table 1), increases DNA methylation at one *Avp* CpG in male mice of two strains, DBA/2J (D2) and B6.[34] This is associated with increased anxiety-like behaviors in both strains but memory deficits only in D2 mice. Interestingly, although there was no differential methylation in the ELA exposed females from either strain, they nevertheless display alterations in anxiety-like behavior. Altogether this suggests that ELA may have an epigenetic impact at the *Avp* locus in male mice but that the outcomes are strain dependent. In addition, it suggests that the mechanism by which ELA impacts anxiety-like behaviors in females is not via DNA methylation at this gene promoter. Much more research is needed to understand how individual factors interact with ELA and determine DNA methylation patterns at the *Avp* locus.

Only one study thus far has focused on ELA-associated epigenetic regulation of ACTH. However, the results are concordant with ELA inducing long-term hyperactivity of the HPA axis. Maternally separated (Table 1) male mice display reduced DNA methylation in the *Pomc* gene.[27] As POMC is the precursor of ACTH, this further suggests that ELA promotes HPA reactivity in a sex-dependent manner. In addition, van Oers *et al.*'s[151] demonstration on the differential impacts of nursing/feeding and tactile stimulation (licking and grooming) on ACTH and GC expression suggest that regulation of ACTH may be sensitive to licking and grooming, but not nursing, and thus may be an interesting area of investigation of the epigenetic effects of specific maternal behaviors.

6.2 Brain-Derived Neurotropic Factor

BDNF plays multiple key roles in neurodevelopment, such as the differentiation and branching of neurons, and synaptogenesis.[152] In addition, it is critical for neurogenesis, synaptic plasticity, and connectivity throughout life.[153,154] Mild stressors such as exercise and positive experiences increase BDNF activity.[155] Chronic stressors, in general, are thought to decrease BDNF activity[156–159]; however, many reports have been mixed. In concordance with individual differences in stress vulnerability, chronic stress has been shown to both increase and decrease BDNF activity depending on a number of individual factors, such as the perceived or objective severity of the stressor, age of exposure and measurement, brain region, sex, and genetic background.[160–162] Alterations in BDNF during development are associated with depression,[163] anxiety disorders,[164] and PTSD.[165]

In an early study, maternal maltreatment (MM) (Table 1) in rats was shown to increase DNA methylation and decrease expression of BDNF in the PFC.[47] In addition, chronic treatments of intracerebroventricular zebularine, a DNMT inhibitor, were able to reverse these impacts. However, these effects may be sex and age dependent in rats. In a subsequent study, the same group demonstrated that although both male and female Long Evans rats exposed to ELA have increased methylation in the ventral hippocampus, the exact location at the *Bdnf* loci differed.[143] In addition, only males display decreased methylation of *Bdnf* in the amygdala. Most recently, this group further confirmed a sex difference by using the same paradigm of MM to conduct a cell-type specific investigation at the *Bdnf* IV locus, finding that only females exposed to ELA have increased DNA methylation and only in the neuronal fraction.[139] Other studies investigating this locus at different ages found opposite patterns between young versus older rats after exposure to MS. For example, a decrease in H3K9me2 and an increase in expression of exon IV were evident in both adolescence and early adulthood, and this was further associated with increased cognitive performance on a hippocampal-dependent task.[28] However, in older adults, the exact opposite patterns in histone PTM, expression, and behavior were found. In agreement with this, MD in rats increases the expression of a miRNA that targets the *Bdnf* locus (miR-16) during mid adulthood and is associated with a decrease in expression.[35] However, rats exposed to chronic unpredictable stress in adulthood do not display an increase in miR-16 or a decrease in BDNF.

In mice, there is evidence for a strain effect as well as a disconnect between epigenetic patterns and expression. Balb/c, but not B6, mice exposed to MS display increased methylation of CpG sites in hippocampal exon IX, but show no differences in expression.[41] Interestingly, B6 mice do show ELA-associated reduction in expression in the hippocampus, and Balb/c mice show ELA-associated changes in expression in the PFC. Together this suggests that methylation of the BDNF gene in Balb/c mice might be transcriptionally silent, and that ELA may alter BDNF expression via other mechanisms in B6 mice. In CD-1 mice, the enrichment paradigm, communal bedding (CN) (Table 1), increases H3Ac in hippocampal *Bdnf* promoters, as well as expression,[50] suggesting that positive early life experiences also alter epigenetic mechanisms.

6.3 GABA and Glutamate Signaling

GABA and glutamate signaling is also involved in stress adaptations such as neurogenesis and plasticity, as well as stress overload in the form of excitotoxicity,[66] and is critical for learning and memory, which is often impacted in

ELA animal models and multiple psychiatric diseases.[166] Glutamate signaling is also currently receiving much needed research attention through the study of the effects of ketamine on depression (for a reviews, see Caddy *et al.* and Niciu *et al.*[167,168]). ELA is associated with alterations in many receptors along these excitatory and inhibitory pathways, such as expression of ionotropic (NMDA GluN2A, GluN2B, and GluN1),[169,170] and metabotropic (mGluR1) glutamate receptors,[137] as well as decreased $GABA_A$ receptors in the locus coeruleus, medial PFC, and amygdala.[11,16] In Long Evans rats, low-LG offspring have increased DNA methylation and H3K9Ac at the mGluR1 promoter along with decreased expression.[137] In addition, the same model displays increased DNA methylation in the promoter region of the GAD1 gene (an important, rate-limiting, enzyme in GABA synthesis), which is associated with decreased NGFI-A binding and expression.[17] Although preliminary, these studies suggest that ELA alters both excitatory and inhibitory signaling, and that this may be mediated, in part, by epigenetic processes.

6.4 Serotonin and Dopamine Signaling

Alterations in serotonin and dopamine signaling have long been associated with increased risk for multiple psychopathologies.[171] Multiple psychotropic drugs target these systems and the *serotonin transporter* gene (*Slc6a4*) has been the topic of intense research for multiple decades, as the *Slc6a4 or 5-Httlpr* short variant (*ss*) has been widely associated with an increased risk for major depression.[172] The associated ELA-induced epigenetic impacts at this locus have largely been investigated in peripheral cells of nonhuman primate models. Rhesus macaques who are both carriers of the *ss* variant and are exposed to MD display increased methylation at the *Slc6a4* promoter and decreased expression, as well as increased stress reactivity later in life,[37,38,173] suggesting that ELA interacts with genetic background here as well. In another study in bonnet macaques and using the variable foraging demand (VFD) paradigm (Table 1), the same group[46] found no overall ELA effect on DNA methylation at the *Slc6a4* gene, or genome wide. Instead, they found that exposure to VFD plus increased DNA methylation at the *Slc6a4* locus was associated with increased reactivity to stressful circumstances in adulthood, again suggesting individual differences in ELA resilience/vulnerability. In a more recent longitudinal study, this same group[24] demonstrated that exposure to high levels of maternal aggression during maternal observations (MO) was predictive of increased methylation at the *Slc6a4* locus and associated with poorer health outcomes up to 8 years later, suggesting that the impacts of ELA on DNA methylation at this locus are stable over long periods of time.

Although it plays many other roles, the heterotrimeric G protein alpha subunit (Gαq) couples to 5HT receptors and so has also been studied as part of the serotonin system. In a cross-generational study, Schmauss *et al.*[144] investigated the effects of ELA on the *Gαq* gene in both male and female mice from two strains, the stress sensitive Balb/c and the resilient B6, and identified strain, sex, and generational effects. In Balb/c mice, both sexes directly exposed to MS have increased H4K5Ac, H4K8Ac, and H4K12Ac at the *Gαq* locus in the forebrain, whereas the F2 females have increased H4K12Ac as well as increased expression. Neither generation of B6 mice displays changes in histone PTMs at this locus. In addition, a cross-fostering condition where B6 mice were raised by Balb/c mothers during MS still showed no changes in *Gαq* methylation or expression in offspring, suggesting that this resilience is not due to differences in maternal behavior between the strains. This study adds a new dimension to the literature on the differences between Balb/c and B6 mice in stress-induced alterations to serotonin signaling, and adds to the growing body of literature detailing the importance of considering genetic background in the study of ELA-induced epigenetic changes.

Sasagawa *et al.*[36] investigated the effects of MS on tyrosine hydroxylase (TH) in the ventral tegmental area (VTA), as well as the dopamine D1 and D2 receptors in the nucleus accumbens (NAc) of B6 mice, and found that MS is associated with an increase in DNA methylation at the D1 promoter in the NAc in female, but not male mice. This was also correlated with a decrease in expression and increased anhedonic behavior, in contrast with many other studies showing resilience of B6 mice to ELA. It is important to note that the authors used chocolate eating to test anhedonia rather than the more common sucrose preference protocol; in addition, they were also using a substrain B6N, rather than 6J. Thus, it would be interesting to test B-6J mice on this specific anhedonic measure. In addition, a study of the prefrontal cortex in rhesus macaques exposed to maternal deprivation (MD) in the form of an inanimate surrogate mother (Table 1) showed that ELA is associated with decreased hydroxymethylation (5hmC) at the *Slc6a4* and *dopamine receptor (D3)* loci.[39] When they examined the genome wide methylome they found a positive correlation between increased 5hmC and expression, at least at loci where there was low CpG 5mC CpG methylation and high levels of promoter 5mC.

In addition to serotonin and dopamine receptors, neurotensin (NTS) and its receptors neurotensin receptor type 1 (NTSR1) and neurotensin receptor type 2 (NTSR2) are important in monoamine signaling pathways, are densely populated in the amygdala and hippocampus, and have been linked to multiple psychiatric disorders, such as schizophrenia (for a review, see Mitra[174]). MS exposure in Sprague Dawley rats is associated with increased methylation of the *Ntsr1* promoter in the amygdala, along with decreased expression.[145] However, there was no impact on *Ntsr2* or NTS expression itself. Lastly, there has been growing research interest in the *monoamineoxidase A* gene (*Maoa*), which encodes a catalyst enzyme for the deamination of dopamine and serotonin. SNPs at this locus have been shown to interact with ELA and increased conduct disorder in human males.[175] However, a cross-generational study in B6 mice found no epigenetic differences at this locus in either sex,[42] although this may be due to the lack of assessment of the *Maoa* risk variants.

6.5 Estrogen and Oxytocin Signaling

Another early finding from the LG-ABN model (Table 1) was the intergenerational transmission of maternal behaviors. Adult female offspring that were exposed to high-LG conditions become high-LG mothers, whereas low LG pups become low LG mothers.[176] Cross-fostering studies have shown that this transmission is not dependent on genetic factors but is associated with increased expression of estrogen (ER) and oxytocin receptors (OXTR) in the medial preoptic area (MPOA) of the hypothalamus.[18] In addition, maternal exposure to stress that decreases maternal care behaviors has been shown to decrease OXTR expression in several brain regions.[177]

In rats, low LG-ABN has been associated with increased DNA methylation in the *estrogen receptor alpha* (*Erα*) promoter and decreased binding of the transcriptional activator stat5b.[19] In a follow-up study, the same group found that low LG-ABN is associated with increased methylation and H3K9me3, as well as decreased H3K4me3 at *Erα* during weaning and later in adulthood.[20] In addition, these epigenetic changes are associated with decreased expression. Interestingly, they also found that these outcomes varied with the timing of ELA exposure and the timing of DNA methylation measurement, suggesting that the ELA-associated epigenetic profile at the *Erα* locus is dependent on the age of the stress exposure and the age of outcome measures.

Beery *et al.*[21] recently investigated the impact of LG-ABN model at the *Oxtr* locus in the blood and brain of adult male rats and found decreased

DNA methylation in the plasma of low LG-ABN adults. However, they did not find any changes in the brain, suggesting that although peripheral epigenetic profiles might serve as useful biomarkers for ELA exposure, this does not necessarily reflect alterations in the brain.

6.6 Other Genes

Multiple other candidate genes have been investigated to examine the wide range of epigenetic impacts associated with ELA. Methyl-CpG-binding protein 2 (MeCP2) is itself involved in the downstream impacts of DNA methylation via the recruitment of other enzymes and transcription factors to methylated CpGs, including histone-modifying enzymes and DNMTs,[178–180] and may also be involved in miRNA processing.[181] Mutations in *Mecp2* are well known to cause Rett syndrome; however, alterations in expression and activity are also associated with other disorders such as autism and schizophrenia.[182] ELA may alter MeCP2 signaling as B6 mice exposed to MUCS show increased methylation of the *Mecp2* promoter in male germ cells. Strikingly this pattern can be transferred to the F2 female population as the same methylation pattern is found in the brain along with decreased expression.[42] The same group found that MS also impacts the *Cannabinoid receptor type 1* (*Cb1*) gene in the same cross-generational pattern leading to an increase in methylation and a decrease in expression in the female F2 offspring. *Cb1* encodes a G protein-coupled receptor that binds endocannabinoids, and THC from marijuana, and may be involved in stress resilience.[6] Altered cannabinoid signaling has been associated with multiple psychopathologies, including mood disorders and schizophrenia.[183]

Another gene with widespread downstream impacts, *Egr1*, has also been shown to be impacted by ELA. MS exposure is associated with increases of H4Ac at the *Egr1* locus and increased expression in the hippocampus of adolescent B6 male mice.[148] As *Egr1* encodes the transcription factor NGF1-A, it plays an important role in the activation of many other genes discussed previously in this review, such as *Nr3c1*. As B6 mice have been shown to be resilient to ELA at the *Nr3c1* locus, this ELA-induced upregulation of NGF1-A may be an important compensatory mechanism. Of course more research is needed to see if the trend is reversed in more ELA vulnerable strains, such as Balb/c mice.

Finally, two genes involved in adult neural plasticity, *Activity-regulated cytoskeleton-associated protein* (*Arc*) and *Reelin* (*Reln*), have both been shown to alter epigenetic marks and expression in ELA exposed rodents. Along with

Egr1, MS is also associated with increased H4Ac and expression of *Arc* in the hippocampus of male B6 mice.[148] However, MM is associated with decreased methylation at the *Reln* promoter in female Long Evans infant rats, and counterintuitively, decreased expression of RELN in adulthood.[138] ARC interacts with neurotropic factors such as BDNF, and dysregulation is associated with autism and Alzheimer's.[184] RELN is known to be important in neuronal migration during neurodevelopment and adult neurogenesis, and dyseregulation is associated with schizophrenia and bipolar disorder.[185] Although preliminary, these results increase the evidence discussed earlier in the *BDNF* section, suggesting that ELA may exert long-term impacts on neurogenesis and synaptic plasticity via epigenetic modifications.

6.7 Genome-Wide Studies

It was suggested, in the early studies, that large areas of the epigenome are likely effected by ELA.[22] Some researchers thus hypothesized that similar to the effects of ELA on the protein synthesis machinery discussed above, ELA may also alter the expression of global epigenetic machinery, such as DNMTs and HDACs. Indeed, this has been confirmed by several studies in animals. For example, Long Evans rats exposed to Low LG-ABN have increased expression of DNMT1 in the hippocampus,[17] whereas Wistar rats exposed to MS have increased expression of DMNT1, DMNT3a, and 3b in the NAc.[29] Multiple studies have found ELA-dependent alterations in HDACs as well. In Long Evans rats, MM decreases HDAC1 expression,[138] and in Sprague Dawley rats, MS alters HDACs 2, 3, 7, 8, 9, as well as a histone H3 lysine-9 (H3-K9)-specific methyltransferase (HMT SUV39H), in an age-dependent manner.[30] Similarly, in Balb/c mice, MS decreases the expression of HDACs 1, 3, 7, 8, and 10.[31]

However, although these machinery alterations appear to have an impact, it is unlikely that this results in uniform increases or decreases of global epigenetic patterns as there are multiple processes involved. The first genome-wide approaches have begun to investigate the complexity of the ELA epigenome and have so far illustrated global alterations that nevertheless encompass complex site-specific bidirectional alterations.

The ELA-induced increased expression of DMNTs has been associated with both a global decrease in DNA methylation and an increase in methylation at specific sites, such as the promoters for *Pp1c* and *A2ar*, which are involved in cocaine addiction.[29] However, the impact of ELA on DMNTs may be sex and age specific. For example, MM increases DNA methylation at the gene loci for *Dnmt1*, *Dnmt3a*, *Mecp2*, and *Gadd45b* (a protein necessary for activity-induced DNA demethylation) in both male and female rats.[48] Yet, these alterations are transcriptionally silent during development. Later, in adult male rats, the methylation patterns correlate with a reduction in expression of all the methylating enzymes. However, in female adult rats there is only reduced expression of GADD45b. In addition, there is evidence for strain and age effects on ELA-induced alterations in HDAC expression. Levine *et al.*[31] found that a number of HDACs are increased in the forebrain of adolescent Balb/c mice exposed to MS, but that these same HDACs are decreased later in adulthood. In addition, this pattern of expression inversely correlates with global levels of the active mark H4K12Ac in the forebrain at both ages. In contrast to this, MS does not have any effects on HDAC expression or global histone acetylation in B6 mice at any age.

McCoy *et al.*[32] also found that species/strain can vary the behavioral outcome of a specific methylation pattern. They used the Wistar Kyoto (WKY) rat strain that is known to be highly vulnerable to adult stress and after exposure exhibits increased depressive and anxiety-like behavior, as well as HPA axis hyperreactivity, stress-induced ulcers, and cardiac issues. Although they found the typical hypermethylated hippocampus in WKY rats exposed to MS, this was not associated with an anxiety or depressive-like phenotype. Instead, MS exposed rats display increased resilience to adult stressors, increased exploratory and social behaviors, and improved cardiovascular health. Most recently, they followed up this study by breeding rats to be either low or high responders to novelty and investigated DMNT expression and global methylation patterns in the amygdala.[186] They not only found differential methylation between the two strains, but were also able to rescue the increased anxiety-like behavior of the low responders by increasing overall methylation through dietary methyl donors. These studies underline the need for caution in interpreting phenotypic outcomes of specific epigenomic alterations without considering genetic background and collecting observational data.

Although only beginning, altogether genome-wide approaches have resulted in somewhat mixed results, possibly also due to the wide range of utilized methods. For example, Doherty *et al.*[49] used a MethylFlash approach, which combines antibody-based enrichment and absorbance measures, to show increased global methylation in the hippocampus of male rats. In agreement with this, McCoy *et al.*[32] also used MethylFlash to show a similar global pattern of hypermethylation in the hippocampus of male Wistar Kyoto rats. However, conversely, using a novel immunohistochemistry staining approach, Brown *et al.*[23] showed no differences in global methylation in the hippocampus of Long Evans rats exposed to low and high LG-ABN. In studies that used methods with the capacity to interrogate site-specific patterns, such as MethylCap-seq, or genome-wide methods paired with bisfulfite sequencing at a targeted loci, the majority have found that ELA is associated with complex patterns of site-specific hyper- and hypomethylation clusters across the genome,[43,49,186] suggesting a complex coordination of alterations in multiple enzymes.

The first genome-wide studies on ELA-associated hydroxymethylation (5hmC) have also been conducted. Maternal unpredictable chronic stress (MUCS) (Table 1) in female B6 mice result in global 5hmC changes that are correlated with changes in corresponding transcript expression.[43] In contrast to this, exposure to MM in Long Evans rats alters the global percentage of 5hmC in the amygdala of males, but not females.[49] At the level of specific loci however, using an amplicon approach, ELA is associated with differential 5mC methylation at *Bdnf* exon IV in multiple brain regions in females, but not in males, once again suggesting that the method of interrogation has a significant impact on results. Together these studies suggest that ELA may impact global 5hmC patterns in a species- and sex-dependent manner.

7. HUMAN STUDIES ON THE EPIGENETIC IMPACT OF ELA

The earliest attempts at translating animal findings into human populations were conducted by our group using postmortem brain tissues from individuals who died by suicide and had a history of child abuse, compared with those who died by suicide without a history of ELA, and those who died by sudden means other than suicide without a history of ELA or psychopathology. Since then, the field has expanded to investigate peripheral cell epigenetics in multiple child and adult populations to determine if these noninvasive techniques can be used as a biomarker for a molecular signature of ELA and psychopathology. Recently, genome-wide and single-cell approaches have started to be conducted, lending more insight into the wide range of impacts of ELA on human cells. Here we will review in detail the current evidence of the ELA-associated epigenetic patterns at multiple sites across the human genome. See Table 3 for a list of the human studies included in this review.

Table 3 Human Studies on ELA-Associated Alterations in Epigenetic Processes.

Reference	ELA Exposure	Gene	Tissue	Sex	Age	Epigenetic Alteration	Epigenetic Method
187	Child abuse + adoption	*SLC6A4*	Lymphoblast cell lines	B	Adults M = ~48	Increased methylation in promoter Increased methylation of CpG 1 and 3 in females No change in individual CpG in males	Bisulfite Sequenom EpiTYPER
187	Sexual abuse	*SLC6A4*	Lymphoblast cell lines	F	Adults	Increased methylation in promoter Methylation associated with antisocial personality disorder and carrying at least one copy of the 5-HTTPLR-*s* risk allele	Bisulfite Sequenom EpiTYPER
188	Low SES	GW	Blood	B	Children 19	Increased methylation at one CpG in females. Interaction with 5-HTTLPR genotype in females only	Illumina 450K Array
189	Low SES	GW	Blood	F	Children M = 11–19	Differential methylation across the genome Increased differential methylation is associated with carriers of at least one	Illumina 450K Array

(*Continued*)

Table 3 Human Studies on ELA-Associated Alterations in Epigenetic Processes.—cont'd.

Reference	ELA Exposure	Gene	Tissue	Sex	Age	Epigenetic Alteration	Epigenetic Method
						copy of the 5-HTTPLR-*s* risk allele	
190	Sexual abuse	*SLC6A4*	Lymphoblast cell lines	F	Adults M = 41.1	Increased methylation associated with sexual abuse, antisocial personality disorder and major depression No association between methylation and biological parental psychopathology (no genetic load)	Bisulfite + PCR
191	Low SES	GW	Blood	M	Adult M = 45	Increased methylation at 666 promoters Decreased methylation at 586 promoters Genes involved in cell signaling, DNA, and metabolic processes	
192	Trauma	GW	Blood	B	Adults	Increased methylation at KITLG	
193	Childhood adversity	*SLC6A4*	Lymphoblast cell lines	B	Adults M = ~48	Increased methylation in promoter at CpG 7 Increased methylation of CpG 2 associated with physical abuse	

194	Child abuse	*FKBP5*	Blood	B	Adult	Decreased methylation in GRE-binding site. Methylation plus SNP associated with PTSD	Bisulfite sequencing
195	Severe adversity before 43 months	GW	Buccal cells	B	Adolescents M = 15	Increased methylation at 9 CpGs in the promoter of cytochrome P450 gene (CYP2EI) Associated with altered social cognition	195
196	Child abuse and suicide	*NR3C1*	HPC ACC	M	Adult	Differential methylation associated with depressed suicides not child abuse. Increased methylation in exon 1B, C, CpGs 6 and 8. Decreased methylation at CpG 11. Associated changes in expression were only in child abuse group.	196
197	Child abuse and suicide	GW	HPC	M	Adult	Increased methylation at 248 (68.5%) of promoters Decreased methylation at 114 (31.5%) of promoters Increased methylation correlated with decreased expression	

(*Continued*)

Table 3 Human Studies on ELA-Associated Alterations in Epigenetic Processes.—cont'd.

Reference	ELA Exposure	Gene	Tissue	Sex	Age	Epigenetic Alteration	Epigenetic Method
198	Child abuse and suicide	*OPRK*	INS	B	Adult	Decreased methylation and hydroxymethylation in 2nd intron Decreased expression	
199	Child abuse and suicide	GW	ACC	B	Adult M = 45.3	RRBS: Differential methylation across the genome Decreased methylation at LINGO3 and POU3F1 Increased DNA methylation at ITGB1 RNAseq: 17.9% differentially expressed genes, including 55 genes related to myelin and oligodendrocytes Targeted BS: Decreased methylation at LINGO3 and POU3F1 in oligodendrocytes	
200	Child abuse	*NR3C1*	Blood	B	Adult M = 29.4	Increased methylation in exon 1F in individuals who had experienced physical abuse only.	

201	Low SES, parental absence, breastfeeding	GW	Blood	B	Adult 20–22	Differential methylation at multiple genes associated with inflammation regulation Increased methylation in body of C1S and decreased methylation at promoter of GNG2 associated with low SES Increased methylation in 3′-UTR of EGR4 associated with parental absence Increased methylation at 5′-UTR of TLR1 and decreased methylation at SULT1C2 5′-UTR associated with breastfeeding duration These above associations were not associated with season of birth, infectious diarrhoea, or exposure to animal feces Differential methylation is associated with differential expression of inflammatory cytokines

(Continued)

Table 3 Human Studies on ELA-Associated Alterations in Epigenetic Processes.—cont'd.

Reference	ELA Exposure	Gene	Tissue	Sex	Age	Epigenetic Alteration	Epigenetic Method
202	Child abuse and suicide	*rRNA*	HPC	M	Adult M = 30	Increased methylation at promoter and 5' regulatory region at 21 out of 26 CpGs Decreased expression	
203	Child abuse and suicide	*NR3C1*	HPC	M	Adult	Increased methylation of exon 1F and NGFI-A-binding site	
204	Child abuse	GW	Blood	B	Adult	Differential global methylation at 14 genes Increased methylation in CA + PTSD by 11.78% Increased methylation in CA + PTSD by 0.8%	
205	Parental loss	*NR3C1* *MAOA*	Saliva	B	Adult M = ~57	NR3C1: Increased methylation at exon 1F, CpG 35 associated with early parental death and carriers of MOAO-L risk allele MAOA: Decreased methylation associated with depression in females, no differences based on CA	
206	Institutional care	GW	Blood	B	Children M = 8.25	Increased methylation at 744 promoters Decreased methylation at 94 promoters	

						Genes involved in cell and immune signaling	
207	Bullying	*SLC6A4*	Buccal cells	B	Children 10	Increased methylation of 5-HTT	
208	Childhood adversity	*NR3C1*	Blood	B	Adult	Increased methylation of promoter. Methylation associated with sexual abuse and severity.	
209	Child abuse, neglect	*NR3C1*	Blood	B	Children 11–14	Increased methylation at CpG 3, 6 and 7, NGFI-A site	
210	Child abuse	GW	Blood	B	Adult	No change based on CA Increased plasma TNFa in CA Increased global methylation in PTSD	
211	Child abuse	*NR3C1*	Blood	F	Adult M = 26	No change in methylation based on CA	
212	Trauma	*NR3C1*	Blood	B	Adult M = 27	Increased methylation at exon 1F, CpG 1 and 3 Increased methylation was correlated with increased post-dex cortisol	Bisulfite sequencing
213	Parental loss Child abuse	*NR3C1*	Blood	B	Adult M = 32.9	Decreased methylation in exon 1F over 13 CpGs in CA Decreased methylation at	Bisulfite sequencing

(*Continued*)

Table 3 Human Studies on ELA-Associated Alterations in Epigenetic Processes.—cont'd.

Reference	ELA Exposure	Gene	Tissue	Sex	Age	Epigenetic Alteration	Epigenetic Method
						CpG 6, 9, 10, 11, 12, 13 No CA + no disorder had the highest levels of methylation Increased methylation was correlated with increased post-dex cortisol	
214	Pre and post natal stress and adolescent stress and trauma comparison	*NR3C1*	Blood	B	Adolescents M = 16	Increased methylation at 11 CpGs in promoter Could be attributed to adolescent exposures	
215	Child abuse + adoption	*SLC6A4*	Lymphoblast cell lines	B	Adults M = ~48	Increased methylation with trauma and risk allele Psychopathology associated with risk allele and low methylation	
216	Pre and post natal trauma and recent life stress	*SLC6A4*	Blood	B	Adults	Increased methylation at 2 CpGs for prenatal stress. No change for postnatal stress, or recent stress	Bisulfite sequencing

217	Childhood abuse, neglect, domestic violence	GW	Saliva	B	NR	Differential methylation at multiple loci, including BDNF, NR3C1, FKBP5. Increased methylation in ID3, GRINI, TPPP associated with depression	
218	Holocaust survivors Offspring child abuse	*FKBP5*	Blood	B	Survivors: M = 77.3 Offspring: Adults	Decreased methylation at promoter in survivors Increased methylation at promoter in offspring associated with CA and carriers of the risk allele (rs1360780 A)	Bisulfite sequencing

Gene: NR3C1, glucocorticoid receptor; FKBP5, FK-binding protein 5; rRNA, ribosomal RNA; SLC6A4, 5-HTT; OPRK, kappa opioid receptor; GW, genome wide. Sex: M, male; F, female; B, both. Age: M, mean. Epigenetic alteration: CA, child abuse. Tissue: HPC, hippocampus; ACC, anterior cingulate cortex; INS, anterior insula. NR, not reported.

7.1 HPA Axis

Paralleling the original findings from rodent studies, our group provided the first human evidence of ELA-associated increases in DNA methylation at exon 1F of the *NR3C1* locus, which is the human homologue of rodent *Nr3c1* exon 17. Depressed suicides with a history of child abuse have increased levels of DNA methylation at 2 CpG sites in the promoter region and decreased NR3C1 expression when compared with both nonabused suicide completers and healthy controls.[203] Follow-up in vitro experiments further suggest that the animal findings translate to humans, by showing that the increased methylation of these CpGs blocked the binding of the transcription factor NGFI-A. Although this original work only investigated one transcript of *NR3C1*, our group also subsequently investigated other highly expressed transcripts: exons 1B, 1C, and 1H[196] and found increased DNA methylation in the promoters of exons 1B and 1C in depressed suicides both with and without a history of child abuse, compared with controls, along with reduced expression in only those with a history of ELA. Exon 1H, in contrast, displays decreased methylation in depressed suicides compared with controls, but this , atypically, was associated with decreased expression, and only in individuals with a history ELA, suggesting that DNA methylation does not always inversely correlate with expression, and that it can even be functionally "silent" as in the case of the depressed suicides without ELA exposure. Although it is well established that DNA methylation is not always repressive, the relevance of transcriptionally silent methylation patterns is still being debated. Some researchers hypothesize that this could be a type of poised promoter that enables increased speed in specific environmental contexts.[219]

There have also been several studies of ELA-associated DNA methylation of the *NR3C1* locus in the peripheral cells, such as blood or saliva. In one of the first studies, the severity of childhood sexual abuse was shown to be predictive of increased *NR3C1* promoter DNA methylation and associated with depression, PTSD, or borderline personality disorder.[208] Other studies also found increased *NR3C1* methylation in the peripheral cells of adults with a history of ELA and a current psychiatric diagnosis, such as depression[205] and borderline personality disorder.[200] One study did not find ELA-associated differential *NR3C1* methylation in adults with eating disorders,[211] although increased methylation was associated with increased borderline personality disorder and/or suicidality. Peripheral studies have also been conducted on children and teenagers with a history of ELA exposure and have largely found the same methylation patterns at the *NR3C1* locus as in adults with psychopathology.[209,214,217] However, one study found that additional stressful life events in adolescents could account for the increased *NR3C1* methylation.[214]

Overall, the consistency of findings on the epigenetic impact of ELA at the *NR3C1* locus between animal and human neural tissues is encouraging.[220] However, peripheral studies are mixed. Although there is evidence that DNA methylation profiles in peripheral cells might serve as biomarkers for alterations in HPA axis function, caution is still needed in interpreting these results and predicting specific mental health outcomes. For example, adults with a history of ELA but without a history of psychiatric disorders also have increased *NR3C1* methylation in leukocytes.[212] Interestingly, increased methylation is still correlated with decreased response to dexamethasone suppression, suggesting that those with increased *NR3C1* methylation have increased HPA axis activity, despite the lack of psychiatric issues. Recently, the same group tested a second but similar cohort and found an opposite ELA-associated decrease in *NR3C1* promoter methylation in those both with, and without, a history of psychopathology.[213] However, they still found a positive correlation between methylation and postdexamethasone cortisol levels, suggesting that HPA axis hyperactivity can occur in the absence of increased *NR3C1* methylation, and is not deterministic of psychiatric illness. Thus, although the overall consistency between peripheral and central *NR3C1* studies is encouraging, more research needs to be done to determine if peripheral epigenetic patterns can be used to evaluate changes in the brain or HPA axis regulation, or predict the risks for adult psychopathology.

Studies investigating ELA-associated epigenetic alterations to the *FKBP5* gene have largely been conducted in human peripheral cells. Multiple single nucleotide polymorphisms (SNPs) have been identified in the *FKBP5* gene that have been shown to interact with ELA to predict depression, PTSD, and suicidal behaviors.[221–223] The first study at this locus explored the interaction between ELA and genetic background in individuals diagnosed with PTSD and found decreased DNA methylation levels in intron 7 of *FKBP5* in ELA exposed individuals who also carry a specific risk allele for adult psychopathology.[194] They used a chromatin capture method to identify an altered chromatin conformation of this risk allele (rs1360780 A). This conformation allowed for increased expression of FKBP5 via a closer interaction between the binding site for GR [the glucocorticoid response element (GRE)] in an enhancer in intron 2 and the promoter region. In the protective allele (rs1360780 G), this interaction between the GRE and the promoter was not as strong. FKBP5 expression was increased in risk allele carriers at baseline as well as after GR binding. Thus, ELA exposure may have a stronger impact on individuals with the risk allele, whereby chronic GR activation triggers a positive feedback loop of DNA demethylation at the GRE in intron 7 and increased *FKBP5* transcription.

A recent study also found differential methylation patterns at the *FKBP5* locus in Holocaust survivors and their unexposed offspring. Although survivors had increased methylation, the offspring had decreased methylation at the *FKBP5* promoter,[218] possibly reflecting the complex cross-generational patterns of epigenetic inheritance seen in some animal studies.[42] However, it was also found that in offspring, decreased methylation was associated with their own experiences of child abuse as well as with the possession of the risk allele (rs1360780 A), paralleling the methylation patterns from the study discussed above. Together these studies suggest that ELA interacts with genetic background to determine methylation patterns at the *FKBP5* locus and highlight the importance of controlling for genetic background in future studies. However, as discussed in the peripheral studies of GR methylation, these studies also suggest caution in predicting mental health risks from epigenetic patterns in peripheral cells. The first study was able to associate the methylation/risk allele interaction with PTSD outcomes. However, despite possessing a similar methylation pattern and risk allele interaction, the majority of the holocaust offspring cohort did not display psychiatric symptoms.

7.2 Brain-Derived Neurotropic Factor

The trends of ELA-associated epigenetic silencing at the *BDNF* locus in animal models are congruent with the first findings in humans using peripheral cells. In individuals diagnosed with borderline personality disorder, the number of childhood traumas is positively correlated with increased DNA methylation of the *BDNF* promoter in circulating leukocytes.[224] However, this is not correlated with BDNF expression in the plasma. Taken with the animal findings, this supports the evidence that epigenetic alterations can be transcriptionally silent at the *BDNF* locus.

7.3 Serotonin Signaling

In human peripheral studies, the majority of studies have shown an impact of ELA. Wang *et al.*[225] used a novel positron emission tomography (PET) approach to show that increased *5-HTT* (*SLC6A4*) methylation is associated with childhood-limited aggression in the adult male orbitofrontal cortex. They were also able to show that methylation is positively correlated with central measures of 5-HTT synthesis in the orbitofrontal cortex.[225] A study using bisulfite sequencing at this locus, however, did not find an effect of postnatal ELA.[216] In a monozygotic twin study, exposure to bullying was associated with increased DNA methylation at the *SLC6A4* promoter in comparison to their unexposed twin.[207] In addition, bullying associated methylation was associated with a decrease in cortisol response to a stress challenge.

Although the findings in animals suggest an interaction between the short *SLC6A4* risk allele (*SS*) and ELA, the evidence is mixed in humans. Beach *et al.*[188,189] conducted two genome-wide studies using peripheral blood samples from children who were carriers of at least one *SS* allele and exposed to cumulative low SES stress and found differential methylation in risk allele carries of both sexes in one study,[189] but only in females in the other.[188]

Multiple studies into ELA-associated *SLC6A4* methylation was also conducted using virus-transformed lymphoblast cell lines derived from children and adults with or without a history of ELA. One study found that increased DNA methylation of the *SLC6A4* promoter region is associated with sexual abuse in females.[187] Although these in vitro results should be interpreted with caution, they were able to replicate these results in other cohorts exposed the multiple types of child abuse.[190,215,226,227] In addition, another group was also able to replicate these findings in cell lines from adults diagnosed with depression, and found that increased methylation at the *SLC6A4* promoter is associated with increased exposure to multiple types of childhood adversities.[193]

ELA-associated epigenetic alterations in other genes involved in serotonin signaling have also been investigated. One study focusing on the *MAOA* locus found that early parental loss is associated with increased DNA methylation at a CpG site near the NGFI-A-binding site at *N53C1*, but only in the saliva from depressed females with the *MAOA-L* gene variant.[205]

7.4 Immune Signaling

It is well established that the immune system interacts extensively with the HPA axis. GCs play a key role in regulating the expression and release of both pro- and anti-inflammatory cytokines, which, in turn, can stimulate the release of GCs.[2] In normal conditions GCs act in an overall anti-inflammatory capacity; however, chronic increased GCs can lead to GC desensitization in immune cells, similar to what occurs in HPA axis dysregulation, and thus can result in chronic proinflammatory signaling.[228,229] ELA exposure is associated with increased levels of proinflammatory cytokines and inflammatory markers, such as interleukin-6 and C-reactive protein, that can persist into adulthood.[230] Indeed, immune disruption is a common factor that may explain the wide variety of ELA-associated outcomes that span both physical and psychiatric diseases as previously discussed. However, despite the strong evidence of ELA alterations in immune signaling, most investigations have interrogated the epigenome of immune cells as a proxy for neural function. The investigation of the ELA-associated epigenetic alterations of immune function genes is still preliminary. Smith *et al.*[210] investigated the methylome in circulating blood cells of African American subjects with or without PTSD and a history of child abuse. Differential methylation was associated with PTSD but not child abuse in multiple immune-related genes; however, the expression of proinflammatory cytokines was associated with a history of child abuse. Recently, McDade *et al.*[201] conducted an interesting study investigating the impact of exposure to early life nutritional, microbial, and psychosocial stressors on DNA methylation in inflammatory genes. Types of ELA, such as low SES exposure and parental loss, were associated with differential expression of multiple inflammatory cytokines. Interestingly, there was little overlap between the inflammatory genes impacted by psychosocial stressors, and those impacted by physiological stressors such as decreased breastfeeding, season of birth, infectious diarrhea, or increased exposure to animal feces. These preliminary studies contribute to the growing evidence that ELA impacts immune regulation in adulthood, and suggests that epigenetic mechanisms may be involved.

7.5 Other Genes

The opioid system has been shown to be involved in social bonding in mammals,[231] and alterations of the kappa opioid receptor (*OPRK*) have been associated with trauma-related psychopathology in humans.[232] Our group found that decreased DNA methylation and hydroxymethylation in the second intron of *OPRK*, as well as decreased expression of the receptor, is associated with a history of child abuse in depressed suicides.[198] In addition, our group also investigated the impacts of ELA on ribosomal RNA using the same approaches and found that ELA is associated with increased hippocampal methylation of the small (40S) ribosomal subunit promoter,[202] which along with the impact of ELA on epigenetic machinery suggests that ELA may profoundly affect both transcription and translation in a global manner.

7.6 Genome-Wide Studies

Our group conducted the first investigation of ELA-associated epigenome alterations in postmortem brain tissues of suicide completers, with or without a history of child abuse using a custom 400K microarray. Suicides with an ELA history had 248 hypermethylated and 114 hypomethylated promoter regions compared with those without an ELA history and nonsuicides.[197] In addition, by separating neurons from nonneurons, we were able to identify neuron-specific DNA methylation changes, which were at multiple loci involved in plasticity, such as *ALS2*, *NR1D1*, and *HIST2H2AB*. These results highlight the benefit of assessing epigenetic adaptations at the cell-type level.[233,234] We also recently showed that child abuse may alter the epigenome in oligodendrocytes, by using similar cell-type isolation methods.[199] We found that a history of child abuse is associated with differential methylation at the promoters of multiple genes involved in myelination These alterations in methylation are also associated with changes in expression and even with changes in the morphology of nearby myelinated neurons, suggesting that child abuse may have a lasting impact on neural connectivity and function.

Multiple investigations on ELA-induced epigenome changes have been conducted in human peripheral cells mostly using the Illumina 450K[188,189,192,195,204,217] or 27K[206,210] microarray. In children raised in institutions, versus those raised by their parents, more than 800 gene loci were found to be differentially methylated, including those involved in immune response and cellular signaling.[206] More than 2600 CpG sites were found to be differentially methylated in saliva samples from children that were removed from their homes due to child abuse as compared with children with no history of maltreatment or psychopathology.[235] Effected loci included genes implicated in cortical development, depression and addiction, diabetes, asthma, and cancer. Houtepen *et al.*[192] recently reported a methylome analysis from three independent cohorts using either leukocytes or buccal cells that had multiple findings, including an association between a history of childhood trauma, DNA methylation in the *KITLG* gene, and cortisol stress reactivity. They later used another cohort with premortem blood samples and postmortem brain samples to compare peripheral versus central *KITLG* methylation in the same individuals and found a positive correlation between the blood and the PFC. In addition, they observed a negative correlation between blood and the superior temporal gyrus, and no association between the blood and the entorhinal cortex or the cerebellum. Although more research is needed comparing peripheral and central cell epigenetic patterns, these findings suggest that peripheral epigenome might be a useful biomarker in some cases.

8. INDIVIDUAL VARIATION IN THE IMPACTS OF ELA

A major theme arising from both the animal and human literature is that individual differences, such as sex, genetic background, age, as well as the type, developmental timing, and duration/severity of the stressor, can vary the impact that ELA has on the epigenome, transcriptome, and behavioral outcomes. It is possible that each of these factors determine an individual's tolerance for stress. Selye's original concepts of distress and eustress[236,237] and the inverted-U dose–response curve[3] discussed at the beginning of this article, as well as the two and three "hit" models of stress vulnerability,[94] all propose that the detrimental effects of stress exposure occur at a threshold unique to the individual. Genetic background appears to be heavily influential. For example, Kundakovic *et al.*[41] found that the same ELA paradigm both increases and decreases anxiety- and depressive-like behaviors based on mouse strain and sex. Studies by McCoy *et al.*[32] also found that ELA induces resilience to later stressors in rats who have a genetic predisposition for anxiousness, and that this resilience is mediated by dramatic reconfigurations of DNA methylation in the hippocampus. McIlwrick *et al.*[238] showed that in mice bred for low HPA reactivity, exposure to ELA is associated with enhanced memory performance in adulthood, whereas those bred for high trait HPA reactivity show deficits. In humans it is well established that many children exposed to early adversities show a remarkable ability to recover and can be resilient to future stressors and the development of adult psychopathologies.[239,240] At the same time, an individual's stress tolerance is not necessarily set and can be shifted by experience. For example, mild levels of early life stress exposure can increase resilience to future stressors[11,32,40] Indeed, resilience in children exposed to ELA has been shown to involve a combination of factors, including possibly innate traits such as increased sensitivity to positive stimuli, mature amygdala-prefrontal circuitry, low negative cognition, easy temperament, increased empathy, and social skills, but also environmental factors, such as supportive relationships, connectedness to school, and many others.[241–243] Altogether these data suggest high levels of individual variability in ELA vulnerability and resilience. Thus, despite the compelling evidence that on average ELA contributes to lasting negative outcomes in both animals and humans, there is strong evidence of resilience at the level of the individual. One of the major challenges of future research, therefore, will be to disentangle how various factors interact with epigenetic machinery to either promote susceptibility or resilience to ELA. As per the current literature, it is also unclear whether the cellular and molecular mechanisms underlying resilience act in a passive or active manner. In the former, epigenetic patterns would not change in response to ELA, thereby resulting in a similar molecular profile as controls. However in the latter, active resilience mechanisms would counteract maladaptive molecular alterations (see Russo *et al.* and Zannas and West[244,245] for in depth reviews). Active mechanisms occur at various physiological levels and may be a promising future target of ELA investigations as a way to help clarify mixed findings in the literature.[244] Another way to explore resilience and differential susceptibility is by identifying resilient and vulnerable subpopulations in both animal (within and between species) and human studies. This approach has already been insightfully incorporated into animal models of adult stress, such as social defeat,[246] and may also benefit future ELA studies. In the next discussion, we will assess how genetic background, sex, type/intensity, and timing of a stressor have impacted the current literature reviewed earlier in this article and discuss ways of addressing the issue of differential susceptibility in future studies.

8.1 Genetic Differences

The recent use of mice in the investigation of ELA-associated epigenetics has identified strain-dependent resilience and susceptibility, with the resilience in B6 mice being especially striking. This may be due in part to the fact that B6 mothers display increased licking and grooming behavior compared with other strains, such as Balb/c.[11] However, adult B6 mice also show a consistent bimodal distribution of either resilience or susceptibility when exposed to social defeat paradigm,[246] and cross fostering of B6 mice with Balb/c mothers during ELA exposure does not alter B6 resilience,[144] after thus it is plausible that in this strain of mouse, genetic factors are involved in conferring increased stress resilience in certain circumstances. In fact, results from Uchida *et al.*[247] suggest that epigenetic regulation over GDNF expression in the ventral striatum and the associated behavioral responses are genotype dependent. After exposure to the chronic unpredictable maternal stress

(CUMS) paradigm, both stress vulnerable (Balb/c) and stress resilient (B6) adult mice show increased methylation of the *Gdnf* promoter and recruitment of MeCP2. However, levels of H3 acetylation are reduced in the susceptible strain but increased in the resilient strain. This difference was found to be due to the recruitment of the transcriptional repressor HDAC2 in susceptible mice versus the transcriptional activator CREB in resilient mice. At the level of expression, stress vulnerable and resilient mice show a decrease and increase of GDNF expression, respectively. Here, B6 mice display a unique resilience profile, suggestive of active resilience. Thus, future studies that use multiple genetic backgrounds and/or bimodal distributions of susceptibility and resilience to ELA might be useful in our attempt to identify genetic vulnerabilities and epigenetic responses to ELA.

One issue arising when attempting to address the same question in humans is the lack of this dissectible bimodal distribution of susceptible and resilient ELA responses seen in some animal models. However, SNP × methylation interaction investigations have shown epigenetic marks to be strongly determined by the underlying genetic code for multiple genes, including those often associated with ELA, such as the cases at the *BDNF*, *SLC6A4*, and *FKBP5 loci*, discussed earlier.[164,194,205,248–250]

8.2 Sex Differences

Sex differences are especially challenging to characterize as they involve multiple factors, including differences in neuroendocrine signaling,[251] neurodevelopment, and brain region specificity.[252] Because of this complexity, ELA studies in rodents have traditionally only used males.[9,253] However, recent work including females suggests that overall males appear more sensitive to ELA,[41,254–257] at least on anxiety-like and cognitive outcomes. In contrast to this, some studies have found that female rodents are more sensitive to ELA on specific behavioral assays, such as sucrose preference,[258,259] or other anhedonia measures.[36] Beyond behavioral differences, ELA causes neurogenesis deficits in females before puberty while increasing neurogenesis in males,[256,260] although this trend is reversed by adulthood. Overall, sex-specific vulnerabilities to ELA, both at the behavioral and neurobiological level, may be the result of sexually dimorphic expression of epigenetic factors.[261] Mechanisms such as DNA methylation, histone modifications, chromatin remodeling, and noncoding RNAs are thought to be responsible for promoting sexual dimorphism in the brain and may thus explain, in part, sex-specific susceptibility or resilience to ELA.[262] For instance, as discussed earlier, Dohertyet *al.*[49] demonstrated that adolescent female Long Evans exposed to ELA have no significant alterations in global 5mC or 5hmC levels but greater methylation of exon IV of the *Bdnf* gene in the amygdala and ventral hippocampus. However for males, ELA resulted in significantly higher 5mC levels in the dorsal hippocampus and lower 5hmC levels in the amygdala, but no significant changes to *Bdnf* methylation. In addition, this sex specificity may be further dependent on cell type. Interestingly, Blaze and Roth[139] revealed that ELA induced neuronal but not nonneuronal hypermethylation of *BdnfIV* in the mPFC of adult female rats, a finding not observed in either cellular fraction for males. Altogether these data suggest that sex-specific sensitivities to ELA may be cell type, brain, and genomic region dependent.

Recent characterization of the transcriptional profiles of multiple brain areas in both males and females diagnosed with major depressive disorder via gene coexpression network analysis demonstrated that there are considerable sexual dimorphisms in the human brain as well.[263] At the behavioral level, results from Alastalo *et al.*[264] also parallel some of the animal studies in observing that parental separation during childhood had a greater impact on adult male physical and psychosocial outcomes. This is counterintuitive, as many stress-related disorders are more prevalent in females,[265] and ELA increases depression and drug use risk in females more than males.[266] As with the aforementioned animal literature, sex-specific vulnerability to ELA in humans may be a reflection of epigenetic differences. For instance, Gouin *et al.*[267] found that ELA was associated with methylation at one CpG site in the first intron of the *Oxtr* gene among females but not males. Another CpG site within the promoter region was significantly related to childhood trajectories of anxiousness, also only in females. Thus, although it is clear that ELA interacts with sex, and there is evidence to support a role for epigenetics at the root of these sex differences, more research is needed to identify a clear pattern and to better tease apart the epigenetic mechanisms involved.

8.3 Stressor Differences

Although exposure to ELA is robustly associated with an increased risk for adult psychopathology, studies linking specific adversities to unique epigenomic patterns have been mixed, likely due in part to inadequate subgroup sample sizes. Despite some evidence that specific types of ELA exposures are associated with specific adult outcomes,[192,193,205,268] large and multinational studies suggest that the often observed increased risk for psychopathology is the result of combined ELA exposures, rather than a specific type of stressor.[72,266,269] This issue is especially difficult to resolve in human studies, as early stressors rarely occur in isolation. For example, children exposed to parental mental illness are often also exposed to conflict and other forms of instability.[72,270] Interestingly, Teicher *et al.*[271] have shown that exposure to multiple forms of ELA results in more deleterious effects than an additive sum of each type of stressor. Another important consideration is the frequency of ELA because exposure to abuse often results in an increased risk for subsequent abuse.[272] Moreover, the severity of ELA may also differentially impact the epigenome and mental health outcomes. Indeed, Perroud *et al.*[208] have shown that sexual abuse in childhood, its severity, and the number of different types of abuse positively correlate with *Nr3C1* methylation. In bipolar disorder subjects, they found that repeated abuse and types of sexual abuse translated into higher *Nr3C1* methylation.[273]

In addition to the aforementioned complexities, psychosocial stressors often cooccur with physiological stressors. Children exposed to low socioeconomic conditions have increased exposures to both physical stressors, such as inadequate nutrition and industrial toxins, and psychosocial stressors, such as instability and violence.[274] Although often categorized and studied separately, physiological stressors can have significant mechanistic overlaps with psychosocial stressors. For example, postnatal malnutrition in mice was shown to result in global changes to the epigenomic landscape of the hippocampus and thalamus, whereby 60% of significantly altered regions are related to neuronal development or psychiatric diseases.[275] In humans, malnutrition in the first year of life results in differential methylation of regions involved in regulating metabolism[276] and genes that have been previously identified as involved in neuropsychiatric risk.[277] Not surprisingly, ELA has also long been associated with metabolic disturbances and obesity.[278] Even in controlled animal models, disentangling the physiological and psychological impacts of ELA is difficult. Animals exposed to significant amounts of stress often show reduced body weight, altered levels of leptin and glucose, and are prone to increased adult obesity compared with controls.[278,279] Furthermore, maternal high-fat and/or high-sugar diets have recently been shown to be surprisingly protective for their offspring in the context of ELA,[280,281] suggesting a complex relationship between stress and metabolic processes. Thus, ELA paradigms in animal models may be exerting some of their impacts via routes that are not entirely "psychological."

However, this may not be an issue. Our definitional distinctions between different types of stressors may be arbitrary. It is clear in humans that what we define as "psychological" stress has significant physiological impacts and vice versa. Both types of stressors have been associated with DNA methylation and gene expression changes as well as increased risks of chronic disease in adulthood.[191,274] Thus, it is unclear how helpful it will be to associate specific stressor types with specific epigenetic outcomes. The current evidence suggests overall that the type of the stressor is not as impactful as the accumulation or severity of stressors,[191,282] or the developmental timing and duration of the exposure.

8.4 Age Differences

The timing of ELA can significantly alter outcomes. In animal studies, the start and duration of ELA can impact the type, direction, and magnitude of specific behavioral, gene expression,[51] and epigenetic profiles.[20,35] The developmental window at which ELA occurs may also impinge upon region-specific neurodevelopmental programs that mature at different rates, thus making particular brain areas more sensitive than others. For instance, hippocampal development in mice undergoes a growth spurt during the second week of life; this developmental pattern is associated with accelerated rRNA synthesis. Correspondingly, levels of DNA methylation at rDNA promoters are lower during the second week of life compared with before or after this window of time. After a peak in hippocampal growth, DNA, RNA, and rRNA content gradually decline into adulthood.[283] In terms of human development, hippocampal growth follows a similar pattern with the largest increase in volume during infant/toddler years followed by a steady decline[284] Interestingly, early child maltreatment has been associated with increased DNA methylation at ribosomal promoters, decreases in rRNA levels, blunted growth of the hippocampus, and reduced hippocampal volumes in humans.[284,285] On a clinical level, there is also some evidence of age-specific effects on psychiatric diagnosis. For example, abuse before age 12 is associated with internalizing symptoms in childhood and adult PTSD, whereas abuse after 12 is associated with externalizing symptoms and adult depression.[285,286] Maercker *et al.*,[287] on the other hand, report that females exposed to political trauma before the age of 13 have an approximately equal risk of developing MDD or PTSD in adulthood. Yet trauma after the age of 13 throughout adolescence is associated with a greater risk for PTSD than for MDD. In another study, parental loss before 9 years of age was associated with a higher risk for depression compared with loss between 9 and 17.[288]

In addition, outcomes from ELA can change dramatically over the life span. Some animal studies suggest that males exposed to ELA may display more resilience in adolescence, yet more deficits in mid-to-late adulthood.[28,30,138,256] This contributes to the in vitro evidence that ELA induces progressive regulatory events that eventually lead to stable epigenetic patterns over time,[150] suggesting that outcomes may be age dependent. However, this may also be due to acute epigenetic dynamics. Not all epigenetic patterns have immediate transcriptional outcomes. Similarly to the poised promoters discussed previously at the Nr3c1, Bdnf, Dnmt1, Dnmt3a, Mecp2, and Gadd45b loci,[35,219,289,290] these transcriptionally silent epigenetic alterations may modulate expression only in response to specific environmental cues, such as brain activity. This epigenetic metaplasticity[219] makes it important for future studies to examine both the timing of ELA exposure and the timing and specific circumstances surrounding outcome measures.

8.5 Cell-Type Differences

Epigenetic findings associated with ELA are clearly dependent on brain region, as every study that has interrogated multiple brain regions has found differences in epigenetic patterns (Table 2). However, more recently it has also become clear that ELA-associated epigenetic patterns are cell-type specific. This is not surprising, as it has long been known that epigenetic processes are tightly coupled with developmental processes underlying cellular differentiation and identity, and that some epigenetic marks are specific to defined cell types, as previously discussed in this review. However, technological limitations until recently have restricted many of the studies reviewed here to the investigation of epigenetic and transcriptomic alterations in tissue homogenates. Arguably, the experience of adversity could specifically impact the molecular biology of some cell types while leaving others unscathed. As such, tissue homogenates introduce a level of background noise that may mask important cell-type specific signals. One solution has been the employment of separation techniques that allow for cell-type specific investigations. Our group has done much of the work in this area showing unique global and targeted epigenetic patterns in isolated neurons[197] and oligodendrocytes,[199] and therefore offers a resolution to precisely determine cell-type specific molecular responses to ELA. Although still in their infancy, new technologies that enable single cell whole genome investigation, (discussed in more detail below), will no doubt increase our ability to identify more specifically the mechanisms by which

ELA interacts with the epigenome in a cell-type specific manner. This approach may even resolve some of the high interindividual variability and conflicting results in the literature, and enable us to more accurately target the alterations directly associated with behavioral outcomes.

9. CURRENT TECHNICAL LIMITATIONS AND POSSIBLE SOLUTIONS

Taken together the candidate and genome-wide studies conducted thus far indicate that ELA modifies epigenetic processes at multiple sites across the genome in many individuals. However, the functional impact of this is not entirely clear. Many of the studies suggest that ELA-induced changes to the epigenome may lead to altered transcriptome and phenotypic outcomes, such as increasing the risk of altered behaviors and psychopathology. However, the evidence on outcomes is mixed and complex and appears to depend on multiple individual factors as discussed earlier. Thus, a big challenge of future research will be to link epigenetic alterations with specific phenotypic outcomes while controlling for these, and possibly many other individual factors. In addition, there are still many other challenges to these investigations that need to be addressed if we are to gain a comprehensive understanding of the molecular impact of ELA on the brain and stress response pathways. Here we will discuss some of the current technical limitations in the field and offer possible solutions for future studies.

9.1 Dynamic Nature of Epigenetic Processes

One of the biggest limitations in the current literature is the necessary retrospective nature of postmortem studies. Although some DNA methylation patterns are stable over long periods of time, animal studies also suggest that ELA can trigger dynamic and sometimes rapid changes in DNA methylation and associated transcriptional activity,[291] as well as progressive changes in epigenetic processes that slowly change over the life span.[141] Differentiating between stable and dynamic epigenetic patterns remains out of the reach of postmortem human studies as they only offer a temporal snapshot of the brain. Peripheral cell studies may be able to serve as biomarkers of epigenetic plasticity in humans. However, results thus far suggest that findings need to be interpreted on a gene-by-gene basis and with caution, as their accurate representation of alterations in central tissues might only apply in specific contexts. For example, Provencal *et al.*[142] looked at both peripheral and central samples from the same rhesus monkeys and although they found an effect of ELA and a small correlation between different cell types, cell type-based differences in DNA methylation were so pronounced that the low amplitude of ELA-associated alterations was difficult to detect. In addition, studies conducted in peripheral blood cells should take into account the acute environmental circumstances of their subjects. Epigenetic processes in immune cells may not be as stable as in brain tissues, due to the dynamic nature of the immune system in responding to environmental cues. Thus, any alterations may be due to recent events. This was illustrated by van der Knaap *et al.*[214] who found that although increased *NR3C1* DNA methylation is associated with ELA exposure in teenagers, this could be entirely accounted for by recent stressful life events. It is possible that individuals who have experienced ELA are also more likely to experience later stressors, thus this overlap may confound peripheral studies of ELA-associated epigenetic patterns. New tools that can noninvasively visualize epigenetic processes in vivo, such as PET ligands for chromatin-modifying enzymes,[33] would enable future studies to investigate how the epigenetic machinery dynamically responds to various environmental cues in humans and thus improve our understanding of the stability of epigenetic profiles.

9.2 Cell-Type Specificity of Epigenetic Mechanisms

Although much progress has been made in developing cell-type isolation methods as discussed earlier, there are still limitations, as these methods rely on current cellular classification systems based off of morphology, or the presence of cellular markers and/or RNA content. Thus, this approach limits investigation to classically defined "cell types" and makes the assumption that all cells of the same "type" share a similar molecular biology that responds to environmental insults in an identical way. Given that a cell's function is derived from its molecular composition, grouping individual cells by their molecular content is a reasonable solution to the conundrum.[292] However, with the advancement of single-cell sequencing technologies, the systematic categorization of individual neurons based off of their molecular properties is now feasible and thus these assumptions can now be directly tested. Future studies may therefore benefit from investigating the impact of ELA on single neurons or glia to explore if epigenetic profiles are consistent within a specific cell type.

The findings from genome-wide characterizations of chromosomal architecture and genomic organization in heterogeneous samples of central cells have provided mechanistic insight into how these features may be epigenetically regulated. Arguably, though, generalizations extrapolated from these bulk analyses have not improved our understanding of the temporal dynamics of the epigenome, nor its causal role in gene expression.[293] Recently developed single-cell isolation and sequencing approaches will inevitably enable the generation of mechanistic models that causally link epigenomic variation with transcriptomic dynamics of cells in development and disease. The roadblocks to achieving this endeavor have been single-cell isolation and library preparation methods requiring the input of biological material from millions of cells. However, these obstacles are beginning to be overcome, with studies conducting single-cell reduced representation bisulfite sequencing approach (scRRBS),[294] single-cell whole genome bisulfite sequencing (scBS-seq),[295–297] and single nuclei RNA-seq.[298] Of note, Luo *et al.*[297] interrogated single nuclei derived from the frontal cortex of both mouse and human subjects and identified 16 mouse and 21 human neuronal subpopulations as well as a unique layer 6 excitatory neuronal subtype not previously described. In partial agreement with this, Lake *et al.*[298] identified 16 neuronal subtypes in the human cortex.

In addition to profiling the methylome and transcriptome of single nuclei or cells, other available single-cell technologies include single-cell ChIP-seq (histone modifications),[299] single-cell MNase-seq (nucleosome positioning),[300] single-cell Hi-C (chromatin conformation),[301] single-cell DamID (nuclear lamina interactions),[302] and single-cell DNaseI-seq and single-cell ATAC-seq (chromatin accessibility)[303,304] to dissect the epigenetic heterogeneity of cellular populations. However, to make direct comparisons of epigenomic variation and transcriptome heterogeneity, combined single-cell sequencing methods are required. Indeed, methods have been developed to simultaneously analyze genome-transcriptome (DR-seq),[305] DNA methylome-transcriptome (M&T-seq),[306] and genome-DNA methylome-transcriptome (Trio-seq)[307] data from single cells. Recently, a multiomic approach was developed to measure multiple layers of epigenetic information from the same single cell. Chromatin Overall Omic-scale Landscape Sequencing (COOL-seq) simultaneously analyzes chromatin state, nucleosome positioning, DNA methylation, copy number variation (CNVs), and ploidy from a single cell.[308]

Advancements in single-cell high-throughput sequencing have laid the foundation for future research oriented at mapping the global epigenome and transcriptome of the brain. Given the feasibility of not only single-cell but also single-nuclei sequencing, investigations can move beyond animal models to postmortem human brain samples as well. In the context of ELA, complete molecular profiles of specific neuronal subtypes will allow us to understand whether certain cell types are more vulnerable to environmental insults than others. In the same vein, these advancements open up the possibility of exploring whether or not cell types negatively impacted by ELA are rescuable through specific pharmacological, optogenetic, and/or chemogenetic interventions. A thorough understanding of cellular subtypes, defined by expression, is essential for the design of paradigms that increase target specificity and therefore give us the capacity to manipulate the activity of subpopulations of neurons,while reducing off target impacts, such as the tools discussed later. Although these approaches are not yet feasible in humans, an increased ability to target pharmacological therapies to specific cells and loci would improve drug efficacy and specificity.

9.3 Issues of Causation

Another major limitation of the current literature is the difficulty of testing causal direction. Conclusions on the causal nature of specific epigenetic patterns were out of reach for all the studies reviewed here because, until recently, methods to manipulate epigenetic marks at specific gene loci were unavailable. Thus, despite the correlative evidence between ELA exposure, specific epigenetic alterations, and specific phenotypic outcomes discussed earlier, we do not yet know if epigenetic changes are sufficient to alter gene and protein expression, nor the associated phenotypic outcomes. For example, to determine if DNA methylation at the *NR3C1* promoter is sufficient to cause decreased expression of GR, a dysregulation of the HPA axis, and increased anxiety-like behaviors, future research needs to be able to manipulate DNA methylation at the *NR3C1* locus alone while keeping other epigenetic and regulatory processes constant. Although it appears likely that the DNA methylation of promoter regions results in decreased expression, this has yet to be confirmed in all cases. Furthermore, the connection between increased DNA methylation at one locus and phenotypic outcomes is even more speculative. Given the sheer number of gene expression regulatory processes, it is likely that multiple other factors are involved in determining the transcriptional outcomes from epigenomic alterations and the downstream impacts as well as the way these processes interact with the individual factors we have previously discussed in this article. In addition, as genome-wide studies suggest that ELA impacts large areas of the epigenome, it is likely that ELA-associated health and behavioral outcomes require the combined effects of widespread alterations in methylation, as well as other epigenetic processes, across multiple brain regions, or even in other organs.

Importantly, tools are now becoming available that will enable this causal investigation of specific epigenetic marks. Zinc fingers (ZFs), transcription activator-like effectors (TALEs), and an adapted form of CRISPR-Cas9 that directs epigenetic machinery (dCas9) are already being used by to manipulate the epigenome. In vivo these tools have been used to both validate[307,308] and challenge[309] the current assumptions about the transcriptional impact of specific epigenetic patterns. In vivo, Heller *et al.*[310] notably used both ZF and TALEs approaches to demonstrate the sufficiency of specific changes to chromatin structure at the *FosB* locus in the NAc to cause altered reward behaviors and an increased vulnerability to social stress. More recently, Liu *et al.*[311] used dCas9-Dnmt3a and dCas9-Tet1 constructs to respectively methylate and demethylate specific sites at multiple gene loci, including *Bdnf*. They used the dCas9-Tet1 construct in vitro to demethylate the *Bdnf* promoter IV in mouse neurons, resulting in an increase in BDNF expression. Importantly, they subsequently showed that this construct could operate in vivo to demethylate two other gene loci. Two other groups have also used a dCas9-Tet1 construct in vitro to demethylate and upregulate gene expression at multiple loci,[312,313] and Vojta *et al.*[314] recently tested a dCas9-Dnmt3a construct in human cells. Although most of the excitement around CRISPR to date has focused on the possibility of genome editing to cure disease, epigenome editing might be more applicable in vivo as it allows for more subtle and reversible modulatory control. Nevertheless, future studies will greatly benefit from these new tools and will allow for increased investigation into the causal relationship between specific epigenetic patterns and behavioral phenotype outcomes.

10. CONCLUSIONS

The ultimate aim in the study of the epigenetic links between ELA and disease outcomes is to contribute to the reduction of the devastating and lasting impacts on individuals, families, and communities. Although ELA is not the only cause of psychiatric disease, it is a significant contributor. For example, the World Health Organization estimates that the elimination of ELA would reduce adult mood, behavioral, and substance disorders by almost 30% globally.[269] Of course eradicating ELA is a complex task requiring effort on multiple fronts, including political, cultural, educational, and economic interventions, not to mention addressing the psychological, physical, and financial health of parents. However, research can also play an important role in this effort through informing social policy by highlighting the impacts of ELA, identifying vulnerable populations and discovering effective interventions and treatments.

Both the animal and human literature suggest that intervention during ELA is possible during development, and can be mitigated even long after exposure. For example, Wang *et al.*[146] reversed the epigenetic impact of ELA exposed rats with adolescent enrichment, not only rescuing the methylation profile but also recovering control levels of plasticity and behavior. Murgatroyd *et al.*[315] also recently showed that increased tactile stimulation, or stroking, can normalize DNA methylation in human infants who had been exposed to mothers diagnosed with major depression. Although the prevention and treatment of ELA may appear daunting, there is growing evidence on the effectiveness of intervention programs (for a review, see Fisher *et al.*[89]). The literature on differential susceptibility and gene by environment interactions strongly suggests that the children who are the most vulnerable to ELA are also the ones who readily respond to these interventions,[316] especially those that occur earlier in development. Thus, it is imperative that we work to identify vulnerable populations early on. In addition to behavioral interventions, multiple effective psychotropic medications are thought to alter epigenetic marks, such as imipramine.[317] However, based on the variable response to ELA illustrated in the current literature, future drug treatments will likely need to be tailored to the individual as well as being cell and loci specific to avoid off-target effects. The new in vivo epigenetic tools being developed based on CRISPR-Cas9 technology might be useful in developing a new generation of highly targeted drug treatments.

The potential benefits of studying and targeting the epigenome in mental illness lie in its capacity for both stability and change. This nature of epigenetic homeostatic plasticity is an encouraging reminder that we are a malleable and resilient species and that neither our genes, nor our negative experiences, are entirely deterministic of our outcomes.

REFERENCES

1. Koolhaas JM, Bartolomucci A, Buwalda B, et al. Stress revisited: a critical evaluation of the stress concept. *Neurosci Biobehav Rev.* 2011;35(5):1291–1301.
2. Chrousos GP. Stress and disorders of the stress system. *Nat Rev Endocrinol.* 2009;5 (7):374–381.

3. Calabrese EJ, Bachmann KA, Bailer AJ, et al. Biological stress response terminology: integrating the concepts of adaptive response and preconditioning stress within a hormetic dose-response framework. *Toxicol Appl Pharmacol.* 2007;222(1):122–128.
4. Selye H. The evolutioin of the stress concept. *Am Sci.* 1973;61(6):692–699.
5. Selye H. Stress without distress.In: *Psychopathology of Human Adaptation. Boston, MA: Springer US;1976.137–146.*
6. McEwen BS, Bowles NP, Gray JD, et al. Mechanisms of stress in the brain. *Nat Neurosci.* 2015;18(10):1353–1363.
7. Sapolsky RM. Stress and the brain: individual variability and the inverted-U. *Nat Neurosci.* 2015;18(10):1344–1346.
8. McLaughlin KA, Sheridan MA, Nelson CA. Neglect as a violation of species-expectant experience: neurodevelopmental consequences. *Biol Psychiatry.* 2017;82 (7):462–471.
9. Weaver ICG, Cervoni N, Champagne Fa. et al. Epigenetic programming by maternal behavior. *Nat Neurosci.* 2004;7(8):847–854.
10. Liu D, Diorio J, Tannenbaum B, et al. Maternal care, hippocampal glucocorticoid receptors, and hypothalamic-pituitary-adrenal responses to stress. *Science.* 1997;277 (5332):1659–1662.
11. Caldji C, Tannenbaum B, Sharma S, Francis D, Plotsky PM, Meaney MJ. Maternal care during infancy regulates the development of neural systems mediating the expression of fearfulness in the rat. *Proc Natl Acad Sci USA.* 1998;95(9):5335–5340.
12. McGowan PO, Suderman M, Sasaki A, et al. Broad epigenetic signature of maternal care in the brain of adult rats. *PLoS One.* 2011;6(2):e14739.
13. Liberman SA, Mashoodh R, Thompson RC, Dolinoy DC, Champagne FA. Concordance in hippocampal and fecal Nr3c1 methylation is moderated by maternal behavior in the mouse. *Ecol Evol.* 2012;2(12):3123–3131.
14. Daniels W, Fairbairn L, van Tilburg G, et al. Maternal separation alters nerve growth factor and corticosterone levels but not the DNA methylation status of the exon 1 7 glucocorticoid receptor promoter region. *Metab Brain Dis.* 2009;24(4):615–627.
15. Kosten TA, Nielsen DA. Litter and sex effects on maternal behavior and DNA methylation of the Nr3c1 exon 17 promoter gene in hippocampus and cerebellum. *Int J Dev Neurosci.* 2014;36:5–12.
16. Caldji C, Diorio J, Meaney MJ. Variations in maternal care alter GABAA receptor subunit expression in brain regions associated with fear. *Neuropsychopharmacology.* 2003;28(11):1950–1959.
17. Zhang T, Hellstrom IC, Bagot RC, Wen X, Diorio J, Meaney MJ. Maternal care and DNA methylation of a glutamic acid decarboxylase 1 promoter in rat hippocampus. *J Neurosci.* 2010;30(39):13130–13137.
18. Champagne FA, Diorio J, Sharma S, Meaney MJ. Naturally occurring variations in maternal behavior in the rat are associated with differences in estrogen-inducible central oxytocin receptors. *PNAS.* 2001;98(22):12736–12741.
19. Champagne FA, Weaver ICG, Diorio J, Dymov S, Szyf M, Meaney MJ. Maternal care associated with methylation of the estrogen receptor-alpha1b promoter and estrogen receptor-alpha expression in the medial preoptic area of female offspring. *Endocrinology.* 2006;147(6):2909–2915.
20. Peña CJ, Neugut YD, Champagne FA. Developmental timing of the effects of maternal care on gene expression and epigenetic regulation of hormone receptor levels in female rats. *Endocrinology.* 2013;154(11):4340–4351.
21. Beery AK, McEwen LM, MacIsaac JL, Francis DD, Kobor MS. Natural variation in maternal care and cross-tissue patterns of oxytocin receptor gene methylation in rats. *Horm Behav.* 2016;77:42–52.

22. Weaver ICG, Meaney MJ, Szyf M. Maternal care effects on the hippocampal transcriptome and anxiety-mediated behaviors in the offspring that are reversible in adulthood. *Proc Natl Acad Sci USA.* 2006;103(9):3480–3485.
23. Brown SE, Weaver ICG, Meaney MJ, Szyf M. Regional-specific global cytosine methylation and DNA methyltransferase expression in the adult rat hippocampus. *Neurosci Lett.* 2008;440(1):49–53.
24. Kinnally EL. Epigenetic plasticity following early stress predicts long-term health outcomes in rhesus macaques. *Am J Phys Anthropol.* 2014;155(2):192–199.
25. Henningsen K, Dyrvig M, Bouzinova EV, et al. Low maternal care exacerbates adult stress susceptibility in the chronic mild stress rat model of depression. *Behav Pharmacol.* 2012;23(8):1.
26. Chen J, Evans AN, Liu Y, Honda M, Saavedra JM, Aguilera G. Maternal deprivation in rats is associated with corticotrophin-releasing hormone (CRH) promoter hypomethylation and enhances CRH transcriptional responses to stress in adulthood. *J Neuroendocrinol.* 2012;24(7):1055–1064.
27. Wu Y, Patchev AV, Daniel G, Almeida OFX, Spengler D. Early-life stress reduces DNA methylation of the *Pomc* gene in male mice. *Endocrinology.* 2014;155(5):1751–1762.
28. Suri D, Veenit V, Sarkar A, et al. Early stress evokes age-dependent biphasic changes in hippocampal neurogenesis, Bdnf expression, and cognition. *Biol Psychiatry.* 2013;73(7):658–666.
29. Anier K, Malinovskaja K, Pruus K, Aonurm-Helm A, Zharkovsky A, Kalda A. Maternal separation is associated with DNA methylation and behavioural changes in adult rats. *Eur Neuropsychopharmacol.* 2014;24(3):459–468.
30. Suri D, Bhattacharya A, Vaidya VA. Early stress evokes temporally distinct consequences on the hippocampal transcriptome, anxiety and cognitive behaviour. *Int J Neuropsychopharmacol.* 2014;17(2):289–301.
31. Levine A, Worrell TR, Zimnisky R, Schmauss C. Early life stress triggers sustained changes in histone deacetylase expression and histone H4 modifications that alter responsiveness to adolescent antidepressant treatment. *Neurobiol Dis.* 2012;45(1):488.
32. McCoy CR, Rana S, Stringfellow SA, et al. Neonatal maternal separation stress elicits lasting DNA methylation changes in the hippocampus of stress-reactive Wistar Kyoto rats. *Eur J Neurosci.* 2016;44(10):2829–2845.
33. Wang C, Schroeder FA, Hooker JM. Visualizing epigenetics: current advances and advantages in HDAC PET imaging techniques. *Neuroscience.* 2014;264:186–197.
34. Kember RL, Dempster EL, Lee THA, Schalkwyk LC, Mill J, Fernandes C. Maternal separation is associated with strain-specific responses to stress and epigenetic alterations to Nr3c1, Avp, and Nr4a1 in mouse. *Brain Behav.* 2012;2(4):455–467.
35. Bai M, Zhu X, Zhang Y, et al. Abnormal hippocampal BDNF and miR-16 expression is associated with depression-like behaviors induced by stress during early life. *PLoS One.* 2012;7(10):e46921.
36. Sasagawa T, Horii-Hayashi N, Okuda A, Hashimoto T, Azuma C, Nishi M. Long-term effects of maternal separation coupled with social isolation on reward seeking and changes in dopamine D1 receptor expression in the nucleus accumbens via DNA methylation in mice. *Neurosci Lett.* 2017;641:33–39.
37. Capitanio JP, Mendoza SP, Mason WA, Maninger N. Rearing environment and hypothalamic-pituitary-adrenal regulation in young rhesus monkeys (Macaca mulatta). *Dev Psychobiol.* 2005;46(4):318–330.
38. Kinnally EL, Capitanio JP, Leibel R, et al. Epigenetic regulation of serotonin transporter expression and behavior in infant rhesus macaques. *Genes Brain Behav.* 2010;9(6):575–582.
39. Massart R, Suderman M, Provencal N, et al. Hydroxymethylation and DNA methylation profiles in the prefrontal cortex of the non-human primate rhesus macaque and the

impact of maternal deprivation on hydroxymethylation. *Neuroscience.* 2014;268:139–148.
40. Meaney MJ, Aitken DH, Bhatnagar S, Sapolsky RM. Postnatal handling attenuates certain neuroendocrine, anatomical, and cognitive dysfunctions associated with aging in female rats. *Neurobiol Aging.* 1991;12(1):31–38.
41. Kundakovic M, Lim S, Gudsnuk K, Champagne FA. Sex-specific and strain-dependent effects of early life adversity on behavioral and epigenetic outcomes. *Front Psychiatry.* 2013;4:78.
42. Franklin TB, Russig H, Weiss IC, et al. Epigenetic transmission of the impact of early stress across generations. *Biol Psychiatry.* 2010;68(5):408–415.
43. Papale LA, Madrid A, Li S, Alisch RS. Early-life stress links 5-hydroxymethylcytosine to anxiety-related behaviors. *Epigenetics.* 2017;12(4):264–276.
44. Dalle Molle R, Portella A, Goldani M, et al. Associations between parenting behavior and anxiety in a rodent model and a clinical sample: relationship to peripheral BDNF levels. *Transl Psychiatry.* 2012;2.
45. Naninck EFG, Oosterink JE, Yam K-Y, et al. Early micronutrient supplementation protects against early stress-induced cognitive impairments. *FASEB J.* 2017;31 (2):505–518.
46. Kinnally EL, Feinberg C, Kim D, et al. DNA methylation as a risk factor in the effects of early life stress. *Brain Behav Immun.* 2011;25(8):1548–1553.
47. Roth TL, Lubin FD, Funk AJ, Sweatt JD. Lasting epigenetic influence of early-life adversity on the BDNF gene. *Biol Psychiatry.* 2009;65(9):760–769.
48. Blaze J, Roth TL. Epigenetic mechanisms in learning and memory. *Wiley Interdiscip Rev Cogn Sci.* 2013;4(1):105–115.
49. Doherty TS, Forster A, Roth TL. Global and gene-specific DNA methylation alterations in the adolescent amygdala and hippocampus in an animal model of caregiver maltreatment. *Behav Brain Res.* 2015;298(pt A):15–19.
50. Branchi I, Karpova NN, D'Andrea I, Castrén E, Alleva E. Epigenetic modifications induced by early enrichment are associated with changes in timing of induction of BDNF expression. *Neurosci Lett.* 2011;495(3):168–172.
51. Molet J, Maras PM, Avishai-Eliner S, Baram TZ. Naturalistic rodent models of chronic early-life stress. *Dev Psychobiol.* 2014;56(8):1675–1688.
52. Baram TZ, Davis EP, Obenaus A, et al. Fragmentation and unpredictability of early-life experience in mental disorders. *Am J Psychiatry.* 2012;169(9):907–915.
53. Ivy AS, Brunson KL, Sandman C, Baram TZ. Dysfunctional nurturing behavior in rat dams with limited access to nesting material: a clinically relevant model for early-life stress. *Neuroscience.* 2008;154(3):1132–1142.
54. Hackman DA, Farah MJ, Meaney MJ. Socioeconomic status and the brain: mechanistic insights from human and animal research. *Nat Rev Neurosci.* 2010;11(9):651–659.
55. Solomon MB. Evaluating social defeat as a model for psychopathology in adult female rodents. *J Neurosci Res.* 2017;95(1–2):763–776.
56. Peña CJ, Kronman HG, Walker DM, et al. Early life stress confers lifelong stress susceptibility in mice via ventral tegmental area OTX2. *Science.* 2017;356 (6343):1185–1188.
57. Dranovsky A, Picchini AM, Moadel T, et al. Experience dictates stem cell fate in the adult hippocampus. *Neuron.* 2011;70(5):908–923.
58. Nilsson M, Perfilieva E, Johansson U, Orwar O, Eriksson PS. Enriched environment increases neurogenesis in the adult rat dentate gyrus and improves spatial memory. *J Neurobiol.* 1999;39:569–578.
59. Kuzumaki N, Ikegami D, Tamura R, et al. Hippocampal epigenetic modification at the brain-derived neurotrophic factor gene induced by an enriched environment. *Hippocampus.* 2011;21(2):127–132.

60. De Kloet ER, Joëls M, Holsboer F. Stress and the brain: from adaptation to disease. *Nat Rev Neurosci.* 2005;6(6):463–475.
61. Million M, Larauche M. Stress, sex, and the enteric nervous system. *Neurogastroenterol Motil.* 2016;28(9):1283–1289.
62. Fiori LM, Turecki G. Implication of the polyamine system in mental disorders. *J Psychiatry Neurosci.* 2008;33(2):102–110.
63. Fagundes CP, Glaser R, Kiecolt-Glaser JK. Stressful early life experiences and immune dysregulation across the lifespan. *Brain Behav Immun.* 2013;27(1):8–12.
64. Hostinar CE, Nusslock R, Miller GE. Future directions in the study of early-life stress and physical and emotional health: implications of the neuroimmune network hypothesis. *J Clin Child Adolesc Psychol.* 2018;47(1):142–156.
65. Zannas AS, Wiechmann T, Gassen NC, Binder EB. Gene–stress–epigenetic regulation of FKBP5: clinical and translational implications. *Neuropsychopharmacology.* 2016;41 (1):261–274.
66. Gold PW. The organization of the stress system and its dysregulation in depressive illness. *Mol Psychiatry.* 2015;20163:32–47.
67. Nusslock R, Miller GE. Early-life adversity and physical and emotional health across the lifespan: a neuroimmune network hypothesis. *Biol Psychiatry.* 2016;80(1):23–32.
68. Gilbert LK, Breiding MJ, Merrick MT, et al. Childhood adversity and adult chronic disease: an update from ten states and the District of Columbia, 2010. *Am J Prev Med.* 2015;48(3):345–349.
69. Rutten BPF, Mill J. Epigenetic mediation of environmental influences in major psychotic disorders. *Schizophr Bull.* 2009;35(6):1045–1056.
70. Gluckman PD, Hanson MA, Beedle AS. Early life events and their consequences for later disease: a life history and evolutionary perspective. *Am J Hum Biol.* 2007;19 (1):1–19.
71. Bernet CZ, Stein MB. Relationship of childhood maltreatment to the onset and course of major depression in adulthood. *Depress Anxiety.* 1999;9(4):169–174.
72. Green JG, McLaughlin KA, Berglund PA, et al. Childhood adversities and adult psychiatric disorders in the national comorbidity survey replication I: associations with first onset of DSM-IV disorders. *Arch Gen Psychiatry.* 2010;67(2):113–123.
73. Teicher MH, Samson JA. Annual research review: enduring neurobiological effects of childhood abuse and neglect. *J Child Psychol Psychiatry Allied Discip.* 2016;57 (3):241–266.
74. Anda RF, Felitti VJ, Bremner JD, et al. The enduring effects of abuse and related adverse experiences in childhood. A convergence of evidence from neurobiology and epidemiology. *Eur Arch Psychiatry Clin Neurosci.* 2006;256(3):174–186.
75. Lutz PE, Turecki G. DNA methylation and childhood maltreatment: from animal models to human studies. *Neuroscience.* 2014;264:142–156.
76. Nemeroff CB. Paradise lost: the neurobiological and clinical consequences of child abuse and neglect. *Neuron.* 2016;89(5):892–909.
77. Hertzman C. The biological embedding of early experience and its effects on health in adulthood. *Ann N Y Acad Sci.* 1999;896(1):85–95.
78. Weinstock M. The long-term behavioural consequences of prenatal stress. *Neurosci Biobehav Rev.* 2008;32(6):1073–1086.
79. Bale TL. Epigenetic and transgenerational reprogramming of brain development. *Nat Rev Neurosci.* 2015;16(6):332–344.
80. Bick J, Nelson CA. Early adverse experiences and the developing brain. *Neuropsychopharmacology.* 2016;41(1):177–196.
81. Teicher MH, Samson JA, Anderson CM, Ohashi K. The effects of childhood maltreatment on brain structure, function and connectivity. *Nat Rev Neurosci.* 2016;17 (10):652–666.

82. Edmiston EE, Wang F, Mazure CM, et al. Corticostriatal-limbic gray matter morphology in adolescents with self-reported exposure to childhood maltreatment. *Arch Pediatr Adolesc Med.* 2011;165(12):1069–1077.
83. Luby J, Belden A, Botteron K, et al. The effects of poverty on childhood brain development. *JAMA Pediatr.* 2013;167(12):1135.
84. Tottenham N, Hare TA, Millner A, Gilhooly T, Zevin JD, Casey BJ. Elevated amygdala response to faces following early deprivation. *Dev Sci.* 2011;14(2):190–204.
85. Pollak SD. Mechanisms linking early experience and the emergence of emotions: illustrations from the study of maltreated children. *Curr Dir Psychol Sci.* 2008;17 (6):370–375.
86. Pollak SD, Tolley-Schell SA. Selective attention to facial emotion in physically abused children. *J Abnorm Psychol.* 2003;112(3):323–338.
87. Aguliar-Valles A, Kim J, Jung S, Woodside B, Luheshi GN. Role of brain transmigrating neutrophils in depression-like behavior during systemic infection. *Mol Psychiatry.* 2014;19(5):599–606.
88. Kim P, Evans GW, Angstadt M, et al. Effects of childhood poverty and chronic stress on emotion regulatory brain function in adulthood. *Proc Natl Acad Sci USA.* 2013;110 (46):18442–18447.
89. Fisher Pa, Beauchamp KG, Roos LE, Noll LK, Flannery J, Delker BC. The neurobiology of intervention and prevention in early adversity. *Annu Rev Clin Psychol.* 2016;12 (1):331–357.
90. Fischer S, Strawbridge R, Vives AH, Cleare AJ. Cortisol as a predictor of psychological therapy response in depressive disorders: systematic review and meta-analysis. *Br J Psychiatry.* 2017;210(2):105–109.
91. Gunnar M, Quevedo K. The neurobiology of stress and development. *Annu Rev Psychol.* 2007;58:145–173.
92. Kvetnansky R, Pacak K, Fukuhara K, et al. Sympathoadrenal system in stress: interaction with the hypothalamic pituitary adrenocortical system. *Ann N Y Acad Sci.* 1995;771 (1):131–158.
93. Monroe SM, Simons AD. Diathesis-stress theories in the context of life stress research: implications for the depressive disorders. *Psychol Bull.* 1991;110(3):406–425.
94. Daskalakis NP, Bagot RC, Parker KJ, Vinkers CH, De Kloet ER. The three-hit concept of vulnerability and resilience: toward understanding adaptation to early-life adversity outcome. *Psychoneuroendocrinology.* 2013;38(9):1858–1873.
95. Parker KJ, Buckmaster CL, Sundlass K, Schatzberg AF, Lyons DM. Maternal mediation, stress inoculation, and the development of neuroendocrine stress resistance in primates. *Proc Natl Acad Sci USA.* 2006;103(8):3000–3005.
96. Bauer AM, Quas Ja, Boyce WT. Associations between physiological reactivity and children's behavior: advantages of a multisystem approach. *J Dev Behav Pediatr.* 2002;23(2):102–113.
97. Belsky J, Pluess M. Beyond diathesis stress: differential susceptibility to environmental influences. *Psychol Bull.* 2009;135(6):885–908.
98. Boyce WT, Ellis BJ. Biological sensitivity to context: I. An evolutionary–developmental theory of the origins and functions of stress reactivity. *Dev Psychopathol.* 2005;17 (2):271–301.
99. Del Giudice M, Ellis BJ, Shirtcliff EA. The adaptive calibration model of stress responsivity. *Neurosci Biobehav Rev.* 2011;35(7):1562–1592.
100. Carabotti M, Scirocco A, Maselli MA, Severi C. The gut-brain axis: interactions between enteric microbiota, central and enteric nervous systems. *Ann Gastroenterol.* 2015;28(2):203–209.
101. Foster JA, Rinaman L, Cryan JF. Stress & the gut-brain axis: regulation by the microbiome. *Neurobiol Stress.* 2017;7:124–136.

102. O'Mahony SM, Clarke G, Dinan TG, Cryan JF. Early-life adversity and brain development: is the microbiome a missing piece of the puzzle? *Neuroscience.* 2017;342:37–54.
103. Lieberman-Aiden E, van Berkum NL, Williams L, et al. Comprehensive mapping of long-range interactions reveals folding principles of the human genome. *Science.* 2009;326(5950):289–293.
104. Kumari S, Swaminathan A, Chatterjee S, Senapati P, Boopathi R, Kundu TK. *Chromatin Organization, Epigenetics and Differentiation: An Evolutionary Perspective.* Dordrecht: Springer; 2013. 3–35.
105. Duncan EJ, Gluckman PD, Dearden PK. Epigenetics, plasticity, and evolution: how do we link epigenetic change to phenotype? *J Exp Zool Part B Mol Dev Evol.* 2014;322 (4):208–220.
106. Levenson JM, Roth TL, Lubin FD, et al. Evidence that DNA (cytosine-5) methyltransferase regulates synaptic plasticity in the hippocampus. *J Biol Chem.* 2006;281 (23):15763–15773.
107. Cameron NM, Champagne FA, Parent C, Fish EW, Ozaki-Kuroda K, Meaney MJ. The programming of individual differences in defensive responses and reproductive strategies in the rat through variations in maternal care. *Neurosci Biobehav Rev.* 2005;29 (4–5):843–865.
108. Singh-Taylor A, Korosi A, Molet J, Gunn BG, Baram TZ. Synaptic rewiring of stress-sensitive neurons by early-life experience: a mechanism for resilience? *Neurobiol Stress.* 2015;1:109–115.
109. Nagy C, Turecki G. Transgenerational epigenetic inheritance: an open discussion. *Epigenomics.* 2015;7(5):781–790.
110. Bronfman ZZ, Ginsburg S, Jablonka E. Shaping the learning curve: epigenetic dynamics in neural plasticity. *Front Integr Neurosci.* 2014;8:55.
111. Feng J, Chang H, Li E, Fan G. Dynamic expression of de novo DNA methyltransferases Dnmt3a and Dnmt3b in the central nervous system. *J Neurosci Res.* 2005;79(6):734–746.
112. Jones PA. Functions of DNA methylation: islands, start sites, gene bodies and beyond. *Nat Rev Genet.* 2012;13(7):484–492.
113. Iurlaro M, von Meyenn F, Reik W. DNA methylation homeostasis in human and mouse development. *Curr Opin Genet Dev.* 2017;43:101–109.
114. Klose RJ, Bird AP. Genomic DNA methylation: the mark and its mediators. *Trends Biochem Sci.* 2006;31(2):89–97.
115. Oda S, Fukami T, Yokoi T, Nakajima M. Epigenetic regulation is a crucial factor in the repression of UGT1A1 expression in the human kidney. *Drug Metab Dispos.* 2013;41 (10):1738–1743.
116. Deng G, Chen A, Pong E, Kim YS. Methylation in hMLH1 promoter interferes with its binding to transcription factor CBF and inhibits gene expression. *Oncogene.* 2001;20 (48):7120–7127.
117. Domcke S, Bardet AF, Adrian Ginno P, Hartl D, Burger L, Schübeler D. Competition between DNA methylation and transcription factors determines binding of NRF1. *Nature.* 2015;528(7583):575–579.
118. Maunakea AK, Nagarajan RP, Bilenky M, et al. Conserved role of intragenic DNA methylation in regulating alternative promoters. *Nature.* 2010;466(7303):253–257.
119. Mellén M, Ayata P, Dewell S, et al. MeCP2 binds to 5hmC enriched within active genes and accessible chromatin in the nervous system. *Cell.* 2012;151(7):1417–1430.
120. Gross JA, Pacis A, Chen GG, et al. Gene-body 5-hydroxymethylation is associated with gene expression changes in the prefrontal cortex of depressed individuals. *Transl Psychiatry.* 2017;7(5):e1119.
121. Kinney SM, Chin HG, Vaisvila R, et al. Tissue-specific distribution and dynamic changes of 5-hydroxymethylcytosine in mammalian genomes. *J Biol Chem.* 2011;286 (28):24685–24693.

122. Li W, Liu M. Distribution of 5-hydroxymethylcytosine in different human tissues. *J Nucleic Acids.* 2011;2011:870726.
123. He Y, Ecker JR. Non-CG methylation in the human genome. *Annu Rev Genomics Hum Genet.* 2015;16(1):55–77.
124. Ziller MJ, Müller F, Liao J, et al. Genomic distribution and inter-sample variation of non-CpG methylation across human cell types. *PLoS Genet.* 2011;7(12):e1002389.
125. Guo JU, Su Y, Shin JH, et al. Distribution, recognition and regulation of non-CpG methylation in the adult mammalian brain. *Nat Neurosci.* 2014;17(2):215–222.
126. Lister R, Mukamel EA, Nery JR, et al. Global epigenomic reconfiguration during mammalian brain development. *Science.* 2013;341(6146):1237905.
127. Luger K, Mäder AW, Richmond RK, Sargent DF, Richmond TJ. Crystal structure of the nucleosome core particle at 2.8 A resolution. *Nature.* 1997;389(6648):251–260.
128. Jenuwein T. Translating the histone code. *Science.* 2001;293(5532):1074–1080.
129. Koch CM, Andrews RM, Flicek P, et al. The landscape of histone modifications across 1% of the human genome in five human cell lines. *Genome Res.* 2007;17 (6):691–707.
130. Clark MB, Amaral PP, Schlesinger FJ, et al. The reality of pervasive transcription. *PLoS Biol.* 2011;9(7):e1000625.
131. Kaikkonen MU, Lam MTY, Glass CK. Non-coding RNAs as regulators of gene expression and epigenetics. *Cardiovasc Res.* 2011;90(3):430–440.
132. Rinn JL, Chang HY. Genome regulation by long noncoding RNAs. *Annu Rev Biochem.* 2012;81(1):145–166.
133. Geisler S, Coller J. RNA in unexpected places: long non-coding RNA functions in diverse cellular contexts. *Nat Rev Mol Cell Biol.* 2013;14(11):699–712.
134. Siomi H, Siomi MC. On the road to reading the RNA-interference code. *Nature.* 2009;457(7228):396–404.
135. Obiols-Guardia A, Guil S. The role of noncoding RNAs in neurodevelopmental disorders: the case of Rett syndrome. *Advances in Experimental Medicine and Biology.* vol. 978. Cham: Springer International Publishing; 2017;23–37.
136. Chen M, Deng J, Yang M, et al. Noncoding RNAs, neurodevelopment, and neurodegeneration.In: *Post-Transcriptional Gene Regulation: RNA Processing in Eukaryotes. Weinheim, Germany: Wiley-VCH Verlag GmbH & Co. KGaA; 2013. 215–238.*
137. Bagot RC, Zhang T-Y, Wen X, et al. Variations in postnatal maternal care and the epigenetic regulation of metabotropic glutamate receptor 1 expression and hippocampal function in the rat. *Proc Natl Acad Sci USA.* 2012;109(suppl 2):17200–17207.
138. Blaze J, Scheuing L, Roth TL. Differential methylation of genes in the medial prefrontal cortex of developing and adult rats following exposure to maltreatment or nurturing care during infancy. *Dev Neurosci.* 2013;35(4):306–316.
139. Blaze J, Roth TL. Caregiver maltreatment causes altered neuronal DNA methylation in female rodents. *Dev Psychopathol.* 2017;29(2):477–489.
140. Desarnaud F, Jakovcevski M, Morellini F, Schachner M. Stress downregulates hippocampal expression of the adhesion molecules NCAM and CHL1 in mice by mechanisms independent of DNA methylation of their promoters. *Cell Adh Migr.* 2008;2 (1):38–44.
141. Murgatroyd C, Patchev AV, Wu Y, et al. Dynamic DNA methylation programs persistent adverse effects of early-life stress. *Nature.* 2009;12(12):1559–1566.
142. Provencal N, Suderman MJ, Guillemin C, et al. The signature of maternal rearing in the methylome in rhesus macaque prefrontal cortex and T cells. *J Neurosci.* 2012;32 (44):15626–15642.
143. Roth TL, Matt S, Chen K, Blaze J. Bdnf DNA methylation modifications in the hippocampus and amygdala of male and female rats exposed to different caregiving environments outside the homecage. *Dev Psychobiol.* 2014;56(8):1755–1763.

144. Schmauss C, Lee-McDermott Z, Medina LR. Trans-generational effects of early life stress: the role of maternal behavior. *Sci Rep.* 2014;4:4873.
145. Toda H, Boku S, Nakagawa S, et al. Maternal separation enhances conditioned fear and decreases the mRNA levels of the neurotensin receptor 1 gene with hypermethylation of this gene in the rat amygdala. *PLoS One.* 2014;9(5):e97421.
146. Wang A, Nie W, Li H, et al. Epigenetic upregulation of corticotrophin-releasing hormone mediates postnatal maternal separation-induced memory deficiency. *PLoS One.* 2014;9(4):e94394.
147. Weaver ICG. Reversal of maternal programming of stress responses in adult offspring through methyl supplementation: altering epigenetic marking later in life. *J Neurosci.* 2005;25(47):11045–11054.
148. Xie L, Korkmaz KS, Braun K, Bock J. Early life stress-induced histone acetylations correlate with activation of the synaptic plasticity genes Arc and Egr1 in the mouse hippocampus. *J Neurochem.* 2013;125(3):457–464.
149. Meaney MJ, Diorio J, Francis D, et al. Early environmental regulation of forebrain glucocorticoid receptor gene expression: implications for adrenocortical responses to stress. *Dev Neurosci.* 2010;18(1–2):61–72.
150. Murgatroyd C, Spengler D. Polycomb binding precedes early-life stress responsive DNA methylation at the Avp enhancer. *PLoS One.* 2014;9(3):e90277.
151. van Oers HJJ, de Kloet ER, Whelan T, Levine S. Maternal deprivation effect on the infant's neural stress markers is reversed by tactile stimulation and feeding but not by suppressing corticosterone. *J Neurosci.* 1998;18(23):10171–10179.
152. Park H, Poo M. Neurotrophin regulation of neural circuit development and function. *Nat Rev Neurosci.* 2012;14(1):7–23.
153. Greenberg ME, Xu B, Lu B, Hempstead BL. New insights in the biology of BDNF synthesis and release: implications in CNS function. *J Neurosci.* 2009;29 (41):12764–12767.
154. Greenough WT, Black JE, Wallace CS. Experience and brain development. *Child Dev.* 1987;58(3):539.
155. Duman RS. Neuronal damage and protection in the pathophysiology and treatment of psychiatric illness: stress and depression. *Dialogues Clin Neurosci.* 2009;11(3):239–255.
156. Boulle F, van den Hove DLA, Jakob SB, et al. Epigenetic regulation of the BDNF gene: implications for psychiatric disorders. *Mol Psychiatry.* 2012;17(6):584–596.
157. Branchi I, Francia N, Alleva E. Epigenetic control of neurobehavioural plasticity: the role of neurotrophins. *Behav Pharmacol.* 2004;15(5–6):353–362.
158. Chatterjee D, Chatterjee-Chakraborty M, Rees S, Cauchi J, de Medeiros CB, Fleming AS. Maternal isolation alters the expression of neural proteins during development: "stroking" stimulation reverses these effects. *Brain Res.* 2007;1158:11–27.
159. Smith MA, Makino S, Kvetnansky R, Post RM. Stress and glucocorticoids affect the expression of brain-derived neurotrophic factor and neurotrophin-3 mRNAs in the hippocampus. *J Neurosci.* 1995;15(3):1768–1777.
160. Marmigère F, Givalois L, Rage F, Arancibia S, Tapia-Arancibia L. Rapid induction of BDNF expression in the hippocampus during immobilization stress challenge in adult rats. *Hippocampus.* 2003;13(5):646–655.
161. Pardon MC, Roberts RE, Marsden CA, et al. Social threat and novel cage stress-induced sustained extracellular-regulated kinase1/2 (ERK1/2) phosphorylation but differential modulation of brain-derived neurotrophic factor (BDNF) expression in the hippocampus of NMRI mice. *Neuroscience.* 2005;132(3):561–574.
162. Roceri M, Cirulli F, Pessina C, Peretto P, Racagni G, Riva MA. Postnatal repeated maternal deprivation produces age-dependent changes of brain-derived neurotrophic factor expression in selected rat brain regions. *Biol Psychiatry.* 2004;55(7):708–714.

163. Monteggia LM, Luikart B, Barrot M, et al. Brain-derived neurotrophic factor conditional knockouts show gender differences in depression-related behaviors. *Biol Psychiatry.* 2006;61(2):140–143.
164. Chen Z-Y, Jing D, Bath KG, et al. Genetic variant BDNF (Val66Met) polymorphism alters anxiety-related behavior. *Science.* 2006;314(5796):140–143.
165. Zovkic IB, Meadows JP, Kaas GA, Sweatt JD. Interindividual variability in stress susceptibility: a role for epigenetic mechanisms in PTSD. *Front Psychiatry.* 2013;4:60.
166. Pechtel P, Pizzagalli DA. Effects of early life stress on cognitive and affective function: an integrated review of human literature. *Psychopharmacology (Berl).* 2011;214(1):55–70.
167. Niciu MJ, Henter ID, Luckenbaugh DA, Zarate CA, Charney DS, Charney DS. Glutamate receptor antagonists as fast-acting therapeutic alternatives for the treatment of depression: ketamine and other compounds. *Annu Rev Pharmacol Toxicol.* 2014;54:119–139.
168. Caddy C, Amit BH, McCloud TL, et al. Ketamine and other glutamate receptor modulators for depression in adults. In: Cipriani A, ed. In: *Cochrane Database of Systematic Reviews. Chichester, UK: John Wiley & Sons, Ltd; 2015.*
169. Bagot RC, Tse YC, Nguyen HB, Wong AS, Meaney MJ, Wong TP. Maternal care influences hippocampal *N*-methyl-D-aspartate receptor function and dynamic regulation by corticosterone in adulthood. *Biol Psychiatry.* 2012;72(6):491–498.
170. Roceri M, Hendriks W, Racagni G, Ellenbroek BA, Riva MA. Early maternal deprivation reduces the expression of BDNF and NMDA receptor subunits in rat hippocampus. *Mol Psychiatry.* 2002;7(6):609–616.
171. Thompson DF, Ramos CL, Willett JK. Psychopathy: clinical features, developmental basis and therapeutic challenges. *J Clin Pharm Ther.* 2014;39(5):485–495.
172. Caspi A. Influence of life stress on depression: moderation by a polymorphism in the 5-HTT gene. *Science.* 2003;301(5631):386–389.
173. Kinnally EL, Tarara ER, Mason WA, et al. Serotonin transporter expression is predicted by early life stress and is associated with disinhibited behavior in infant rhesus macaques. *Genes Brain Behav.* 2010;9(1):45–52.
174. Mitra SP. Neurotensin and neurotensin receptors in health and diseases: a brief review. *Indian J Biochem Biophys.* 2015;54:7–23.
175. Caspi A, McClay J, Moffitt TE, et al. Role of genotype in the cycle of violence in maltreated children. *Science.* 2002;297(5582):851–854.
176. Francis D. Nongenomic transmission across generations of maternal behavior and stress responses in the rat. *Science.* 1999;286(5442):1155–1158.
177. Champagne FA, Meaney MJ. Stress during gestation alters postpartum maternal care and the development of the offspring in a rodent model. *Biol Psychiatry.* 2006;59(12):1227–1235.
178. Tao J, Hu K, Chang Q, et al. Phosphorylation of MeCP2 at Serine 80 regulates its chromatin association and neurological function. *Proc Natl Acad Sci USA.* 2009;106(12):4882–4887.
179. Kimura H, Shiota K. Methyl-CpG-binding protein, MeCP2, is a target molecule for maintenance DNA methyltransferase, Dnmt1. *J Biol Chem.* 2003;278(7):4806–4812.
180. Dong E, Guidotti A, Grayson DR, Costa E. Histone hyperacetylation induces demethylation of reelin and 67-kDa glutamic acid decarboxylase promoters. *Proc Natl Acad Sci USA.* 2007;104(11):4676–4681.
181. Woo JS, Kim VN. MeCP2 caught moonlighting as a suppressor of microRNA processing. *Dev Cell.* 2014;28(5):477–478.
182. Shibayama A, Cook EH, Feng J, et al. MECP2 structural and 3′-UTR variants in schizophrenia, autism and other psychiatric diseases: a possible association with autism. *Am J Med Genet Part B Neuropsychiatr Genet.* 2004;128:50–53.

183. Rubino Ti. Zamberletti E, Parolaro D. *Endocannabinoids and Mental Disorders*. Cham: Springer; 2015. 261–283.
184. Korb E, Finkbeiner S. Arc in synaptic plasticity: from gene to behavior. *Trends Neurosci.* 2011;34(11):591–598.
185. D'Arcangelo G, Lossi L, Merighi A. Editorial: reelin-related neurological disorders and animal models. *Front Cell Neurosci.* 2017;10:299.
186. McCoy CR, Jackson NL, Day J, Clinton SM. Genetic predisposition to high anxiety- and depression-like behavior coincides with diminished DNA methylation in the adult rat amygdala. *Behav Brain Res.* 2017;320:165–178.
187. Beach SRH, Brody GH, Todorov AA, Gunter TD, Philibert RA. Methylation at 5HTT mediates the impact of child sex abuse on women's antisocial behavior: an examination of the Iowa adoptee sample. *Psychosom Med.* 2011;73(1):83–87.
188. Beach SRH, Dogan MV, Brody GH, Philibert RA. Differential impact of cumulative SES risk on methylation of protein-protein interaction pathways as a function of SLC6A4 genetic variation in African American young adults. *Biol Psychol.* 2014;96(1):28–34.
189. Beach SRH, Brody GH, Lei MK, Kim S, Cui J, Philibert RA. Is serotonin transporter genotype associated with epigenetic susceptibility or vulnerability? Examination of the impact of socioeconomic status risk on African American youth. *Dev Psychopathol.* 2014;26(2):289–304.
190. Beach SRH, Brody GH, Lei MK, Gibbons FX, Gerrard M, Simons RL. Impact of child sex abuse on adult psychopathology: a genetically and epigenetically informed investigation. *J Fam Psychol.* 2014;27(1):1–16.
191. Borghol N, Suderman M, Mcardle W, et al. Associations with early-life socio-economic position in adult DNA methylation. *Int J Epidemiol.* 2012;41(1):62–74.
192. Houtepen LC, Vinkers CH, Carrillo-Roa T, et al. Genome-wide DNA methylation levels and altered cortisol stress reactivity following childhood trauma in humans. *Nat Commun.* 2016;7:10967.
193. Kang HJ, Kim JM, Stewart R, et al. Association of SLC6A4 methylation with early adversity, characteristics and outcomes in depression. *Prog Neuropsychopharmacol Biol Psychiatry.* 2013;44:23–28.
194. Klengel T, Mehta D, Anacker C, et al. Allele-specific FKBP5 DNA demethylation mediates gene–childhood trauma interactions. *Nat Neurosci.* 2012;16(1):33–41.
195. Kumsta R, Marzi S, Viana J, et al. Severe psychosocial deprivation in early childhood is associated with increased DNA methylation across a region spanning the transcription start site of CYP2E1. *Transl Psychiatry.* 2016;6(6):e830.
196. Labonte B, Yerko V, Gross J, et al. Differential glucocorticoid receptor exon 1B, 1C, and 1H expression and methylation in suicide completers with a history of childhood abuse. *Biol Psychiatry.* 2012;72(1):41–48.
197. Labonté B, Suderman M, Maussion G, et al. Genome-wide epigenetic regulation by early-life trauma. *Arch Gen Psychiatry.* 2012;69(7):1097–1105.
198. Lutz PE, Gross JA, Dhir SK, et al. Epigenetic regulation of the kappa opioid receptor by child abuse. *Biol Psychiatry.* 2017. 10.1016/j.biopsych.2017.07.012.
199. Lutz P-E, Tanti A, Gasecka A, et al. Association of a history of child abuse with impaired myelination in the anterior cingulate cortex: convergent epigenetic, transcriptional, and morphological evidence. *Am J Psychiatry.* 2017;174(12):1185–1194.
200. Martín-Blanco A, Ferrer M, Soler J, et al. Association between methylation of the glucocorticoid receptor gene, childhood maltreatment, and clinical severity in borderline personality disorder. *J Psychiatr Res.* 2014;57(1):34–40.
201. McDade TW, Ryan C, Jones MJ, et al. Social and physical environments early in development predict DNA methylation of inflammatory genes in young adulthood. *Proc Natl Acad Sci.* 2017;114(29):7611–7616.

202. McGowan PO, Sasaki A, Huang TCT, et al. Promoter-wide hypermethylation of the ribosomal RNA Gene promoter in the suicide brain. *PLoS One.* 2008;3(5): e2085.
203. McGowan PO, Sasaki A, D'Alessio AC, et al. Epigenetic regulation of the glucocorticoid receptor in human brain associates with childhood abuse. *Nat Neurosci.* 2009;12 (3):342–348.
204. Mehta D, Klengel T, Conneely KN, et al. Childhood maltreatment is associated with distinct genomic and epigenetic profiles in posttraumatic stress disorder. *Proc Natl Acad Sci USA.* 2013;110(20):8302–8307.
205. Melas PA, Wei Y, Wong CCY, et al. Genetic and epigenetic associations of MAOA and NR3C1 with depression and childhood adversities. *Int J Neuropsychopharmacol.* 2013;16 (7):1513–1528.
206. Naumova OY, Lee M, Koposov R, Szyf M, Dozier M, Grigorenko EL. Differential patterns of whole-genome DNA methylation in institutionalized children and children raised by their biological parents. *Dev Psychopathol.* 2012;24(1):143–155.
207. Ouellet-Morin I, Wong CCY, Danese A, et al. Increased serotonin transporter gene (SERT) DNA methylation is associated with bullying victimization and blunted cortisol response to stress in childhood: a longitudinal study of discordant monozygotic twins. *Psychol Med.* 2013;43(9):1813–1823.
208. Perroud N, Paoloni-Giacobino A, Prada P, et al. Increased methylation of glucocorticoid receptor gene (NR3C1) in adults with a history of childhood maltreatment: a link with the severity and type of trauma. *Transl Psychiatry.* 2011;1(12):e59.
209. Romens SE, Mcdonald J, Svaren J, Pollak SD. Associations between early life stress and gene methylation in children. *Child Dev.* 2015;86(1).
210. Smith AK, Conneely KN, Kilaru V, et al. Differential immune system DNA methylation and cytokine regulation in post-traumatic stress disorder. *Am J Med Genet Part B Neuropsychiatr Genet.* 2011;156(6):700–708.
211. Steiger H, Labonte B, Groleau P, Turecki G, Israel M. Methylation of the glucocorticoid receptor gene promoter in bulimic women: associations with borderline personality disorder, suicidality, and exposure to childhood abuse. *Int J Eat Disord.* 2013;46 (3):246–255.
212. Tyrka AR, Price LH, Marsit C, Walters OC, Carpenter LL. Childhood adversity and epigenetic modulation of the leukocyte glucocorticoid receptor: preliminary findings in healthy adults. *PLoS One.* 2012;7(1):e30148.
213. Tyrka AR, Parade SH, Welch ES, et al. Methylation of the leukocyte glucocorticoid receptor gene promoter in adults: associations with early adversity and depressive, anxiety and substance-use disorders. *Transl Psychiatry.* 2016;6(7):e848.
214. van der Knaap LJ, Riese H, Hudziak JJ, et al. Glucocorticoid receptor gene (NR3C1) methylation following stressful events between birth and adolescence. The TRAILS study. *Transl Psychiatry.* 2014;4(4):e381.
215. Vijayendran M, Beach SRH, Plume JM, Brody GH, Philibert RA. Effects of genotype and child abuse on DNA methylation and gene expression at the serotonin transporter. *Front Psychiatry.* 2012;3:55.
216. Wankerl M, Miller R, Kirschbaum C, Hennig J, Stalder T, Alexander N. Effects of genetic and early environmental risk factors for depression on serotonin transporter expression and methylation profiles. *Transl Psychiatry.* 2014;4(6):e402.
217. Weder NM, Zhang HP, Jensen KP, et al. Child abuse, depression, and methylation in genes involved with stress, neural plasticity, and brain circuitry. *J Am Acad Child Adolesc Psychiatry.* 2014;53(4):417–424.
218. Yehuda R, Daskalakis NP, Bierer LM, et al. Holocaust exposure induced intergenerational effects on FKBP5 methylation. *Biol Psychiatry.* 2016;80(5):372–380.

219. Baker-Andresen D, Ratnu VS, Bredy TW. Dynamic DNA methylation: a prime candidate for genomic metaplasticity and behavioral adaptation. *Trends Neurosci.* 2013;36 (1):3–13.
220. Turecki G, Meaney MJ. Effects of the social environment and stress on glucocorticoid receptor gene methylation: a systematic review. *Biol Psychiatry.* 2015;79:87–96.
221. Appel K, Schwahn C, Mahler J, et al. Moderation of adult depression by a polymorphism in the fkbp5 gene and childhood physical abuse in the general population. *Neuropsychopharmacology.* 2011;36(10):1982–1991.
222. Binder EB. Association of FKBP5 polymorphisms and childhood abuse with risk of posttraumatic stress disorder symptoms in adults. *JAMA.* 2008;299(11):1291.
223. Roy A, Gorodetsky E, Yuan Q, Goldman D, Enoch M-A. Interaction of FKBP5, a stress-related gene, with childhood trauma increases the risk for attempting suicide. *Neuropsychopharmacology.* 2010;35(8):1674.
224. Perroud N, Salzmann A, Prada P, et al. Response to psychotherapy in borderline personality disorder and methylation status of the BDNF gene. *Transl Psychiatry.* 2013;3(1):e207.
225. Wang D, Szyf M, Benkelfat C, et al. Peripheral SLC6A4 DNA methylation is associated with in vivo measures of human brain serotonin synthesis and childhood physical aggression. *PLoS One.* 2012;7(6):e39501.
226. Beach SRH, Ph D, Brody GH, Todorov AA, Gunter D, Philibert RA. Methylation at SLC6A4 is linked to family history of child abuse: an examination of the Iowa adoptee sample. *Am J Med Genet B Neuropsychiatr Genet.* 2011;153B(2):710–713.
227. Beach SRH, Brody GH, Lei MK, et al. Impact of child sex abuse on adult psychopathology: a genetically and epigenetically informed investigation. *J Fam Psychol.* 2013;27 (1):3–11.
228. Schreier HMC, Roy LB, Frimer LT, Chen E. Family chaos and adolescent inflammatory profiles. *Psychosom Med.* 2014;76(6):460–467.
229. Miller GE, Chen E. Harsh family climate in early life presages the emergence of a proinflammatory phenotype in adolescence. *Psychol Sci.* 2010;21(6):848–856.
230. Miller GE, Chen E, Fok AK, et al. Low early-life social class leaves a biological residue manifested by decreased glucocorticoid and increased proinflammatory signaling. *Proc Natl Acad Sci.* 2009;106(34):14716–14721.
231. Donaldson ZR, Young LJ. The neurobiology and genetics of affiliation and social bonding in animal models. *Animal Models of Behavior Genetics.* New York, NY: Springer; 2016. 101–134.
232. Pietrzak RH, Naganawa M, Huang Y, et al. Association of in vivo κ-opioid receptor availability and the transdiagnostic dimensional expression of trauma-related psychopathology. *JAMA Psychiatry.* 2014;71(11):1262.
233. Akbarian S, Halene T. The neuroepigenetics of suicide. *Am J Psychiatry.* 2013;170 (5):462–465.
234. Turecki G. The molecular bases of the suicidal brain. *Nat Rev Neurosci.* 2014;15 (12):802–816.
235. Yang BZ, Zhang H, Ge W, et al. Supplement: child abuse and epigenetic mechanisms of disease risk. *Am J Prev Med.* 2013;44(2):101–107.
236. Selye H. Forty years of stress research: principal remaining problems and misconceptions. *Can Med Assoc J.* 1976;115(1):53–56.
237. Kloet ERDE, Vreugdenhil E, Oitzl MS, Pharmacology M, Amsterdam L. Brain corticosteroid receptor balance in health and disease. *Endocr Rev.* 1998;19(3):269–301.
238. McIlwrick S, Pohl T, Chen A, Touma C. Late-onset cognitive impairments after early-life stress are shaped by inherited differences in stress reactivity. *Front Cell Neurosci.* 2017;11:9.

239. Boyce WT. Differential Susceptibility of the developing brain to contextual adversity and stress. *Neuropsychopharmacology.* 2016;41(1):142–162.
240. Masten AS. Ordinary magic-resilience processes in development. *Am Psychol.* 2001;56 (3):227–238.
241. McLaughlin KA, Lambert HK. Child trauma exposure and psychopathology: mechanisms of risk and resilience. *Curr Opin Psychol.* 2017;14:29–34.
242. Agnafors S, Svedin CG, Oreland L, Bladh M, Comasco E, Sydsjö G. A Biopsychosocial approach to risk and resilience on behavior in children followed from birth to age 12. *Child Psychiatry Hum Dev.* 2017;48(4):584–596.
243. Happer K, Brown EJ, Sharma-Patel K. Children's resilience and trauma-specific cognitive behavioral therapy: comparing resilience as an outcome, a trait, and a process. *Child Abuse Negl.* 2017;73(suppl C):30–41.
244. Russo SJ, Murrough JW, Han M, Charney DS, Nestler EJ. Neurobiology of resilience. *Nat Neurosci.* 2012;15(11):1475–1484.
245. Zannas AS, West AE. Epigenetics and the regulation of stress vulnerability and resilience. *Neuroscience.* 2014;264:157–170.
246. Krishnan V, Han M-H, Graham DL, et al. Molecular adaptations underlying susceptibility and resistance to social defeat in brain reward regions. *Cell.* 2007;131(2):391–404.
247. Uchida S, Hara K, Kobayashi A, et al. Epigenetic status of Gdnf in the ventral striatum determines susceptibility and adaptation to daily stressful events. *Neuron.* 2011;69 (2):359–372.
248. Kerkel K, Spadola A, Yuan E, et al. Genomic surveys by methylation-sensitive SNP analysis identify sequence-dependent allele-specific DNA methylation. *Nat Genet.* 2008;40(7):904–908.
249. Schalkwyk LC, Meaburn EL, Smith R, et al. Allelic skewing of DNA methylation is widespread across the genome. *Am J Hum Genet.* 2010;86(2):196–212.
250. Alexander N, Wankerl M, Hennig J, et al. DNA methylation profiles within the serotonin transporter gene moderate the association of 5-HTTLPR and cortisol stress reactivity. *Transl Psychiatry.* 2014;4(9):e443.
251. Desantis SM, Baker NL, Back SE, et al. Gender differences in the effect of early life trauma on hypothalamic-pituitary-adrenal axis functioning. *Depress Anxiety.* 2011;28 (5):383–392.
252. Nugent BM, McCarthy MM. Epigenetic underpinnings of developmental sex differences in the brain. *Neuroendocrinology.* 2011;93(3):150–158.
253. Champagne FA, Francis DD, Mar A, Meaney MJ. Variations in maternal care in the rat as a mediating influence for the effects of environment on development. *Physiol Behav.* 2003;79(3):359–371.
254. Llorente R, Gallardo ML, Berzal AL, Prada C, Garcia-Segura LM, Viveros M-P. Early maternal deprivation in rats induces gender-dependent effects on developing hippocampal and cerebellar cells. *Int J Dev Neurosci.* 2009;27(3):233–241.
255. Frodl T, Carballedo A, Hughes M, et al. Reduced expression of glucocorticoid-inducible genes GILZ and SGK-1: high IL-6 levels are associated with reduced hippocampal volumes in major depressive disorder. *Transl Psychiatry.* 2012;2(3):e88.
256. Loi M, Koricka S, Lucassen PJ, Joëls M. Age- and sex-dependent effects of early life stress on hippocampal neurogenesis. *Front Endocrinol (Lausanne).* 2014;5:13.
257. Sutanto W, Rosenfeld P, De Kloet ER, Levine S. Long-term effects of neonatal maternal deprivation and ACTH on hippocampal mineralocorticoid and glucocorticoid receptors. *Dev Brain Res.* 1996;92(2):156–163.
258. Gracia-Rubio I, Moscoso-Castro M, Pozo OJ, Marcos J, Nadal R, Valverde O. Maternal separation induces neuroinflammation and long-lasting emotional alterations in mice. *Prog Neuropsychopharmacol Biol Psychiatry.* 2016;65:104–117.

259. Fuentes S, Daviu N, Gagliano H, et al. Sex-dependent effects of an early life treatment in rats that increases maternal care: vulnerability or resilience? *Front Behav Neurosci.* 2014;8:56.
260. Oomen CA, Girardi CEN, Cahyadi R, et al. Opposite effects of early maternal deprivation on neurogenesis in male versus female rats. *PLoS One.* 2009;4(1):e3675.
261. Jessen HM, Auger AP. Sex differences in epigenetic mechanisms may underlie risk and resilience for mental health disorders. *Epigenetics.* 2011;6(7):857–861.
262. Qureshi IA, Mehler MF. Genetic and epigenetic underpinnings of sex differences in the brain and in neurological and psychiatric disease susceptibility. *Prog Brain Res.* 2010;186 (C):77–95.
263. Labonté B, Engmann O, Purushothaman I, et al. Sex-specific transcriptional signatures in human depression. *Nat Med.* 2017;23(9):1102–1111.
264. Alastalo H, von Bonsdorff MB, Räikkönen K, et al. Early life stress and physical and psychosocial functioning in late adulthood. *PLoS One.* 2013;8(7):e69011.
265. Becker JB, Monteggia LM, Perrot-Sinal TS, et al. Stress and disease: is being female a predisposing factor? *J Neurosci.* 2007;27(44):11851–11855.
266. Macmillan HL, Fleming JE, Streiner DL, et al. Childhood abuse and lifetime psychopathology in a community sample. *Am J Psychiatry.* 2001;158(11):1878–1883.
267. Gouin JP, Zhou QQ, Booij L, et al. Associations among oxytocin receptor gene (OXTR) DNA methylation in adulthood, exposure to early life adversity, and childhood trajectories of anxiousness. *Sci Rep.* 2017;7(1):7446.
268. Heim CM, Mayberg HS, Mletzko T, Nemeroff CB, Pruessner JC. Decreased cortical representation of genital somatosensory field after childhood sexual abuse. *Am J Psychiatry.* 2013;170(6):616–623.
269. Kessler RC, McLaughlin KA, Green JG, et al. Childhood adversities and adult psychopathology in the WHO world mental health surveys. *Br J Psychiatry.* 2010;197 (5):378–385.
270. Chapman DP, Whitfield CL, Felitti VJ, Dube SR, Edwards VJ, Anda RF. Adverse childhood experiences and the risk of depressive disorders in adulthood. *J Affect Disord.* 2004;82(2):217–225.
271. Teicher MH, Samson JA, Polcari A, McGreenery CE. Sticks, stones, and hurtful words: relative effects of various forms of childhood maltreatment. *Am J Psychiatry.* 2006;163 (6):993–1000.
272. LaPrairie J, Heim CM, Nemeroff C. Neuroendocrine effects of early life trauma. In: Lanius R, Vermetten E, Pain C, eds. In: *The Impact of Early Life Trauma on Health and Disease: The Hidden Epidemic. Cambridge, UK: Cambridge University Press; 2010:157–165.*
273. Perroud N, Zewdie S, Stenz L, et al. Methylation of serotonin receptor 3A in ADHD, borderline personality, and bipolar disorders: link with severity of the disorders and childhood maltreatment. *Depress Anxiety.* 2016;33(1):45–55.
274. Olden K, Lin Y-S, Gruber D, Sonawane B. Epigenome: biosensor of cumulative exposure to chemical and nonchemical stressors related to environmental justice. *Am J Public Health.* 2014;104(10):1816–1821.
275. Weng X, Zhou D, Liu F, et al. DNA methylation profiling in the thalamus and hippocampus of postnatal malnourished mice, including effects related to long-term potentiation. *BMC Neurosci.* 2014;15(1):31.
276. Fischer L, Peter C, Garg P, et al. The effects of childhood malnutrition on DNA methylation in adulthood. *The FASEB Journal.* 2015;29(1_supplement):749.2.
277. Peter CJ, Fischer LK, Kundakovic M, et al. DNA methylation signatures of early childhood malnutrition associated with impairments in attention and cognition. *Biol Psychiatry.* 2016;80(10):765–774.

278. Joung KE, Park KH, Zaichenko L, et al. Early life adversity is associated with elevated levels of circulating leptin, irisin, and decreased levels of adiponectin in midlife adults. *J Clin Endocrinol Metab.* 2014;99(6):E1055–E1060.
279. Levin BE. Metabolic imprinting: critical impact of the perinatal environment on the regulation of energy homeostasis. *Philos Trans R Soc B Biol Sci.* 2006;361 (1471):1107–1121.
280. Viveros MP, Diaz F, Mateos B, Rodriguez N, Chowen JA. Maternal deprivation induces a rapid decline in circulating leptin levels and sexually dimorphic modifications in hypothalamic trophic factors and cell turnover. *Horm Behav.* 2010;57(4–5):405–414.
281. Moussaoui N, Larauche M, Biraud M, et al. Limited nesting stress alters maternal behavior and in vivo intestinal permeability in male Wistar pup rats. *PLoS One.* 2016;11(5):e0155037.
282. Felitti VJ, Anda RF, Nordenberg D, et al. Relationship of childhood abuse and household dysfunction to many of the leading causes of death in adults. *Am J Prev Med.* 1998;14 (4):245–258.
283. Wei L, Hao J, Kaffman A. Early life stress inhibits expression of ribosomal RNA in the developing hippocampus. *PLoS One.* 2014;9(12):e115283.
284. Andersen SL, Teicher MH. Stress, sensitive periods and maturational events in adolescent depression. *Trends Neurosci.* 2008;31(4):183–191.
285. Vythilingam M, Heim C, Newport J, et al. Childhood trauma associated with smaller hippocampal volume in women with major depression. *Am J Psychiatry.* 2002;159 (12):2072–2080.
286. Schoedl AF, Costa MCP, Mari JJ, et al. The clinical correlates of reported childhood sexual abuse: an association between age at trauma onset and severity of depression and PTSD in adults. *J Child Sex Abus.* 2010;19(2):156–170.
287. Maercker A, Michael T, Fehm L, Becker ES, Margraf J. Age of traumatisation as a predictor of post-traumatic stress disorder or major depression in young women. *Br J Psychiatry.* 2004;184(6):482–487.
288. Agid O, Shapira B, Zislin J, et al. Environment and vulnerability to major psychiatric illness: a case control study of early parental loss in major depression, bipolar disorder and schizophrenia. *Mol Psychiatry.* 1999;4(2):163–172.
289. LaSalle JM, Powell WT, Yasui DH. Epigenetic layers and players underlying neurodevelopment. *Trends Neurosci.* 2013;36(8):460–470.
290. Guo JU, Ma DK, Mo H, et al. Neuronal activity modifies DNA methylation landscape in the adult brain. *Nat Neurosci.* 2011;14(10):1345–1351.
291. Sweatt JD. Experience-dependent epigenetic modifications in the central nervous system. *Biol Psychiatry.* 2009;65(3):191–197.
292. Fishell G, Heintz N. The neuron identity problem: form meets function. *Neuron.* 2013;80(3):602–612.
293. Schwartzman O, Tanay A. Single-cell epigenomics: techniques and emerging applications. *Nat Rev Genet.* 2015;16(12):716–726.
294. Guo H, Zhu P, Wu X, Li X, Wen L, Tang F. Single-cell methylome landscapes of mouse embryonic stem cells and early embryos analyzed using reduced representation bisulfite sequencing. *Genome Res.* 2013;23(12):2126–2135.
295. Smallwood SA, Lee HJ, Angermueller C, et al. Single-cell genome-wide bisulfite sequencing for assessing epigenetic heterogeneity. *Nat Methods.* 2014;11(8):817–820.
296. Farlik M, Sheffield NC, Nuzzo A, et al. Single-cell DNA methylome sequencing and bioinformatic inference of epigenomic cell-state dynamics. *Cell Rep.* 2015;10 (8):1386–1397.
297. Luo C, Keown CL, Kurihara L, et al. Single-cell methylomes identify neuronal subtypes and regulatory elements in mammalian cortex. *Science.* 2017;357(6351):600–604.

298. Lake BB, Ai R, Kaeser GE, et al. Neuronal subtypes and diversity revealed by single-nucleus RNA sequencing of the human brain. *Science.* 2016;352 (6293):1586–1590.
299. Rotem A, Ram O, Shoresh N, et al. Single-cell ChIP-seq reveals cell subpopulations defined by chromatin state. *Nat Biotechnol.* 2015;33(11):1165–1172.
300. Small EC, Xi L, Wang J-P, Widom J, Licht JD. Single-cell nucleosome mapping reveals the molecular basis of gene expression heterogeneity. *Proc Natl Acad Sci USA.* 2014;111 (24):E2462–E2471.
301. Nagano T, Lubling Y, Stevens TJ, et al. Single-cell Hi-C reveals cell-to-cell variability in chromosome structure. *Nature.* 2013;502(7469):59–64.
302. Kind J, Pagie L, de Vries SS, et al. Genome-wide maps of nuclear lamina interactions in single human cells. *Cell.* 2015;163(1):134–147.
303. Cusanovich DA, Daza R, Adey A, et al. Multiplex single cell profiling of chromatin accessibility by combinatorial cellular indexing. *Science.* 2015;348(6237):910–914.
304. Jin W, Tang Q, Wan M, et al. Genome-wide detection of DNase I hypersensitive sites in single cells and FFPE tissue samples. *Nature.* 2015;528(7580):142–146.
305. Dey SS, Kester L, Spanjaard B, Bienko M, van Oudenaarden A. Integrated genome and transcriptome sequencing of the same cell. *Nat Biotechnol.* 2015;33(3):285–289.
306. Angermueller C, Clark SJ, Lee HJ, et al. Parallel single-cell sequencing links transcriptional and epigenetic heterogeneity. *Nat Methods.* 2016;13(3):229–232.
307. Hou Y, Guo H, Cao C, et al. Single-cell triple omics sequencing reveals genetic, epigenetic, and transcriptomic heterogeneity in hepatocellular carcinomas. *Cell Res.* 2016;26(3):304–319.
308. Guo F, Li L, Li J, et al. Single-cell multi-omics sequencing of mouse early embryos and embryonic stem cells. *Cell Res.* 2017;27(8):967–988.
309. O'Geen H, Ren C, Nicolet CM, et al. dCas9-based epigenome editing suggests acquisition of histone methylation is not sufficient for target gene repression. *Nucleic Acids Res.* 2017;45(17):9901–9916.
310. Heller EA, Cates HM, Peña CJ, et al. Locus-specific epigenetic remodeling controls addiction- and depression-related behaviors. *Nat Neurosci.* 2014;17(12):1720–1727.
311. Liu XS, Wu H, Ji X, et al. Editing DNA methylation in the mammalian genome. *Cell.* 2016;167(1). 233-247.e17.
312. Xu X, Tao Y, Gao X, et al. A CRISPR-based approach for targeted DNA demethylation. *Cell Discov.* 2016;2:16009.
313. Choudhury SR, Cui Y, Lubecka K, Stefanska B, Irudayaraj J. CRISPR-dCas9 mediated TET1 targeting for selective DNA demethylation at BRCA1 promoter. *Oncotarget.* 2016;7(29):46545–46556.
314. Vojta A, Dobrinić P, Tadić V, et al. Repurposing the CRISPR-Cas9 system for targeted DNA methylation. *Nucleic Acids Res.* 2016;44(12):5615–5628.
315. Murgatroyd C, Quinn JP, Sharp HM, Pickles A, Hill J. Effects of prenatal and postnatal depression, and maternal stroking, at the glucocorticoid receptor gene. *Transl Psychiatry.* 2015;5(5):e560.
316. Brody GH. Using genetically informed prevention trials to test gene × environment hypotheses.In: *Gene-Environment Transactions in Developmental Psychopathology. Cham: Springer International Publishing; 2017. 211–233.*
317. Boks MP, de Jong NM, Kas MJH, et al. Current status and future prospects for epigenetic psychopharmacology. *Epigenetics.* 2012;7(1):20–28.

FURTHER READINGS

Rincel M, Lépinay AL, Delage P, et al. Maternal high-fat diet prevents developmental programming by early-life stress. *Transl Psychiatry.* 2016;6(11):e966.

Evans GW, Schamberg MA. Childhood poverty, chronic stress, and adult working memory. *Proc Natl Acad Sci USA.* 2009;106(16):6545–6549.

Kaplow JB, Widom CS. Age of onset of child maltreatment predicts long-term mental health outcomes. *J Abnorm Psychol.* 2007;116(1):176–187.

Snowden AW, Gregory PD, Case CC, et al. Gene-specific targeting of H3K9 methylation is sufficient for initiating repression in vivo. *Curr Biol.* 2002;12(24):2159–2166.

Laganiere J, Kells AP, Lai JT, et al. An engineered zinc finger protein activator of the endogenous glial cell line-derived neurotrophic factor gene provides functional neuroprotection in a rat model of Parkinson's disease. *J Neurosci.* 2010;30(49):16469–16474.

INDEX

E

M

N

O

P

S

T

V

Kolibri 360 GmbH
Ferdinand-Jühlke-Straße 7
99095 Erfurt, Deutschland
produktsicherheit@kolibri360.de